PLUMBING ENGINEERING SERVICES DESIGN GUIDE

Compiled and published by:

The Institute of Plumbing

64 Station Lane, Hornchurch, Essex RM12 6NB. Telephone: 04024 72791. Fax: 04024 48987

ISBN 0 9501671 8 5

FOREWORD

by Colin Moynihan
Parliamentary Under Secretary, Department of the Environment

PARLIAMENTARY UNDER SECRETARY OF STATE

DEPARTMENT OF THE ENVIRONMENT
2 MARSHAM STREET LONDON SW1P 3EB
01-212 3434

28 March 1988

I am particularly pleased to have been asked to contribute the foreword to this edition of the Plumbing Engineering Services Design Guide.

Since the Guide was launched in 1977, it has proved to be an indispensable reference source for designers, engineers and craftsmen. It has now been completely revised and updated.

There has been a constant flow of requests for the first edition, both from home and abroad. The market for information of this kind is clearly there. Given the essential nature and growing volume and complexity of plumbing services, that should come as no surprise.

Plumbing services have a long history as a respected craft, and the Institute of Plumbing is dedicated to promoting plumbing as a skilled and professional service which delivers the highest standards of workmanship.

The second edition of the Design Guide provides a key contribution to the achievement of the Institute's objectives. I commend it to the plumbing industry.

Colin Moynihan

COLIN MOYNIHAN

CONTENTS

Section A Hot and Cold Water Supplies Pages 1-49

Pipe sizing; hot water storage; sizing of primary and secondary circulations; pumped hot water systems; water quality; outward thrust in pipelines due to internal pressure; acoustic and vibration control; water hammer; inspection and testing of water mains and water supply systems; prevention of contamination of water by backsiphonage or cross connection; backflow prevention devices; protection of draw-off taps, hose connections, bidets, clothes and dishwashing machines and cisterns; unvented hot water systems; corrosion causes, effects and preventative measures; effects of corrosive environments; prevention of corrosion; welding, brazing, soft soldering and adhesive bonding; references; British Standards and Codes of Practice.

Section B Domestic Central Heating 50-77

Design considerations; heating controls; systems of control; electric central heating storage boilers; domestic electric water heating; references; British Standards and Codes of Practice.

Section C Sanitary Plumbing and Drainage 78-136

Design of sanitary pipework systems; kitchen and laboratory drainage; materials for above ground drainage systems; inspection and testing of discharge pipes; vacuum systems; methods of waste collection; sanitary accommodation; the design of drainage systems; sewage pumping systems; below ground drain renovation; methods of pipe renovation; rainwater systems; solid waste refuse disposal; references; British Standards and Codes of Practice.

Section D Pumps and Pumping 137-142

Definitions and descriptions; pump types; systems; pump selection; noise; pump installation; British Standards.

Section E Piped Gas Services 143-169

Natural gas; liquefied petroleum gas installations; compressed air; vacuum; medical gases; references; British Standards and Codes of Practice.

Section F Fire Protection Services 170-188

Principal causes of fires in buildings; classification of fire risks; sprinkler installations; automatic sprinklers; risk categories; classes of system; water supplies; grading systems; pressure and flow requirements; proving water supplies; fire brigade inlets; sizing of sprinkler heads; area covered by sprinklers; multiple controls; other installations using water; systems using steam; hose reel installations; dry risers; water risers; foam systems; carbon di-oxide systems; total flooding systems; halogenated extinguishing agents systems; private hydrants; references; British Standards and Codes of Practice.

Section G Steam Supplies 189-200

Steam; condensate; steam traps and separators; British Standards and Codes of Practice.

Section H Pipework Expansion 201-207

Considerations for design and installation; cold draw; anchor design; pipework systems working under test conditions; pipe guides and supports; expansion bellows; plastic soil and waste pipes; British Standards and Codes of Practice.

Section I Designing for the Disabled 208-216

Approaches to buildings; inside buildings; WC compartments in public buildings; access through lobbies; WC facilities for wheelchair disabled; WC facilities for ambulant disabled; facilities for private use and nursing homes; advanced WC cubicle design for wheelchair disabled; central heating; references; British Standards and Codes of Practice.

Section J Legionnaires Disease 217-220

Engineering implications; outbreak control; cold water storage cisterns and tanks; continuous chlorination plant; testing outlets; water softening and hot water storage systems; stagnant water positions; domestic hot water calorifiers; cooling towers; humidifiers; references.

Section K Fountain and Water Display Techniques 221-224

Design considerations; types of fountain and water display; the art of water display; construction and maintenance; cascades.

Section L Airflow in Ducts 225-231

Duct sizing; duct fittings resistance; ideal system; fans; British Standards and Codes of Practice.

Section M Electrical Earthing and Bonding in Water Installations 232-235

Protective multiple earthing; bonding; problems with plastics; supplementary bonding; drainage systems; new regulations; plumber's responsibility; British Standards.

Section N Conversion Factors and Miscellaneous Data 236-259

Conversion factors; identification of pipelines; building drawings; symbols; maximum spacing of pipe supports; spacing of pipes; head and pressure of water; hydrostatic tables; relation to services to each other; approximate weight of lead required for caulked joints; corrosion of metals; coefficients of linear expansion; definition of plumbing terms; useful addresses.

PREFACE

Since publication of the previous Design Guide in 1977, plumbing in the United Kingdom has changed considerably in both scope and technology. Revised water byelaws and codes of practice have introduced a new set of operating conditions for both the designer and the craftsman. The manufacturing industry's advance research into sophisticated control systems is more akin to space age technology than the plumbing industry.

These changes, which allow the use of unvented domestic hot water storage systems, relate to standards and practices that have operated in Europe, North America and other parts of the world for some years. It is essential that such systems are correctly designed and installed to protect public health and safety.

This new Guide seeks to recognise many recent developments in UK plumbing technology by extending many of the existing sections and also by including new subjects such as basic design in steam supplies; pipeline expansion; facilities for the disabled; protection against legionnaires disease; fountain and water display techniques; airflow in ducts and electrical earth bonding.

The water services section has been extended to cover current regulations in backsiphonage and unvented domestic hot water storage systems as well as vibration and noise control.

The drainage and sanitation section includes drain renovation; design for seismic disturbances and a completely new method for the sizing of underground drainage systems.

In order to include this new data, the new Guide now contains over 100 pages more than the previous edition.

An enormous amount of voluntary time and energy has gone into the production of this Guide in the form of research and drafting, particularly from the section authors. I would like to take this opportunity to thank all who have been associated with the Guide – in particular the Review Panel, the Institute's Technical Officer Kevin Wellman who has acted as Secretary to the Panel and the Editor, Colin Sturgeon.

I know their efforts will be appreciated by all who use this Guide in the advancement of plumbing technology for the service and protection of the public.

J G Rodgers IEng FIWEM MIP RP

Chairman
Design Guide Review Panel
Past President
The Institute of Plumbing
December 1988

ACKNOWLEDGEMENTS

The Institute pays tribute to Jack Rodgers (Chairman) and members of the Design Guide Review Panel for their dedicated efforts in the mammoth task of compiling this book. Kevin Wellman and Colin Sturgeon deserve special mention for their meticulous attention to detail in bringing this project to fruition.

We greatly appreciate the information supplied by Government Departments, Agencies, other organisations and companies as well as the part played by the individuals listed below.

The competence of all those involved is reflected in the finished work which is a credit to the plumbing industry in the United Kingdom and one of which we can all be proud.

Andrew Watts Eng Tech MIP RP MASSE MASPE
Chief Executive and Secretary
The Institute of Plumbing

British Standards Institution
Building Research Establishment
Calor Gas Limited
Confederation for the Registration of Gas Installers
Department of Environment
The Electricity Council
Engineering Appliances Limited
Health and Safety Executive
Home Office Fire Service Inspectorate
Marley Extrusions Limited
Ove Arup & Partners
Raychem Limited
Rice Hayward Limited
Severn Trent Water Authority
Thomas Telford Publications
Water Techniques
Water Research Centre
Wavin Building Products Limited

D Butler BSc MSc DIC
P B Caplin BSc CEng MICE MIWEM
B Pinkerton BSc CEng MICE MIWES
B Webb FIFireE

Word processing:
Mrs G S Rust and Miss A J Skipworth

Drawings: Guy Smith

Design and artwork: Richard Bonwick

Typesetting: Crowborough Typesetters

Produced by Colin Sturgeon
and Project Marketing Services Limited

Printed by Saunders and Williams

Members of the Plumbing Engineering Services Design Guide Review Panel

B Armstrong CEng FIGasE MInstE MIP
D A Baggs BSc CEng MIDHE MCIBSE MInstE
L Bassett IEng MIHospE MIDHE ACIBSE MIP
M F Bennett IEng FIWEM MRSH MIP
J Byfield EngTech MIP RP
A W Clarke IEng FIOP RP
C Daniels IEng MIElecIE
J P Finnegan BSc ARCS
A J Goodger BSc MSc CChem CEng MPRI MRSC MIM MICorrST MIP
Dr D Harper PhD BSc IEng MIHospE MIWSoc MASEE
H A Howick IEng FIWEM FIOP
V H J Jackson MIP RP
K Kettle IEng MIPHE MInstBE MIWEM FWOA MIP
G A J Knight MIP
R J Oakley EngTech MIDHE MIP
J Prestage IEng FIOP FRSH MIWEM RP
W T Millar IEng MIElecIE MRSH MASHRAE MIP
P J Reeves IEng FIOP MCIOB
J G Rodgers IEng FIWEM MIP
P A Sterlini MCIBSE
R G Towse EngTech AMIGasE
D C Walker IEng FIOP MIWEM RP
S A Walsh CEng FIWEM FIOP MCIBSE FInstPet

Secretary & Co-ordinator
Kevin Wellman EngTech MIP RP
Technical Officer, The Institute of Plumbing

Editor
Colin G Sturgeon Comp IP

PIPE SIZING

Assessment of Required Flow Rate

The minimum recommended rates of flow at various appliances and fittings are set out in Table A1. Except under very special circumstances, these are rates which are found to be generally acceptable to consumers and the design of a system should provide for these delivery rates.

Table A1 Recommended minimum rates of flow of various appliances or fittings

Outlet fitting	Rate of flow (litres/second)
WC flushing cistern, float operated valve	0.10
WC flushing trough, float operated valve	0.15 per WC served
Spray tap or spray mixer tap	0.04 per tap
Wash basin tap	0.15 hot or cold
Bath tap nominal size ¾in	0.30 hot or cold
Bath tap nominal size 1in	0.60 hot or cold
Shower head	0.10 hot or cold
Sink tap nominal size ½in	0.20 hot or cold
Sink tap nominal size ¾in	0.30 hot or cold
Sink tap nominal size 1in	0.60 hot or cold
Urinal flushing cisterns served	0.004 per position

Notes

1. The draw-off rate from mixer fittings is normally less than that from two separate taps. The difference may be disregarded when pipe sizing.

2. The flow rate of shower heads can vary depending upon the manufacturer. Refer to supplier when in doubt.

3. Where the frequency of usage of WCs is likely to be more than at 60 second intervals, flushing troughs are advised. Urinal demand is very low and is normally disregarded.

4. Water demand for industrial and process applications should be designed for their full flow rate. The domestic flow rate obtained by using loading units should, where appropriate, be added to the required industrial flow.

Assessment of Probable Demand

In hot and cold water installations, it rarely happens that all the appliances installed are in simultaneous use.

For economic reasons it is the usual design practice to provide for a simultaneous usage less than the maximum possible.

Loading Units

In most installations, the simultaneous demand can be determined with an adequate degree of accuracy by using the loading unit concept.

Although the flow rates at hot and cold water draw-off points may not be identical, for practical purposes, the same loading units can be applied to both types of draw-off.

In certain industrial situations where peak usage is possible, the judgement of the designer must prevail and a flow rate demand provided in excess of that which, in a domestic environment, would be considered an adequate simultaneous flow rate.

Table A2 sets out the 'loading unit' rating for various appliances. A loading unit has no precise value in terms of litres per second. By multiplying the total number of each type of appliance by the appropriate loading unit number and adding the resultant totals together, the recommended flow rate can be read from Graph A1 or Graph A2.

Table A2 Loading units

Appliance	Loading units Lu
WC flushing cistern (9 litres)	2
Wash basin domestic use	1½
Wash basin public use	2
Wash basin concentrated use	3
Bath tap nominal size ¾in	10
Bath tap nominal size 1in	22
Shower	3
Sink tap nominal size ½in	3
Sink tap nominal size ¾in	10
Spray tap	½

Notes

1. WC flushing cisterns with single or dual flush control have the same loading unit.

2. The wash basin loading unit is for pillar taps, and the larger unit is applicable to situations such as schools and those offices where there is a peak period of use. Where spray taps are installed, an equivalent continuous demand of 0.04 l/s/tap should be assumed.

Equivalent Pipe Lengths for Copper, Plastics and Stainless Steel Pipework

The diameter of pipe necessary to give a required flow rate will depend upon the head available, the smoothness of the pipe used (i.e. type of material) and the effective length of pipe run.

To the loss of head in the actual length of pipe must be added that caused by pipe fittings such as elbows, tees, etc. For convenience, the latter is usually expressed as the loss in an equivalent length of pipe as set out in Table A3.

Table A3 Equivalent pipe length

Pipe size (OD) copper mm	Elbow m	Tee m
15	0.5	0.6
22	0.8	1.0
28	1.0	1.5
35	1.4	2.0
42	1.7	2.5
54	2.3	3.5
66.5	3.0	4.5
76	3.4	5.8
108	4.5	8.0

Notes

1. The losses through tees should be assumed to occur at changes of direction only.

2. In major hot water distribution systems, special fittings with significant head losses are often used. For information on these, reference should be made to manufacturers' literature.

PIPE SIZING

Table A4 Loss of head through draw-off taps expressed as equivalent pipe lengths

Fitting (BS 1010)*	Discharge rate tap fully open l/s	Equivalent length of pipe of same diameter as tap (metres) Copper	Galvanised mild steel
½in (15mm nominal) bibtap or pillar tap	0.20	2.70	4.00
¾in (20mm nominal) bibtap or pillar tap	0.30	8.50	5.75
1in (25mm nominal) bib tap or pillar tap	0.60	20.00	13.00

*Some patented draw-off taps have a loss of head much higher than BS 1010 fittings.

Loss of Head at Point of Delivery

In calculating the diameter of a pipe to supply an individual fitting, the loss of head through the draw-off or delivery valve should be taken into account. A table of head losses through draw-off taps is given in Table A5. For easy calculation, a table of the loss of head through draw-off taps expressed as equivalent pipe lengths is given in Table A4.

Loss of Head through Ballvalves and Stopvalves

The loss of head through ballvalves, which can be very high, and stopvalves varies with the rate of flow. It is usually more convenient therefore to deduct the loss of head through these fittings from the available head. The loss of head through ballvalve orifices of various sizes is given by Graph A3 and that through stopvalves by Graph A4.

Table A5 Loss of head through draw-off taps (metres)

Fittings (BS 1010)	Discharge rate in l/s 0.075	0.12	0.15	0.20	0.22	0.30	0.35	0.47	0.55	0.60
½in bibtap	0.185	0.365	0.52	0.75	1.1	1.8				
¾in bibtap			0.21	0.33	0.45	0.75	1.1	1.4		
1in bibtap						0.45	0.70	0.90	1.2	1.5
½in pillar tap	0.24	0.40	0.52	0.70	0.80					
¾in pillar tap			0.24	0.30	0.40	0.50	0.70			

Table A6 Loss of head through taps and equivalent pipe lengths

Tap	Flow rate l/s	Loss of head m	Equivalent pipe length m
BS 5412/3 Nominal size ½in	0.16	1.0	7.0
Nominal size ¾in	0.32	1.0	13.5

*Head loss for stated flow rates are typical and may vary with taps of different manufacture.

Relative Discharging Power of Pipes

From the pipe sizing graph it can be seen that at H/L = 1, the flow rates shown in Table A7 occur.

Table A7 Flow rates at H/L = 1

Pipe size (OD) copper mm	Flow rates l/s
15	0.44
22	1.40
28	2.70
35	4.90
42	8.00
54	16.00
76	45.00

Expressed as a ratio of the 15mm draw-off rate we obtain the following tabulations:

Table A8 Number of outlets at 100% demand at full flow conditions

Pipe size (OD) mm	Flow rate l/s	Ratio	No.of outlets to be supplied
15	0.44	1.0	1
22	1.40	3.18	3
28	2.70	6.13	6
35	4.90	11.13	11
42	8.00	18.18	18
54	16.00	36.36	36
76	45.00	102.27	100

Table A9 Average water consumption of domestic and commercial appliances

Appliance	Average water consumption (hot and cold) litres/operation
WC (normal flush)	9
Wash basin	6
Bath	80-90
Shower	4 per minute
Clothes washing machine	60-180
Dishwashing machine	30-70
Domestic waste disposal unit	10-15 per minute
Bidet	6
Installations Domestic	litres/day/person served
WC flushing	35
Personal washing and bathing	25
Laundry	13
Washing up	10
Car washing and garden use	3
Drinking, food preparation and cooking	20
Offices	litres/day/person served
WC flushing women – normal flush	12
WC flushing women – dual flush	9
WC flushing men – urinals provided	5
Urinal flushing: 3 flushes/hour – 24 hour day	14
Urinal flushing: 3 flushes/hour – 8 hour day	5
Sanitary towel disposal by macerator	2
Hand washing: normal taps	11
Hand washing: spray taps	5

Graph A1 Pipe sizing chart, copper tubes to BS 2871 Table X

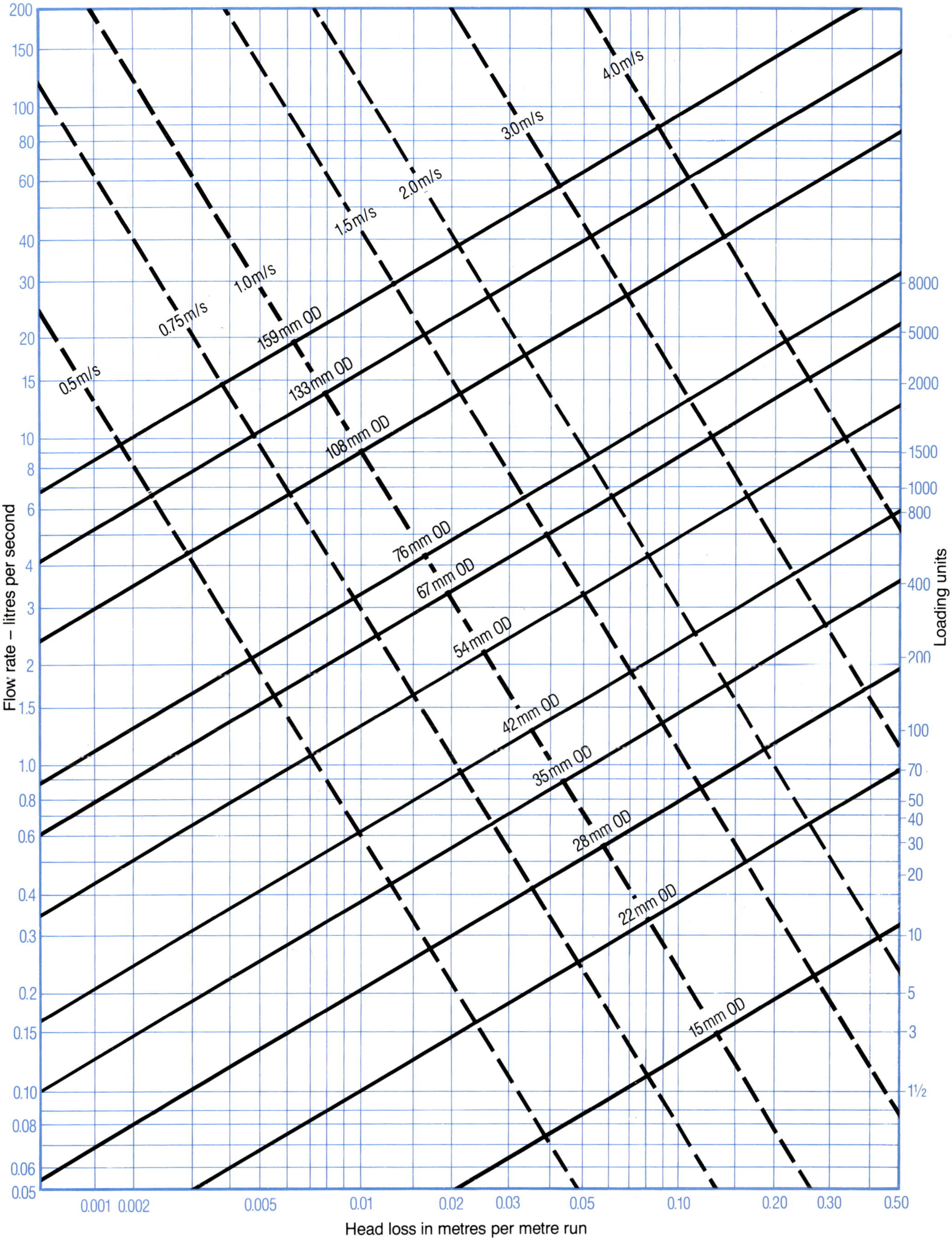

Graph A2 Pipe sizing graph, heavy galvanised steel tubes to BS 1387

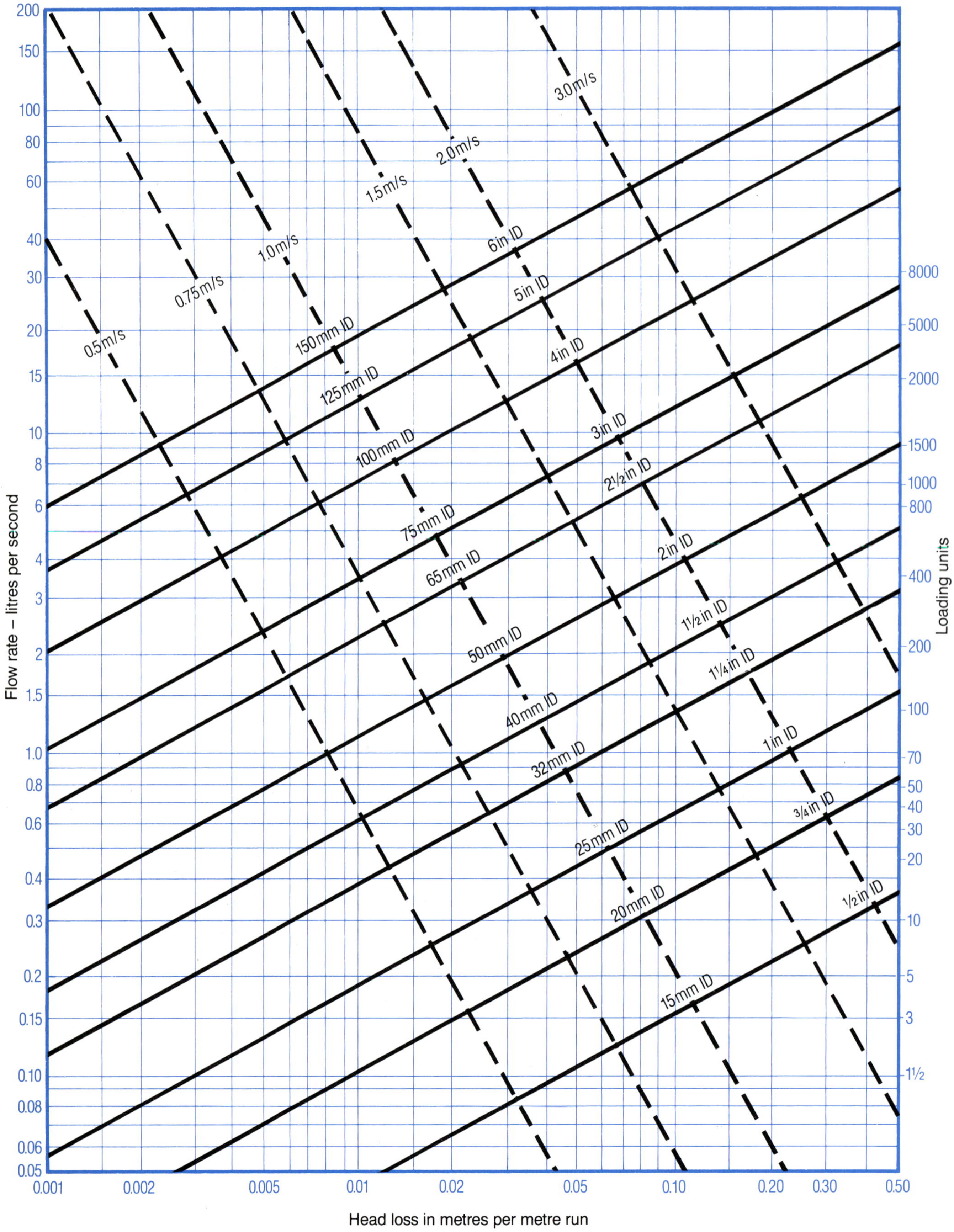

Graph A3 Head loss through ballvalve orifices

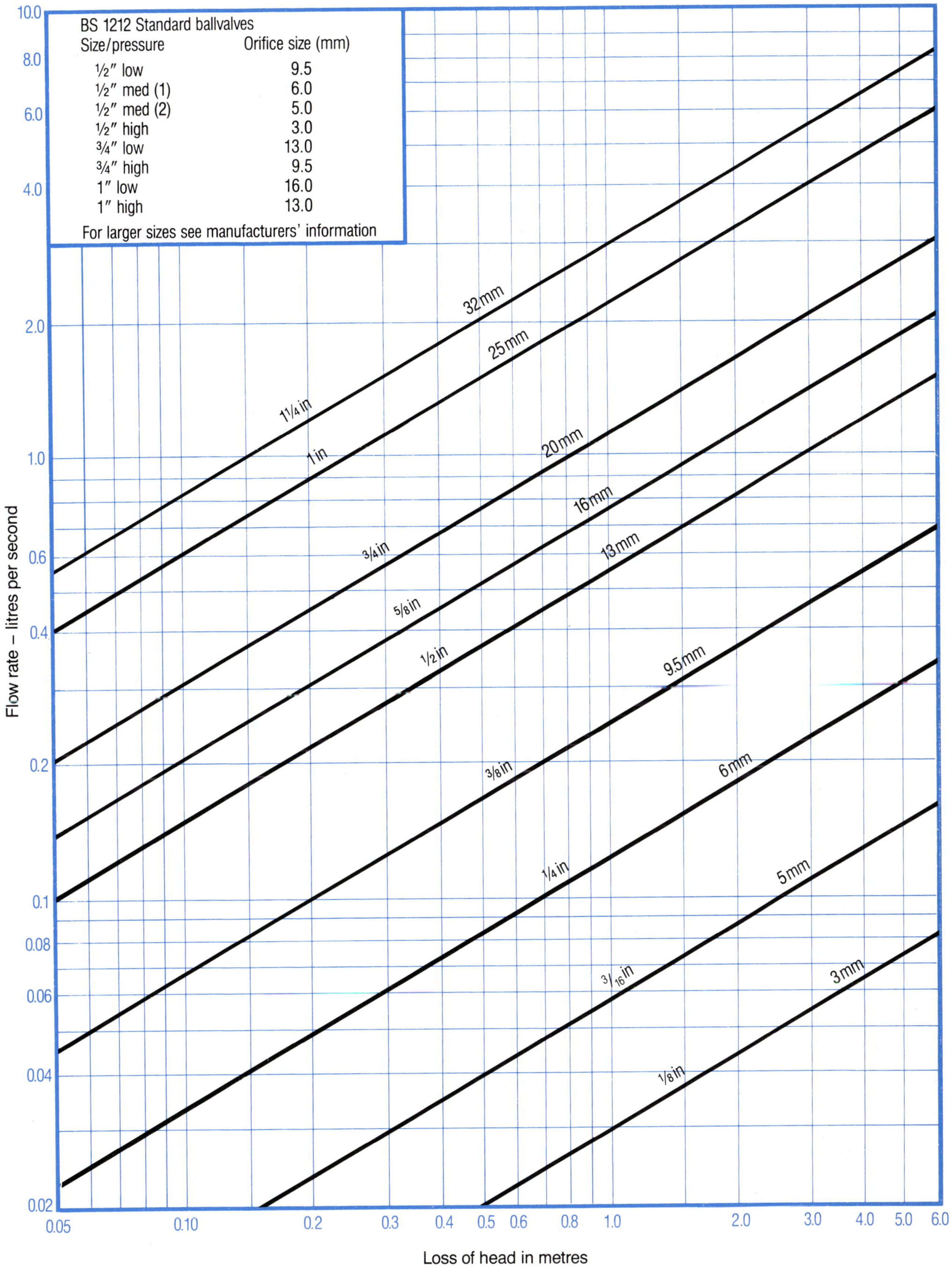

Graph A4 Head loss through stopvalves

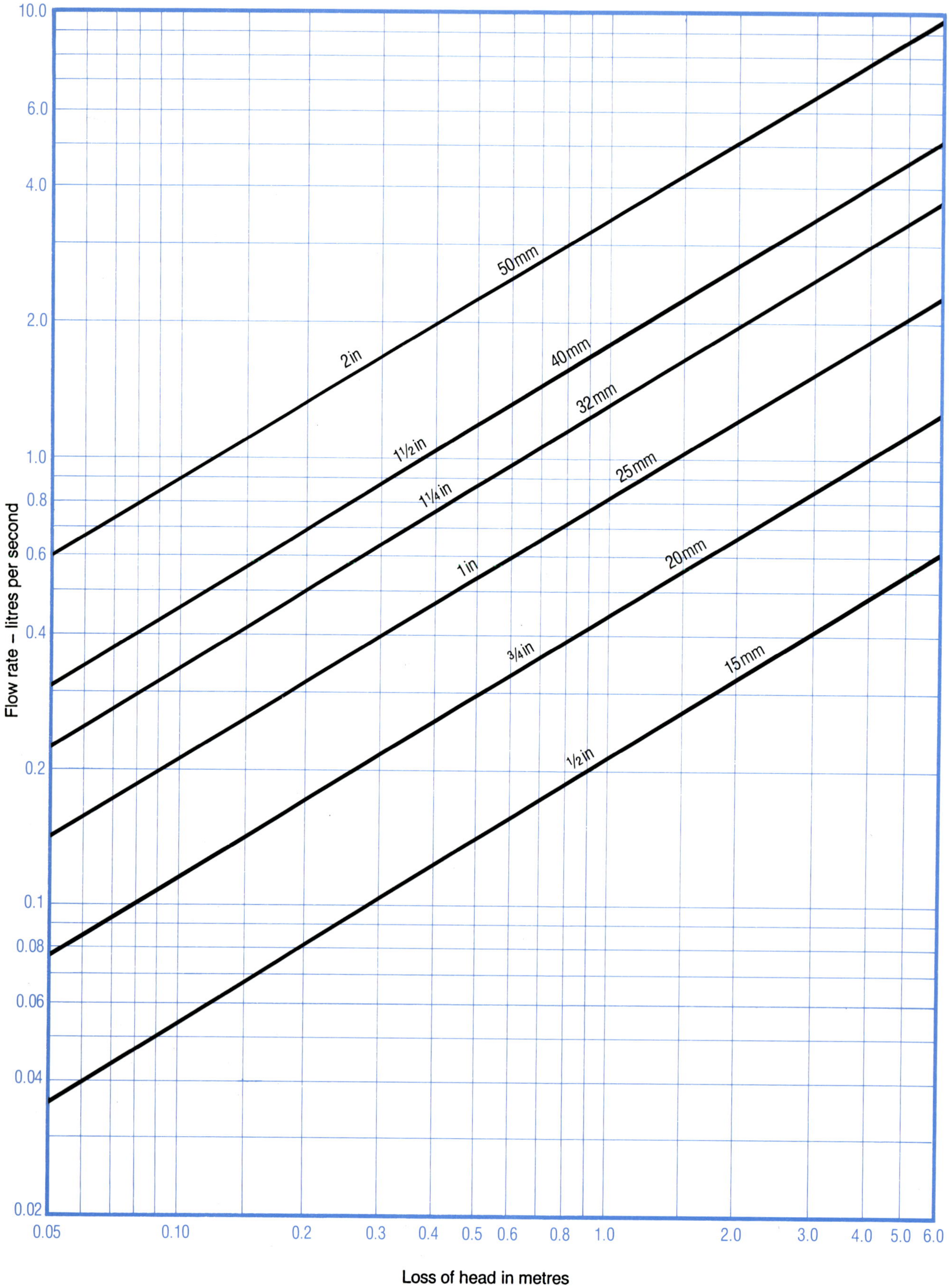

HOT WATER STORAGE

Except in cases of special hot water demand, the requirements for the storage of hot water may be assessed from Tables A10 and A11.

Table A10 Cold water storage in buildings

Type of building	Storage (litres)
Dwelling up to 4 bedrooms	120 per bedroom
Dwelling more than 4 bedrooms	100 per bedroom
Hostel	90 per bed space
Hotel	135 per bed space
Office premises with canteen facilities	45 per employee
Office premises without canteen facilities	40 per employee
Restaurant	7 per meal
Day School Nursery, Primary	15 per pupil
Secondary, Grammar, Comprehensive, Technical	20 per pupil
Boarding School	90 per pupil
Children's Home or Residential Nursery	135 per bed space
Nurses Home	120 per bed space
Nursing or Convalescent Home	135 per bed space

Table A11 Assessment of hot water demand and storage requirements (based on day of heaviest demand during week)

Type of building	Max. daily demand per person	Storage requirement per person
Colleges and schools	litres	litres
Boarding	114	23
Day	14	4.5
Dwelling houses		
Low rental	114	see note* below
Medium rental	114	45
High rental	136	45
Flats (blocks)		
Low rental	68	23
Medium rental	114	32
High rental	136	32
Factories	14	4.5
Hospitals§		
General	136	27
Infectious	225	45
Infirmaries	68	23
Infirmaries (with laundries)	90	27
Maternity	225	32
Mental	90	23
Nurses' homes	136	45
Hostels	114	32
Hotels		
First class	136	45
Average	114	36
Offices	14	4.5
Sports pavilions (with spray-type showers)	36	36
Restaurants	Per meal	6 litres

*Storage normally a minimum of 114 litres with a 4-hour heat-up period.
§Refer also to relevant Hospital Building Notes (DHSS) HMSO.

Method of Assessing Recovery/Storage Relationships for Hot Water Storage Vessels

When sizing hot water storage generators, it is important to determine not only the correct size but also the correct relationship between recovery and storage to enable the most efficient selection of plant to be made. If too high a storage volume is selected and the recovery rate determined as a time function of the volume, then not only is the vessel and heat exchanger oversized but likewise the primary generating plant, pumps, pipework and controls.

The effect of oversizing will vary depending on the degree and type of building, but where the domestic hot water load represents a substantial proportion of the total heating load, unnecessary expenditure, both in capital and running costs, is involved.

Where inadequate storage and recovery are provided, then the system will not meet the user requirements, with the result that costly modifications may be necessary when the building is occupied and the engineering services in daily use.

Whatever the building under consideration, a pattern of hot water usage can be projected, the actual usage being largely a function of the building population and the type of activity that takes place within. In determining the pattern of usage, it is important to differentiate between the maximum daily demand and the average daily demand, so that the implications of the system not meeting the demands made upon it can be recognised, and the maximum requirements designed for where necessary.

To project the demand pattern over the operating period an hour by hour analysis of hot water usage should be made. Taking into account the building population, the level and type of activity and all other factors which affect hot water demand, a quantitive assessment of hourly usage should be tabulated for every hour of the operating period.

For offices and schools, a continuous 12 hour operating period may be sufficient, but for other buildings such as hospitals and factories a 24 hour operating period may be necessary. The projected demand pattern should be recorded in the form of a histogram profile as illustrated in Figure A1.

Figure A1 Demand pattern histogram

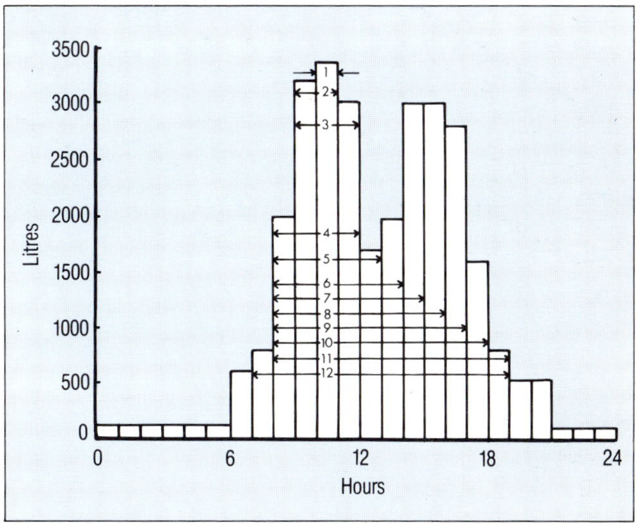

From the histogram, a series of sequential peak hourly loads should be calculated for groups of hours, from 1 hour through to the maximum number of hours in the operating period. The total hourly loads should be recorded as in Table A12 together with the average hourly load for each group of hours under consideration.

On completion of Table A12, a graph should be drawn using Column 3 co-ordinates as the 'y' axis and Column 2 co-ordinates as the 'x' axis as in Figure A2. Linking the co-ordinates for each

HOT WATER STORAGE

group of hours will produce a curve, any point of which establishes a specific relationship of recovery to storage which satisfies the system requirements, assuming of course these have been predicted with sufficient accuracy in the demand profile.

Table A12 Peak hourly loads

1 Peak hours	2 Litres used	3 Litres per hour	1 Peak hours	2 Litres used	3 Litres per hour
1	3400	3400	13	27000	2130
2	6600	3300	14	28300	2021
3	9600	3200	15	28800	1920
4	11600	2900	16	28900	1806
5	13300	2660	17	29000	1705
6	16300	2716	18	29100	1616
7	19300	2757	19	29200	1536
8	22000	2750	20	29300	1465
9	24000	2666	21	29400	1400
10	25600	2560	22	29500	1340
11	26400	2400	23	29600	1286
12	27200	2266	24	29700	1237

Figure A2 Storage make-up ratio curve

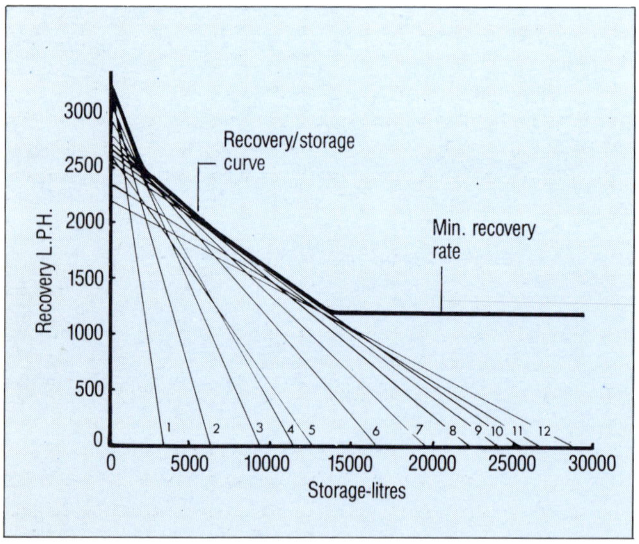

It should be noted that for groups of buildings in which similar types of activity take place and which have the same operating period, the shape of the curves will be closely related, the only variable being the quantitive values of the storage and recovery co-ordinates. It should also be noted that the storage volumes represent net usable volumes to which an allowance should be added to compensate for stratification and heat losses.

The method described is suitable for calculating all types of hot water storage vessels, including off-peak electric heated vessels. The method is not however, applicable to non-storage generators because the maximum performance of these is determined by the peak system requirement per minute not per hour.

SIZING OF PRIMARY AND SECONDARY CIRCULATIONS

Table A13 Density of water

Temperature °C	Density kg/m³	Temperature °C	Density kg/m³
0	999.8	62	982.1
4	1000.0	64	981.1
10	999.7	66	979.9
12	999.4	68	978.9
14	999.2	70	977.7
16	998.9	72	976.6
18	998.6	74	975.4
20	998.2	76	974.3
22	997.7	78	973.1
24	997.2	80	971.8
26	996.7	82	970.6
28	996.2	84	969.3
30	995.6	86	968.0
32	995.0	88	966.7
34	994.3	90	965.3
36	993.6	92	964.0
38	993.0	94	962.7
40	992.2	96	961.2
42	991.4	98	959.8
44	990.6	100	958.3
46	989.8	102	956.9
48	988.9	104	955.5
50	988.0	106	954.0
52	987.2	108	952.6
54	986.2	110	951.0
56	985.2	120	943.1
58	984.3	130	934.8
60	983.2	140	926.1

Circulating Pressure

In a gravity heating circuit it is necessary to calculate the circulating pressure in the system in order to determine the allowable friction loss per metre run of effective pipe length.

For any system, this is obtained as follows:

CP in N/m² = 9.81 (D_r – D_f) × circulating height in metres
where D_r = density of water in return pipe
and D_f = density of water in flow pipe.

As low pressures are usually expressed in millibars and 1m/bar = 100 N/m², the expression becomes:

$$CP(\text{mbar}) = \frac{9.81\,(D_r - D_f) \times \text{circulating height}}{100}$$

Example

Determine the circulating pressure in a system designed for a flow temperature of 82°C and a return temperature of 60°C if the circulating head is 5m.

$$CP = \frac{9.81\,(983.2 - 970.6) \times 5}{100} = \frac{9.81 \times 12.6 \times 5}{100}$$

CP = 6.18 mbar = 618.00 Pa.

Primary Circulation Pipes

When the primary circuit to a hot water supply system operates by natural circulation, the sizing of the primary circulation pipes is carried out in the same manner as that used for sizing gravity heating circuits. The circulating head is the vertical distance between the centre of the boiler and the centre of the hot water storage vessel.

Normally, the temperature of the stored water should not exceed 65°C and the temperature drop between the primary flow and return pipes should not exceed 22°C. For any indirect system with a pump assisted circulation, the temperature drop should not exceed 11°C.

Secondary Circulation Pipes

Secondary circuits may be designed for pumped or natural circulation, but where practicable they should be designed for natural circulation. Circulation depends upon the heat loss from the circuit, the circulation pressure available and the frictional resistance of the circuit.

The flow pipe of a secondary circuit is also the hot water distributing pipe and should be sized accordingly. The return pipe should be sized so that the temperature drop on the circuit is not less than 8°C and not more than 22°C, dependent upon the temperature required at the draw-off points.

Hot Water Distributing Pipes

Dead legs to hot water draw-off points should be as short as possible and in no case must they exceed those given below.

Maximum Permissible Lengths of Uninsulated Hot Water Draw-off Pipes

BS 6700 gives the following restrictions on the lengths of pipe supplying a hot water draw-off tap, measured along the axis of the pipe from the cylinder or tank or from a secondary circuit. These are:

Largest internal diameter of copper pipe:

(1) Not exceeding 12mm	20.0m length
(2) Over 12mm, up to and including 22mm	12.0m length
(3) Over 22mm, up to and including 28mm	8.0m length
(4) Exceeding 28mm	3.0m length

In the case of a compound pipe of differing diameters the largest diameter is to be taken.

PUMPED HOT WATER SYSTEMS

In order to allow for a rapid response to user demands, many large hot water installations include a circulating pump within the secondary circuit.

Owing to the large amount of air that is present in the systems, care should be taken with the pump positioning.

The pump should not be fitted in the lowest point of the secondary circulation pipework because sediment builds up and when passed through the pump could easily damage the bearings and also cause the pump to seize.

To ensure quiet running, the pipework should be generously sized to reduce friction losses and, the pump head kept to a minimum.

Adequate space to permit free ventilation around the pump should be allowed. This will also facilitate servicing should this be necessary.

To render servicing of the pump easier, isolating valves should be fitted on both the inlet and delivery side so that the pump can be inspected without the need to drain down the system.

To prevent strain to the pump body the pipework on either side should be adequately supported and care taken to line up with the inlet and delivery connectors.

The type of pump most suited to this work is of canned rotor, bronze or stainless steel construction with water lubricated bearings having no seals to replace. These units will normally give years of trouble free service.

Owing to their form of construction, this type of pump can be refitted into the existing main pump housing when a replacement motor and rotating element is required.

The power output of these pumps varies between 40W – 120W, and with a running current of 0.40 to 1.9Amp.

The recommended position for a secondary H W circulating pump is shown in Figure A3.

Figure A3 Recommended pump positioning for hot water circulations

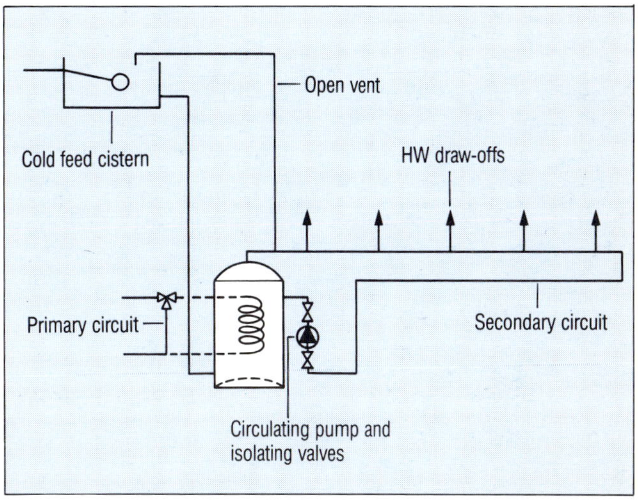

WATER QUALITY

The quality of water is defined by chemical and bacterial analysis and where the end usage is directly or indirectly for human consumption. It should comply with recognised standards to ensure freedom from harmful bacteria, acute and long term toxic substances and in addition, the water should be clear, odourless, tasteless and wholesome.

The 'International Standards for Drinking Water' produced by the World Health Organisation define the toxic limits for substances frequently found in water supplies. In the UK, the Ministry of Housing Report No. 71 (4th edition 1969), recommends the bacterial purity standards for water supplied by public water authorities in addition to establishing procedures to be followed in obtaining samples.

Waters which satisfy the quality standards for human consumption are not always suitable for certain medical, industrial and commercial uses. They may contain levels of trace elements, high total dissolved solids contents, non-pathogenic bacteria, gases and suspended matter which necessitate some form of treatment before the water can be used in activities such as food and pharmaceutical preparation, brewing, research, medicine and in many areas where heat generation and transfer take place.

Chemically pure water has no free ions; its hydrogen ion concentration, usually known as pH value, is equivalent to the hydroxyl ion concentration and on the pH scale has a value of 7 which represents neutrality. This figure indicates the concentration of hydrogen ions in a litre of solution and for pure water the concentration for both types of ions being equal at 10^{-7} gram molecules per litre.

WATER QUALITY

The pH value usually being expressed as $pH = \log \frac{1}{H+}$

where H+ is the hydrogen ion concentration, hence

the pH of pure water is $\log_{10} \frac{1}{10^{-7}} = 7$

Water having a hydrogen ion concentration in excess of 10^{-7} gram molecules per litre will have a pH of less than 7 and will therefore tend to acidity and conversely, waters with a pH of greater than 7 will tend to alkalinity. From the above it follows that pure water having no free ions has poor conductivity and high resistance to electrical current. The conductivity, or its reciprocal resistivity, of a sample of water is a good measure of its purity where this is concerned with dissolved ionisable solids or gases, which is largely the case in research and laboratory applications.

Deionised Water

Where water is required to a higher biological and chemical purity than that supplied by the local Water Authority, single or mixed bed deionisers are capable of producing pure and ultra pure water approaching theoretical H_2O. Purity is usually measured by a conductivity monitor which measures the conductivity in microsiemens.

The operation of deionisation is based on using twin bed or mixed bed columns of positive and negative charged ionised resins so that the incoming water disassociates into positively charged ions such as magnesium and calcium which exchange with the hydrogen ions of the acidic resins.

Likewise, the negative ions of sulphate and bicarbonates are exchanged with the hydroxyl ions of the alkaline resins, the resultant hydrogen and hydroxyl ions combining to form theoretical H_2O. The exhausted resins are periodically replaced or regenerated with acid and alkaline regenerates.

As the name of the process implies, only ionisable dissolved solids and gases can be removed and the treated water is not necessarily pure or sterile. Centralised recirculating systems, local units, and both in combination where terminal polishing is required, can produce water up to 18 megohms at 25°C which is close to theoretical maximum.

Figure A4 shows a two bed recirculation system suitable for providing 3 M^3/HR of deionised water up to 1 microsiemen centimetre quality.

Installations should comply with local Water Authority requirements to prevent backflow contamination.

Softened Water

Water softening in large quantities for commercial use and distribution is usually carried out using one or more of the lime or lime/soda processes. For domestic application however, the base exchange system of softening is more commonly used and only this method is described. Base exchange softening operates on a similar principle to the ion exchange described in Deionised Water and functions by the process of exchanging sodium salts for those of calcium.

The action of softening can be expressed chemically as follows:

SODIUM ZEOLITE + CALCIUM CARBONATE exchanges to CALCIUM ZEOLITE + SODIUM CARBONATE

and that of re-generation:

CALCIUM ZEOLITE + SODIUM CHLORIDE exchanges to SODIUM ZEOLITE + CALCIUM CHLORIDE

Because the base exchange system produces water of zero hardness, it is often only necessary to soften part of the total water requirement and blend the raw and softened water together.

Operation of a Base Exchange Water Softener

The raw/hard water supply is connected into the top of the softener tank and flows downwards under pressure through the bed of softening mineral (zeolite) where the hardness is removed by the process of 'ion exchange'. At the same time, sediment in the water is filtered out and retained by the mineral bed. Clear soft water leaves the bottom of the tank into the water distribution system.

The softening mineral eventually becomes saturated with hardness and no longer softens the water. Re-generation of the softening mineral is then necessary and the three stage process is as follows:

Figure A4 Recirculation deionisation system

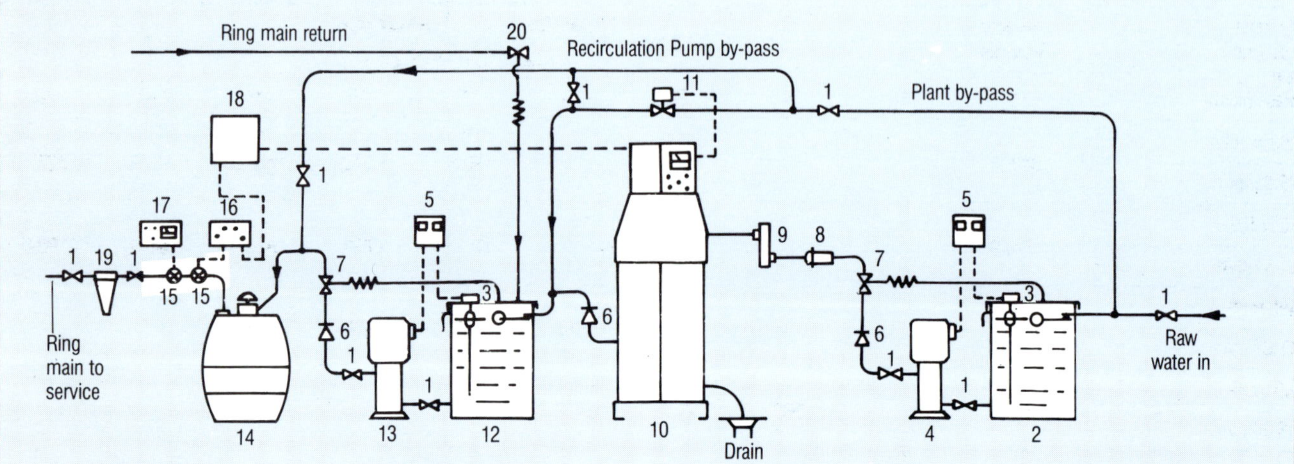

1. Hand isolation valve
2. Break tank for raw water
3. Low level float switch
4. Delivery pump
5. Pump motor starter
6. Non return valve
7. Pressure relief valve
8. Integrating water meter
9. Rate of flow meter
10. Two bed automatic deioniser
11. Solenoid valve
12. Break tank for recirculation water
13. Recirculation pump
14. Deioniser cylinder
15. Line cell
16. Conductivity controller
17. Conductivity meter
18. Water condition alarm
19. 0.2μ Particle filter
20. Pressure regulating/relief valve

(a) Backwash

Reverse the water flow through the mineral bed to remove accumulated sediment and wash to drain. A backwash controller is necessary to limit the water flow and thus prevent washing out the mineral bed.

(b) Brine Rinse

Ordinary salt (sodium chloride) has the ability to fully restore the softening capacity of the mineral. A measured amount of salt-brine is drawn from the brine tank through the brine injector and is rinsed slowly down through the mineral bed to remove the hardness which is rinsed to drain.

(c) Flushing

The water flow is again reversed to re-pack the mineral bed and any trace of sediment not removed by the backwash is flushed to drain. The softener can now be returned to normal service.

Distributed Loads on Steel Joists

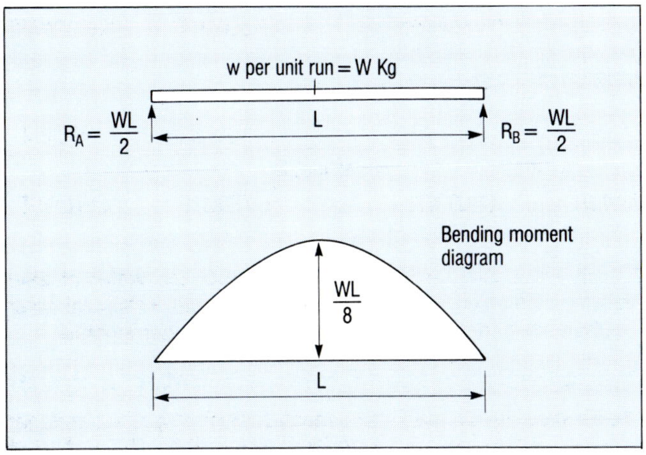

(1) Bending moment $= \frac{WL}{8}$

(2) Bending moment $=$ stress $\times$ section modulus
$= f \times z$

Load distributed (kg) = 134.55 × z/m

$$\text{Equating } \frac{WL}{8} = f \times z$$

and when $f = 165\,\text{N/mm}$

$l = 1$ metre

$$W = \frac{8 \times 165 \times z}{m} = 1320.z/m$$

$$W\text{kg} = \frac{13200 \times z}{9.81 \times m} = 134.55.z/m$$

Distributed loads in kg which can be carried by some Standard beams as RSJs are shown in Tables A14 and A15.

Table A14

	Beam mm	kg/m	Section modulus (z) cm^3
(1)	203 × 102	25.33	225.8
(2)	178 × 102	21.54	170.9
(3)	152 × 89	17.69	115.6
(4)	127 × 76	13.36	74.94
(5)	102 × 64	9.65	42.84
(6)	76 × 51	6.67	21.67

Table A15

	Calculations		W(kg/m)
(1)	134.55 × 225.8	=	30381.39
(2)	134.55 × 170.9	=	22994.59
(3)	134.55 × 115.6	=	15553.98
(4)	134.55 × 74.94	=	10083.18
(5)	134.55 × 42.84	=	5764.00
(6)	134.55 × 21.67	=	2915.69

Refer to Table A16 for loads at clear spans, 1–3 metres.

Table A16 Distributed loads for spans up to 3 metres

Nominal Size (mm) and Mass (kg/m)	Section Modulus cm^3 'z'	Distributed Loads in Kilogrammes								
		Clear Spans in Metres								
		1	1.25	1.50	1.75	2.0	2.25	2.50	2.75	3.00
203 × 102 25.33 kg/m	225.8					15190	13500	12150	11050	10130
78 × 102 21.54 kg/m	170.9					11470	10197	9180	8340	7650
152 × 89 17.69 kg/m	115.6			10370	8890	7780	6910	6225		
127 × 76 13.36 kg/m	74.94		8066	6720	5760	5040				
102 × 64 9.65 kg/m	42.84	5764	4610	3840						
76 × 51 6.67 kg/m	21.67	2916	2330							

WATER QUALITY

The nominal sizes of pipes and fittings used for hot and cold water installations in buildings are normally decided on the basis of two main criteria:

(a) that the resistance of the pipework is low enough to give the required outflow at the various draw-offs within the pressure head available.

(b) that the flow velocities are kept low enough to ensure, as far as is possible, that the system does not generate excessive noise when the appliances are being supplied with water.

Another factor, which in special cases, must be borne in mind is the effect of erosion (due to cavitation) and corrosion in pipework whether external or internal.

Experiments carried out at BRE some years ago indicated that acceptable noise levels would be when flow velocities reached about 3 m/s.

The noise which arose when water was travelling along the pipework was affected, to some extent, by the pipework configuration, i.e. whether sharp pattern elbows were used or large radius bends.

It would appear from the BRE experiments that the time taken for the sound level to be reduced to acceptable levels was affected by the method of pipe fixing and the degree of absorbence of the wall surface finish.

Very high water velocities create noise when cavitation is occurring, generally at abrupt changes of direction. Cavitation is a condition arising when pockets of vapour in the water suddenly collapse and the in-rush of pressure causes a distinct noise to emanate from the pipework.

Cavitation is also well known to hydraulic engineers as a destructive agent which causes severe erosion and pitting of the material in contact with water where cavitation is occurring.

It would appear that limiting the water velocities to about 3 m/s, placing the pipework on brackets which keep the pipework out of contact with the wall surface and using sound absorbent finishes to the wall surfaces wherever possible are important design considerations. A smooth waterway through even small elbows also reduces substantially the degree of noise created at these sharp changes of direction.

OUTWARD THRUST IN PIPELINES DUE TO INTERNAL PRESSURE

When water under pressure flows round any form of bend there is an outward thrust at the bend. The force which exists is created by the static internal pressure and the dynamic force created by a change in direction of the flowing water.

For most normal velocities as used in water supply the dynamic force is small and can be disregarded.

All pipelines subject to internal pressure must therefore be provided with firm anchorage at any change of direction and be restrained from movement at any blanked off ends.

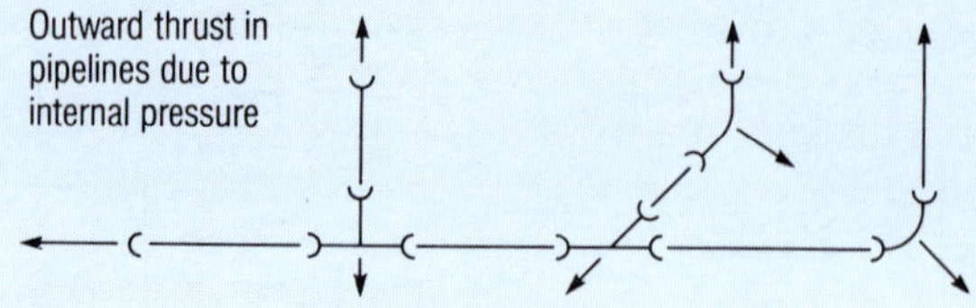

The thrust P is given by the basic expression

$$P = p.A.2\,\text{Sin}\,\frac{\theta}{2}$$

where

P = thrust (Newtons)
p = internal pressure (bar) = $10^5\,N/m^2$
A = cross sectional area of pipe (m^2)
θ = angle of bend (degrees)

For ready application to most practical problems the basic expression given can be simplified down to:

$$P = 8.10^{-2}.K.d^2(\text{mm})$$

where

K (90°) = 1.42
K (45°) = 0.765
K (22½°) = 0.390
K (11¼°) = 0.180
K (blank end) = 1.00

Example

Estimate the thrust which will be transmitted to an anchor block restraint at a 90° bend if the bore of the pipeline is 100 mm and the internal water pressure is 1 bar.

$$P = 8.10^{-2} \times 1.42 \times 100^2$$
$$= 8 \times 1.42 \times 10^2$$

P = 1136 Newtons = 1.136 kN

The outward thrust for various bores and angle of bend at a pressure of 1 bar can be read from Graph A5 based on the simplified formula.

For higher pressures, multiply by the system pressure in bars. Where test pressures are in excess of system working pressures the greatest pressure must be used in the calculation.

Values of K for Various Bends

90°

$$2\,\text{Sin}\,\frac{90^\circ}{2} = 2 \times \text{Sin}\,45^\circ = K = 2 \times 0.7071 = 1.414 = 1.42$$

45°

$$2\,\text{Sin}\,\frac{45^\circ}{2} = 2 \times \text{Sin}\,22\tfrac{1}{2}^\circ = K = 2 \times 0.3827 = 0.7654$$

22½°

$$2\,\text{Sin}\,\frac{22\tfrac{1}{2}^\circ}{2} = 2 \times \text{Sin}\,11\tfrac{1}{4}^\circ = K = 2 \times 0.1951 = 0.3902$$

11¼°

$$2\,\text{Sin}\,\frac{11\tfrac{1}{4}^\circ}{2} = 2 \times \text{Sin}\,5\tfrac{1}{8}^\circ = K = 2 \times 0.0895 = 0.180$$

K = (Blank End) = 1.0

Graph A5 External thrusts developed in pipelines

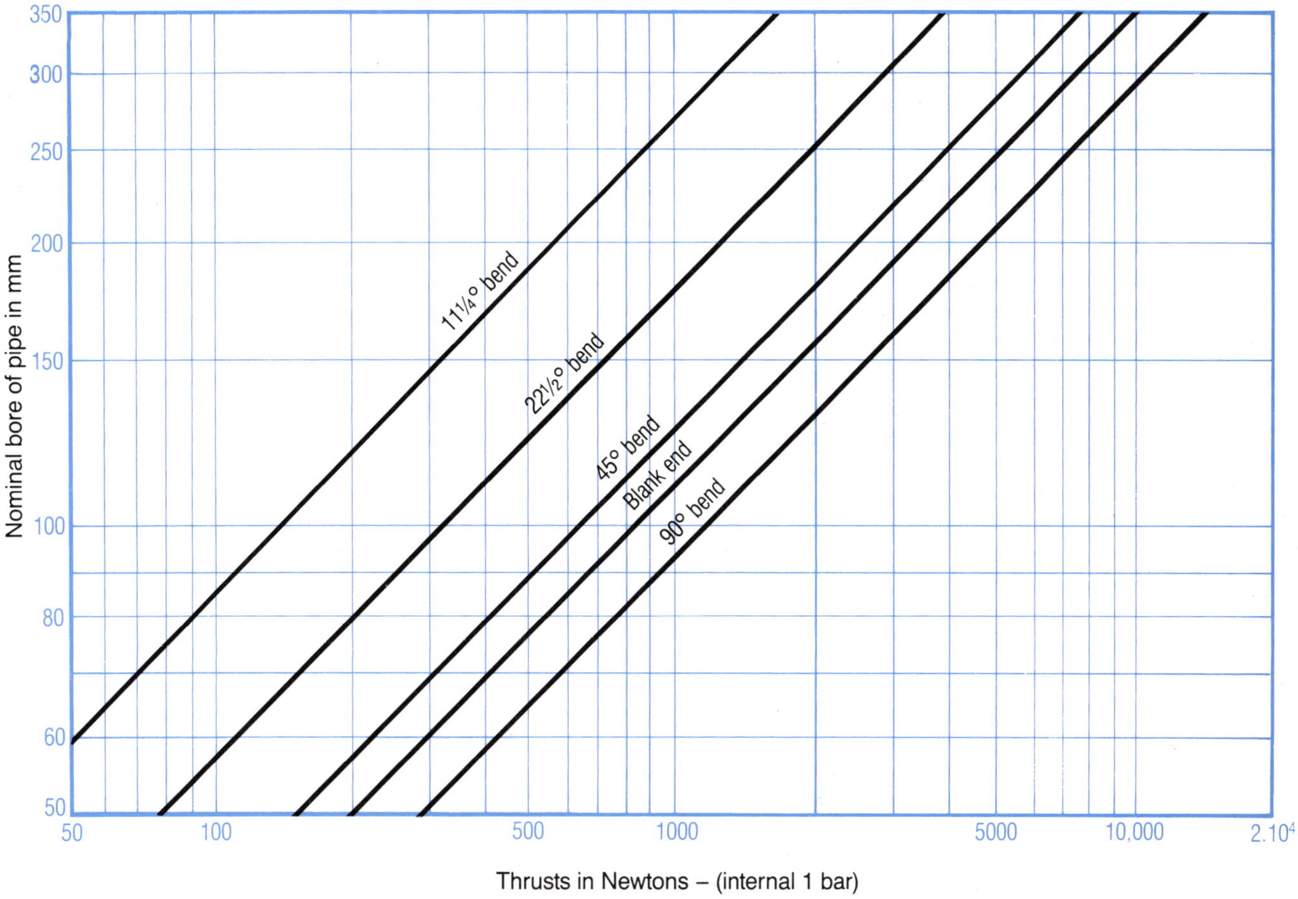

The size and shape of thrust blocks (concrete) will be determined by the load bearing capacity of the ground where the thrust takes place. The following information is given as a guide, but on- site checks should be made, and if necessary, an analysis of the ground condition should be obtained.

Table A17 Bearing capacity of soils

Soil type	Safe bearing load kN/m
Soft clay	24
Sand	48
Sandstone and gravel	72
Sand and gravel bonded with clay	96
Shale	240

ACOUSTIC AND VIBRATION CONTROL

All pipes should be securely fixed to rigid structures rather than to lightweight flexible panels. When pipes are rigidly fixed to flexible structures with expansion joints, the pipe must be provided with movement joints. To reduce plumbing noises to a minimum, the pipes should be installed in substantial and reasonably airtight ducts or enclosures, and adequately supported clear of the enclosing walls. The pipes should never touch the structure.

Pipes should be run so that air is expelled through terminal fittings, ballvalves and space heating elements or vent pipe connections. Pipes must be supported to prevent sagging and the formation of high points which will collect air. This is particularly important for pipework where the circulation is by gravity.

Pipes should be fitted clear of timber joists, floorboards and all building components. Where this is not possible, pads resistant to damage by insects and vermin should be fitted between the pipe and structure to minimise noise.

Protective saddles to prevent damage by nails when laying floor boards and carpets should be installed around all pipes located in positions where they might be subject to damage.

Acoustic

The pipework should be installed and secured in such a manner which will prevent contact with electric cables.

Pipework should not be buried in solid concrete. Where unavoidable, only short lengths of pipe should be installed in solid concrete floors or in walls without using ducts or chases. Then, the pipes must be installed to comply with Water Byelaws and British Standard Codes of Practice.

The minimum acceptable requirements for pipes installed with solid and concrete floors are:

(1) The pipe must be in a continuous length without joints along the whole length of the embedded pipe.

(2) The pipes must be adequately protected from damage and corrosion by a specifically designed and manufactured plastics covering or be products specifically manufactured for the purpose.

(3) The pipe must be adequately thermally insulated to meet the

ACOUSTIC AND VIBRATION CONTROL

requirements of Water Byelaws and British Standard Codes of Practice.

(4) The pipe must be free to move along its entire embedded length to allow thermal movement without damage to itself or other members or property and must not cause, create, emit or transmit an audible noise.

(5) Pipes passing through brickwork, masonary, concrete, or other building components must not be 'built-in' but be provided with a neatly drilled hole or sleeve of suitable diameter and material.

(6) Pipe fixings and supports must be fitted at intervals not exceeding l.8 metres and where the pipes are fixed to skirtings or in other vulnerable positions, the interval must not exceed 1.2 metres. All pipes should be supported to allow free movement for expansion and contraction. The support brackets should be of the screw fixing type with the provision of suitable plugs for masonary locations.

Water Flow Noise

Noise from pipework is generally caused when the pipe surfaces are set into vibration and/or oscillation by the action of the water flowing throught the system. Water flow noises do not become significant below a water velocity of 3 m/s.

Cavitation will greatly increase noise. Although cavitation can occur in outlet fittings, it is not common in pipework because at normal pressures, an average velocity of 7 m/s to 8 m/s is needed to produce cavitation in an elbow fitting. Reduced pressures at the upper parts of water systems incorporating long drop pipes can cause cavitation to begin at lower flow rates and such pipework should be avoided if possible. It is essential that cavitation is prevented for noise levels to be kept within acceptable limits.

Water Terminal Fittings

The sudden changes of direction and minimal downstream pressures that occur at the seating of an outlet fitting, such as a tap or float operated valve, can create cavitation. This phenomenon is the major cause of noise in such fittings, which are the main sources of flow noise in water systems. Cavitation can be reduced by lowering the pressure at the point of inception and by reducing the water velocity. In terminal fittings, the pressure at the point of cavitation may be increased by restricting the outlet.

The water velocity at the seating, the usual location of cavitation inception, may be reduced by design changes or by lowering the supply pressure. When such a pressure reduction results in a tap being opened further to maintain the rate of flow, the water velocity through the seating/washer clearance will be further reduced allowing the tap to operate more quietly. Similarly, a float operated valve will operate more quietly at a lower supply pressure, with the delivery rate maintained by fitting a larger diameter seat.

Water Hammer (Impulsive noise)

Stopvalve Operation

When a tap or valve is closed rapidly, the operation is sometimes accompanied by a backing noise like that of a hammer blow.

Possible prevention and alleviation of 'water hammer' may be achieved by preventing the sudden closing of the valves, by absorbing the pressure peaks; by increasing the attenuation of pressure waves when transmitted through the pipework; by designing the pipework to avoid long straight pipe runs and by restricting the water velocities.

Solenoid-operated valves and self-closing taps are frequently the cause of 'water hammer' noise. When fittings of this type are specified, non-concussive types should be used and properly maintained to ensure continuance of their non-concussive qualities.

An air vessel may be used to absorb the pressure wave at a point close or adjacent to its formation.

It may be possible to reduce the pressure peak wave as it travels along the pipe. One solution is a bellows-type vibration isolator to reduce pressure peak waves, another and more effective being a rubber hose-type isolator. By analogy with the sound transmission situation, the use of 'plastic' pipes may be expected to provide increased attenuation of the pressure pulse but with some materials, exposure to excessive pressure surges should be avoided. Reference to British Standard CP312: Part 2 gives guidance regarding the permissible limits of pressure surges in unplasticized PVC pipework.

The sizing of system pipework to limit water velocities to a maximum of 3 m/s will not necessarily prevent the occurrence of 'water hammer' but will be helpful in limiting the magnitude of the pressure peak waves produced.

Float Operated Valve Oscillation

In the event of an operating arm and float of a float operated valve being set into oscillation, the consequent rapidly repeated opening and closing of the valve usually causes a loud and violent noise in the system which can shear the operating arm. There are many reasons for this oscillation, the most common being the formation of waves on the water surface in the water storage cistern/tank. This wave system is maintained by the intermittent entry of water through the float operated valve. Because of the extended duration of the 'water hammer' noise caused in this way, compared with that of hammer noise resulting from valve closure, it constitutes a very serious risk of damage to pipework and fittings should this oscillation continue.

When the float operated valve is correctly installed, maintained, and is suitable for the supply pressure, the prevention of oscillation is obtained by installing a damping plate immersed in the water of the storage cistern/tank in such a position as will give the greatest resistance to vertical motion. This plate should easily be fitted to the float or arm.

An alternative solution is to fit baffle plates in the water storage cistern/tank to prevent the surface waves affecting the float. Surface waves can also be magnified when the length of the cistern tank is in sympathy with the wave length and this can be eliminated by fixing baffle plates across the width of the storage structure.

Tap Washer Oscillation (Valve bounce)

Renewing the washer normally cures the problem but in obstinate cases, a change of washer material may be necessary. For float operated valves, proper maintenance of the tap mechanism will reduce the tendency to oscillate (valve bounce).

Pump Noise

A correctly designed pumping system used under the correct conditions should not generate excessive noise. When the flow is much greater than the rated flow or the static pressure on the pump is insufficient, there is a risk of cavitation and turbulence, which could result in noise and vibration.

Noise transmission from pumps and pumping sets can be reduced by using rubber vibration isolators between the pump and the main pipework. Suitable isolators can restrict transmission of water-borne noise and vibration of the pipes. The pipework should be isolated from the building structure by using resilient inserts in the pipe brackets. In severe cases of noise transmission, a hydraulic-acoustic filter, tuned to the unwanted frequency, should be used and at this point the advice of a specialist acoustic engineer should be obtained.

Acoustic Transmission

In practice, the transmission process from the noise source to the hearer is complex involving multiple transmission paths, including possible direct airborne transmission; transmission through the pipework; transmission through the building structure and possible various combinations. To minimize audible noise from fittings it is necessary to take all possible transmission paths into consideration.

Sound is transmitted along metal pipes with little loss, and outlet fitting noise can be mistakenly diagnosed as pipe flow noise. Plastics pipes attenuate noise appreciably. The reduction is not linear proportionally to the pipe length. BS 6700 gives lengths between 5 m and 20 m generating the range 1.0 db/m to 2.5 db/m, depending on the pipe material and wall thickness, and states that insertion of the metal-bellows type of vibration isolator in a metal pipe run can reduce the noise transmission by 5 db to 15 db, but a reinforced rubber hose type of vibration isolator can usually give better attenuation values.

Structure-borne noise should be reduced at source. The weight of the structure and the stiffness of the pipe are major factors in eliminating the noise. A 15 mm diameter copper pipe firmly fixed to a 230 mm thick solid brick wall will not induce any appreciable vibrations in it while a 25 mm diameter nominal bore galvanised steel pipe due to its stiffer structure, can generate vibration and so transmit noise. Lightweight structures are relatively easy to set into vibration and readily transmit structure-borne noise. When pipes have to be secured to lightweight structures, the use of flexible and vibration-isolating clips or brackets are required. Where a design choice exists, the use of small diameter copper or plastics rather than steel pipes should be used to reduce the transfer of sound energy from the pipework to the structure.

Where noise is likely to be transmitted to large surfaces such as storage cisterns and tanks which rest directly on ceiling joists, thereby giving a direct acoustic coupling with the air in the habitable rooms, isolation of the structure from the plumbing component is essential. Resilient mountings can be helpful in isolating a water storage cistern/tank from its supporting structure and the advice of the manufacturer of the mountings should be sought to ensure the correct type are selected for the specific application.

Flow noise originating in outlet fittings and transmitting to the pipe runs can be reduced by running the pipes in closed ducts when sound radiation from the pipework is a potential problem.

Float Operated Valve Water Noise

When the stream of water issuing from a float operated valve strikes the water surface in the water storage cistern/tank or the cistern/tank itself, the resulting noise can cause serious disturbance to the occupants of the building. Previously it was common practice to suply water from a float operated valve at a level below the normal water surface by attaching a 'silencing tube' to the valve outlet. Due to difficulty in meeting the backsiphonage protection requirements of BS 1212 when silencing tubes are fitted, this practice is no longer used.

Backsiphonage protection is usually achieved by the float operated valve discharging above the surface of the water in the water storage cistern/tank with silencers available for some types of BS 1212 valves. These reduce flow noise generation at the seat and also control the emergent water stream to reduce splash noise.

Float operated valves using collapsible silencer tubes in conjunction with built-in anti-vacuum devices are available and are listed by the United Kingdom Water Research Centre (WRc) Fittings Scheme which comply to the byelaws but do not comply with BS 1212. The isolation of storage cisterns from the building structure is essential and this is necessary to reduce noises caused by splashing.

Impact Noise in Sanitary Fittings

Taps can generate noise by the impact of the discharging water stream with the surfaces of the sanitary fitting appliance. The use of flow aerators is helpful in reducing such noise as the most disturbance from this type of noise can occur with metal sinks and similar appliances.

The underside of stainless steel bowls and draining boards complying with BS 1244 must be treated to minimize vibration and noise transmission. This treatment could be applied to other types of lightweight appliance when water impact noise is a problem, and in some instances, additional treatment of stainless steel appliances could be worthwhile. It is good practice to isolate lightweight appliances from lightweight structural elements.

Thermal Movement Noise

Noise caused by the thermal movement of pipes can be disturbing by generating creaks, squeaks and banging noises which can occur after the water has been drawn off. Significant thermal movement occurs in hot water pipes as a result of temperature changes. Expanding pipes when installed correctly with proper expansion allowances do not create noise. The use of resilient pipe clips or brackets, or resilient pads between pipes and fixings, will, in many cases, introduce sufficient flexibility to take up thermal movements. Where very long straight runs of pipe are involved, expansion loops or joints may be needed. In such cases, refer to pipe thrust section of this Guide.

Aeration of Water Pipes

Bubbles in the water can cause a significant increase in water flow noise which normally occurs in the hot water system and arises from poor design or operation of the system. Bubbles forming in or passing into water heaters of hot water cylinders are particularly noisy. Domestic hot water systems must be designed and operated so as to avoid general and/or localized boiling, with the whole system being designed to facilitate removal of air during filling of the system after maintenance.

Acoustic Bridge

In addition, noise generated by the water service installation may introduce problems by acting as a noise transmission path bridging structural sound insulation. For sensitive habitat areas such as recording studios, the insertion of flexible vibration isolators in pipe runs may be necessary and should provide the desired degree of acoustic isolation.

WATER HAMMER

Water hammer is a pulsating type of noise which may, on occasion, be heard emanating from a pipework installation. The noise is caused mainly by shock waves which are set up when water, which is flowing at high velocity, is suddenly arrested.

The practical conditions which can create this adverse effect are the too rapid closure of manual valves, or the virtual instantaneous closure of automatic control valves.

Pumped piped water supplies are also liable to water hammer if provision is not made at the design stage. Plunger type pumps for example give a pulsating type of discharge which may require damping out.

Cases have occurred of water hammer damage although little audible sound was present. Premature failure of valves, joints and the loosening of supports and clips are common examples of such damage.

Gravity water installations do not as a rule give rise to the problem

WATER HAMMER

but when water velocities exceed 3 m/s, water hammer conditions may exist.

The increasing use of pressurized domestic water systems in the UK could well lead to an increase in the incidence of water hammer if proper precautions are not taken at the design stage.

System Protection

Shock pressures in pipework supplied at high pressure can be reduced to acceptable limits by incorporating hydro-pneumatic accumulators or water hammer arrestors in the pipeline at suitable points.

The earliest form of water hammer absorber was the simple air vessel or air bottle. In service, these usually became water-logged and ineffective unless means were available to renew the air cushion.

Modern hydro-pneumatic accumulators incorporate a pre-charged flexible gas bag in their design which is filled with air or nitrogen gas (see Figure A5). Loss of air or gas in service is thereby prevented. Consequently, their continuing efficiency in service is satisfactory and reliable.

When used for this application, hydro-pneumatic accumulators absorb the kinetic energy of the moving water and much of the initial shock pressure; this occuring when the nitrogen gas is compressed in the flexible bag.

Basic Design Principles

The shock pressure waves which may be generated in pipes, travel at a velocity V which is affected both by the compressibility of the water and the elasticity of the pipe walls.

It can be shown mathematically that the velocity of the shock wave is close to the velocity of sound under the same conditions.

The time interval, in seconds, for a shock pressure wave to form at the back of a valve or other arrest point is given by:

$$t = \frac{2 \times \text{length of pipe}}{\text{velocity of sound}}$$

$$t = \frac{2 \times L}{V_s}$$

The velocity of sound is 1450 m/s under perfect conditions. In the case of shock pressure waves, this is usually reduced by the conditions prevailing which must be taken into account.

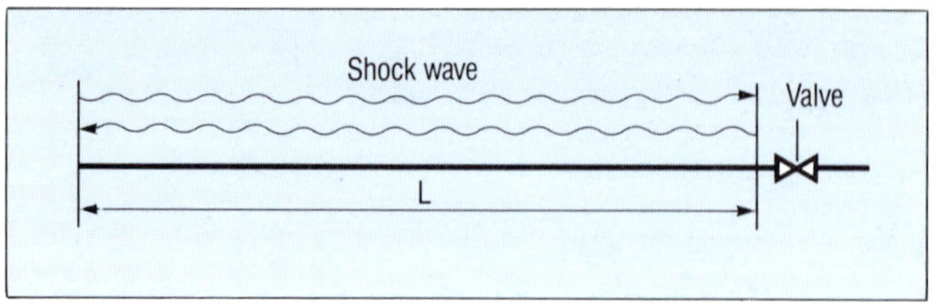

Example

Calculate the time required for a shock wave to form at a rapid closing valve in a pipe 14.5 metres long.

$$V = 1450\,\text{m/s}$$

$$t = \frac{2L}{1450}$$

$$t = \frac{2 \times 14.5}{1450}$$

$$t = \frac{1}{50} \text{ seconds}$$

Solenoid valves of the type frequently fitted to automatic washing machines etc, have closure times of a few milliseconds i.e, 5/1000 or 2/200 seconds. This type of valve can therefore be regarded as rapid closing and liable, in some circumstances, to cause shock waves to be set up when in operation, particularly when they are installed in high pressure water installations.

It can be shown that, if the time taken to close a valve is less than that for the pressure wave to travel the length of the pipe, the pressure wave produced is equal to that if the valve was closed instantaneously.

Shock Pressure Rise

When considering water hammer problems, it is necessary to take into account the initial water velocities, the pipe material and the disposition of the pipe runs.

Figure A5 Hydro-pneumatic accumulator

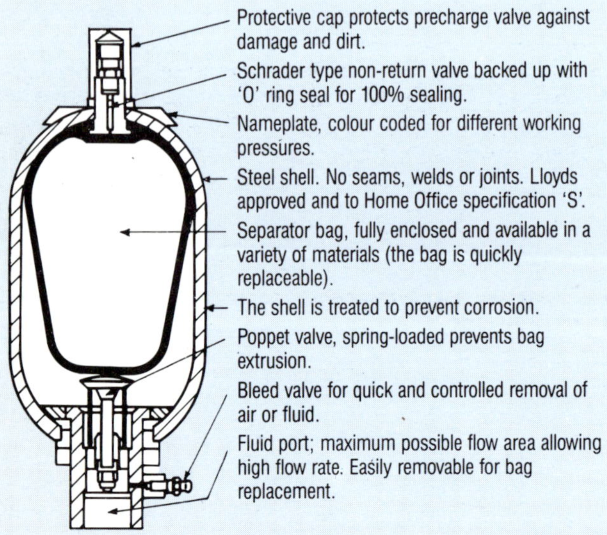

Operation
The accumulator is first charged with nitrogen at a predetermined pressure. Fluid from the pump enters through the fluid port, compressing the gas in the rubber bag, the bag floating between gas and fluid in equilibrium. The gas maintains pressure in the accumulator and forces out the system fluid under pressure when the circuit requires it.

1. *Accumulator neither pressurised nor charged.*
2. *Accumulator pre-loaded with gas.*
3. *Gas in bag compressed by fluid from pump.*
4. *Bag expanded after discharging fluid.*

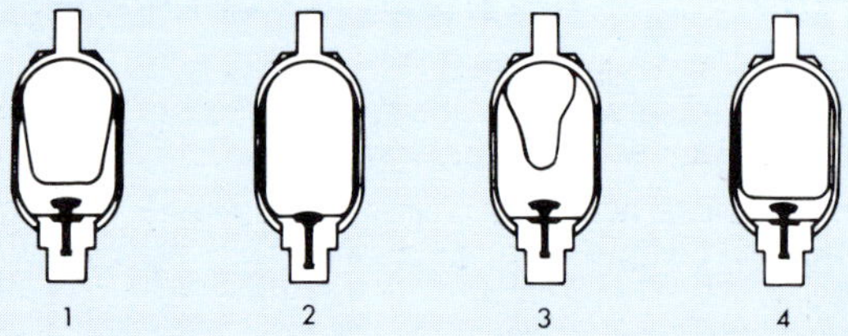

Graph A6 shows that the velocity of the sound wave is affected by the main conditions:

(i) The pipe material, i.e. value of 'E'.
(ii) The diameter/wall thickness ratio.
(iii) The bulk modulus of water or 'K'.

These variables are taken into account in the basic formula below:

$$V_s = \sqrt{\frac{1}{p\left(\frac{1}{K} + \frac{d}{tE}\right)}}$$

where:

V_s = velocity of sound pressure waves (m/s)
p = density of water in (kg/m^3)
K = bulk modulus of water = 2.07 GN/m^2
E = modulus of elasticity of the pipe material

E = for steel = 200 GN/m^2
E = for cast iron = 100 GN/m^2
E = for copper = 110 GN/m^2
$\frac{d}{t}$ = pipe diameter/wall thickness ratio (independent of the units)

Graph A6 is based on the above formula.

The shock pressure rise in a pipe can be shown to be given by the product of the initial water velocity and the sound pressure wave velocity from which:

$P_r = V_s.V.$ kN/m^2

where V = initial water velocity in m/s

P_r = pressure rise kN/m^2

Example 1

Calculate the shock pressure rise when water is flowing through a 150 mm nominal bore copper pipe at 3 m/s when the water is suddenly arrested by the closure of a valve.

For 150 mm pipe $\frac{d}{t}$ = 80 (see Graph A6)

From Graph A6 V_s = 925 m/s. V = 3 m/s

Pressure rise P_r = 925 × 3
P = 2775 kN/m^2

Example 2

Calculate the shock pressure rise when water is flowing through a 15 mm nominal bore copper pipe at 3 m/s when the water is suddenly arrested by the closure of a valve.

For 15 mm pipe $\frac{d}{t}$ = 20 m (see Graph A6)

From Graph A6 V_s = 1225 m/s. V = 3 m/s

P_r = 1225 × 3
P = 3675 kN/m^2

It will be seen from these two examples that thinner wall pipes reduce the pressure rise more than rigid thick wall pipes.

Sizing of Hydro-Pneumatic Accumulators

In addition to the shock pressure rise, the plumbing designer must also calculate or otherwise evaluate, the kinetic energy released by the moving water when it is brought to rest.

In a hydro-pneumatic accumulator, the kinetic energy is absorbed in doing work by compressing nitrogen gas.

kinetic energy destroyed on impact = work done on gas in accumulator

$$KE = \tfrac{1}{2} mV^2 = P_2 . \frac{V_2 - P_3V_3}{n - 1} = \text{work done}$$

where:

for KE	work done on gas
m = mass kg	P_1 = gas charge pressure
g = 9.81 m/s^2	V_1 = gas charge volume
V = Velocity m/s	P_2 = initial pressure
	V_2 = initial gas volume due to P_2
	P_3 = shock pressure
	V_3 = shock volume
	n = 1.4 (gas index for nitrogen)
	P = N/m^2
	V = cm^3

Example 3

Water is flowing at 3 m/s through a 15 mm nominal bore (13.6 mm actual) pipe when the water is suddenly arrested by the closing of a valve. Calculate the KE of the moving mass of water if the pipe is 15 m long.

KE = ½ m V^2

mass of water moving at 3 m/s

$$\text{mass} = 1000 \times \frac{\pi d^2}{4} \times \frac{L}{10^6}$$

$$= 0.7854 \times 13.6^2 \times \frac{15}{10^6}$$

mass = 2.18 kg

KE = ½ × 2.18 × 3^2

KE = 9.8 Nm

To provide a small working margin to allow for variations in actual pipe bores, say KE = 10 Nm.

Note

(i) If the pipe length remains constant then,

$$\text{new KE} = \text{original KE} \times \left(\frac{D}{d}\right)^2$$

(ii) If the pipe length is altered then,

$$\text{new KE} = \text{original KE} \times \frac{L}{l}$$

Example 4

Calculate the volume of nitrogen gas in cm^3 required to be provided in a hydro-pneumatic accumulator which is required to absorb the kinetic energy of water which is flowing at 3 m/s through a 15 mm diameter pipe 15 m long. The line pressure is 414 kN/m^2 and the shock pressure rise is not to exceed a further kN/m^2.

Note

Initial fill pressure of the gas is to be 90% of line pressure.

Absolute pressures have been obtained by adding 100 kN/m^2 to gauge pressures.

WATER HAMMER

Figure A6 Accumulator sizing

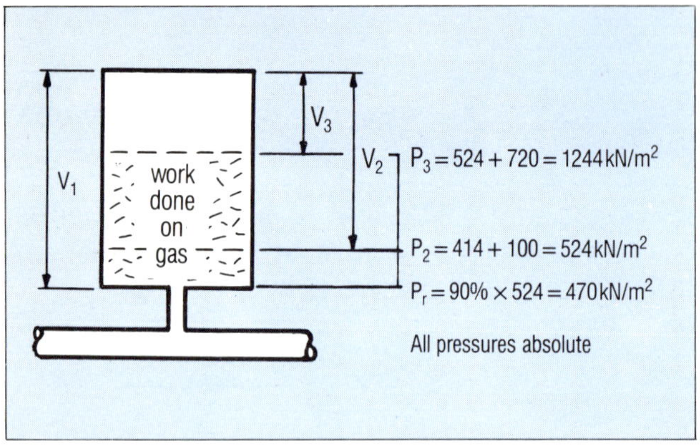

(i) Initial compression = $\dfrac{P_1V_1}{T_1} = \dfrac{P_2V_2}{T_2}$

(ii) Compression due to shock pressures = $P_2.V_2^n = P_3.V_3^n$

(iii) Work done on gas = $\dfrac{P_2V_2 - P_3V_3}{n-1}$

Allowing for initial compression due to line pressure, and assuming $T_1 = T_2$ (constant temperature)

$$P_1V_1 = P_2.V_2 \quad \text{or} \quad V_2 = \frac{P_1}{P_2}V_1$$

$$= \frac{470}{524} \times 100$$

$$V_2 = 90\,cm^3$$

Taking n = 1.4 (nitrogen)

Main compression due to shock pressure

$$P_2.V_2^n = P_3.V_3^n$$

$$V_3 = V_2\left(\frac{P_2}{P_3}\right)^{\frac{1}{n}} = 90 \times \left(\frac{524}{1244}\right)^{0.71}$$

$$= \frac{90}{2.38^{0.71}} = \frac{90}{1.854}$$

$$V_3 = 48.0\,cm^3$$

Work done on the nitrogen gas during compression stage

$$WD = \frac{P_2V_2 - P_3V_3}{n-1}$$

$$= \frac{(524 \times 90 - 1240 \times 48.0)}{(1.4 - 1) \qquad 10^3}$$

$$= \frac{(5.24 \times 9 - 12.4 \times 4.80)}{0.4}$$

$$= \frac{47.2 - 59.5}{0.4}$$

$$= -\frac{12.3}{0.4}$$

$$WD = -30.6\,Nm \text{ per } 100\,cm^3$$

from Example 3, KE for 15m run of 15mm pipe, V = 3m/s

KE = 10Nm

Volume of nitrogen for 10Nm of energy

$$\text{Volume} = 10 \times \frac{100}{30.6}$$

$$= 32.6\,cm^3$$

Volume say = 33cm³

Example 3 can be used as a basis for other conditions.

It will be observed that three variables can be isolated as affecting the kinetic energy to be absorbed under any given practical conditions.

(i) KE can vary with the square of the diameter (d^2).
(ii) KE can vary with the square of the velocity (V^2).
(iii) KE can vary with the length of the branch.

From which we get the following relationships:

(i) New KE = KE × $\dfrac{D^2}{d^2}$ = KE × $\left(\dfrac{D}{d}\right)^2$

(ii) New KE = KE × $\left(\dfrac{V_2}{V_1}\right)$ = KE × $\left(\dfrac{P_2}{P_1}\right)$

(iii) New KE = KE × $\dfrac{L_2}{L_1}$

These relationships have been applied to obtain the volume of nitrogen gas for other pipe diameters, velocities, pressures and lengths.

Hydro-Pneumatic Accumulator Sizing

KE of 13.6mm dia (15mm NB) copper pipe when the velocity of the water is 3m/s and length of 15m.

KE = 10Nm

For 20mm NB
20.2mm (actual)

$$KE = 10 \times \left(\frac{20.2}{13.6}\right)^2 = 22.2\,Nm$$

For 25mm NB
26.2mm (actual)

$$KE = 10 \times \left(\frac{26.2}{13.6}\right)^2 = 37.0\,Nm$$

For 32mm NB
32.6mm (actual)

$$KE = 10 \times \left(\frac{32.6}{13.6}\right)^2 = 57.5\,Nm$$

For 40mm NB
39.6mm (actual)

$$KE = 10 \times \left(\frac{39.6}{13.6}\right)^2 = 85.0\,Nm$$

For 50mm NB
51.6mm (actual)

$$KE = 10 \times \left(\frac{51.6}{13.6}\right)^2 = 144.0\,Nm$$

Table A18 Accumulator sizing 1

Calculated Copper Pipe	42m head Kinetic Energy	v = 3m/s Volume IOP	Volume USA/PDI	PDI 42m
15	10.0	32.5	33	A
22	22.2	72.0	68	B
28	37.0	120.0	119	C
35	57.5	186.0	187	D
42	85.0	276.0	267	E
54	144.0	465.0	475	F

Table A19 Accumulator sizing 2

Calculated Copper Pipe	60m head KE × 1.5	v = 3.5m/s Volume	USA PDI 60m Head
15	15.0	48.6	B
22	33.2	108.0	C
28	55.2	180.0	D
35	86.0	280.0	E
42	122.0	470.0	F
54	216.0	700.0	DF

Table A20. Flow rates from copper tubes at 3m/s and equivalent IOP loading units

Diameter mm	Flow rate l/s	KE Nm	IOP Loading units	USA vol.	PDI size
15	0.52	10.0	25	33	A
22	1.10	22.2	75	68	B
28	1.60	37.0	140	119	C
35	2.80	57.5	400	187	D
42	4.00	85.0	500	267	E
54	6.60	144.0	900	475	F
67	10.00	214.0	1600	800	DF
76	14.00	288.0	2500	900	FF

In the USA where hydro-pneumatic accumulators or water hammer arrestors are widely used on high pressure water installations, a method of rationalizing the capacities of the units has been adopted.

Six standard sizes from 15 mm to 54 mm are given code letters A to F. Size A, which is intended for ½in or 15mm pipes is the smallest (see Graph A7). For applications outside the largest standard size, various combinations of the six basic units can be used (see Graph A8).

Note

The basic pipe length is 15m and the velocity of the water is up to 3m/s. The basic line pressure used is 42m head and the shock pressure 1 MN/m² or 102m head.

The adoption of a 3m/s limit on the velocities used with the basic sizes also provides a rational basis for sizing the units in a l/s rating or in IOP loading units.

A table giving the alternative sizing is shown in Table A20.

Hydro-pneumatic accumulator sizing based on the kinetic energy of the water flowing in the branch can be conveniently carried out by referring to the Graphs A7 and A8.

For sizing units where the branches are of variable lengths use Graph A9.

Example 5

Calculate the KE to be dissipated by a pumping main at the point of switch-off and the resultant shock pressure rise.

Data

Static head 50m — Diameter of main 75mm

Pump power output 4kW

$$\text{Mass of water} = \rho \times \frac{\pi}{4}\,\frac{d^2}{10^6} \times L$$

$$= 10^3 \times 0.7854 \times \frac{76^2}{10^6} \times 50$$

$$= 0.7854 \times 7.6^2 \times 5$$

Mass = 227 kg

Force = m × a = 227 × 9.81 = 2220N

4kW = 4000N.m/s

$$\text{Velocity of water} = \frac{4000\,\text{N m/s}}{2220\,\text{N}} = 1.8\,\text{m/s}$$

KE = ½.mV² = ½.227 × 1.8²

KE = 368Nm

Pressure rise at switch off

For 75mm steel pipe, $\frac{d}{t} = 15$

from the graph A6

$V_S = 1315\,\text{m/s}$

$P_r = V_S\,V = 1315 \times 1.8 = 2370\,\text{kN/m}^2$

$P_r = 2.370\,\text{M N/m}^2 = 2.37\,\text{M.Pa}$

Example 6

A pump is supplying overhead storage tanks with water under the following hydraulic conditions:

Static head 50m

Effective pipe length including fitting resistance 150m

Pipe diameter 50m

$$\frac{\text{head}}{\text{length}} = \frac{50}{150} = 0.33\text{m per metre run}$$

From IOP pipe sizing chart, Graph A1, a 54mm pipe at 0.33m/m run will give a flow velocity of 4m/s

$$\text{mass} = \rho \,.\, \frac{\pi}{4} \,.\, \frac{d^2}{10^6}.L$$

$$= 10^3 \times \frac{0.7854}{10^6} \times 50^2 \times 125$$

$$= 10 \times 0.7854 \times 5^2 \times 1.25$$

= 246kg

KE = ½m.V² = ½ × 246 × 4²

KE = 1960Nm

50mm pipe, $\frac{d}{t} = 12$

from graph A6

V_S = 1330m/s

$P_r = V_S.V\,\text{kN/m}^2$

= 1330 × 4

= 5320kN/m²

$P_r = 5.32\,\text{M N/m}^2 = 5.32\,\text{M.Pa}$

WATER HAMMER

Example 7

Calculate the capacity of a hydro-pneumatic accumulator required to be provided for a 6m branch pipe supplying 10 wash basins concentrated use.

Velocity of flow 3 m/s

From Table A2

10wb = 10 × 3 = 30 LU

and from pipe sizing chart, Graph A1

30 LU = 0.60 l/s

From the pipe sizing chart Graph A1, a 15mm pipe will convey 0.6 l/s at 3 m/s and from the selection chart for hydro-pneumatic accumulators, Table A18, a 15mm pipe carrying water at 3 m/s requires a 15 cm^3 accumulator.

The nearest standard size is 33 cm^3 or PDI 'A'.

Example 8

A branch pipe 11 m long, is under a static head of 40m. The estimated flow rate is 3.2 l/s. From the IOP pipe sizing chart a 42mm pipe would be required for the branch.

3.2 l/s is equivalent to 350 LU.

From the hydro-pneumatic sizing graph A9 a 42mm diameter branch 11 m long would require an accumulator of 190 cm^3 capacity.

The nearest single size is 267 cm^3 or PDI size 'E'.

In practice two smaller units coupled in tandem would probably be used.

From which $\frac{190}{2} = 95\,cm^3$

Nearest size to 95 cm^3 is 119 cm^3.

Therefore two 119 cm^3 accumulators could be used coupled PDI size 'C'.

Graph A6 Velocity of shock wave/pipe wall thickness graph

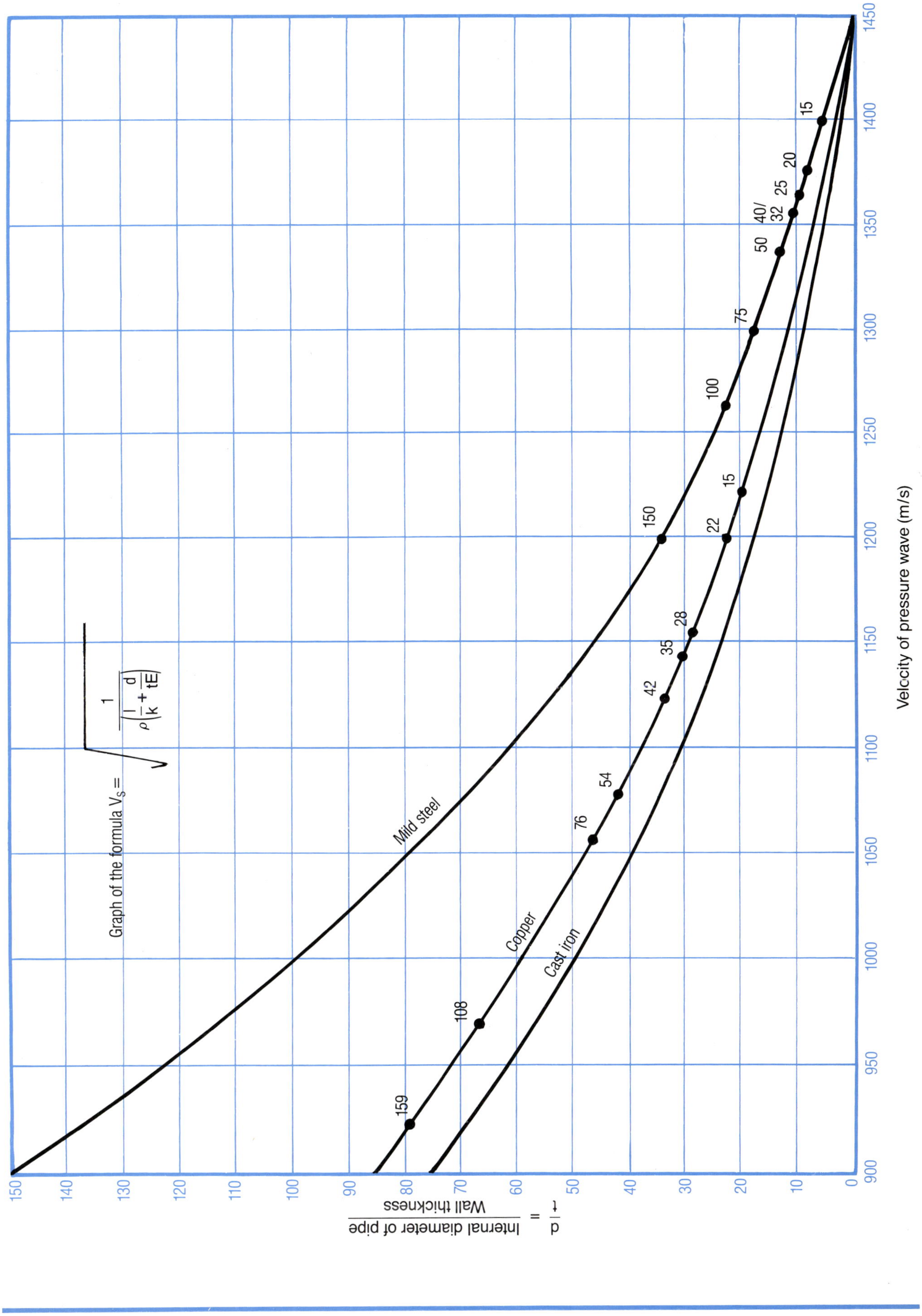

Graph A7 Hydro-pneumatic accumulator selection chart (single units)

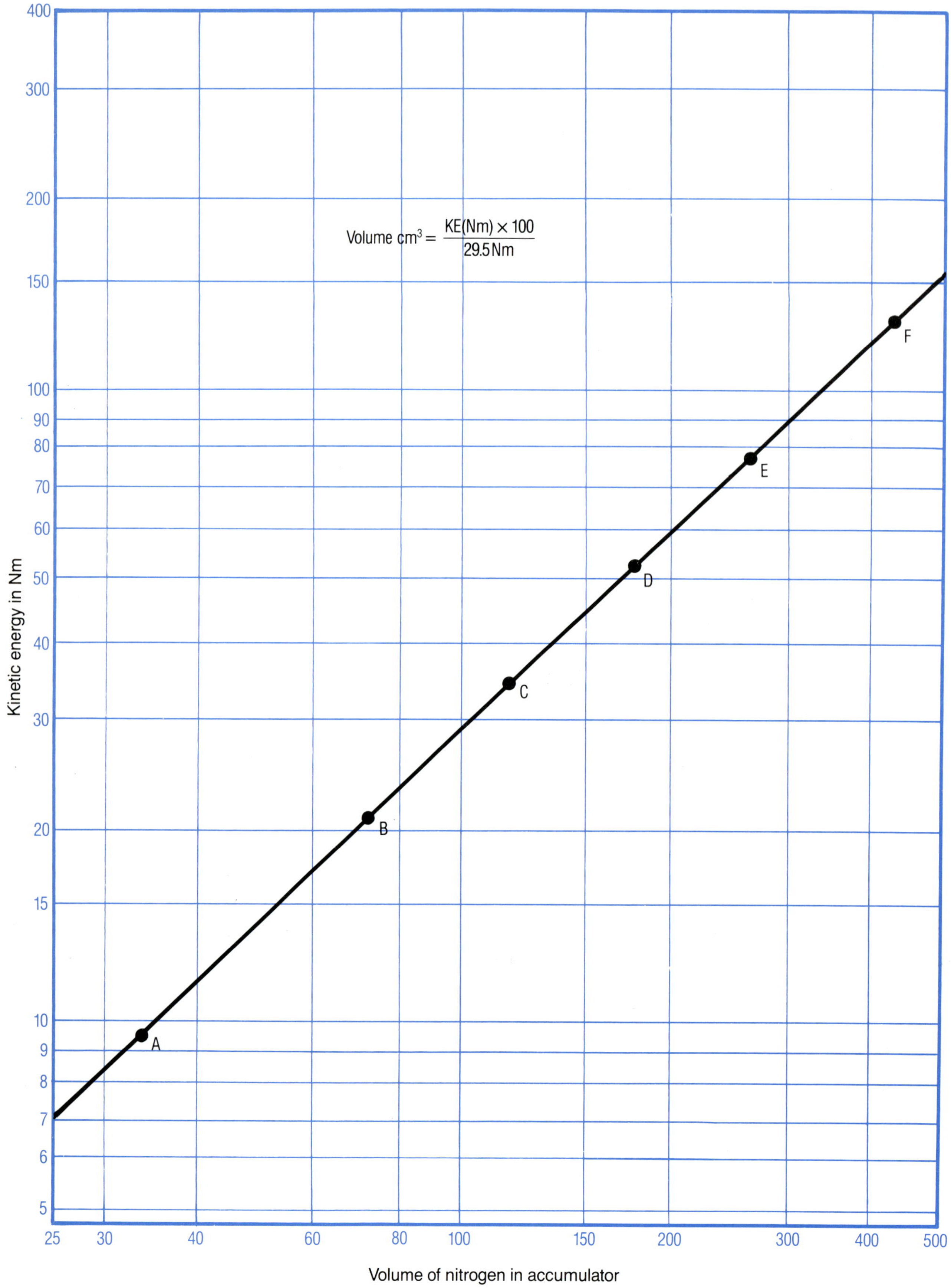

Graph A8 Hydro-pneumatic accumulator selection chart (multiple units)

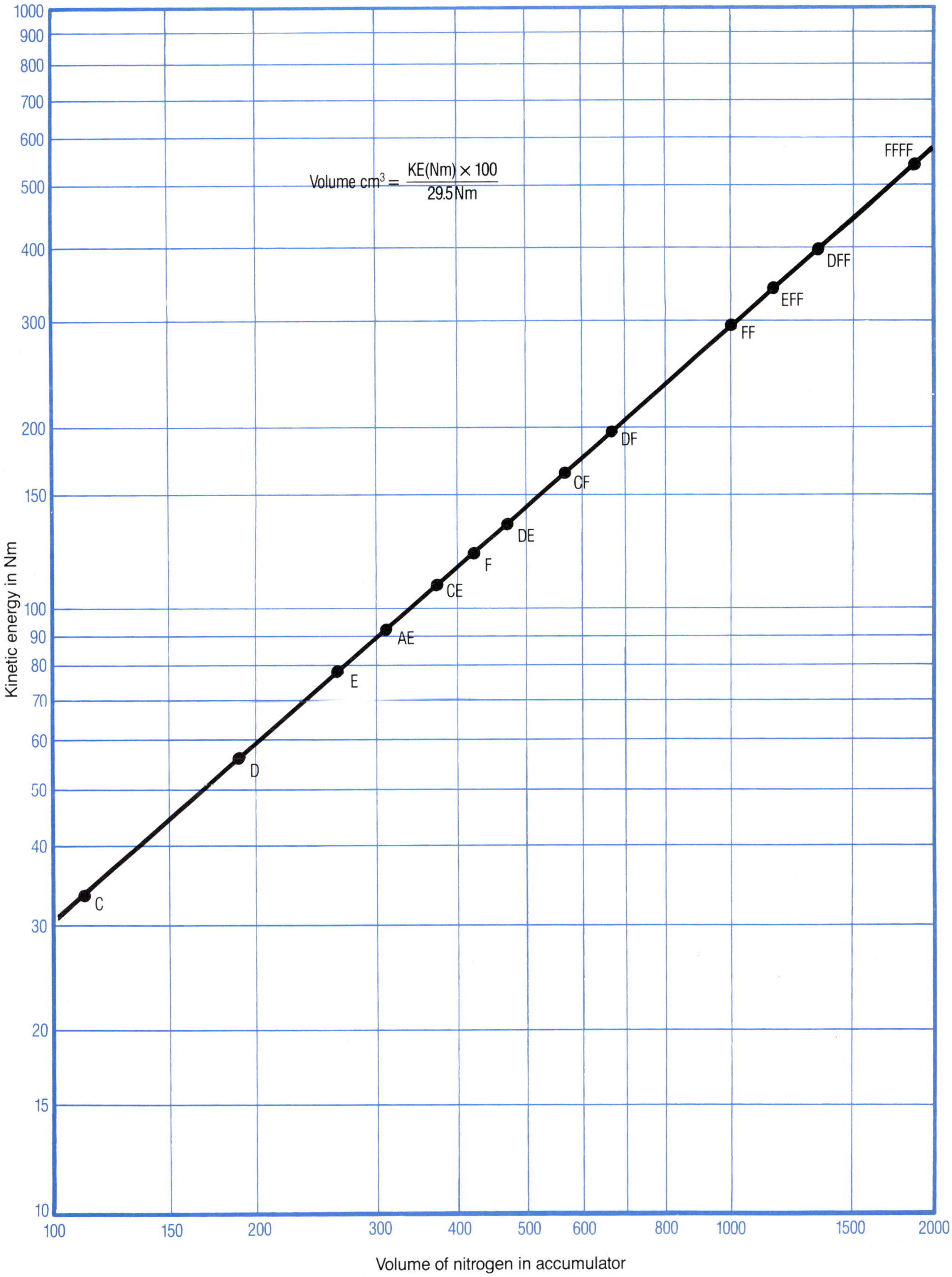

Graph A9 Hydro-pneumatic accumulator selection chart (5-50 m pipe run)

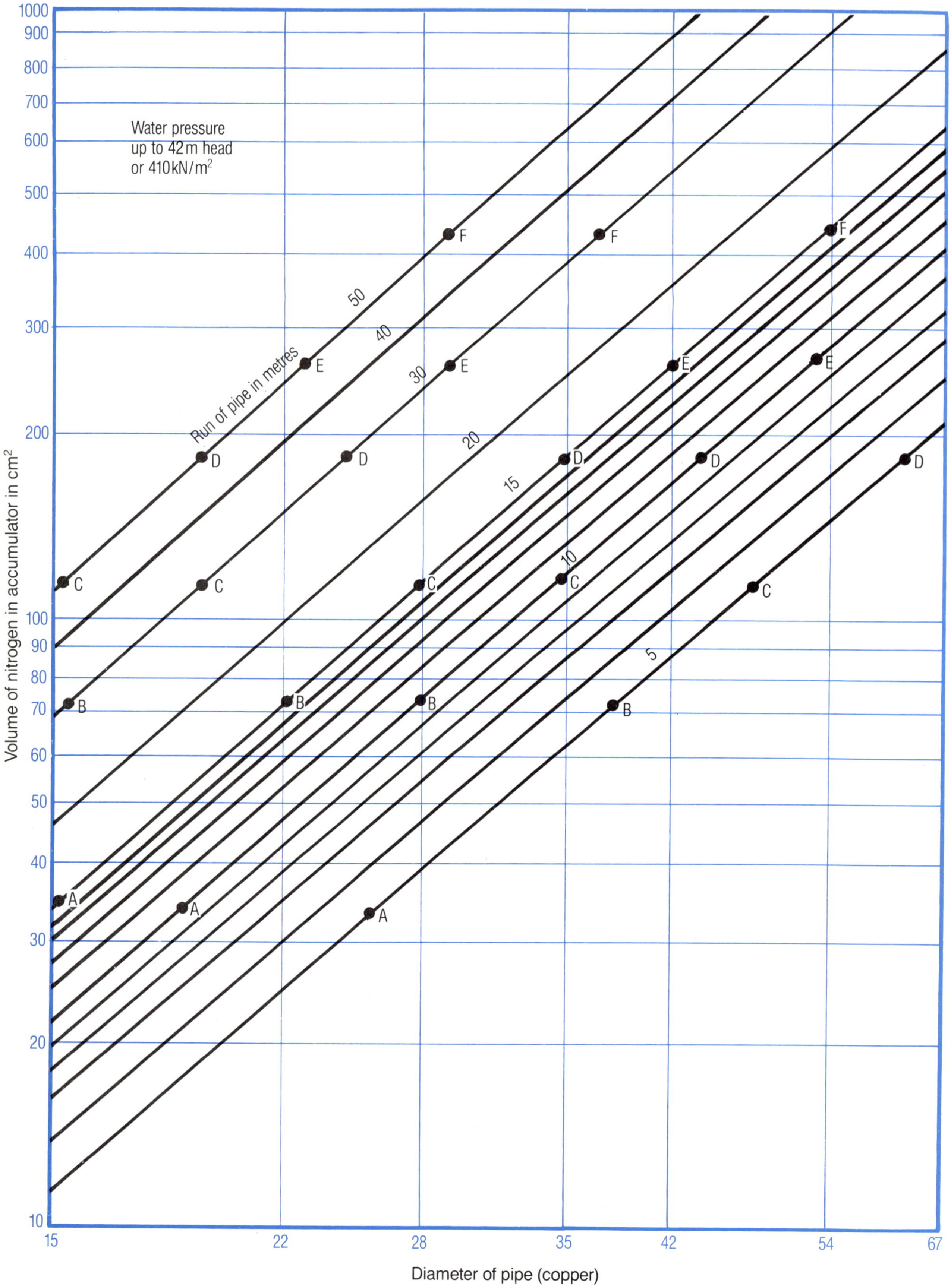

INSPECTION AND TESTING OF WATER MAINS AND WATER SUPPLY SYSTEMS

Water Mains

The mains should be tested in sections as the work of laying proceeds and joints should be left exposed for inspection during testing. After completion of each section, the main should be carefully and slowly charged with water so that all air is expelled, allowed to stand full of water for 1–2 days if possible and then tested under pressure. The test pressure should be 90 m head or the maximum working pressure plus 50%, whichever is the greater.

The pressure should be applied by means of a manually operated test pump or, in the case of long mains or mains of large diameter, by a power-driven test pump provided that the pump is not left unattended. Precaution must always be taken to see that the test pressure is not exceeded. Pressure gauges must be accurate and if necessary should be re-calibrated before the test.

After the pump has been stopped, the test pressure should be maintained as long as is necessary to inspect the whole of the pipework under test and in any event not less than half-an-hour.

Open ends of mains should be temporarily closed for testing under moderate pressure by fitting watertight expanding plugs. The end of the main and any test plug must be well secured to resist the end thrust of the water pressure in the main, i.e. maximum test pressure x cross sectional area of pipe. If the section of main terminates with a sluice valve, the wedge of the valve should never be used to retain the water because this might lead to permanent distortion of the working parts of the valve. Instead, the valve should be fitted with a blank flange or socket plug and the valve left in the open position whilst testing. End support should be provided as explained previously.

Cold Water Systems

When the installations are complete they should be slowly and carefully charged with water, allowing all air to escape thus avoiding shock or water hammer. The systems should be inspected under working conditions of pressure and flow and when all draw-off taps are closed, should be absolutely watertight. Each draw-off tap should be opened and tested for rate of flow.

Certain specifying authorities may require pressure testing of internal pipework in which case, systems should be tested in accordance with the pressure test previously described. In such cases, it may be necessary to isolate items of equipment from the pressure test if they are not capable of withstanding the test pressure. Where these items are removed, blanking flanges or plugs must be used or a make-up piece of pipework installed temporarily.

All piping, fittings and appliances should be inspected and checked for satisfactory support and protection from physical damage, corrosion and frost.

Because of the possibility of damage in transit, it is always advisable to re-test cisterns, tanks and cylinders for water-tightness on arrival at site and before fixing.

Hot Water Systems

Hot water systems should be thoroughly flushed out and then tested in the same manner as described for cold water systems. Where thermal insulation is used, the hydraulic test should be made before the insulation work is completed and whilst all joints are exposed.

Where a pressure test is employed, boiler and calorifier relief valves should be removed and these valves should be tested later. The test pressure should be one and a half times the normal working pressure and this should be maintained for thirty minutes after making good any leaks.

It is necessary to carry out the hydraulic pressure test on sections of pipework prior to completion of the whole installation where these are fixed in ducts, chases, trenches, etc, and are concealed from view. If rectification of faulty materials or workmanship on such sections is likely to involve disturbance to finished structural features, the test pressure should be twice the normal working pressure.

Sterilization of Cold Water Systems

The whole of the system should be sterilized to eliminate possible traces of bacteria.

Sterilization of public water mains is carried out by the Water Authority who may also carry out the sterilization of new private mains. Where this is not standard practice, the plumbing contractor should carry out the sterilizing process as described below.

After cleaning the cistern of all debris, the cistern and pipework should be filled with water and the whole thoroughly flushed out. The system should then be filled with water a second time, but as the cistern is filling, a sterilizing chemical containing chlorine should be added to ensure thorough mixing of the chemical and water. The dose should be such as to give 50 parts of chlorine to one million parts of water. If ordinary bleaching powder is used, the proportion should be 150 g of powder to 1000 litres of water, the powder first being mixed with water to a creamy consistency before being added. Proprietary brands of sterilizing chemicals should be added in the proportions as instructed by the manufacturers.

After filling the system, the incoming water supply should be shut-off and each tap on the distributing pipes opened successively, starting with that nearest the cistern. As the water which issues from each tap begins to smell of chlorine, the tap should be closed. The cistern should then be filled again to the water-line with water to which has been added the correct dose of chemical.

The whole system should then be allowed to stand charged with treated water for a period of at least 3 hours, after which a test should be made by smell for residual chlorine. If none is found, the sterilization should be repeated.

Before any water is used for domestic purposes, the whole system must be emptied and thoroughly flushed out with clean water.

PREVENTION OF CONTAMINATION OF WATER BY BACKSIPHONAGE OR CROSS CONNECTION

Water Byelaw requirements for protecting mains supplied water services from the risks of contamination arising from backsiphonage, backflow or cross-connection are concerned with three principles of protection:

1. Water supplied from the mains must be kept separate from any other water, such as that supplied from a private source (e.g. a well or borehole) and water that is otherwise non-potable. This includes mains supplied water that has been drawn-off for use, even if its quality has not deteriorated.

2. There must be no cross-connection within a water installation, such as between mains-fed supply pipes and cistern-fed distributing pipes, and between cistern-fed primary and secondary hot water circuits. Although a closed circuit must not be connected to a mains-fed supply pipe, the primary heating circuit of a sealed system may be filled or topped up from a temporary

PREVENTION OF CONTAMINATION OF WATER BY BACKSIPHONAGE OR CROSS CONNECTION

connection made to a supply pipe, through an approved and effective backflow prevention device, provided that the temporary connection is removed after use. The connection of any pump or other similar device into a supply pipe, so as to cause backflow into the pipe, is prohibited.

3. Every point of water use, such as a draw-off tap, WC flushing cistern or washing machine, must be protected against backflow by the use of a suitable backflow prevention device.

Requirements for backflow protection at every point of use of water are based upon the recommendations of the 1974 Report of the Committee on Backsiphonage in Water Installation, for the three categories of risk shown in Table A21, and illustrated in Figures A7–A12.

Table A21 Requirements for backflow protection

Category of risk	Source	Examples of recommended protection
Class 1 Risk of contamination which is likely to be harmful to health, from a substance continuously or frequently present	WC pan Bidet	Approved flushing cistern correctly installed Type A air-gap
Class 2 Risk of contamination which is, or is likely to be harmful to health, from a substance which may be present	Taps at sinks etc Hose union taps. Clothes washing machines	Type A air-gap Combination in-line-vacuum and check valve. Type B air-gap.
Class 3 Risk of contamination which is not, or is not likely to be, harmful to health	Hot & cold mixer taps Domestic	Any of the above devices, or a check valve or an in line anti-vacuum valve. Any of the above devices.

BACKFLOW PREVENTION DEVICES

The following protective devices are those considered acceptable and mentioned in the Model Water Byelaws, June 1986. As other 'no less effective devices' are developed and accepted, they will be listed in the 'Water Fittings Directory' published twice-yearly by the Water Byelaws Advisory Service of the Water Research Centre.

Figures A7–A12, Examples of backflow protection

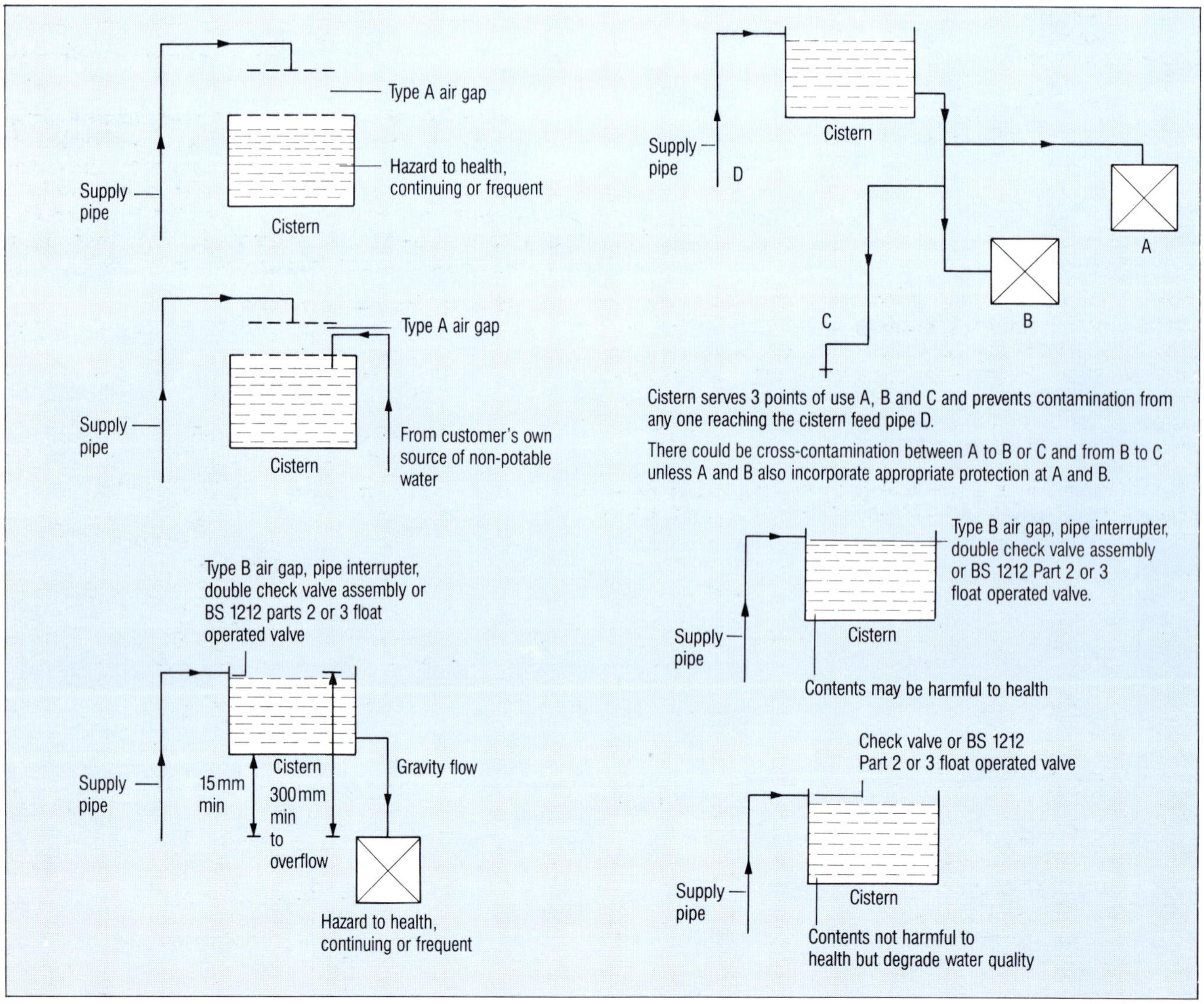

Type A Air-Gap

A visible, unobstructed and complete gap measured vertically between the lowest point of discharge of the water inlet pipe or fitting and the spill over level of the cistern, vessel or other fitting into which the flow is discharged.

Table A22 Required dimensions of Type A air-gaps

Bore of inlet or feed pipe mm	Minimum vertical distance of point of outlet above spill over level mm
Not exceeding 14	20
Exceeding 14 but not exceeding 21	25
Exceeding 21 but not exceeding 41	70
Exceeding 41	Twice the bore of the outlet

Figure A13 Example of a Type A air-gap at a draw-off tap on a sink, wash basin or bath

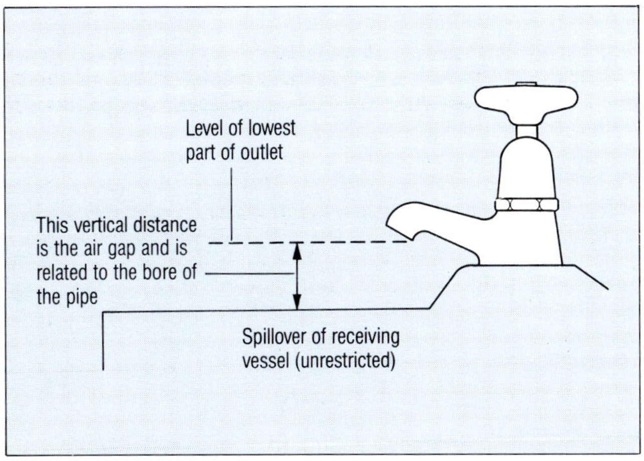

Type B Air-Gap

A vertical distance between the lowest point of discharge into a cistern, vessel or other fitting and its steady water level when there is a maximum inflow of water and all outlets, except any overflow pipe, are closed. The necessary air-gap must either be sufficient to prevent backsiphonage should a vacuum occur in the discharge pipe or fitting, or comply with the appropriate dimensions given in Table A22.

Figure A14 Example of a Type B air-gap provided at a cold water storage cistern fitted with a BS 1212 Part 2 or Part 3 float operated valve

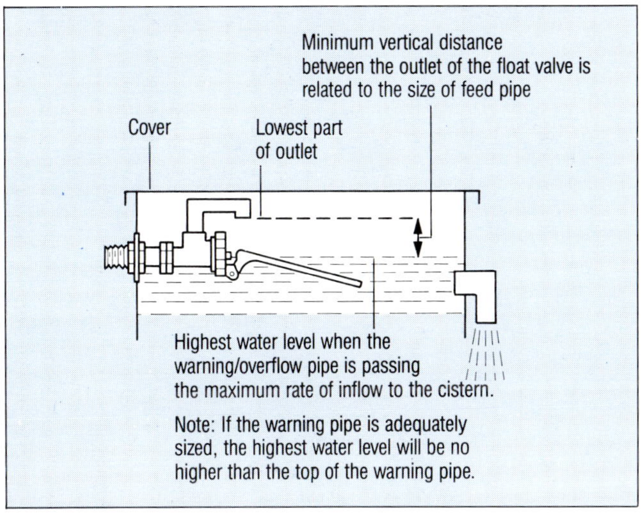

Check Valve

A mechanical device incorporating resilient elastic and positively tight seals designed to permit water to flow in one direction only. The installation of two check valves with an intervening draining tap, known as a 'double check valve assembly', is an acceptable form of protection at certain points of use or delivery of water.

Figure A15 Example of a spring-loaded check valve

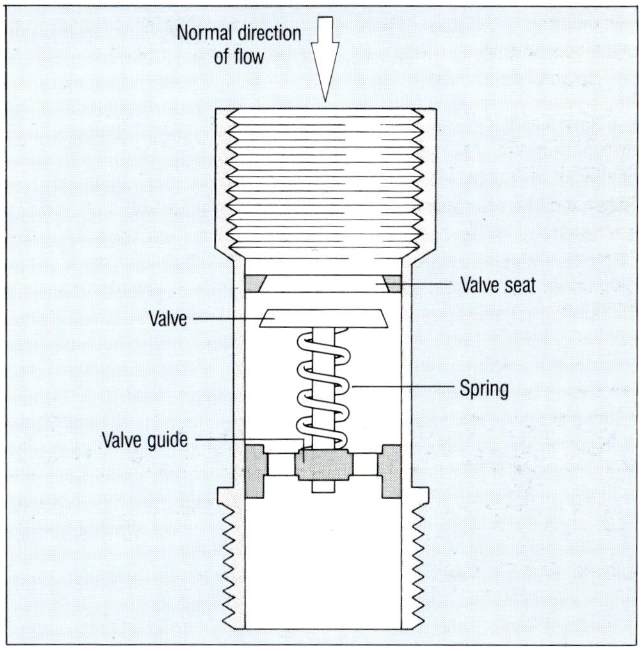

Pipe Interrupter

A non-mechanical device through which water passes and into which air can enter through an unobstructed annular aperture, or apertures. When a vacuum occurs on the inlet side of the device, a corresponding vacuum is produced on the outlet side, thus preventing backsiphonage. Pipe interrupters need to be correctly located and installed to avoid any restriction downstream (eg, a stopvalve) that could impose backpressure on the device.

Figure A16 Example of a pipe interrupter

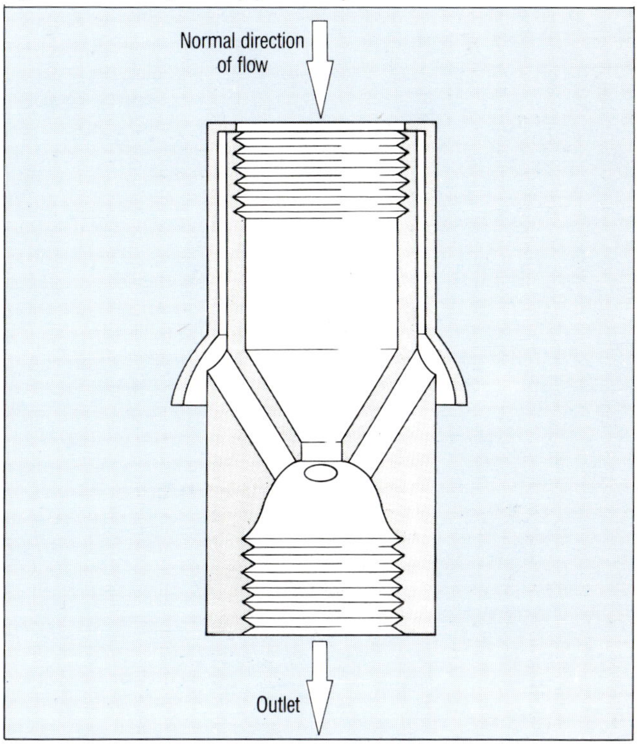

BACKFLOW PREVENTION DEVICES

Vacuum Breaker

A mechanical device with an air-inlet which is closed when water flows past it at or above atmospheric pressure, opening to admit air if a vacuum occurs in the pipe to which the device is connected and closing so as to be watertight when the flow of water is resumed at normal pressure. The combination of a check valve and an in-line vacuum breaker is an acceptable alternative to a double check valve assembly if:

(a) the vacuum breaker element is situated downstream of the check valve; and

(b) no control valve is installed downstream of the vacuum breaker; and

(c) the vacuum breaker is installed on a suitable upstand.

Figure A17 Example of a vacuum breaker (atmospheric type)

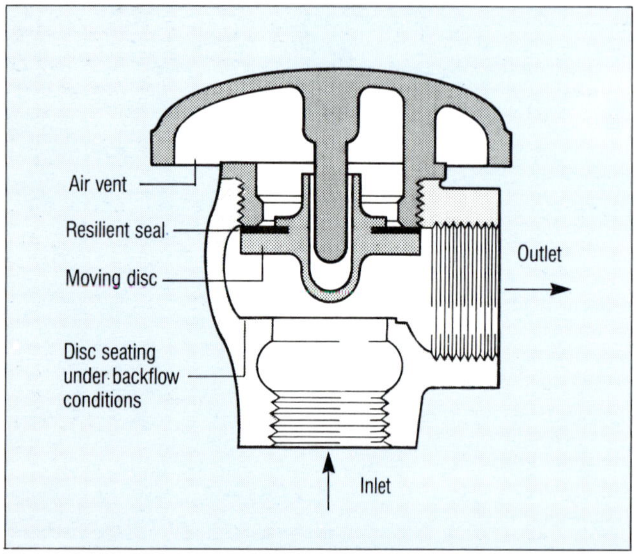

As the atmospheric type needs to draindown and admit air after use to remain operable, no control valve should be fitted downstream of the device.

The pressure type has a spring-loaded, air-inlet and check element, and will operate satisfactorily under continuous pressure for long periods.

Both types must be installed at least 150mm above the highest level of any contaminated water, and neither type will operate under backpressure backflow.

Upstands on Supply and Distributing Pipes
(see Figures A18 and A19)

An upstand is an arrangement of pipework which ensures that a lateral distribution pipe is connected with the supply pipe, or draw-off pipe from a cistern, at a safe distance above the highest free water level in any appliance, draw-off point or fitting served from it. This is to prevent contaminated liquid being drawn from an appliance into the supply or distributing pipe should the required backflow protection at the point of water use fail.

Figures A18 & A19 Examples of backflow protection by the use of upstands

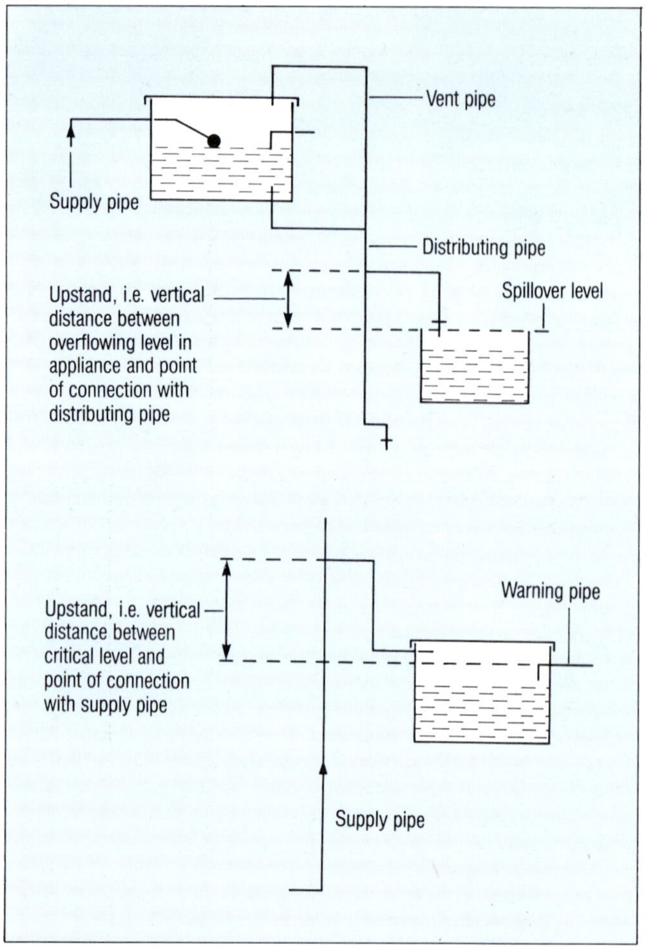

PROTECTION OF DRAW-OFF TAPS, HOSE CONNECTIONS, BIDETS, CLOTHES AND DISHWASHING MACHINES AND CISTERNS

Note

The following examples of backflow protection requirements at various points of water use are based on the Model Water Byelaws, June 1986. To ensure compliance with local requirements, designers and installers must refer to the actual byelaws made and operated by Water Authorities and Water Companies within their respective areas of supply.

For further guidance on particular byelaw requirements, the 'Water Supply Byelaw Guide' published by the Water Research Centre should be consulted.

Draw-off Taps

All draw-off taps or similar water fittings (excluding shower hosepipes) discharging into sinks, wash basins, baths or similar appliances (other than bidets) require one of the following forms of protection:

(a) a double check valve assembly on the hot and cold water feed pipes; or

(b) some other equally effective backflow or backsiphonage

prevention device fitted as close as practicable to the point of draw-off or use; or

(c) air-gaps complying with the dimensions given in Table A23.

Table A23

Size of tap or combination fitting	Vertical distance of point of outlet above spill-over level of appliance
Not exceeding ½ in	20mm
Exceeding ½ in but not exceeding ¾ in	25mm
Exceeding ¾ in	70mm

Note

Requirements (a), (b) and (c) do not apply to draw-off taps or other water fittings which are:

(a) supplied by gravity only from a cistern, cylinder or tank, constantly vented to the atmosphere; and

(b) supplied by a pipe connected to a cistern, cylinder or tank not less than 25mm above the spill-over level of the associated appliance; and

(c) connected to a pipe which does not supply water to any other tap or fitting (other than a draining tap) situated at a lower level.

Shower Hose Connections

All draw-off taps or other water fittings incorporating shower hosepipes (whether or not operated by manual or automatic diverters) discharging into sinks, wash basins, baths, shower trays or similar appliances (other than bidets) require one of the following forms of protection:

(a) a double check valve assembly; or

(b) a check valve and a vacuum breaker; or

(c) some other equally effective backflow or backsiphonage prevention device fitted as close as practicable to the point of draw-off or use.

Note

Requirements (a), (b) and (c) do not apply to draw-off taps or other water fittings which are:

(1) supplied by gravity only from a cistern, cylinder or tank constantly vented to atmosphere; and

(2) supplied by a pipe connected to a cistern, cylinder or tank not less than 25mm above the spill-over level of the associated appliance; and

(3) connected to a pipe which does not supply water to any other tap or fitting (other than a draining tap) situated at a lower level; or

(4) where the movement of a showerhead attached to a hosepipe is restricted by a fixed or sliding attachment, preventing it from discharging water below the spill-over level of the associated appliance; or

(5) where the air-gap between the showerhead of a freely movable shower hosepipe and the spill-over level of the associated appliance complies with the dimension given in Table A23.

Other hosepipes for garden watering, car washing etc. must not be connected to draw-off taps or other similar fittings for use either inside or outside any premises, unless:

(a) the draw-off tap or other similar fitting is supplied by gravity from a cistern, through a pipe which does not supply water to any other tap or fitting (other than a draining tap) situated at a lower level; or

(b) where situated on domestic premises (or elsewhere with the written consent of the water undertaker) the tap or other similar fitting is protected by either a double check valve assembly or some other equally effective backflow or backsiphonage prevention device fitted as close as practicable to the point of draw-off or use.

Bidets

The principal requirement of Water Byelaws is that bidets equipped with any type of submersible spray, or any draw-off fitting allowing the attachment of a hand-held flexible spray, shall not be connected to any mains-fed supply pipe. It is also a contravention of the Byelaws to attach a hand-held flexible spray or similar fitting to any draw-off fitting on a bidet connected to any mains-fed supply pipe.

An over-rim water feed type of bidet can, however, be connected to a mains-fed supply pipe, provided that the vertical distance between the outlet point of any draw-off tap or similar fitting is not less than the air-gap dimension for the size of tap or fitting given in Table A23.

Pipes supplying hot and cold water to a bidet must not also supply any other draw-off tap which can discharge water at a point below the spill-over level of the bidet (other than a draining tap), except in the case of:

(a) a mains-fed supply pipe delivering water to an over-rim water feed type of bidet incorporating approved air-gaps at the hot and cold water draw-off points; or

(b) a cistern-fed distributing pipe supplying cold water only to a flushing cistern or to a urinal; or

(c) a hot water pipe supplying only a bidet and which either incorporates a check valve and vent pipe arranged to prevent backflow or backsiphonage of water from the bidet, or is fitted with some other no less effective backflow prevention device.

Note

Because there are serious contamination risks associated with bidets, it is necessary to ensure that their proposed installation and methods of backflow protection comply with the particular Byelaw requirements of the local water undertaker.

Washing Machines and Dishwashers

Clothes washing machines, dishwashers or tumbler driers permanently or temporarily connected to a water service in any premises are required to incorporate either a Type B air-gap or a pipe interrupter, the removal of which makes it impossible to operate the machine. In addition, every such machine similarly connected to a water service elsewhere than in a domestic dwelling is required to draw water by gravity only from a storage cistern, except where the machine incorporates a Type A air-gap at the supply inlet.

An ion-exchange common salt regenerated type water softener used in conjunction with a clothes washing machine or a dishwasher is required to have its inlet pipe fitted with a check valve and vacuum breaker, or some other no less effective backflow prevention device, except where water supplied to such a softener passes first through a backflow prevention device (eg, a Type B air-gap or a pipe interrupter) incorporated with the washing machine or dishwasher.

Storage Cisterns

Supply pipes conveying water to cisterns (whether or not fitted with float operated valves) are required to incorporate a Type A air-gap at the water inlet if the cistern could receive or contain any substance likely to be harmful to health.

Where a cistern supplies water to a domestic primary heating circuit, or is a flushing cistern, the supply pipe must incorporate either a Type B air-gap, pipe interrupter or a double check valve assembly, except in the case of:

(a) a cistern especially designed and installed in compliance with Water Byelaw requirements preserving the quality of stored water (see Figure A20); or

PROTECTION OF DRAW-OFF TAPS, HOSE CONNECTIONS, BIDETS, CLOTHES AND DISHWASHING MACHINES AND CISTERNS

(b) a cistern fitted with a reducing flow type of float operated valve which will prevent backsiphonage if a vacuum occurs in the feed pipe (eg, a float operated valve complying with either Part 2 or Part 3 of BS 1212).

Cisterns storing water for domestic use are required to:

(1) be installed in a place or position where polluted or contaminated water cannot enter the cistern; and

(2) be thermally insulated; and

(3) if made of materials likely to contaminate stored water, be effectively lined or coated to prevent such contamination; and

(4) be equipped with a rigid, close-fitting cover which is not airtight, excludes light and insects from the cistern and is made of shatterproof material which will not contaminate any water condensing on the underside of the cover; and

(5) be provided with warning and/or overflow pipes constructed and arranged to exclude insects.

In the case of a cistern storing more than 1000 litres of water, the cover is to be constructed so that it does not have to be completely removed to obtain access for maintenance of the float operated valve and inspection and cleansing of the interior of the cistern. It must also fit closely around any vent or expansion pipe conveying water into the cistern.

Figure A20 Arrangement of a domestic cold water storage cistern complying with Water Byelaws requirements for preventing contamination of stored water.

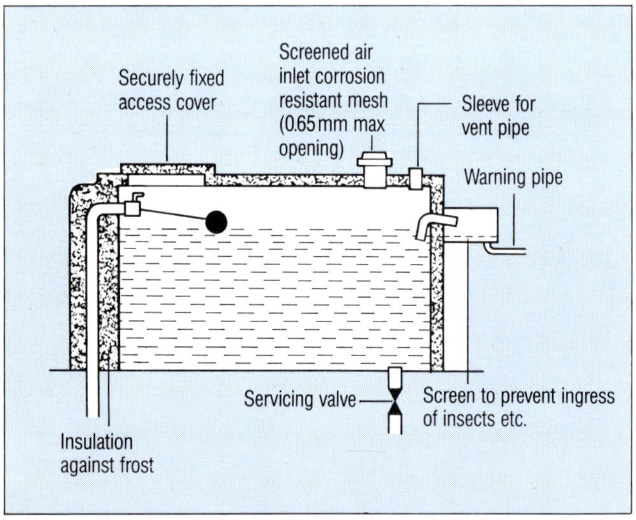

Primary Heating Circuits and Feed Cisterns

The Water Byelaws also require the following precautions to be taken to prevent contamination of domestic storage cisterns and secondary hot water systems from primary heating circuits and feed cisterns.

(a) No vent pipe connected to a cistern-fed primary circuit is to be installed so that it can convey water into a cistern supplying water to a secondary system.

(b) No warning or overflow pipe from any cistern connected to a vented primary circuit is to be installed so that it can convey water to any cistern supplying water for domestic purposes.

(c) The internal pressure in a primary heater within a double feed indirect hot water cylinder must not exceed the pressure of the stored water under normal operating conditions, except where the primary heater has no joints or is constructed so that any joints will withstand normal working pressures.

(d) No single feed indirect cylinder in which secondary hot water is separated from primary heating water eg, by an air pocket is to be directly connected to a mains-fed supply pipe. Every such cylinder must be permanently vented to the atmosphere, and be so constructed and installed that water in the primary circuit is prevented from mixing with water in any secondary circuit or associated central heating system when operating at a sustained temperature not exceeding 80°C. The use of corrosion inhibitors or similar additives in single feed indirect hot water cylinders is therefore prohibited.

Secondary Backflow Protection

Because the risks of backflow and backsiphonage may be increased within water supply installations serving multi-occupied buildings, it is necessary to provide a second line of defence against possible contamination. In these cases, the Water Byelaws require supply and distributing pipes to be provided with secondary backflow protection by being fitted with a combination (or combinations) of check valves, vacuum breakers, double check valve assemblies or some other no less effective backflow prevention device, sufficient to prevent backflow or backsiphonage of water from one premises to another, or from any floor or level of a building to a lower floor or level. These requirements do not however, apply in the case of separately occupied premises served by an individual supply pipe.

Examples of where secondary backflow protection would be required are:

(1) On a common supply pipe system serving two or more separately occupied houses, flats and maisonettes, and on a common distributing pipe system serving two or more dwellings in a block of flats (see Figures A21 and A22).

(2) On a common supply pipe conveying an intermittent supply of water to premises which, under any special enactment, are required to provide a storage cistern capable of holding sufficient water for not less than 24 hours ordinary use.

Similar measures for secondary backflow protection are necessary to limit the risks of contamination occuring in industrial premises, schools, hospitals etc.

Acceptable Methods of Backflow Protection at Points of Use or Delivery of Water

Tables A24, A25 and A26 list some common usages of water, and the appropriate backflow protection requirements for the three classes of contamination risk defined in the 1974 Report of the Committee on Backsiphonage in Water Installations.

Table A24 Class 1 risk of contamination from a substance continuously or frequently present.

Required backflow protection: Type A air-gap or an interposed cistern
WC pans
Urinal bowls
Bedpan washers
Dental sputum bowls
Slaughter house equipment, butchery and meat trades
Laboratories
Sprinkler systems (containing anti-freeze)
Non-domestic central heating systems (sealed)
Dairies
Industrial chemical baths (eg, cyanides)
Poultry and animal drinking troughs
Bottle washing apparatus
Water treatment plant
Dye works and similar applications
Non-domestic hose union taps (alternatively protected by either a double check valve assembly or some other no less effective backflow protection device incorporated in the tap, as close as is practicable to the point of draw-off or use)
Sewage works
Supplementary sources of water supply (eg; fire reserve, agriculture)
Cisterns receiving or containing substances which are, or are likely to be harmful to health (eg; cisterns connected to non-domestic central heating systems; fire fighting reserve storage, where supplemented from non-potable sources, agricultural storage.

Table A25 Class 2 risk of contamination from a substance which may be present

Required backflow protection: Type A or Type B air-gap, a pipe interrupter, a combination of check valve and vacuum breaker*, a double check valve assembly or some other no less effective backflow prevention device
Commercial softening plant (common salt regenerated)
Home dialysing machine (with integral membrane washing).
Fire sprinkler systems (gravity supply from uncovered storage cistern; pumped from storage.
Primary hot water circuits and their associated cisterns.
Standpipes and marinas (serving small boats).
Standpipes on mains supplying water to mobile apparatus.

Note

As vacuum breakers alone do not afford protection against backpressure backflow, a combination of check valve and vacuum breaker is unacceptable if the vacuum breaker cannot be installed on a suitable upstand (see Figure A18 and A19 for examples of upstands).

Table A26 Class 3 risk of contamination where backflow or back-siphonage is not, or is not likely to be harmful to health at points of use or delivery of water

Required backflow protection: A check valve or some other no less effective backflow prevention device
Single outlet hot and cold water mixing taps in domestic premises (hot water from vented apparatus; cold water from the supply pipe) This is additional to the appropriate protection provided at the point of discharge of the mixed water
Home dialysing machine (without integral membrane washing)
Domestic water softening plant (common salt regenerated)
Fire sprinkler systems (certain types, eg, mains fed system – no water draw-off for other uses; point of draw-off for normal use before sprinkler system)

Figure A21 Secondary backflow protection on a common supply pipe serving two or more dwellings

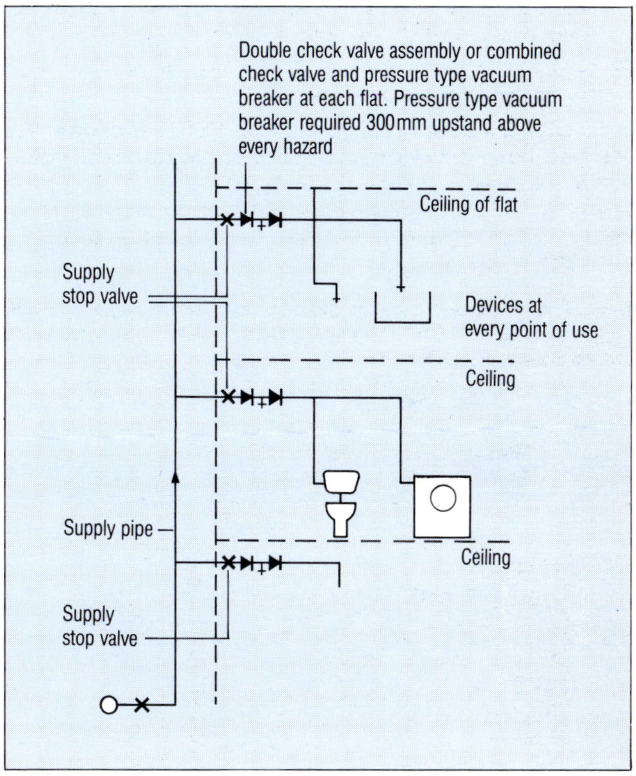

PROTECTION OF DRAW-OFF TAPS, HOSE CONNECTIONS, BIDETS, CLOTHES AND DISHWASHING MACHINES AND CISTERNS

Figure A22 Secondary backflow protection on a common distributing pipe serving two or more dwellings

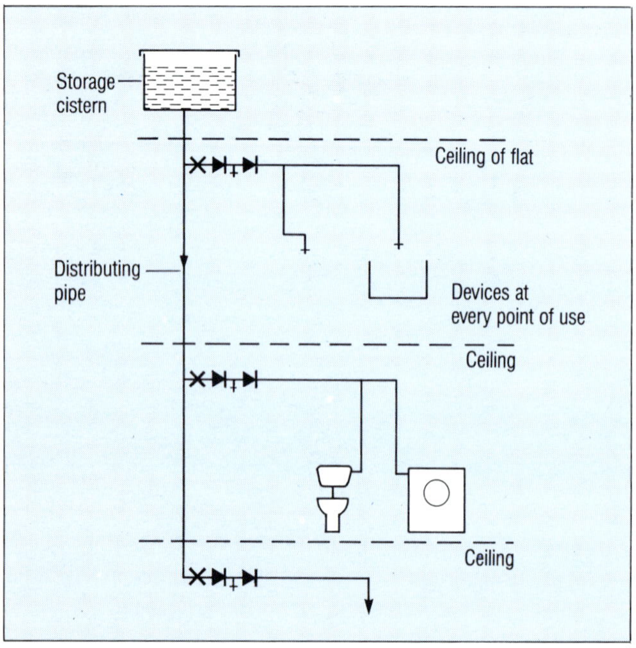

UNVENTED HOT WATER SYSTEMS

Water Byelaws in the UK have for many years permitted hot water systems up to 15 litres capacity to be fed from the mains, whilst generally requiring larger capacity systems to be fed by cold water storage cisterns. However, new Byelaws made and operated by statutory water undertakers now permit unvented hot water systems above 15 litres capacity to be connected directly to mains-fed supply pipes.

There are many possible arrangements of unvented systems, but these can be conveniently grouped into two categories:

(1) Systems with a maximum water capacity of l5 litres, heated on demand (i.e. instantaneous gas or electric point of use water heaters).

(2) Systems containing larger volumes of water heated and stored at a controlled pressure.

Figure A23 shows the schematic layout of a mains-fed system fitted with the following control and safety devices:

Stopvalve

Required to isolate the system from the water main for maintenance and repairs.

Line-strainer

Recommended to ensure the efficient functioning of other control devices with moving parts.

Figure A23 Schematic layout of an unvented domestic hot water installation

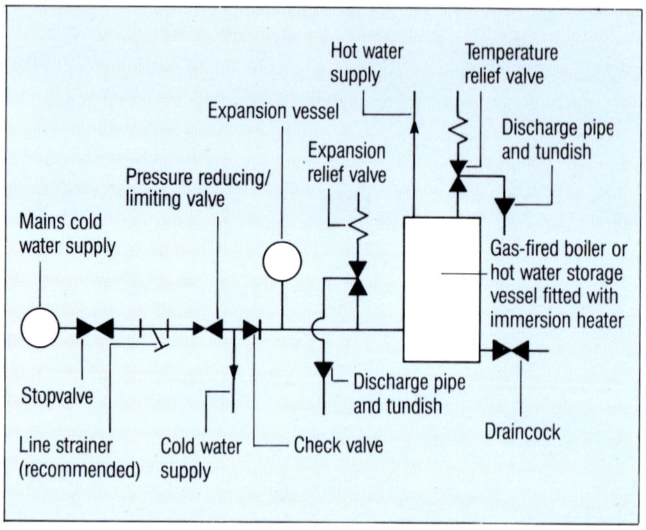

Note

The cold water supply to draw-off points can be taken off the incoming supply pipe before or after the Pressure Reducing Valve (PRV). However, due consideration should be given to the capacity and outlet pressure of the PRV, the mains supply pressure and number of terminal draw-off fittings. An additional PRV installed on the branch cold water supply is sometimes an advantage.

Inlet Pressure Reducing/Limiting Valve

(see also Figure A24)

Protects the system against over-pressurisation and pressure surges by limiting the cold water inlet pressure to a pre-determined level, thereby maintaining a constant performance flow rate if the mains supply pressure exceeds the set operating pressure.

Figure A24 Example of pressure reducing and limiting valves

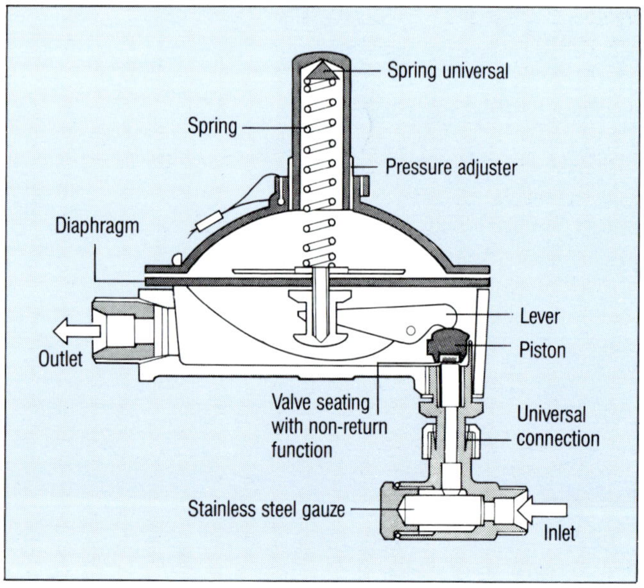

Figure A24 Example of pressure reducing and limiting valves cont'd.

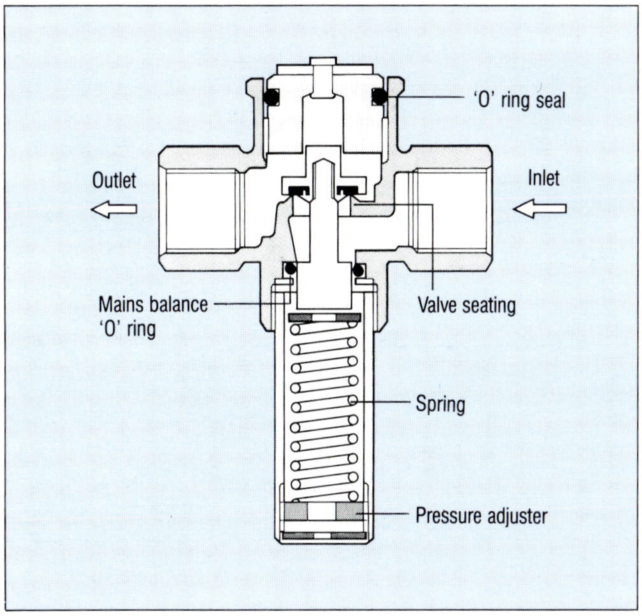

Check Valve

Prevents crossflow of heated water into the mains-fed cold water supply pipe.

Expansion Vessel

Accommodates the increase in volume of water when heated, where such expansion cannot otherwise be accommodated either within the secondary system pipework or the heating apparatus.

Expansion Relief Valve (see also Figure A25)

Automatically releases water from the system during the heat-up period in the event of malfunctioning or failure of the expansion vessel.

Figure A25 Example of an expansion relief valve

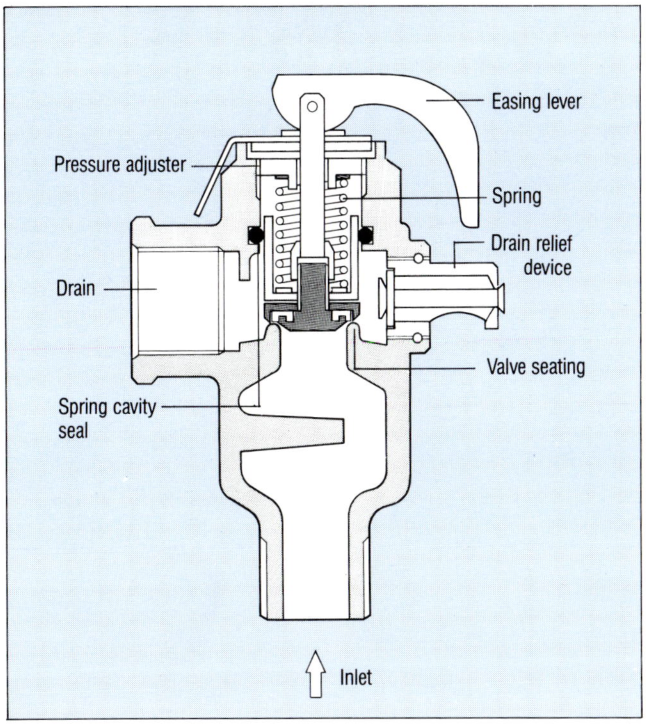

Temperature Relief Valve (see also Figure A26)

Necessary to prevent the water temperature in the system from exceeding 100°C in the event of failure of the control thermostat and high limit energy cut-out, and is normally combined with a pressure relief function.

Figure A26 Example of a temperature and pressure relief valve (usually combined with a pressure relief function)

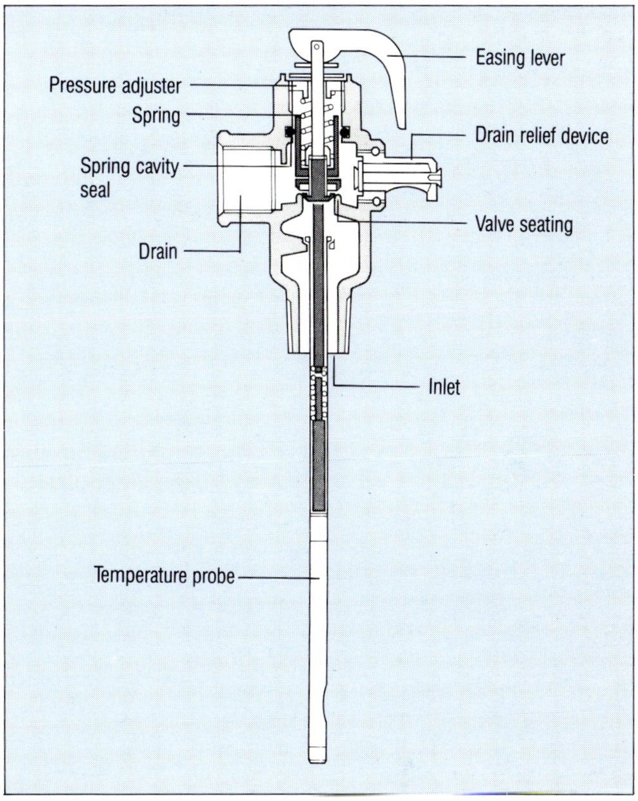

Note

Temperature relief valves may also be used to provide similar protection for traditional vented hot water systems, in which temperatures higher than 100°C can be attained under run-away heating conditions.

Application of Water Byelaws

There are a number of important Water Byelaw requirements for preventing waste and contamination of water from unvented hot water systems.

Every mains-fed unvented water heater must either itself be capable of accommodating any expansion water or be connected to a separate expansion vessel or expansion cistern. Alternatively, it should be installed so that any expansion water can be accommodated within the secondary system pipework, provided that hot water is prevented from entering any communication pipe or supply pipe to which a cold water draw-off tap is connected.

Every cistern-fed unvented system must likewise either be capable of accommodating any expansion water or be connected to a separate expansion vessel or expansion cistern. Alternatively, the system should be so arranged that expansion water can pass back through a feed pipe to any storage cistern to which the unvented apparatus or hot water cylinder is connected.

Air or gas loaded expansion vessels using an internal diaphragm are acceptable for the purposes of the Water Byelaws and should be of a size sufficient to accept at least 4% of the total system water content, to prevent unacceptably high pressures developing within the system. Careful note should therefore be taken of the manufacturer's specification when installing expansion vessels.

Figures A27–A29 show example interpretations of Water Byelaw

UNVENTED HOT WATER SYSTEMS

requirements for accommodating expansion water fed installations.

A further requirement of the Water Byelaws is that every mains-fed or cistern-fed unvented hot water storage cylinder or tank fitted with a non-mechanical (passive) safety device should also be fitted with a temperature relief valve which operates or is designed to operate at a temperature not less than 5°C below that set for the valve to perform its temperature relief function.

All expansion relief valves and temperature/pressure relief valves fitted to unvented water heating apparatus and hot water storage cylinders are required to close automatically after discharging water and to remain watertight when closed. They must also be constructed and installed so that any water discharged from the valve, or from any pipe connected to it, is readily visible.

***Figure A27** Accommodation of expansion water in an expansion vessel, unvented installation*

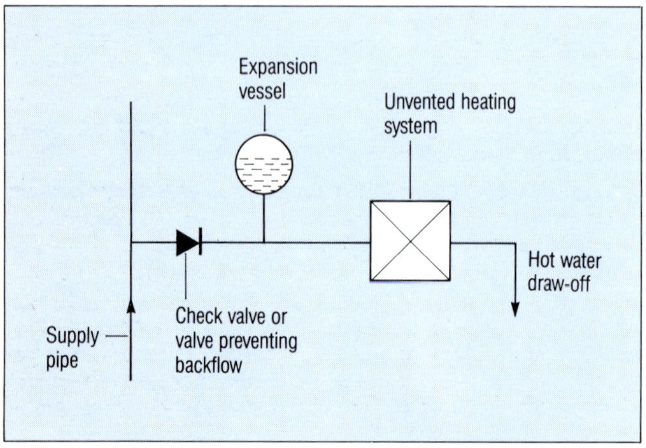

***Figure A28** Accommodation of expansion water in a cistern*

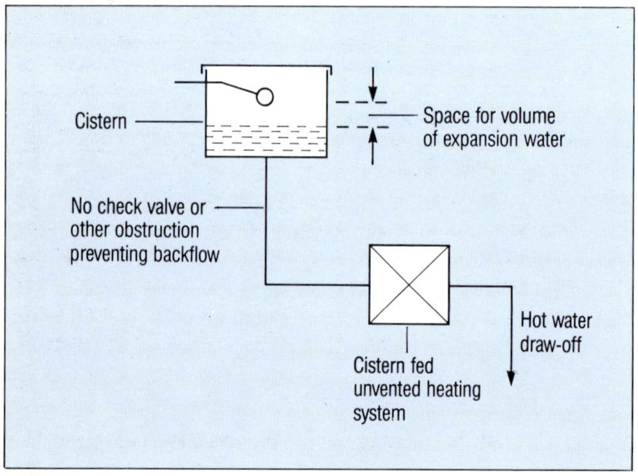

***Figure A29** Accommodation of expansion water in an expansion vessel*

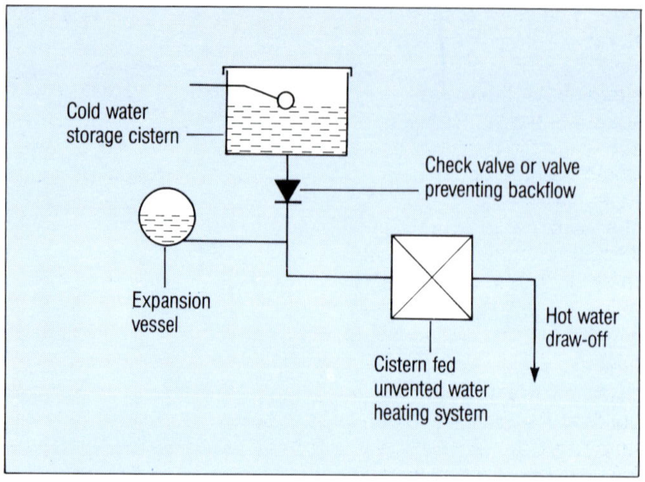

Discharge Pipework from Relief Valves

To avoid injury to persons and damage to property, the very hot water that may be released from expansion relief valves and temperature/pressure relief valves under fault conditions must be discharged through a drainage relief pipe to a visible and safe position outside the building. The discharge warning pipe should be of suitable heat-resistant metal, equal in size to the discharge outlet of the relief valve and fall continuously throughout its length. It must not exceed 9 metres in length without forming an air-break – by discharging into a tundish – and contain no more than three 90° bends. If a safe point of discharge cannot be achieved within a length of 9 metres, a drainage pipe one size larger than the discharge outlet connection of the relief valve may be used. Figure A30 illustrates one acceptable arrangement of discharge pipework from a temperature relief valve.

***Figure A30** Example method of providing safe disposal of hot water discharge from safety devices*

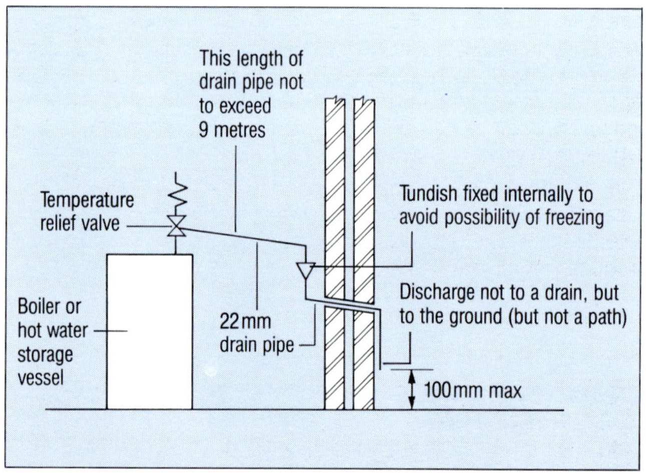

Application of Building Regulations

The minimum safety requirements for unvented hot water storage systems are set out in Part G3 of Schedule 1 of the 1985 Building Regulations, as follows:

Requirements.

If hot water is stored and the storage system does not incorporate a vent pipe to the atmosphere, there shall be adequate precautions to:
a) prevent the temperature of the stored water at any time exceeding 100°C; and
b) ensure that the hot water discharged from safety devices is safely conveyed to where it is visible but will cause no danger to persons in or about the building.

Limits of application.

This requirement does not apply to:
a) a system having a storage capacity of 15 litres or less;
b) a space heating system;
c) a system which heats or stores water for the purposes of an industrial process.

Practical guidance on how to comply with these requirements is given in an Approved Document supporting Part G of the l985 Building Regulations, which states that any unvented hot water storage system should be in the form of a proprietary Unit or Package bearing a current British Board of Agrement Certificate.

A Unit is defined as a system equipped and assembled by the manufacturer with all of the necessary safety devices and other controls.

A Package is defined as a system equipped and assembled by the manufacturer with all necessary safety devices, and supplied by the manufacturer with a kit containing any other devices to prevent backflow or crossflow, control working pressure, relieve excess pressure and accommodate expansion, all to be fitted to the system by the installer.

The main purposes of the Building Regulation requirements for unvented domestic hot water systems are to ensure the health and safety of people in or about buildings. When water is heated in a closed and pressurised system to temperatures above l00°C it becomes superheated. If water at this temperature is discharged into the atmosphere all of its superheat is suddenly released, causing some of the water to vapourise or flash into steam, with the risk of causing severe personal injury or damage to property.

For example, if the water contained in a 136 litre hot water cylinder is heated to 107°C (i.e. 7°C above atmospheric boiling point) any sudden rupturing of the cylinder would instantaneously discharge about 1.8 kilograms (4 lb) weight of steam into the atmosphere. The rapid release of this considerable volume of steam in a confined space could be as damaging as the controlled detonation of an explosive device.

It is therefore of paramount importance that the design and installation of an unvented hot water system should prevent potentially hazardous situations arising as a result of the production of steam under pressure.

CORROSION CAUSES, EFFECTS AND PREVENTATIVE MEASURES

Introduction

Corrosion is one of the major causes of premature failures in plumbing services. Not only is it responsible for increased maintenance costs but losses in efficiency, particularly in heating systems, can increase running costs.

Corrosion may be defined as the reaction of a metal with its environment resulting in damage which impairs the function of a component or system. Whilst the mechanisms of corrosion are common in different plumbing applications, methods of control and/or prevention have to be tailored to suit that particular environment.

CAUSES OF CORROSION: BASIC THEORY

Before discussing the causes of corrosion in detail it is useful to define and clarify in simple terms some of the nomenclature used in the text.

A salt (not to be confused with common salt, NaCl) is formed by the replacement of acidic hydrogen in an acid by a metal or basic group.

$$\underset{\text{metal}}{2Na} + \underset{\text{acid}}{H_2SO_4} \rightarrow \underset{\text{a salt}}{Na_2S_4} + \underset{\text{hydrogen}}{H_2\uparrow}$$

When most salts are dissolved in water they form ions which are atoms or groups of atoms carrying positive or negative charges. The resulting solution is called an electrolyte because of its ability to conduct an electric current.

$$\underset{\text{a salt}}{Na_2SO_4} \rightleftharpoons \underset{\text{positive ion}}{2Na^+} + \underset{\text{negative ion}}{SO_4^{--}}$$

Some salts are much more soluble than others, and their solubility in water often depends to a large extent on solution temperature.

In practical terms, pH is used to measure the acidity or alkalinity of a solution. Values range from O to 14, those less than 7 being acid and those greater are alkaline.

Corrosion is electrochemical in nature and can occur by oxidation (dry corrosion) or wet corrosion. Most corrosion problems in plumbing applications are caused by wet corrosion which requires the presence of an electrolyte to allow the passage of an electric current and some agency to maintain a difference in potential. Many corrosion processes also require the presence of oxygen. Natural and supply waters contain dissolved salts which make them into electrolytes and therefore capable of carrying an electric current. The factors which may cause and will affect the rate of corrosion are as follows:

1. Solution Potential

If a metal is placed in an electrolyte, corrosion in the form of a chemical reaction may occur accompanied by the passage of an electric current. That part of a metal system where current, i.e. positive charges in the form of metal ions, leaves and enters the solution is called the anode while the resulting electrons (negative charges) migrate to an area of higher potential, the cathode, where they react with other ions or oxygen.

In the corrosion process it is always the anode which is dissolved. The difference in potential between the anode and cathode at equilibrium is termed the solution potential. This potential has different values for different metals and conditions, e.g. temperature, electolyte concentration and the surface or metallurgical condition of the metal, and has a profound effect on corrosion.

2. Temperature

In general, rates of chemical reaction and therefore corrosion increase with a rise in temperature. However, some corrosion reactions, for example the corrosion of steel in aerated solution, may be controlled by the solubility and diffusion rate of dissolved gases which vary appreciably with temperature.

As the rate of corrosion of steel is partly determined by the availability of oxygen at the surface, it is found that this rate has a maximum in the range 75°C to 85°C which is approximately 4 times that at ambient temperatures.

With other metals, for example zinc, the variation of corrosion rate with temperature is related to the nature of the corrosion product. At temperatures which produce a continuous/adherent corrosion product, the corrosion rate is low. However, when a granular, non-adherent product is formed, the corrosion rate increases considerably.

CAUSES OF CORROSION: BASIC THEORY

Differences in surface temperature on the same metal component can create areas differing in potential resulting in increased corrosion.

3. Differential Aeration

If part of a metal surface is shielded from air, the natural oxide film can break down. Where the oxygen availability is low, these areas become anodic to those areas to which there is a greater supply. This results in localised corrosion which can occur particularly in crevices and also underneath surface deposits, for example mill scale. This type of corrosion is called differential aeration.

4. Dissolved Salts

The influence on corrosion of salts dissolved in natural or supply waters is determined both by their concentration and, more importantly, by the type of ion produced in solution. In general, a high dissolved solids content would be expected to exacerbate corrosion due to the higher conductivity if conditions favour it. However, some ions, for example carbonates, can be protective due to their scale forming ability on metal surfaces. Others, in particular chloride and sulphate ions, are aggressive because they interfere with the development of protective films and also allow passive films to be broken down more readily. In addition, sulphate ions can allow certain bacteria to proliferate in anaerobic conditions.

Although in certain conditions the overall rate of corrosion may not be increased, the attack may be more localised and therefore corrosion pits tend to be deeper. In the absence of aggressive ions the corrosion rate of steel at different oxygen concentrations reaches a maximum however, where appreciable levels of chlorides are present ($>$ 200 mg/litre), the corrosion rate continues to increase with increasing oxygen availability.

Two indices have been developed to predict whether a carbonate scale will be deposited from a supply water of given composition. These are called the Langelier and Ryznar indices respectively, a positive Langelier index or a Ryznar index of less than 6 indicating that the water is scale forming.

Since the deposition of a carbonate scale can stifle corrosion, it often follows that these waters are less corrosive than those that tend to dissolve calcium carbonate. Their behaviour, however, depends greatly on the form in which the scale is laid down, a discontinuous layer providing relatively little protective value.

When a natural water runs over a metal surface, it can take up traces of that metal which are later deposited on another metal. This may form a bimetallic couple and if the deposited metal is cathodic to the substrate, rapid corrosion will ensue. Typical examples are copper pipework upstream from galvanized tanks and run-off from copper roofs into aluminium gutters. More recently, the high acidity of rainfall in certain areas due to the dissolved sulphur and nitrogen oxide pollutants has generally resulted in reduced service life of exposed metal surfaces.

5. Surface Effects

Certain surface films are cathodic to steel and under wet conditions where there are breaks in the film, the underlying steel surface becomes the anode and will corrode preferentially.

6. Ion Concentration

As discussed above, the solution potential is affected by the concentration of ions in the electrolyte; the higher the concentration, the more cathodic the metal becomes. If, therefore, a metal surface is in contact with an electrolyte which varies in concentration, those areas of metal in contact with dilute solution will become anodic to those areas in contact with the more concentrated solution and corrosion will be accelerated.

7. pH

Many corrosion processes depend on the acidity/alkalinity (pH) of the environment as well as the anode/cathode potential differences. The Pourbaix diagram attempts to account for the influence of both pH and potential on the corrosion of metals. The diagram for the Iron – Water system is illustrated in Figure A31. It is calculated from the thermodynamics of the reactions involved and refers to hydrogen gas being evolved or oxygen gas being available at a standard pressure.

The possible reactions which can occur between iron and water are shown adjacent to the diagram. The electrode potential assumed by iron when immersed in water of pH 7 is approximately at point X on the diagram and corrosion will occur initially by reaction (a). Pourbaix diagrams are obviously applicable to predicting means of corrosion prevention but they have certain limitations. They give no indication as to the probable rate of a corrosion reaction and the 'passivity' regions need to be interpreted carefully since the precipitates of metal hydroxide do not always provide a protective film.

Between pH values of 4.0 and 9.5 the rate of attack on iron is reasonably constant. However, below a pH of 4.0, the acid conditions cause iron to dissolve freely and the corrosion rate increases gradually with decrease in pH.

8. Dissolved Gases

The two most important gases to consider are oxygen and carbon dioxide. In near neutral solution, oxygen reduction is the most important reaction at the cathode and within a certain range, the rate of attack is dependent upon its concentration.

Figure A31 Pourbaix diagram of Fe – H_2O system

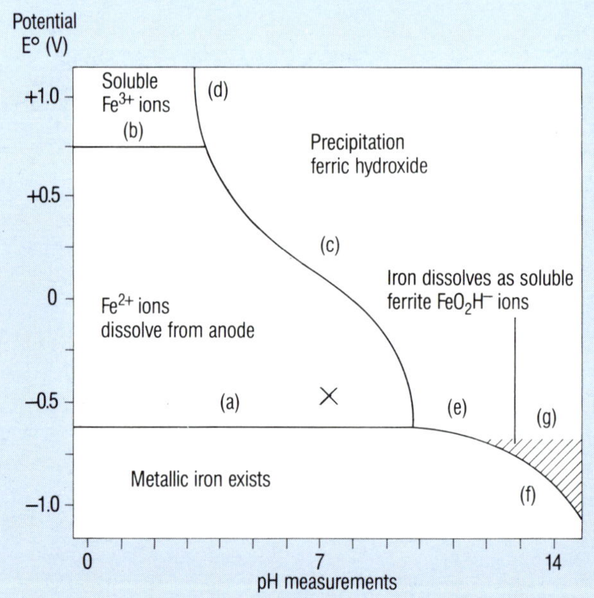

(a) $Fe \rightarrow Fe^{2+} + 2e$. Solution potential effect
(b) $Fe^{2+} \rightleftharpoons Fe^{3+} + e$. Oxidation of dissolved ferrous ions into ferric ions
(c) $Fe^{2+} + 3OH^- \rightleftharpoons Fe(OH)_3 + e$. Precipitation of ferric hydroxide
(d) $Fe^{3+} + 3H_2O \rightleftharpoons Fe(OH)_3 + 3H^+$. Precipitation of ferric hydroxide
(e) $Fe + 3H_2O \rightleftharpoons Fe(OH)_3 + 3H^+ + 3e$. Precipitation of ferric hydroxide
(f) $Fe + 2H_2O \rightleftharpoons FeO_2H^- + 3H^+ + 2e$. Formation of soluble ferrites in solutions of high pH.
(g) $FeO_2H^- + H_2O \rightleftharpoons Fe(OH)_3 + e$. Precipitation of hydroxide from ferrites.

However, with some metals, oxygen may cause passivity when present in high concentrations. Carbon dioxide is present in some supply waters, particularly those derived from deep wells and has the effect of lowering the pH and can simulate rapid acid attack, dissolving any protective scales that are present. It may also be produced by the breakdown of bicarbonates in the feed water of steam boilers and can cause rapid corrosion of condensate pipes particularly if oxygen is also present.

It must be borne in mind that plastics materials, unlike metals, are permeable to gasses. This can exacerbate oxidative corrosion of iron/steel components in closed systems where plastics pipework is used for distribution or underfloor heating.

9. Contact with Dissimilar Materials

(a) Metals

If different metals are immersed in an electrolyte, after a period of time each will attain a characteristic potential. In this way it is possible to arrange metals in what is termed a galvanic series, however the precise order may differ slightly depending on the particular metal/alloy composition and the nature of the electrolyte.

Table A27 gives the galvanic series for metals and alloys when in contact with natural waters.

Table A27 Galvanic Series of Metals and Alloys in Natural Waters

Metal	Normal Electrode Potential (vs. Standard Hydrogen Electrode, mV) 'Noble' or Cathodic, i.e. protected end
Graphite (carbon)	+200
Titanium	
Silver	
Nickel	
Monel	+100
Austenitic stainless steel (types 304/316)	
70/30 Cupro nickel	
Copper	
Ferritic stainless steel (type 416)	
67/33 Nickel copper	
Martensitic stainless steel (type 431)	
Aluminium bronze	
70/30 Brass	0
Gunmetals, phosphor bronzes and tin bronzes	
60/40 Brass	
Chromium	−100
Tin	−200
Tin-lead solder	
Lead	−300
Steel	−400
Cast iron	
Cadmium	−500
Aluminium and aluminium alloys	−700
Galvanised iron	−800
Zinc	
Magnesium and magnesium alloys	−1300
'Base' or anodic, i.e. corroded end	

Corrosion occurring when two differing metals/alloys are in electrical contact is termed bimetallic corrosion. There are three other factors which can influence corrosion at a bimetallic junction.

The first is the relative areas of the two metals in contact. If the area of the cathodic metal is large compared with that of the anodic (corroding) metal, corrosion will be concentrated resulting in rapid attack. Conversely with a large anode connected to a small area of cathodic metal, although some corrosion will be localised adjacent to the contact area, most corrosion will be distributed over the remaining area and the relative thickness loss will be less.

The second is the conductivity of the electrolyte. In an electrolyte of low conductivity, attack will be confined adjacent to the joint area and may be relatively intense. Where the electrolyte conductivity is high, the attack will spread out more and the level of general corrosion will be greater.

The third factor concerns electrical contact between the dissimilar metals. Some metals, in particular aluminium and stainless steel, form coherent oxide films in air which are non-conductive. If they prevent a current flow between the metals then bimetallic corrosion will be prevented. In the absence of dissolved oxygen, bimetallic contact will not cause iron and steel surfaces to corrode.

More detailed information on corrosion at bimetallic contacts is given in British Standard PD 6484: 1979.

(b) Other Materials

Many non-metallic materials found in or around buildings can be responsible for corrosive attack on metals. Some woods are acidic, and under damp conditions can cause attack on steel, cadmium, zinc and lead. If the wood has been treated with a copper/chrome arsenate preservative, corrosion can be accelerated and in particular aluminium alloys and zinc are attacked.

Fresh concrete is very alkaline (pH 12.6–13.5) and while this is responsible for protecting steel reinforcement against corrosion, other metals, for example zinc, aluminium and lead will be attacked. Some concretes also contain chlorides which will increase the risk of corrosion.

Bleach and some soldering fluxes contain high levels of chloride ions which will cause rapid attack on many metals including copper and stainless steels. Furthermore, some plastics on heating or degradation also give off acids and/or chlorides which in the presence of moisture will cause corrosive attack.

Soils are complex in nature and therefore are extremely variable in corrosive activity. The most widespread form of corrosion is due to sulphate reducing bacteria (see below). Other problems are caused by acidic conditions due to natural constituents or contamination by industrial waste, for example, ashes and clinker.

Although not strictly corrosion, it is worth mentioning that other industrial wastes, for example tar oils, can cause failures in buried plastic pipework due to environmental stress cracking. They may also attack protection on metal pipework, thus exposing the underlying surfaces to corrosive attack.

10. Bacteria

In the absence of oxygen, corrosion may continue by the cathodic reduction of sulphate which is present in most soils and natural waters. This is facilitated by anaerobic bacteria (called sulphate reducing bacteria, disulpho-vibrio disulphoricans) which are able to use cathodic hydrogen in their living process and convert sulphate to sulphide. This type of corrosion is responsible for the smell of 'bad eggs' (hydrogen sulphide) sometimes observed when venting radiators.

11. Flow, Erosion, Impingement and Cavitation

Corrosion can be aggravated by the nature of water flow across a metal surface. The overall rate will be controlled both by the rate of reaction and how quickly reactants and products can approach and leave the metal surface. These are termed chemical and diffusion control respectively. Flow may be stagnant, laminar or turbulent which will all affect the rates of diffusion.

Where the flow is turbulent, the impingement of either gas, liquid or solids or a combination of any three can cause mechanical damage which removes or prevents formation of a protective film.

CAUSES OF CORROSION: BASIC THEORY

If the metal or alloy is corrodible in that environment, localised attack of the surface will occur.

Cavitation is a particular type of attack which occurs when a metal surface is exposed to a high velocity, low pressure liquid.

In areas where the pressure is sufficiently low, pockets of vapour form which may suddenly collapse when they pass to an area of higher pressure. The shock generated is sufficient to cause mechanical damage to the metal surface at that point and exacerbate corrosive attack. This mechanism is responsible for water hammer.

12. Stray Current Corrosion

Stray earth currents, in particular DC, can produce cathodic and anodic areas where they enter and leave buried pipelines or other metallic structures causing severe corrosion.

13. Organic Matter

Organic matter, derived from either natural or industrial sources, may lower the pH and increase corrosion rates as well as preventing scale deposition. Bacteria are also often present in organic matter and these may be responsible for fouling and increased corrosive attack under both aerobic and anaerobic conditions.

It can be seen from the foregoing that corrosion phenomena are very complex in nature and are affected by many factors. Most of these factors can have a profound affect on corrosion rates and highlight the difficulties in predicting corrosion behaviour and time to first maintenance or failure. However, many techniques are now available for identifying, monitoring and controlling corrosion to acceptable limits.

EFFECTS OF CORROSIVE ENVIRONMENTS

Corrosion can manifest itself in numerous ways, depending on the metal or alloy and particular environment in which it is in contact. It is convenient to look at the various metals likely to be encountered in plumbing applications and the environments that can be detrimental to their service life.

Some of these metals are also used as protective coatings and their performance together with that of the substrate is also considered.

Aluminium and Aluminium Alloys

Although aluminium is a very reactive metal, it has a high resistance to corrosion because of the tenacious, inert oxide film which forms on the surface. Pure aluminium is generally stable in the pH range 4.5–8.5, but outside this range attack can be rapid. Aluminium alloys, particularly those containing copper in significant quantities (Duralumin), are less resistant to corrosion.

The most commonly encountered form of aluminium corrosion is pitting, but the rate of propagation is very dependent on the alloy composition and nature of the solution with which it is in contact. While many neutral or weakly alkaline solutions allow self passivation others, particularly those containing chlorides or copper ions, cause rapid propagation of corrosion pits in some alloys.

Aluminium alloys may also suffer a preferential attack at the boundaries between the grain structure which is termed intercrystalline corrosion.

Aluminium and its alloys have very good resistance to atmospheric corrosion providing the surfaces are regularly washed by rain, etc. Where deposits of, for example, acidic sulphates, are allowed to build up on sheltered surfaces, moisture from condensation is sufficient to give an increased rate of attack.

In all environments, aluminium is very sensitive to bimetallic corrosion. Contacts with copper and copper alloys, and to a lesser extent, iron and iron alloys, should be avoided.

Cadmium

The main application of cadmium is as a protective coating on steel. It is particularly useful as a plated finish on high strength steels where zinc finishes can heighten the risk of cracking due to hydrogen embrittlement. Its corrosion behaviour is similar to zinc and although it is less protective to steel, unlike zinc, it is stable in alkali solution.

Cast Iron

Cast irons are ferrous alloys containing generally 2–4% carbon and frequently have high silicon contents. A wide variety of compositions are available and additions of silicon, chromium and nickel can be used to improve corrosion resistance. Typical applications include boiler sections, pump housings and pipe fittings (malleable iron). Although corrosion rates are similar to those of steel, above ground corrosion is not normally a problem due to the thick sections used for cast iron components.

When in contact with some natural waters, particularly those that are slightly acidic or contain chlorides (salt), cast iron may suffer a form of attack called graphitisation. The iron corrodes leaving a weak, porous structure composed of graphite and iron oxides. This can result in catastrophic failures in underground pipes where soil or ground movement causes large sections where graphitisation has occurred to burst without warning.

Although not directly associated with corrosion, it is worthwhile mentioning a common cause of premature failure in gas-fired sectional boilers. If hard-water scale is allowed to build up inside the cast iron sections, and flame impingement occurs on the outside, overheating can cause a change in the cast iron structure to a weaker 'undercooled' form. Subsequent thermal stress can result in fracture and consequent leakage from the boiler.

Copper and Copper Alloys

Copper and its alloys comprise a versatile range of materials which are used in a wide variety of plumbing applications.

Copper is used extensively for pipework and heat exchangers due to its excellent ductility and thermal conductivity. A wide range of copper alloys are available and in addition to the aforementioned applications, they are used for valve and pump components, pipe fittings, etc.

Most problems occur when these materials are in contact with water or steam containing dissolved oxygen or carbon dioxide and/or acids or chlorides, although corrosion can also result from contact with aggressive atmospheres e.g. flue gases or certain bacteria.

These environments cause a breakdown in the protective oxide film formed on the surface by contact with oxygen and water, thus allowing corrosion to proceed. Not only will this eventually result in failure of the component, but dissolved copper can cause accelerated corrosion if it is deposited on other metals that are more anodic.

Copper

Copper is used extensively in the form of tube for pipework and the manufacture of hot water cylinders. When in contact with natural waters corrosion can occur by a variety of mechanisms.

Acidity can be a particular problem in steam condensate pipework and is often caused by a high free carbon dioxide content. It results in general dissolution of the metal surface and can be prevented by raising the pH or substituting a more resistant alloy, e.g. aluminium bronze or cupro-nickel.

Two types of localised (pitting) corrosion can cause premature failure due to perforation. 'Type 1' pitting is characterized by fairly large well-defined pits containing loose corrosion products and can occur in areas with moderately hard well waters free from organic matter which will inhibit this type of attack.

The corrosion needs to be initiated by surface deposits, crevices or stagnant conditions, a common cause being carbon films. Carbon films can be produced in the bore of tubes by the breakdown of drawing lubricant residues during manufacture. These films are cathodic to copper, and breaks in the film serve to initiate attack. The British Standard for copper tube has now been amended to include a clause to ensure that deleterious films are not present in tube bores.

'Type 2' pitting is characterized by small deep pits containing hard corrosion products and occurs in certain soft water areas, for example those containing minute quantities of manganese, and generally at temperatures above 60°C.

Both these types of pitting can occur in copper hot water cylinders (as well as pipes), particularly in the base sections where near stagnant conditions exist in crevices or underneath debris. This problem can be prevented by fitting aluminium 'protector rods' in electrical contact with the copper cylinder. The differences in potential cause the aluminium rod to corrode rapidly during which time a permanently protective film is formed on the internal copper surfaces.

Compared with its alloys, copper generally has a lower resistance to impingement attack in waters containing oxygen and/or carbon dioxide. Where dissolved gases are absent, for example in closed heating systems this type of attack is not a problem. Table A28 gives recommended maximum water velocities for copper pipework carrying fresh aerated water of pH not less than 7.

Table A28 Maximum recommended water velocities for copper tube at different temperatures (metres/second)

Temperature °C	10	50	70	90
Pipes that can be replaced	4.0	3.0	2.5	2.0
Pipes that cannot be replaced	2.0	1.5	1.3	1.0
For short connections which are used intermittently, e.g.taps	16.0	12.0	10.0	8.0

Impingement attack has a distinctive appearance, the pits formed having a water swept appearance, sometimes in the shape of horse-shoes, and being free from corrosion products. In closed heating systems where the oxygen content is low, this type of attack does not occur.

Copper can fail by stress corrosion cracking (the combination of constant stress together with a specific corrodent, see brasses) and corrosion fatigue. Corrosion fatigue results from the conjoint action of fluctuating stresses which may be caused by expansion and contraction in restrained pipework, vibration of equipment or poorly supported tubes, or pressure variations, and a corrosive environment.

Occasionally, corrosion can result in copper calorifiers where large temperature differences exist. This may be between a coil carrying steam, high temperature water under pressure or immersion heater sheaths and the calorifier shell. The large differences in temperature are sufficient to generate a potential difference and cause preferential attack on the hotter surface. This may be prevented by electrically insulating the shell and heater.

Brasses

The common brasses consist of a range of copper alloys containing from 10 to 50% zinc and other minor alloying elements. Alloys containing up to 39% zinc from a single phase alloy with copper which is termed brass α and in the range 47-50% zinc another single phase alloy termed β brass. Alloys containing 39-47% zinc contain both phases and are termed $\alpha + \beta$ or duplex brasses. The latter alloys are more suitable for hot pressing or diecasting small components.

More recently an alloy has been developed having a controlled composition so that it is suitable for processing as a duplex brass, but heat treated to attain the dezincification resistance of inhibited brass (see below). This material is termed DZR (dezincification resistant) brass and fittings are marked by the Water Research Centre recognised symbol CR. It should be noted that brazing alloys containing zinc can suffer from similar forms or attack to brasses.

Brasses can suffer a form of selective corrosion called dezincification. Zinc is preferentially removed from the alloy leaving a porous mass of copper having little strength. Where the zinc corrosion products are not washed away, they may form bulky hollow shells which readily block waterways. This is termed 'meringue' dezincification. In the single phase brasses the whole of the metal is corroded either uniformly over the surface or more commonly in the form of a 'plug'. With duplex brasses the phase which contains a higher proportion of zinc is preferentially attacked.

In general, the risk and/or rate of dezincification is increased by higher proportions of zinc in the alloy, high temperature, high chloride content of the water, low pH, low temporary hardness, low water speed and the presence of surface deposits or crevices.

Where leakage or breakage of a fitting has occurred, dezincification may be suspected if the defective areas have a dull, coppery appearance.

Brasses may be inhibited against dezincification by small additions of arsenic. Duplex brasses cannot be inhibited in this way and where there is a risk of dezincification, either DZR brass or gunmetal must be used.

Where brasses have a high degree of internal stress which may be induced during manufacture, fabrication or assembly, certain corrosive environments, in particular those containing ammonia or ammonium compounds (inorganic or organic) can cause failure by a mechanism called stress corrosion cracking (season cracking). The corrosive environment is normally external and may result from ammonical compounds used as lightweight concrete foaming agents, in rubber lattices used as cement additives or for bonding floor tiles, and from fertilizers either as residues in aggregate or airborne pollution.

In common with copper, brasses can suffer attack by impingement or cavitation. This may cause particular problems downstream from valves which, when partially open, cause a pressure drop sufficient to create air bubbles from dissolved gases and intitiate attack. It is important therefore that valves are carefully positioned in plant so that possible damage from this type of attack is minimised.

Bronzes (Gunmetals, Phosphor Bronzes, and Tin Bronzes)

Technically the term 'bronze' should refer to copper-tin alloys, but in practice a variety of copper alloys are termed bronzes irrespective of whether they contain tin. This wide variation in composition makes corrosion behaviour of a component difficult to predict unless the specific composition is known.

In general, copper-tin alloys have a good resistance to corrosion when in contact with both water (natural and sea) and steam and rarely suffer from the selective attack (dezincification) that brasses do.

EFFECTS OF CORROSIVE ENVIRONMENTS

Aluminium Bronze

These materials generally have a high resistance to corrosion, impingement attack and cavitation erosion. They are however surprisingly prone to pitting corrosion in natural waters and can suffer selective corrosion analogous to dezincification.

Cupro Nickel Alloys

Alloys containing copper and nickel have excellent resistance to corrosion in many environments, including those containing chlorides. They are more stable than brasses under flow conditions, less susceptible to stress corrosion and are used in heat exchanger applications.

Lead

Lead and lead alloys have a good corrosion resistance due to the formation of insoluble adherent carbonate or sulphate corrosion products. These films are protective and may be formed over a wide range of pH values (3 to 11) in most natural and treated waters. However, in soft waters with a low carbon dioxide content, a less protective oxide film is formed which allows corrosion to proceed slowly.

Lead is generally resistant to atmospheric corrosion but may be attacked by organic acids or free alkali. Run-off from roofs bearing organic growths, acetic acid derived from wood and fresh cement, can all cause localised corrosive attack.

In its pure form, lead has poor mechanical properties and without small alloying additions is susceptible to creep and fatigue failures induced by thermal movement.

Magnesium

The only plumbing application in which magnesium is likely to be met is that of cathodic protection. The greater negative potential in relation to steel when compared to zinc or aluminium makes magnesium alloys efficient for protecting the interior of steel (either unprotected, painted or galvanised) water cisterns or the exterior of buried steel structures and pipelines.

Soft Solders (Tin Alloys)

Soft solders (i.e. tin alloyed with lead, antimony, silver and/or copper) are anodic to copper and therefore corrosion of solder used in capillary joints could be expected. However, due to the nature of the protective film formed, corrosion only occurs in waters of high conductivity, for example those containing chlorides. Most problems with capillary joints may be attributed to the use of soldering fluxes which leave aggressive residues at, or adjacent to, the joint area.

Stainless Steels

The term 'stainless steel' covers a variety of alloys which may be simply and conveniently divided into the following three categories:

Martensitic Stainless Steels

These alloys generally contain 12–14% chromium and from 0.1 to 2.0% carbon which confers hardenability and controls the mechanical properties. They are used for applications requiring high strength and wear resistance combined with considerable corrosion resistance.

Ferritic Stainless Steels

Alloys containing 16 to 18% chromium and having a low carbon content are termed ferritic stainless steels. They are not hardenable and are used for flue and sink components.

Austenitic Stainless Steels

Often termed 18/8 stainless steels, the common austenitic alloys contain 15–22% chromium, 6–11% nickel and 0.05–0.15% carbon and have the highest resistance to corrosion (grades 302, 304). The addition of molybdenum (2–4%) confers even greater corrosion resistance (grade 316). They are non-magnetic and not hardenable unless heavily cold worked.

The main applications in plumbing services are for pipework, boiler flues, sinks and urinals.

Stainless steels owe their high corrosion resistance to the presence of self healing oxide films on the metal surface. However, certain environments may cause these films to break down and allow corrosion rates comparable to those of mild steel. The most common environments to cause attack are those containing chlorides, and in crevices or underneath surface debris where the absence of oxygen prevents repair of the protective oxide film.

Corrosion of stainless steels is often localised. The presence of chlorides can cause rapid pitting, particularly at grain boundaries and can initiate intergranular attack and stress-corrosion cracking.

Steels (Low Alloy)

Low alloy steels, due to their ease of fabrication and the range of mechanical properties attainable by alloying additions, are used in a wide variety of plumbing applications. When compared with other metals in most environments, they are found to have a much lower corrosion resistance. Under certain conditions, small additions of copper, chromium and nickel improve corrosion resistance but the overall effect is often small and variable.

Steel can exhibit two entirely different types of corrosion behaviour depending on the nature of the environment with which it is in contact. In the passive state, for example when in contact with neutral or alkaline natural water free from bacteria and oxygen, the corrosion rate is negligible. However, in the active state widely differing rates of corrosion, either general or localised, can occur depending on the many factors which have been discussed earlier.

The most common cause of corrosion is exposure to water containing dissolved oxygen. The rate of attack is greatest if the water is soft and/or acidic and the corrosion products often form bulky mounds on the surface called tubercles. These overlie areas where localised attack is occuring and can seriously reduce the carrying capacity of pipes. In severe cases, iron oxides can cause contamination leading to complaints of 'red water'. In hard, neutral or slightly alkaline waters, calcium carbonate (lime scale) can deposit on steel surfaces and provide a protective coating.

Although carbon dioxide has no specific influence on the corrosion of steel, it will increase the corrosion rate by lowering the pH and also preventing the formation of protective calcium carbonate films.

In the absence of oxygen, corrosion can occur in waters where sulphate-reducing bacteria are active.

The effect of dissolved gases and calcium carbonate in fresh waters has already been discussed. Water may also contain a variety of other dissolved material present in the supply or as a result of subsequent contamination. Ions such as chloride and sulphate can interfere with the development of protective films and lead to more localised attack whereas other ions such as calcium and bicarbonate have inhibitive properties. Small concentrations of organic matter can improve the protective qualities of carbonate films, however, organic acids resulting from decomposition of vegetation can increase corrosion rates by lowering the pH. Where the pH is reduced to values below 4 as a result of acid contamination then the corrosion rate will progressively increase with a further decrease in pH. Other surface contaminants such as oil, mill scale or deposits may not increase the overall rate of corrosion, but can localise the attack causing pitting and pinhole corrosion.

In order to minimise corrosion in steel boilers it is necessary to keep the water alkaline, higher pH values being desirable for higher pressures. This can be carried out using caustic soda, or sodium carbonate which at high temperatures, hydrolyses to give caustic soda and carbon dioxide. Where leaks are present, caustic soda can concentrate in the crevice and cause a form of stress corrosion cracking called caustic cracking. This problem is less common now as welding has largely replaced riveting as a method of fabrication.

Steel pipes are not normally suitable for use in wet soils unless protected against external corrosion.

Zinc

Zinc is used almost exclusively in plumbing applications as a protective coating on steel. It may be applied by a variety of methods which give widely varying coating weights and consequent degrees of protection. Both zinc and zinc alloys have good resistance to corrosion under conditions of exterior exposure and when in contact with most natural waters. This resistance is due to the protective layers of zinc oxide and hydroxide, or other basic salts depending on the nature of the environment, which are formed on the metal surface.

Under atmospheric conditions, the rate of corrosion depends on the degree of pollution. Heavily polluted industrial areas containing sulphur dioxide can increase the rate of corrosion by a factor of up to 10 as compared with rural areas. Where new zinc surfaces are stored under damp conditions, an unsightly white corrosion product may be formed. This is termed 'white rust' and can be prevented by oiling or a chromate treatment.

In natural waters, the rate of corrosion is governed by the presence of dissolved salts and gases. The presence of carbon dioxide together with calcium and magnesium salts will form a basic carbonate film which is protective to the base metal. Zinc is anodic to steel and will provide protection in the form of a coating even if small areas of substrate are exposed due to damage or in fabrication.

Temperature has a marked effect on the corrosion of zinc in natural waters. Due to a change in the nature of the protective films, the corrosion rate increases in the range of 55°C to 95°C, giving a maximum increase of some 100 times at 70°C. In some natural waters where zinc is used as a protective coating on steel, a reversal of potential occurs above 65°C. The zinc then becomes cathodic to steel, stimulating attack which results in localised (pitting) corrosion and subsequent perforation.

In general zinc and copper or copper alloys are not compatible when used in natural water plumbing systems. Where waters which are cupro-solvent (e.g. those containing free carbon dioxide) flow through copper pipework, traces of copper are dissolved which can subsequently deposit as metallic copper on zinc surfaces and stimulate attack due to bimetallic corrosion. Brass fittings on galvanized steel tanks and cisterns, however, rarely cause problems due to the relatively small area of brass to zinc.

Where unprotected galvanized pipes are used below ground the most severe attack occurs in soils which are poorly aerated and/or have high acid and soluble salt contents.

In conclusion, galvanized tanks, calorifiers and pipes should not be used at temperatures above 65°C or downstream from copper pipework. Below ground, galvanized pipework is best protected using proprietary tape wrap systems.

Plastics and Rubber Components

Although polymeric materials do not corrode in the accepted sense, they may degrade when exposed to certain environments. These environments include ultraviolet light, heat, organic compounds and inorganic and metallic ions, but the wide range of materials and compositions available make specific guidance impossible.

PREVENTION OF CORROSION

Before deciding on the most suitable course of action required, if any, to control or prevent corrosion, it is often beneficial to assess the nature of the environment to which the metal will be exposed or the rate at which corrosion will occur. In certain cases, metal loss due to corrosion may be economically acceptable providing the required service life or life to first maintenance is exceeded. Furthermore, where corrosion has occured, there are now several non-destructive techniques available which may be used to quantify the extent of the attack and metal loss.

Assessing the Corrosivity of the Local Environment

External Surfaces

The external surfaces of pipework, etc. may suffer attack if buried in soil, located in ductwork subject to condensation or flooding, or where exposed to rain, spray, etc.

The aggressiveness of soils to buried metals, particularly ferrous, may be assessed using resistivity data to measure the likelihood of oxidative corrosion and redox-potentials to indicate the risk of bacterial corrosion. Soils may be considered aggressive if the resistivity is less than 2000 Ωcm at the specified depth or the mean redox-potential is less than +0.400 volts (standard hydrogen scale) when corrected to pH = 7 (+0.430 volts if the soil is predominantly clay).

Accelerated corrosion may also be caused by the presence of stray electrical earth currents from for example, DC railway systems or rectified induced AC from high voltage transmission systems, and these may be identified from the measurement of voltage differences in the soil.

Where metal surfaces may be exposed to natural waters, either due to immersion, condensation or rainfall, corrosion can be severe particularly if salts or other pollutants are present. The likely corrosivity of these environments can be assessed using a variety of electrical and electrochemical techniques.

Internal Surfaces

The nature of the water in contact with the metal surfaces may be most conveniently assessed or monitored by chemical analysis of representative samples. The analysis required depends to a certain extent on the particular situation in which the supply or circulating water is being used.

It must be stressed that conclusions from analytical results should be drawn very carefully and expert consideration needs to be given to the many factors involved. It is only then possible to formulate the most appropriate treatment to control corrosion and also scale deposition which may affect the efficient operation of a system.

The data given below should only be used as a guide and no indications have been given as to the relative importance of the different factors.

Potable (supply), Hot and Cold Domestic Waters

The ability of these waters to deposit a protective film of calcium carbonate scale on internal metal surfaces can play an important part in controlling corrosion. Two indices may be calculated from analytical data (alkalinity, calcium hardness, total solids, temperature and pH) which indicate whether a particular water will dissolve or deposit a protective scale. These are called the Langelier and Ryznar indices.

A negative Langelier index indicates that the water will tend to dissolve scale and promote corrosion while a positive index indicates that a protective film will be deposited.

The Ryznar index is a modified form of the Langelier index which

PREVENTION OF CORROSION

is claimed to be more reliable. Using this index, a water is considered to be corrosive when the index exceeds approximately 6.0 and scale forming when it is less than 6.0. Where the water is used to feed heating plant, then the hardness value will determine the need for chemical treatment/softening to minimise scale deposition.

Primary and Secondary Heating Waters, Chilled Water Systems

These systems are essentially closed systems with water recirculation and there should be little or no make-up.

Problems usually arise when there are excessive water losses, air ingress, or aggressive residues are present from fabrication or cleaning operations. Table A29 indicates how analytical data would be assessed in an area having a supply water with a total hardness of 300 mg/litre as calcium carbonate.

Cooling Waters

Cooling for air conditioning systems is most commonly achieved by evaporation of water, normally 1–1.5% of the circulation rate being used to cause a temperature drop of 5–8°C. This evaporation leaves behind both dissolved solids which build up until deposition occurs and aggressive ions, for example chlorides, which will exacerbate corrosion. In addition the cooling water may become contaminated by atmospheric pollution resulting in acid corrosion due to the reduced pH.

Table A30 gives a guide to the limits that would be used when operating a typical open recirculating cooling system.

Table A30 Recommended water characteristics for evaporative cooling tower waters

Analysis		Value
pH value		7.5–9.0
Chlorines mg/l	max	100
Dissolved solids mg/l	max	700–1000
Calcium hardness mg/l	max	600
Conductivity μS/cm	max	1000–1500

Steam Raising Plant

BS2486 : 1978, 'Treatment of water for land boilers' gives a comprehensive guide to the operation of industrial steam-raising boiler plant and the preparation of feed water for such plant. This standard recommends limiting values for the concentrations of principal chemical constituents in the feed and boiler waters for the following types of steam raising plant:

a) Shell boilers at operating pressures up to 25 bar.

Table A29 Assessment of condition of recirculating systems from analytical data

Analysis	Assessment from analytical result Satisfactory	Fair	Poor	Bad	Units	Comments
Nitrate (as N)	0–0.24	0.25–0.79	0.80–1.20	>1.2	mg/litre	Used as an indication of water losses/water make-up.
Chlorions	<50	50–74	75–100	>100	mg/litre	A high level may indicate the presence of residues from acid descaling or chloride based fluxes. Flushing may be necessary to reduce the risk of corrosion.
pH	>9.0	8.0–9.0	7.0–7.9	<7.0	—	Minimum corrosion rates are achieved with the pH (a measure of acidity/alkinity) above 9.0. Where aluminium is present, the pH should be kept below 8.5.
Sulphate	>50	10–50	10–50	<10	mg/litre	A low level may indicate the presence of sulphate reducing bacteria.
Sulphate Reducing Bacteria						
@ 30°C	0	1	2–5	>5	C.F.U.	These bacteria can cause internal corrosion in the heating system.
@ 55°C	0	1	2–5	>5	C.F.U.	
Iron						
Total	<1.0	1.0–4.9	5.0–10.0	>10.0	mg/litre	Used as an indication of the level of corrosion in a heating system and the need for remedial action and/or water treatment.
Dissolved	<0.5	0.5–1.9	2.0–5.0	>5.0		
Copper						
Total	<0.02	0.02–0.05	0.06–0.20	>0.2	mg/litre	Total values relate to corrosion which has occurred in the past while dissolved levels indicate current corrosion rates
Dissolved	<0.02	0.02–0.05	0.06–0.20	>0.2		
Zinc						
Total	<0.25	0.25–0.49	0.50–1.00	>1.00	mg/litre	
Dissolved	<0.10	0.10–0.29	0.30–0.60	>0.60		
Hardness						
Total	<50	50–100	101–250	>250	mg/litre	High levels may indicate water losses and cause internal scaling. Where low water content boilers are being used, installation of water softening plant may be beneficial.
Water loss	<0.5	0.5–1.0	1.1–10.0	>10.0	%/week	High water losses can cause problems of internal scaling and corrosion.

b) Water-tube boilers operating up to the limit imposed by the critical pressure of steam.

c) Once-through boilers at operating pressures up to 250 bar.

The relevant values are reproduced in Tables A31, A32 and A33.

Table A31 Recommended water characteristics for shell boilers

Total hardness in feed water, mg/l in terms of $CaCO_3$ max	2	20	40	Comments
Feed water				
pH value	7.5 to 9.5	7.5 to 9.5	7.5 to 9.5	—
Oxygen	§	§	§	Should be reduced to lowest practicable level before adding oxygen scavenger
Total solids, alkalinity, silica	§	§	§	
Organic matter	§	§	§	
Boiler water				
Total hardness, mg/l in terms of $CaCO_3$ max. Sodium phosphate, mg/l as Na_3PO_4	50 to 100	50 to 100	50 to 100	—
Caustic alkalinity, mg/l in terms of $CaCO_3$ min.	350	300	200	—
Total alkalinity, mg/l in terms of $CaCO_3$ max.	1200	700	700	—
Silica, mg/l as SiO_2 max.	Less than 0.4 of the caustic alkalinity			
Sodium sulphite, mg/l as Na_2SO_3	30 to 70	30 to 70	30 to 70	—
or				
Hydrazine, mg/l as N_2H_4	0.1 to 1.0	0.1 to 1.0	0.1 to 1.0	—
Suspended solids, mg/l max.	50	200	300	Theoretical values
Dissolved solids, mg/l max.	3500	3000	2000	—

Table A32 Recommended water characteristics for water-tube boilers

Pressure at boiler outlet bar	20	40	60	80	100	120	Above 130	Comments
Feed water at economizer inlet								
Total hardness, mg/l in terms of $CaCO_3$ max	10	2	0.5	Not detectable				—
pH value	8.5 to 9.5	8.5 to 9.5	8.5 to 9.5	8.5 to 9.5	8.5 to 9.5	8.5 to 9.5	8.5 to 9.5	8.5 to 9.2 if copper present after physical de-aeration
Oxygen, mg/l as O_2 max.	0.05	0.02	0.01	0.01	0.005	0.005	0.005	—
Iron + copper + nickel, mg/l max.	—	—	0.02	0.02	0.01	0.01	0.01	—
Total solids, alkalinity, silica	§	§	§	§	§	§	§	—
Oil	Not detectable							
Organic matter	§	§	§	§	§	§	§	—
Boiler water								
Sodium phosphate, mg/l as Na_3PO_4	50 to 100	30 to 70	20 to 50	10 to 40	5 to 20	3 to 10	§	
Caustic alkalinity, mg/l in terms of $CaCO_3$ min.	300	150	60	30	10	5	2	
Total alkalinity, mg/l in terms of $CaCO_3$ max.	700	500	300	200	100	40	10	
Silica§, mg/l as SiO_2 max.	Less than 0.4 of the caustic alkalinity		20	10	§	§	§	
Sodium sulphite, mg/l as Na_2SO_3	30 to 50	20 to 40	15 to 30	10 to 20	§	§	—	
or								
Hydrazine, mg/l as N_2H_4	0.1 to 1.0	0.1 to 0.5	0.05 to 0.3	0.05 to 0.1	§	§	§	
Suspended solids, mg/l max.	200	50	—	—	—	—	—	
Dissolved solids, mg/l max.	3000	2000	1200	700	350	100	15	
Chloride, mg/l as Cl^- max.	—	—	—	—	10	5	1	

§ Numerical values depend upon circumstances.

PREVENTION OF CORROSION

Table A33 Recommended water characteristics for once-through boilers

Feed water at economizer inlet		
Total solids	mg/l max	0.05
Conductivity at 20°C after hydrogen cation exchange	μS/m max.	20
Oxygen after physical de-aeration	mg/l as O_2 max.	0.005
Silica	mg/l as SiO_2 max.	0.02
Iron	mg/l as Fe max.	0.01
Copper plus nickel	mg/l as Cu + Ni max.	0.002
pH value with copper or copper alloy feed heaters		8.5 to 9.2
pH value without copper or copper alloy feed heaters		9.2 to 9.5

Corrosion Control and Prevention

External Surfaces

A variety of methods are available for protecting pipework, etc. when their external surfaces are exposed to a corrosive environment.

Organic coatings such as paints, plastic coatings or tape wrap systems may be applied to the metal surface which has been prepared by grit blasting and/or the application of a suitable primer. In general the more severe the conditions of service, the thicker and more resistant the coating needs to be. These types of coatings can be very susceptible to mechanical damage and this should be borne in mind during installation.

Both tanks and pipes can be repaired and protected using either glass reinforced plastic (GRP) or specially formulated concrete linings.

Metal coatings, for example zinc and aluminium, may be applied to steel surfaces by hot dipping or spraying, which gives the thickest coating, diffusion or in the case of zinc, electro-deposition. These metals will corrode, albeit at a rate much slower than steel, and therefore the coating thickness required depends on the aggressiveness of the environment. In severe cases, an additional paint coating may be applied.

Another method of protecting underground pipelines is that of cathodic protection. Two methods are available each of which produce a counter current sufficiently large to neutralise the currents responsible for corrosion. The first uses a more electro-negative metal, the most common being either magnesium, aluminium or zinc, in the form of sacrificial anodes which are connected electrically to the metal requiring protection. The second uses an impressed current from a generator in conjunction with auxiliary anodes of iron, steel, graphite, lead or platinised titanium.

Internal Surfaces

All internal surfaces that store or carry water, for example pipes, tanks, calorifiers, heat exchangers, radiators, etc. may be subject to corrosive attack. There are many different possible approaches to preventing corrosion and it is of utmost importance that the correct method is chosen to suit each particular situation. In most cases, proprietary blends of chemicals are available which fulfil some or all of the required preventative treatment.

Other factors may also need to be considered, for example toxicity when treating potable waters or if there is a risk of crossover between primary heating and domestic supplies.

Both polyphosphates and sodium silicates may be used as a non-toxic, tasteless, odourless and colourless treatment for domestic waters. They both prevent iron corrosion discolouring the supply water (commonly known as 'red water') and inhibit the deposition of scale by forming a thin film on the metal surfaces which acts as a barrier. They are dosed either in liquid form using a proportional dosing pump or as slowly dissolving crystals from a suitable dispenser.

Storage tanks may be protected internally using cathodic protection. Sacrificial anodes, normally magnesium for steel or aluminium for copper, are connected to and suspended in the tank. While aluminium anodes in copper tanks only have to last for sufficient time to form a permanent protective film, magnesium anodes in steel tanks need to be replaced when consumed.

Primary and Secondary Heating Waters, Chilled Water Systems

These systems, because they are essentially closed recirculating systems, may be treated with either oxygen scavengers, which act by the removal of oxygen responsible for the corrosion process, or inhibitors which slow down the corrosion reactions to acceptable levels, sometimes in conjunction with pH control.

Inhibitors need to be chosen carefully to suit the system under consideration. Some inhibitors may reduce corrosion in one environment while increasing it in another, while others only work effectively in a certain concentration range, actually intensifying corrosion outside this range. Sulphate reducing bacteria may be controlled by maintaining a specified level of biocide in the system.

With the current trend towards very low water content boilers, it is also advisable to use sludge conditioners to prevent scale deposition inside boiler sections and tubes. These materials should be used with caution in older systems as they may dislodge scale and corrosion products highlighting weaknesses at joints and causing leaks. Where larger systems suffer continuous water losses due to the presence of leaks, then it can be beneficial to use softened water for make-up to reduce the need for frequent scale removal.

Cooling Waters

Corrosion may be controlled by dosing the system with a suitable inhibitor. This needs to be carried out continuously at a rate proportional to the bleed-off required to prevent scale deposition and excessive levels of aggressive ions building up. The treatment should also include sludge conditioners to prevent localised scale deposition and biocides to control bacteria and algae (slime) growth.

In hard water areas, the cost of water and treatment chemicals can be high and therefore consideration should be given to using a softened make-up supply. Disposal of bleed-off water also needs to be considered as some treatment chemicals are toxic and cannot be drained into local sewerage systems.

Steam Raising Plant

Steam boilers are treated by ensuring that the boiler water is alkaline at all times and by removing dissolved oxygen to maintain a protective film of magnetite (magnetic iron oxide) on steel surfaces. In addition, treatment may also be necessary to prevent corrosion in condensate lines which is most commonly caused by the presence of dissolved oxygen and carbon dioxide gases.

Dissolved oxygen in boiler water may be removed by means of a physical de-aerator or oxygen scavenger, the chemicals most commonly used being catalysed sodium sulphite, certain selected tannins or hydrazine. Condensate corrosion may be controlled by the addition of volatile amines which either neutralise the acidity caused by the presence of carbon dioxide, or form a protective film on the metal surfaces.

Caustic cracking, which used to be a fairly common phenomenon particularly in riveted boilers, can occur if concentrations of sodium hydroxide greater than 5% exist. This may be prevented by either:

a) maintaining the ratio of sodium nitrate to total alkalinity (in terms of calcium carbonate) at a minimum value of 0.32 at all times, or
b) maintaining the ratio of sodium sulphate to caustic alkalinity (in terms of calcium carbonate) at a minimum value of 2.0 at all times.

In high pressure boilers, the level of chloride ions needs to be limited to minimise corrosion. In addition to treatment for the prevention of corrosion, sludge and scale deposition also needs to be carefully controlled to maintain efficiency and steam purity, particularly as the trend in modern boilers is towards a lower water content and higher heat transfer rates. In brief, the concentration of solids in the boiler water may be controlled by the following methods depending on the quality of supply water:

a) Softening and filtration of the feed water to remove dissolved and suspended solids.
b) Precipitation of hardness salts as a mobile sludge in the boiler by maintaining a controlled reserve of either carbonate or phosphate in solution in the boiler water. In addition, organic sludge conditioners may be added to ensure that the precipitated salts are non-adherent and mobile.
c) Maintaining the total dissolved solids below the maximum level applicable to the particular boiler and operating conditions (usually below 3500 mg/l) by 'blowing down'. Blowdown may be either intermittent or continuous and is estimated from standard formulae given in BS 2486.
d) Cleaning the boiler when shut down.

Cleaning and Descaling of Boilers and Associated Plant

In newly installed systems, it is often necessary to remove internal contaminants, for example flux residues, metal filings and other builders' debris by flushing. Even if carried out in accordance with recommended procedures, it is unlikely that all will be removed due to either their insolubility or the internal geometry of the system. This highlights the need to adopt working procedures that negate the need for flushing, for example using non-aggressive fluxes and cutting rather than sawing copper tubes.

In older systems, particularly where there have been water losses or air ingress, scale and sludge may have deposited which will affect the efficiency. In the worst case, particularly where low water content boilers are used, overheating will lead to eventual failure of sections or splitting of tubes.

If inspection indicates that descaling is necessary, the nature of the deposits should be determined to ascertain the best method of removal.

Loose deposits may be removed manually or mechanically, but hard deposits will normally require chemical removal. Table A34 indicates chemicals that may be used to remove the most common types of deposit. If scaling is present on a confined part of the system, for example the boiler, then treatment should be isolated to that part of the system. Progress of the cleaning should be strictly monitored, care being taken to ensure that all aggressive chemical residues are removed after completion of the descaling operation. Throughout the treatment, adequate safety precautions need to be observed and the effluent disposed of in accordance with statutory and other requirements.

Plant Shutdown

When plant and associated equipment is taken out of service for any length of time, steps need to be taken to prevent damage due to internal corrosion.

In the case of heating and chilled water systems, they should be left completely filled with water containing oxygen scavengers or corrosion inhibitors and a biocide, and maintained at a high pH. In this situation, precautions may also need to be taken to prevent frost damage. Plant should only remain in a drained down condition if left and maintained completely dry internally otherwise rapid localised attack, particularly in for example steel panel radiators, can occur resulting in premature failure.

Cooling towers need to be drained and where necessary dismantled to prevent seizure of pumps, etc. When they are brought back on line, treatment with suitable biocides is recommended to control bacteria.

Table A34 Cleaning and descaling treatments

Treatment	Deposit X indicates Effective Treatment								
	Carbonate	Phosphate	Sulphate	Silicate	Copper & oxide	Millscale	Magnetite	Oil	Grease
Hydrochloric acid (inhibited)	X	X				X	X		
Hydrochloric/hydrofluoric acids (inhibited)	X	X		X		X	X		
Sulphuric acid (inhibited)	X					X	X		
Sulphamic acid	X								
Citric acid						X	X		
Ammonium citrate with oxidising agent (pH 9.5)					X				
Formic acid	X	X				X	X		
Tetra sodium salt of EDTA (ethylene diamine-tetra acetic acid)	X	X	X						
Tetra-ammonium salt of EDTA	X	X	X		X	X	X		
Sodium hydroxide/sodium phosphate								X	X

WELDING/BRAZING AND SOFT SOLDERING/ADHESIVE BONDING

Metal joints may be fabricated using any one of the above processes, and both welding and adhesive bonding may be used as an alternative to rubber 'O' rings to join plastic pipes. All these processes are used extensively in plumbing applications, however it is important to ensure that the technique chosen performs satisfactorily in terms of strength, ductility and durability.

WELDING

Welding is a process where the surface of the parent metal is brought to melting temperature and is fused together with or without a filler rod of similar composition to produce a homogenous joint.

The technique of welding is widely used because it is an efficient method of producing strong joints free from crevices that would require sealing to prevent either leakage or corrosion.

The processes involved in welding may be conveniently divided into two main categories. These are solid phase or pressure welding, which involves pressing parts together sufficiently to deform and subsequently bond the surfaces either at ambient or elevated temperatures and includes butt, cold pressure, explosive, flash butt, forge, friction and ultrasonic welding, and fusion welding.

Fusion welding consists of melting the surface of the parent metal using either a gas flame (oxy-acetylene) or an electric arc and normally adding additional metal from a filler rod to form the joint. The manual metal arc process is the most widely used and involves striking a controlled electric arc between the filler rod and surfaces to be joined. During the fusion process it is necessary to protect the deposited metal as contamination by atmospheric gases can affect the quality of the weld. This is accomplished by using filler rods coated with a consumable flux, or shielding the arc with an inert gas which may be helium, argon or carbon dioxide (or a mixture), depending on the metals being joined. The latter processes are termed metal inert gas (MIG) or tungsten inert gas (TIG) welding.

Both the gas and inert gas welding methods are especially suited to jointing thin sheet metal while for heavy plates the arc processes using either flux or gas shielding are most suitable.

Sheet metal may also be joined using electric resistance welding where an electric current is passed through a lapped joint, the heat generated being sufficient to fuse the surfaces together.

Welding Defects

The most common defects in welds are caused by faulty technique and these may be summarised as follows:

1) Undercut

The reduction in cross-sectional area causes a weakness in the joint and may allow slag to be trapped.

2) Slag Inclusions

Non-metallic particles trapped in the weld metal during welding may seriously reduce the strength of joint.

3) Incomplete Penetration

Produced when the weld metal fails to fill the root of the weld and leaves a gap.

4) Lack of Fusion

Where there is incomplete fusion between the metal surfaces or joint edges.

5) Cracking

Cracks occurring at low temperatures (heat affected zone or cold cracking) or high temperatures (solidification or hot cracking) may be caused by material composition, cooling rate, stress or the presence of hydrogen.

Quality Control

In order to ensure a satisfactory performance, welds must be of adequate quality for the intended application. As the quality of the weld depends to a great extent on the skill of the welder it is important that individual abilities conform with certain laid down minimum standards. These are covered by performance tests laid down in BS 4870, 4871 and 4872 respectively.

The quality of welds may be assessed by several non-destructive techniques such as visual examination, dye penetrant, magnetic particle, radiographic and ultrasonic inspection. These techniques require high levels of operator skill and experience to correctly assess the integrity of a given weld or joint.

Plastic materials may also be joined by welding. The temperatures of the surfaces to be bonded are raised using either heated contact plates, friction, hot gas (with filler rod), a moving heated wire, induction, high frequency or ultrasonic techniques and held or pressed together. In all these processes care must be taken to avoid oxidation/degradation of the polymer as it will result in a weak joint.

BRAZING AND SOLDERING

Both brazing (which encompasses silver-soldering, sometimes termed hard-soldering) and soft soldering employ filler metals with a lower melting point than the parent metal. These are melted and drawn by capillary action into the joint. Providing the surfaces to be joined are clean and the correct flux has been used to remove the surface oxide films, the filler metal will wet the adjoining surfaces and form a metallurgical bond. Brazing generally refers to the use of fillers with melting points above 500°C and soft soldering to those below this temperature.

Brazing

Brazing may be carried out using a wide variety of filler metals which range in melting points from approx 600°C to over 1200°C (see BS 1845: 1984). For most plumbing applications the metals that require joining are likely to be copper, copper alloys, steel and stainless steel for which the lower melting fillers are suitable. Compositions vary widely, but most are based on copper with additions of silver and/or zinc and cadmium (silver brazing alloys), phosphorus or zinc.

Fluxes are generally based on boric acid and alkaline borates sometimes together with small additions of other compounds to improve activity. The copper-phosphorus alloys are self-fluxing when brazing copper to copper in air. Both the composition of the filler metal and flux need to be chosen carefully. The choice depends on the metals to be joined, the size of the joint gap, whether a fillet is required at the joint, whether the environment to which the joint may be exposed is corrosive and the cost and toxicity of the filler metal.

Manufacturers of brazing materials have available comprehensive data sheets to assist with the correct selection of both filler and flux materials and these should be referred to before the final choice is made.

It is essential to remove all flux residues after the brazing operation as corrosive attack may occur when the flux hydrolyses on exposure to moist air. Manufacturers usually give recommended procedures for the removal of a particular composition of flux.

The most common method used for heating is that of a gas/air torch. For the lower melting point filler metals, propane is entirely suitable providing the nozzle size is selected to match the required heat input. This depends very much on the mass and thermal conductivity of the metal components being joined. The aim is to supply sufficient heat to raise the temperature rapidly and evenly up to the filler melting point without causing localised overheating.

The use of oxy-acetylene is not recommended due to the intensely hot localised nature of the flame. This can cause rapid overheating generating fumes, which in the case of cadmium bearing silver solders, can prove fatal.

Soft Soldering

Soft solders used in plumbing applications were mainly alloys of tin and lead having melting points in the range 183–255°C. However to comply with the Water Byelaws, lead free solders must now be used for potable water services and alloys of tin with additions of antimony, copper or silver have been developed as alternatives.

The composition of a solder is chosen to give flow characteristics matching the intended application. For example a 30–70 tin-lead alloy which has a long 'pasty' range i.e. range between the solidus and liquidus temperatures, would be used for wiping lead joints, whereas alloys containing 40–60% tin are used in capillary applications requiring an increased flow and wetting capability.

At temperatures above 80°C, tin-lead solders rapidly lose their strength and therefore at higher service temperatures the lead free alloys are preferred. The properties of the most common alloys are given in Table A35.

Table A35 Composition of typical soft solders

BS 219	Nominal Composition %				Solidus	Liquidus
Grade	Tin	Lead	Antimony	Other	°C	°C
K	60	40.0	0.5 max	—	183	189
F	50	50.0	1.5 max	—	183	212
B	50	47.2	2.8	—	185	203
G	42	58.0	0.4 max	—	183	232
C	40	57.8	2.2	—	185	228
H	35	65.0	0.3 max	—	183	246
L	32	66.2	1.8	—	185	245
D	30	68.5	1.5	—	185	248
95A	95	—	5.0	—	236	243
96S	96	—	—	4 silver	221	235
99C	99	—	—	1 copper	230	335

Ideally, fluxes used for soft soldering should have the following properties:

1. They must be capable of removing oxide films at the chosen temperature of joining.
2. They should be easy to apply and have the correct viscosity so that they flow when displaced by the molten filler alloy but still maintain a reactive liquid cover to prevent further surface oxidation.
3. The film of flux must remain active for the length of the joining operation.
4. Residues remaining after the soldering operation should be non-corrosive and/or easily removable, even where access is limited, e.g. the bore of pipes.

A variety of compounds are used to manufacture commercial fluxes, including inorganic acids and halides, organic acids and halides and rosin. The flux may contain one or a mixture of these made up into a paste using either a petroleum derivative or water miscible base.

Due to the large variety of compositions available, choosing the best flux for a particular application can be difficult. Recommendations are often given to flush pipework after joints have been made using either capillary or end feed fittings to reduce the risk of subsequent corrosion in primary heating systems. This is often impractical for the following reasons:

1. It can be difficult to carry out effectively thus ensuring all residues are removed.
2. On heating, some fluxes produce insoluble residues in the pipe bore which subsequently induce pitting corrosion and eventual perforation.
3. The time and consequent cost of effective flushing is a deterrent against it being carried out.
4. It may not be possible to carry out flushing until some considerable time after fabricating the joints.

For these reasons, the use of organic based, water miscible fluxes is to be preferred for copper services if the risk of corrosion is to be minimised.

When soldering stainless steels, it is essential that only phosphoric acid-based fluxes are used as halide-based fluxes cause rapid pitting corrosion of these alloys.

ADHESIVE BONDING

By far the most common application of adhesives in plumbing is that of joining pipework. Metal pipes with capillary joints can be bonded using anaerobic adhesives specially developed for the purpose. These adhesives are single part products which only cure when air is excluded and they are in contact with a metal surface. These materials are also now available as replacements for the traditional hemp or PTFE tape used for joints with screwed fittings.

Some plastic materials used for plumbing applications (notably ABS and UPVC) may be joined using a solvent welding technique. A suitable adhesive is applied to both joint surfaces and the two halves pushed together. The action of the solvent in the adhesive then rapidly forms a coherent bond.

Health and Safety

Particular care should be taken to ensure adequate ventilation during all welding and brazing operations. The high temperatures involved can generate dangerous levels of toxic vapours, particularly those from zinc or cadmium contained in coatings or filler metals, which can be fatal in the short or long term.

In the case of adhesives and other jointing compounds where it is proposed to use them on potable water services, it should first be ascertained whether they are approved for this application as toxicity can sometimes be a problem.

SECTION A
Hot and Cold Water Supplies

REFERENCE

Water Supply Byelaws Guide, ISBN 0 946653 30 5; WRc Water Byelaws Advisory Service, 660 Ajax Avenue, Slough, Berkshire, SL1 4BG.

BRITISH STANDARDS AND CODES OF PRACTICE

BS 219: 1977 (1984)	Specification for soft solders.
BS 499:	Welding terms and symbols.
BS 499: Part 1: 1983 (1984)	Glossary for welding, brazing and thermal cutting.
BS 499: Part 2C: 1980	Welding symbols.
BS 499: Part 2: 1980	Specification for symbols for welding.
BS 638	Arc welding plant, equipment and accessories.
BS 638: Part 1: 1979	Specification for oil cooled power sources for manual, semi-automatic and automatic metal-arc welding and for TIG welding.
BS 638: Part 2: 1979	Specification for air cooled power sources for manual metal-arc welding with covered electrodes and for TIG welding.
BS 638: Part 3: 1979	Specification for air cooled power sources for semi-automatic and automatic metal-arc welding.
BS 638: Part 4: 1979 (1983)	Specification for welding cables.
BS 638: Part 6: 1984	Specification for safety requirements for construction.
BS 638: Part 7: 1984 (1985)	Specification for safety requirements for installation and use.
BS 638: Part 8: 1984 (1986)	Specification for electrode holders and hand held torches and guns for MIG, MAG and TIG welding.
BS 639: 1986	Manual metal arc welding.
BS 679: 1959 (1977)	Specification for filters for use during welding and similar operations.
BS 807: 1965	Specification for spot welding electrodes.
BS 1010:	Specification for draw-off taps and stop valves for water services (screw-down pattern).
BS 1010: Part 2: 1973 (1984)	Draw-off taps and above ground stopvalves.
BS 1212:	Specification for float operated valves (excluding floats).
BS 1212: Part 1: 1953 (1979)	Piston type.
BS 1212: Part 2 1970 (1980)	Diaphragm type (brass body).
BS 1212: Part 3 1979	Diaphragm type (plastics body), for cold water services.
BS 1244: Part 2: 1982	Metal sinks for domestic purposes.
BS 1295: 1987	Code of Practice for training in arc welding skills.
BS 1377: 1975	Methods of test for soils for civil engineering purposes.
BS 1387: 1985	Specification for screwed and socketed steel tubes and tubulars and for plain steel tubes suitable for welding or for welding or for screwing to BS 21 pipe threads.
BS 1394:	Stationary circulation pumps for heating and hot water service systems.
BS 1394: Part 1 1971	Glanded and glandless pumps.
BS 1394: Part 2 1971	Domestic glandless pumps.
BS 1453: 1972 (1987)	Specification for filler materials for gas welding.
BS 1542: 1982	Specification for equipment for eye, face and neck protection against non-ionising radiation arising during welding and similar operations.
BS 1723: Part 1: 1986	Specification for brazing.
BS 1723: Part 2: 1986	Guide to brazing.
BS 1724: 1959	Specification for bronze welding by gas.
BS 2493: 1985	Specification for low alloy steel electrodes for manual metal arc welding.
BS 2633: 1987	Specification for Class 1 arc welding of feritic steel pipework for carrying fluids.
BS 2653: 1955 (1978)	Specification for protective clothing for welders.
BS 2871: Part 1: 1971 (1976)	Specification for copper tubes for water, gas and sanitation.
BS 2901:	Specification for filler rods and wires for gas shielded arc welding.
BS 2901: Part 1: 1983	Ferritic steels.
BS 2901: Part 2: 1983	Austenitic stainless steels.
BS 2901: Part 3: 1983	Copper and copper alloys.
BS 2901: Part 4: 1983	Aluminiun and aluminium alloys and magnesium alloys.
BS 2901: Part 5: 1983	Nickel and nickel alloys.
BS 2926: 1984	Specification for chromium and chromium nickel steel electrodes for manual metal arc welding.
BS 2971: 1977 (1986)	Specification for Class 2 arc welding of carbon steel pipework for carrying fluids.
BS 3019:	TIG welding.
BS 3019: Part 1: 1984	Specification for TIG welding of aluminium, magnesium and their alloys.
BS 3019: Part 2: 1960	Austenitic stainless an heat-resisting steels.
BS 3571:	MIG welding.
BS 3571: Part 1: 1985	Specification for TIG welding of aluminium, magnesium and their alloys.
BS 3943: 1979 (1985)	Specification for plastic waste traps.
BS 4213: 1986	Specification for cold water storage and feed expansion cisterns.

BS 4215: 1967	Specification for spot welding electrodes and electrode holders.
BS 4346:	Joints and fittings for use with unplasticized PVC pressure pipes.
BS 4346: Part 1 1969 (1987)	Injection moulded unplasticised PVC fittings for solvent welding for use with pressure pipes, including potable water supply.
BS 4346: Part 2: 1970 (1987)	Mechanical joints and fitting principally for unplasticized PVC.
BS 4514: 1983 (1984)	Specification for unplasticized PVC soil and ventilating pipes, fittings and accessories.
BS 4515: 1984	Specification for process of welding of steel pipelines on land and off-shore.
BS 4677: 1984 (1987)	Specification for arc welding of austenitic stainless steel pipework for carrying fluids.
BS 4872: Part 1: 1982	Fusion welding of steel.
BS 4872: Part 2: 1976	TIG and MIG welding of aluminium and its alloys.
BS 5135: 1984	Specification for arc welding of carbon and carbon manganese steels.
BS 5245: 1975 (1985)	Specification for phosphoric acid based flux for soft soldered joints in stainless steel.
BS 5254: 1976 (1983)	Specification for polypropylene waste pipe and fittings.
BS 5255: 1976 (1984)	Specification for plastics waste pipes and fittings.
BS 5412:	Specification for the performance of draw-off taps with metal bodies for water services.
BS 5412: Part 1: 1976 (1984)	Dimensional and design characteristics.
BS 5412: Part 2: 1976 (1980)	Water tightness and pressure resistance characteristics.
BS 5412: Part 3: 1976 (1980)	Hydraulic characteristics.
BS 5412: Part 4: 1976 (1981)	Mechanical and endurance characteristics.
BS 5412: Part 5: 1976 (1984)	Physio-chemical characteristics: materials, coatings.
BS 5413: 1976	Specifications for the performance of draw-off taps with plastics bodies for water services.
BS 5503:	Specification for vitreous china washdown WC pans with horizontal outlet.
BS 5503: Part 1: 1977 (1980)	Connecting dimensions.
BS 5503: Part 2: 1977 (1980)	Materials, quality, performance and dimensions other than connecting dimensions.
BS 5505:	Specification for bidets.
BS 5505: Part 1: 1977	Pedestal bidets over rim supply only. Connecting dimensions.
BS 5505: Part 2: 1977	Wall hung bidets over rim supply only. Connecting dimensions.
BS 5505: Part 3: 1977	Vitreous china bidets over rim supply only. Quality, workmanship and functional dimensions other than connecting dimensions.
BS 5506:	Specification for wash basins.
BS 5506: Part 1: 1977	Pedestal wash basins. Connecting dimensions.
BS 5506: Part 2: 1977	Wall hung basins. Connecting dimensions.
BS 5506: Part 3: 1977	Wash basins (one or three tap holes). Materials, quality, design and construction.
BS 5572: 1978 (1983)	Code of Practice for sanitary pipework.
BS 5627: 1984	Specification for plastics connectors for use with horizontal outlet vitreous china WC pans.
BS 6209: 1982	Specification for solvent cement for non-pressure thermoplastics pipe systems.
PD 6484: 1979 (1984)	Commentary on corrosion at bimetallic contacts and its alleviation.
BS 6678: 1986	Specification for tungsten electrodes for inert gas shielded arc welding and for plasma cutting and welding.
BS 6700: 1987	Specification for design installation testing and maintenance of services supplying water for domestic use within buildings and their curtilages.
CP 312:	Code of Practice for plastics pipework (thermoplastics material).
CP 312: Part 1: 1973	General principles and choice of material.
CP 312: Part 2: 1973 (1977)	Unplasticized PVC pipework for the conveyance of liquids under pressure.
CP 312: Part 3: 1973	Polyethylene pipes for the conveyance of liquids under pressure.
DD 39: 1974	Steel pipelines under pressure.

Note

(1) Year after the Standard's number indicates latest date of issue.

(2) Year in brackets gives date of latest amendment at the time of publication of this Guide.

DESIGN CONSIDERATIONS

Central heating systems should be designed to meet the calculated heat losses. These losses should be based on an outside air temperature no higher than –1°C and should be the total of the structural and ventilation heat losses. A base figure lower than –1°C may be necessary for a building in a very exposed or coastal situation.

Table B1 Temperature and ventilation rates

Room	Room Temperature °C	Ventilation Rate per hour
Living room*	21	1½
Dining room*	21	1½
Bedsitting room*	21	1½
Bedroom**	18	1½
Hall & landing	18	1½
Kitchen	18	2
Bathroom	22	2
Toilet	18	2

*Where open flue appliances are installed the rate of air change should be increased.

**When bedrooms are used as bedsitting rooms or for study purposes a higher room temperature may be required.

Different temperatures and air change rates may be applicable if standards laid down for Gas Warm Homes, All Electric or Medallion Homes are to be met.

In order to reduce unnecessary heat losses and thus reduce operating costs, attention should be given to:

(a) Reducing air leaks through badly fitting external doors and windows, open fire places and open joints in boarded floors.

(b) Insulation of walls, floors, ceilings and windows.

Heat losses can be reduced by double glazing, filling of cavity walls with insulating materials and by internal insulation linings.

Where heating systems are to be operated intermittently, an approximate margin may be added to the calculated heat losses of 10–20% in the case of gas, oil and solid fuel gravity feed and hopper fed boilers, and 20–30% in the case of solid fuel sectional and pot-type boilers.

Heat Emission from Pipes

Where pipework is exposed in heated rooms, the heat emission from the pipes can be usefully employed. The amount of heat emitted can be determined from Graphs B1 and B2.

These graphs give the theoretical heat emission per hour from single horizontal uninsulated steel tubes and painted copper tubes when exposed in still air at temperatures between 10°C and 20°C, but allowing for natural convection currents.

The emission will be greater in conditions of draught, depending upon the velocity of the air. For example, with an air velocity of 1 m/sec there will be an increase of up to 100% for pipes up to 25 mm diameter. Where the air velocity is 0.5 m/s, the increase will be 50% for pipes up to 50 mm diameter.

The emissivity by radiation of a surface will also depend upon the surface condition, e.g. tarnished copper tube will emit approximately 75% of the values given in Graph B1 for painted copper tubes. Where pipes are banked one above the other at close pitch, the emission will be reduced by between 5% and 35%.

Vertical Pipes

Pipes fixed vertically differ in their heat emission from those fixed horizontally. Whereas small diameter pipes emit less heat, large diameter pipes emit more heat. Table B2 gives examples of the percentage of heat emission from vertical pipes as compared with horizontal pipes.

Table B2 Percentage of heat emission from vertical copper pipes as compared with horizontal pipes

Nominal Bore of pipe mm	Temperature Difference 40–45°C	50–130°C
	Percentage of heat emission	
8	70	73
10	73	75
15	75	77
22	78	80
28	80	83
35	83	85
42	85	87
54	87	89
66	89	91
76	91	93

Graph B1 Heat emission from single horizontal uninsulated copper tubes to BS 2871 when painted with oil or water based paints of light colours.
Dark colours and metallic paints will give different emissions.

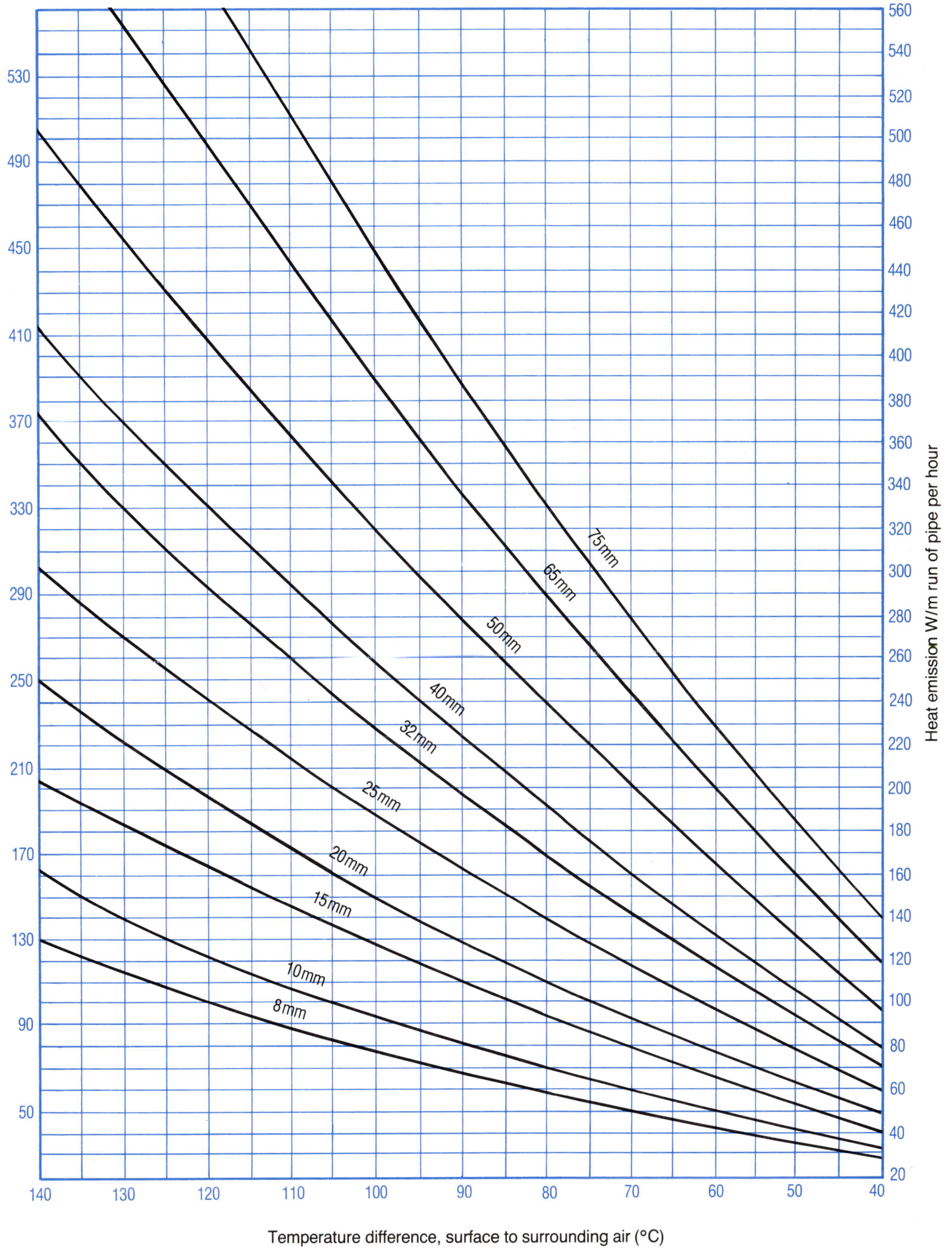

Graph B2 Heat emission from single horizontal uninsulated unpainted steel tubes to BS 1387

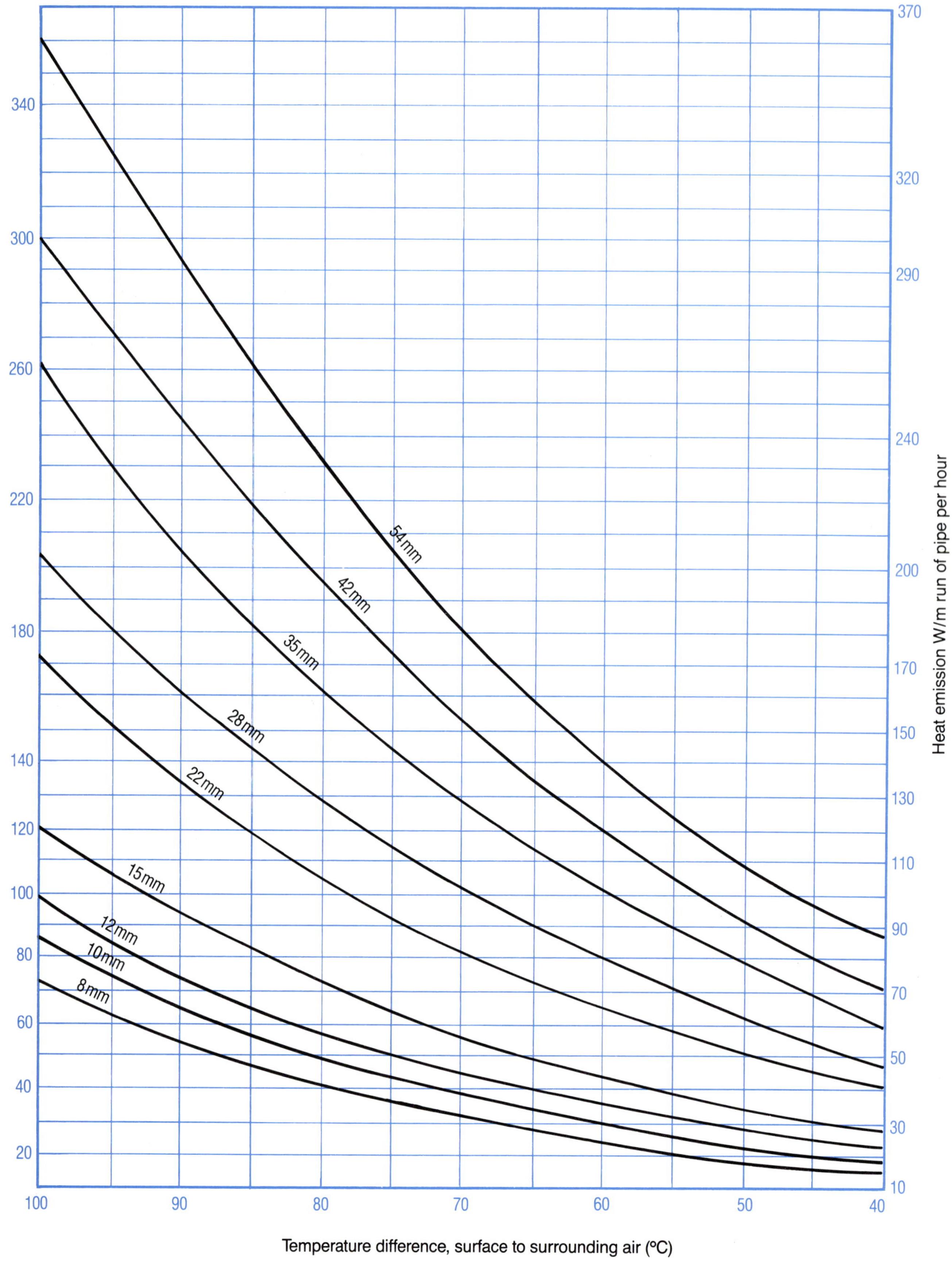

Heat Emission from Insulated Pipes

The heat emission from insulated pipes will vary according to the conductivity of the insulating material and its thickness. Insulation material must be kept dry, as damp insulation can increase heat emission by up to 400%.

Table B3 Heat emission from insulated copper pipes

	Heat emission W/m run of pipe per °C (in still air at 20°C) Thermal conductivity of insulating material (W/m°C)														
	0.040					0.055					0.070				
	Thickness of insulation in mm														
Pipe OD	20	25	30	40	50	20	25	30	40	50	20	25	30	40	50
15	0.20	0.19	0.18	0.16	0.14	0.27	0.25	0.24	0.21	0.19	0.34	0.32	0.30	0.26	0.24
22	0.24	0.23	0.22	0.18	0.16	0.31	0.29	0.28	0.25	0.21	0.38	0.36	0.34	0.30	0.27
28	0.26	0.25	0.24	0.21	0.18	0.35	0.33	0.31	0.28	0.24	0.45	0.42	0.39	0.34	0.30
35	0.31	0.29	0.28	0.24	0.20	0.41	0.39	0.37	0.32	0.28	0.52	0.49	0.46	0.39	0.33
42	0.34	0.32	0.30	0.26	0.22	0.44	0.42	0.40	0.35	0.31	0.57	0.53	0.49	0.42	0.37
54	0.40	0.37	0.34	0.29	0.25	0.52	0.49	0.45	0.39	0.34	0.65	0.61	0.57	0.49	0.42
66	0.47	0.44	0.41	0.35	0.29	0.62	0.58	0.52	0.46	0.39	0.77	0.72	0.68	0.57	0.47
76	0.54	0.50	0.47	0.39	0.33	0.71	0.66	0.60	0.51	0.43	0.87	0.81	0.75	0.63	0.54

Flow of Water in Pipes

The flow of water through pipes can be determined from Table B4 which indicates the pressure loss in N/m² per metre run for given flow rates in kg/s. The velocity of the water in m/s is shown in the Table, and the maximum velocity in any domestic heating system should not exceed 1.5 metres/second.

Table B4 Resistance to hot water flow through copper pipes

Pressure losses per metre run in the flow of hot water through heating circulation pipework where copper tubing to BS 2871: Part 1: Table X is used with a limiting water velocity of 1.5 m/s.

Where microbore tubing to BS 2871: Part 1: Table W is used, manufacturer's tables should be used.

	Pressure loss per metre run due to flow of hot water through copper tubes							
Tube size	6 mm	8 mm	10 mm	12 mm	15 mm	22 mm	28 mm	35 mm
Flow kg/s	N/m²	N/m²	N/m²	N/m²	N/m²	N/m²	N/m²	N/m²
0.00175	59.0	—	—	—	—	—	—	—
0.00180	62.0	—	—	—	—	—	—	—
0.00185	65.0	—	—	—	—	—	—	—
0.00190	68.0	—	—	—	—	—	—	—
0.00195	71.0	—	—	—	—	—	—	—
0.0020	74.0	—	—	—	—	—	—	—
0.0021	80.5	—	—	—	—	—	—	—
0.0022	87.0	—	—	—	—	—	—	—
0.0023	94.0	—	—	—	—	—	—	—
0.0024	101	—	—	—	—	—	—	—
0.0025	108	—	—	—	—	—	—	—
0.0026	116	—	—	—	—	—	—	—
0.0027	124	—	—	—	—	—	—	—
0.0028	132	—	—	—	—	—	—	—
0.0029	140	—	—	—	—	—	—	—
0.0030	148	—	—	—	—	—	—	—
0.0031	156	—	—	—	—	—	—	—
0.0032	165	—	—	—	—	—	—	—
0.0033	174	—	—	—	—	—	—	—
0.0034	183	—	—	—	—	—	—	—

DESIGN CONSIDERATIONS

Table B4 (Cont'd)

	Pressure loss per metre run due to flow of hot water through copper tubes							
Tube size	6mm	8mm	10mm	12mm	15mm	22mm	28mm	35mm
Flow kg/s	N/m^2	N/m^2	N/m^2	N/m^2	N/m^2	N/m^2	N/m^2	N/m^2
0.0035	192	37.8	—	—	—	—	—	—
0.0036	201	39.6	—	—	—	—	—	—
0.0037	211	41.6	—	—	—	—	—	—
0.0038	221	43.5	—	—	—	—	—	—
0.0039	231	45.5	—	—	—	—	—	—
0.0040	241	47.5	—	—	—	—	—	—
0.0042	262	51.5	—	—	—	—	—	—
0.0044	283	55.8	—	—	—	—	—	—
0.0046	305	60.2	—	—	—	—	—	—
0.0048	328	64.8	—	—	—	—	—	—
0.0050	352	69.5	—	—	—	—	—	—
0.0052	376	74.2	—	—	—	—	—	—
0.0054	400	79.0	—	—	—	—	—	—
0.0056	426	84.5	—	—	—	—	—	—
0.0058	452	87.5	—	—	—	—	—	—
0.0060	478	95.0	28.0	—	—	—	—	—
0.0062	505	100	29.6	—	—	—	—	—
0.0064	535	106	31.3	—	—	—	—	—
0.0066	565	112	32.9	—	—	—	—	—
0.0068	595	117	34.6	—	—	—	—	—
0.0070	625	123	36.5	—	—	—	—	—
0.0072	655	129	38.3	—	—	—	—	—
0.0074	685	136	40.2	—	—	—	—	—
0.0076	715	142	42.0	—	—	—	—	—
0.0078	745	148	44.0	—	—	—	—	—
0.0080	780	155	46.0	—	—	—	—	—
0.0084	850	169	50.0	—	—	—	—	—
0.0089	920	183	54.0	20.3	—	—	—	—
0.0092	990	197	58.5	21.9	—	—	—	—
0.0096	1070	212	63.0	23.6	—	—	—	—
0.0100	1150	228	67.3	25.3	—	—	—	—
0.0105	1250	247	73.5	27.6	—	—	—	—
0.0110	1360	268	80.0	30.0	—	—	—	—
0.0115	1460	290	86.0	32.4	—	—	—	—
0.0120	1580	311	92.5	34.8	—	—	—	—
0.0125	1700	333	99.0	37.4	—	—	—	—
0.0130	1820	356	106	40.0	—	—	—	—
0.0135	1940	381	113	42.7	—	—	—	—
0.0140	2060	405	120	45.5	15.3	—	—	—
0.0145	2190	430	128	48.3	16.2	—	—	—
0.0150	2330	455	136	51.0	17.2	—	—	—
0.0155	2460	482	144	54.1	18.2	—	—	—
0.0160	2600	510	153	57.2	19.2	—	—	—

Table B4 (Cont'd)

	Pressure loss per metre run due to flow of hot water through copper tubes							
Tube size	6mm	8mm	10mm	12mm	15mm	22mm	28mm	35mm
Flow kg/s	N/m²	N/m²	N/m²	N/m²	N/m²	N/m²	N/m²	N/m²
0.0165	2740	537	161	60.4	20.3	—	—	—
0.0170	2880	565	170	63.5	21.4	—	—	—
0.0175	3030	595	178	67.0	22.5	—	—	—
0.0180	3180	625	187	70.5	23.6	—	—	—
0.0185	3340	654	196	74.0	24.7	—	—	—
0.0190	3500	684	205	77.5	25.8	—	—	—
0.0195	3600	714	215	81.0	27.0	—	—	—
0.020	3810	746	224	84.5	28.3	—	—	—
0.021	4150	810	244	92.0	30.8	—	—	—
0.022	4500	880	264	100	33.4	—	—	—
0.023	4860	950	286	108	36.1	—	—	—
0.024	5230	1030	307	117	38.8	—	—	—
0.025	5600	1100	330	125	41.6	—	—	—
0.026	6000	1180	353	134	44.6	—	—	—
0.027	—	1260	376	143	47.6	—	—	—
0.028	—	1340	400	152	50.7	—	—	—
0.029	—	1420	425	162	54.0	—	—	—
0.030	—	1500	452	171	57.0	—	—	—
0.031	—	1590	478	182	60.5	9.1	—	—
0.032	—	1680	505	192	64.0	9.6	—	—
0.033	—	1770	532	202	67.5	10.1	—	—
0.034	—	1870	560	213	71.0	10.6	—	—
0.035	—	1960	587	224	74.5	11.2	—	—
0.036	—	2050	615	234	78.5	11.8	—	—
0.037	—	2160	645	246	82.0	12.4	—	—
0.038	—	2260	675	257	86.0	13.0	—	—
0.039	—	2370	705	269	90.0	13.6	—	—
0.040	—	2480	740	281	94.0	14.2	—	—
0.042	—	2690	805	306	103	15.5	—	—
0.044	—	2920	870	332	112	16.8	—	—
0.046	—	3150	940	358	120	18.2	—	—
0.048	—	3380	1010	385	129	19.5	—	—
0.050	—	3620	1080	413	138	21.0	—	—
0.052	—	3870	1160	442	148	22.5	6.48	—
0.054	—	4120	1240	472	158	24.0	6.9	—
0.056	—	—	1320	503	168	25.6	7.35	—
0.058	—	—	1400	535	180	27.2	7.85	—
0.060	—	—	1480	568	191	28.8	8.3	—
0.062	—	—	1570	600	202	30.5	8.8	—
0.064	—	—	1660	634	213	32.3	9.3	—
0.066	—	—	1750	668	224	34.1	9.8	—
0.068	—	—	1840	702	236	35.8	10.3	—
0.070	—	—	1940	736	248	37.8	10.8	—
0.072	—	—	2030	772	261	39.8	11.4	—
0.074	—	—	2130	810	273	41.7	12.0	—

DESIGN CONSIDERATIONS

Table B4 (Cont'd)

	Pressure loss per metre run due to flow of hot water through copper tubes							
Tube size	6mm	8mm	10mm	12mm	15mm	22mm	28mm	35mm
Flow kg/s	N/m²	N/m²	N/m²	N/m²	N/m²	N/m²	N/m²	N/m²
0.076	—	—	2230	848	286	43.7	12.6	—
0.078	—	—	2330	890	298	45.6	13.2	—
0.080	—	—	2430	930	312	47.7	13.8	4.85
0.084	—	—	2640	1010	341	52.0	15.0	5.3
0.088	—	—	2850	1100	368	56.3	16.3	5.75
0.092	—	—	—	1180	397	60.8	17.6	6.23
0.096	—	—	—	1270	430	65.5	19.0	6.70
0.100	—	—	—	1370	462	70.5	20.4	7.20
0.105	—	—	—	1490	502	77.0	22.3	7.55
0.110	—	—	—	1620	545	83.5	24.2	8.55
0.115	—	—	—	1740	588	90.5	26.1	9.25
0.120	—	—	—	1880	633	97.0	28.2	9.95
0.125	—	—	—	2020	680	104	30.2	10.7
0.130	—	—	—	2160	728	113	32.4	11.5
0.135	—	—	—	2310	775	121	34.6	12.3
0.140	—	—	—	—	828	128	36.8	13.1
0.145	—	—	—	—	880	136	39.3	13.9
0.150	—	—	—	—	930	144	41.7	14.7
0.155	—	—	—	—	980	153	44.2	15.5
0.160	—	—	—	—	1040	162	46.6	16.4
0.165	—	—	—	—	1090	171	49.3	17.4
0.170	—	—	—	—	1150	180	52.0	18.3
0.175	—	—	—	—	1210	189	54.8	19.3
0.180	—	—	—	—	1270	199	57.5	20.3
0.185	—	—	—	—	1340	209	60.3	21.3
0.190	—	—	—	—	1400	218	63.0	22.3
0.195	—	—	—	—	1460	229	66.0	23.4
0.20	—	—	—	—	1530	240	69.0	24.4
0.21	—	—	—	—	1670	261	75.5	26.5
0.22	—	—	—	—	1810	283	82.0	28.8
0.23	—	—	—	—	—	305	88.5	31.2
0.24	—	—	—	—	—	330	95.5	33.7
0.25	—	—	—	—	—	355	103	36.2
0.26	—	—	—	—	—	380	110	38.7
0.27	—	—	—	—	—	405	118	41.3
0.28	—	—	—	—	—	432	126	44.1
0.29	—	—	—	—	—	457	134	47.0
0.30	—	—	—	—	—	485	142	49.7
0.31	—	—	—	—	—	513	150	52.8
0.32	—	—	—	—	—	543	159	55.8
0.33	—	—	—	—	—	574	168	59.0
0.34	—	—	—	—	—	606	177	62.0
0.35	—	—	—	—	—	637	186	65.0

Table B4 (Cont'd)

Tube size	6mm	8mm	10mm	12mm	15mm	22mm	28mm	35mm
	Pressure loss per metre run due to flow of hot water through copper tubes							
Flow kg/s	N/m²	N/m²	N/m²	N/m²	N/m²	N/m²	N/m²	N/m²
0.36	—	—	—	—	—	668	195	68.5
0.37	—	—	—	—	—	700	205	72.0
0.38	—	—	—	—	—	735	215	75.5
0.39	—	—	—	—	—	770	225	79.0
0.40	—	—	—	—	—	805	235	82.5
0.42	—	—	—	—	—	875	256	90
0.44	—	—	—	—	—	950	278	97
0.46	—	—	—	—	—	1030	300	105
0.48	—	—	—	—	—	—	324	114
0.50	—	—	—	—	—	—	348	123
0.52	—	—	—	—	—	—	372	132
0.54	—	—	—	—	—	—	398	140
0.56	—	—	—	—	—	—	425	149
0.58	—	—	—	—	—	—	453	158
0.60	—	—	—	—	—	—	480	168
0.62	—	—	—	—	—	—	507	178
0.64	—	—	—	—	—	—	533	189
0.66	—	—	—	—	—	—	562	199
0.68	—	—	—	—	—	—	595	211
0.70	—	—	—	—	—	—	628	222
0.72	—	—	—	—	—	—	660	233
0.74	—	—	—	—	—	—	690	245
0.76	—	—	—	—	—	—	725	256
0.78	—	—	—	—	—	—	755	268
0.80	—	—	—	—	—	—	—	280
0.84	—	—	—	—	—	—	—	304
0.88	—	—	—	—	—	—	—	331
0.92	—	—	—	—	—	—	—	358
0.96	—	—	—	—	—	—	—	385
1.00	—	—	—	—	—	—	—	415
1.05	—	—	—	—	—	—	—	453
1.10	—	—	—	—	—	—	—	490
1.15	—	—	—	—	—	—	—	490
1.20	—	—	—	—	—	—	—	570
1.25	—	—	—	—	—	—	—	—
1.30	—	—	—	—	—	—	—	—

SECTION B
Domestic Central Heating

DESIGN CONSIDERATIONS

When calculating circulating pump duties, the total resistance to flow should be calculated taking into account the resistance to flow through either compression and capillary fittings used with copper tube to BS 2871: Part 1: Table X, as shown in Table B5.

Table B5 Resistance to hot water flow at 0.61 m/s through fittings (Compression and capillary fittings used with BS 2871: Part 1: Table X)

Nominal size (mm)	Equivalent lengths in metres												
	Tee (compression or capillary)			Reducing tee (compression or capillary)		Pitcher tee (compression)			Pitcher tee (capillary)			Elbow (compression)	Elbow (capillary)
8	0.021	0.31	0.27	0.35	0.22	0.024	0.19	0.17	0.026	0.22	0.22	0.24	0.16
10	0.030	0.42	0.37	0.45	0.30	0.032	0.26	0.24	0.036	0.29	0.29	0.33	0.21
12	0.039	0.55	0.49	0.59	0.39	0.042	0.32	0.31	0.046	0.38	0.38	0.42	0.28
15	0.050	0.69	0.62	0.75	0.50	0.056	0.43	0.41	0.059	0.50	0.47	0.56	0.37
22	0.085	1.1	1.0	1.2	0.84	0.090	0.71	0.69	0.10	0.76	0.75	0.90	0.60
28	0.12	1.6	1.4	1.7	1.2	0.12	0.98	0.98	0.14	1.0	1.0	1.2	0.83
35	0.15	2.0	1.7	2.2	1.5	0.15	1.2	1.2	0.17	1.2	1.3	1.5	1.0

Heat Losses Through Building Structures

Table B6 lists heat transmittance coefficients for various types of building structure.

Table B6 Heat transmittance coefficients ('U' values)

Construction	W/m²/°C
External walls solid brickwork unplastered	
121 mm	3.35
236 mm	2.39
351 mm	1.87
Solid brickwork dense plastered 13 mm thick	
236 mm	2.10
351 mm	1.70
Solid brickwork light plastered 13 mm thick	
236 mm	1.90
351 mm	1.50

Cavity Walls (outside to inside)

Brick : Cavity : Brick (+13 mm plaster)

Brick	Air gap	U F Foam	Brick	W/m²/°C
105 mm	25 mm	—	105 mm	1.50
220 mm	25 mm	—	220 mm	1.00
105 mm	—	50 mm	105 mm	0.55

Brick : Cavity : Block (+13 mm plaster)

Brick	Air gap	Mineral Fibre	U F Foam	Concrete Heavy	Block Light	W/m²/°C
105 mm	25 mm	—	—	100 mm	—	1.60
105 mm	—	50 mm	—	100 mm	—	0.49
105 mm	25 mm	—	—	—	100 mm	0.92
105 mm	—	50 mm	—	—	100 mm	0.43
105 mm	—	—	50 mm	—	100 mm	0.46

Table B6 (Cont'd)

Block : Cavity : Block (+13 mm plaster)

Concrete Block Heavy	Concrete Block Light	Air Gap	U F Foam	Concrete Block Heavy	Concrete Block Light	W/m²/°C
100 mm	—	25 mm	—	100 mm	—	1.80
100 mm	—	—	50 mm	100 mm	—	0.61
—	100 mm	25 mm	—	—	100 mm	0.67
—	100 mm	—	50 mm	—	100 mm	0.39

Timber framed walls

Inner leaf: Timber studwork with bracing plywood incorporating 100 mm minimum gap, or wider according to thickness of mineral fibre insulation mat, vapour check and 12.7 mm of plaster internally.

Outer leaf:	Depth of mat mm	W/m²/°C
Concrete tiles hung externally	60	0.47
	80	0.38
	100	0.34
	120	0.29
	140	0.25
Shiplap boarding fixed externally	60	0.44
	80	0.36
	100	0.32
	120	0.28
	140	0.24
105 mm Brick outer cladding	60	0.43
	80	0.36
	100	0.32
	120	0.28
	140	0.24

Table B6 (Cont'd)

Construction		W/m²/°C
Windows		
Frame occupying 20% of window area		
Wooden frame, single glazed		5.0
Wooden frame, double glazed		2.9
Aluminium frame, no thermal break, single glazed		6.4
Aluminium frame, no thermal break, double glazed		4.3
Aluminium frame, with thermal break, single glazed		5.8
Aluminium frame, with thermal break, double glazed		3.7
Partition and party walls		
105mm brick, plastered both sides		1.9
220mm brick, plastered both sides		1.5
50mm breeze block, plastered both sides		2.3
75mm breeze block, plastered both sides		2.0
Pitched roofs		
30° pitched roof of timber trussed rafters, covered with concrete tiles or slates on wooden battens. Joists finished with 9.5mm plasterboard. Mineral fibre mat laid between ceiling joists.		
	Depth of mat	W/m²/°C
	60mm	0.55
	80mm	0.43
	100mm	0.36
With sarking felt underlay	60mm	0.51
	80mm	0.40
	100mm	0.34
With felt laid on sarking board	60mm	0.48
	80mm	0.39
	100mm	0.32
Flat roofs		
3 layer felt on boarded joists, plasterboard ceiling		1.59
3 layer felt on strawboard or chipboard, joists and plasterboard		1.05
Asphalt on 104mm concrete, plaster ceiling		3.41
Asphalt on 150mm concrete, plaster ceiling		3.12

HEATING CONTROLS

The controls included in this section apply mainly to domestic and small commercial installations. More sophisticated control systems will be required for larger commercial, industrial and residential schemes.

The term 'controls' when used in relation to a heating system refers to a number of components which may be used by themselves or together to regulate the system automatically. The cost of heating fuels makes it imperative that efficient control systems are used. These, together with good heating system designs and high insulation standards will give enhanced comfort conditions in the building together with economy in running costs.

Controls usually comprise of the following items:

1. Time Controls

These switch all or part of the system ON and OFF according to the user's requirements. The time switch can control the whole system or the part of the system that is specifically wired to it. A heating programmer enables the heating and hot water to be controlled independently.

2. Sensing Elements

These are usually thermostats which react to changes of temperature to control space heating and stored water temperatures.

(a) Space Heating Controls

These mainly take the form of a room thermostat which will control the temperature of the whole building or a zone of that building. It is important that the room thermostat is fitted in a representative position to give the correct comfort conditions away from any extraneous sources of heat and draughts. It is also important that all the radiators are correctly sized on each circuit being controlled by a single room thermostat.

A fixing height of 1500mm is usually considered satisfactory. Some room thermostats are fitted with a 'heat accelerator'. This is a small electric heater within the thermostat, normally wired so that if the thermostat is calling for heat, a small amount of heat is artificially applied to the sensing element. This reduces the response time of the thermostat. Space temperatures can also be controlled by non-electric controls, usually thermostatic radiator valves. It is important to ensure that thermostatic radiator valves are not installed in the same room as the room thermostat otherwise the sensors will oppose each other and the system will tend to 'hunt'.

(b) Water Temperature Controls

A cylinder thermostat fitted approximately ⅓rd up from the base of the cylinder can control the stored water temperature of the cylinder by controlling a motorised valve.

Alternatively, a non electric 2- or 3-port control valve with a sensing element fitted to the side of the cylinder can also be used.

3. Valves

Valves are used in conjunction with time controls and sensing elements to regulate the flow of water to the various parts of the system as required. Valves on more simple systems may be operated manually, or if used in conjunction with sensing elements, control the flow of water to individual radiators or zones automatically.

SYSTEMS OF CONTROL

Figures B1 to B5 show examples of the control systems that can be used. It must be noted that the systems shown will not necessarily apply to all situations and it is important that the control system chosen is compatible and appropriate to the boiler and fuel being used. In all cases the boiler manufacturer's literature should be carefully consulted before designing a system. Some boilers are not suitable for gravity primary installations, some require the fitting of a pumped by-pass arrangement, especially those fitted with a pump over-run in order to maintain a minimum flow rate through the boiler.

Solid fuel boilers usually require a permanent 'heat-leak' arrangement, with not only the gravity flow and return to the cylinder unobstructed at all times, but also one or two radiators or towel rails connected to a gravity circuit.

Partial Control of System with Gravity Primary Hot Water and Pumped Heating Circuits

This system, shown in Figure B1 is very basic and should be considered as the minimum standard acceptable. A programmer or timeswitch controls the switching of the heating and hot water times with a single room thermostat controlling the heat output from the radiators by turning the pump ON and OFF.

A cylinder thermostat may be added to the system to ensure that boiler 'cycling' does not take place once the cylinder is up to temperature, at any time when the heating pump is switched off. In this case the boiler will not fire again until the cylinder thermostat or room thermostat is calling for heat. The disadvantage of this arrangement is that the cylinder thermostat is ineffective at all times when the heating pump is in operation. When the heating pump is 'ON' the stored water temperature is determined by the thermostat setting on the boiler. This can result in excessive hot water temperatures especially in winter months.

Full Control of System with Gravity Primary Hot Water and Pumped Heating Circuits

For full control at all times of a system with a gravity primary hot water circuit, a two-port valve is required, as illustrated in Figure B2. This prevents the gravity primary flow of water to the cylinder when the stored secondary water is at the temperature set on the cylinder thermostat. It is important that the two-port valve is positioned so as not to obstruct the cold feed connection to the boiler.

With this system both circuits can be fully controlled by use of a programmer. A non-electrically controlled valve may also be used if required to control the cylinder temperature. These are available in a 3-port pattern with bypass or with an auxiliary switch to operate the system via the programmer.

Fully Pumped Systems

Fully pumped systems can be controlled more efficiently than systems employing either full or partial gravity circulation. Most modern heating systems are fully pumped, the advantages being the use of smaller pipe sizes, improved heat up and recovery time of the cylinder temperature, greater flexibility of layout and improved operating efficiencies. It is recommended that gravity systems are not used.

Figure B1 A programmer or time switch controls the boiler and pump timing. Note that the boiler thermostat provides the only control of the stored hot water temperature.

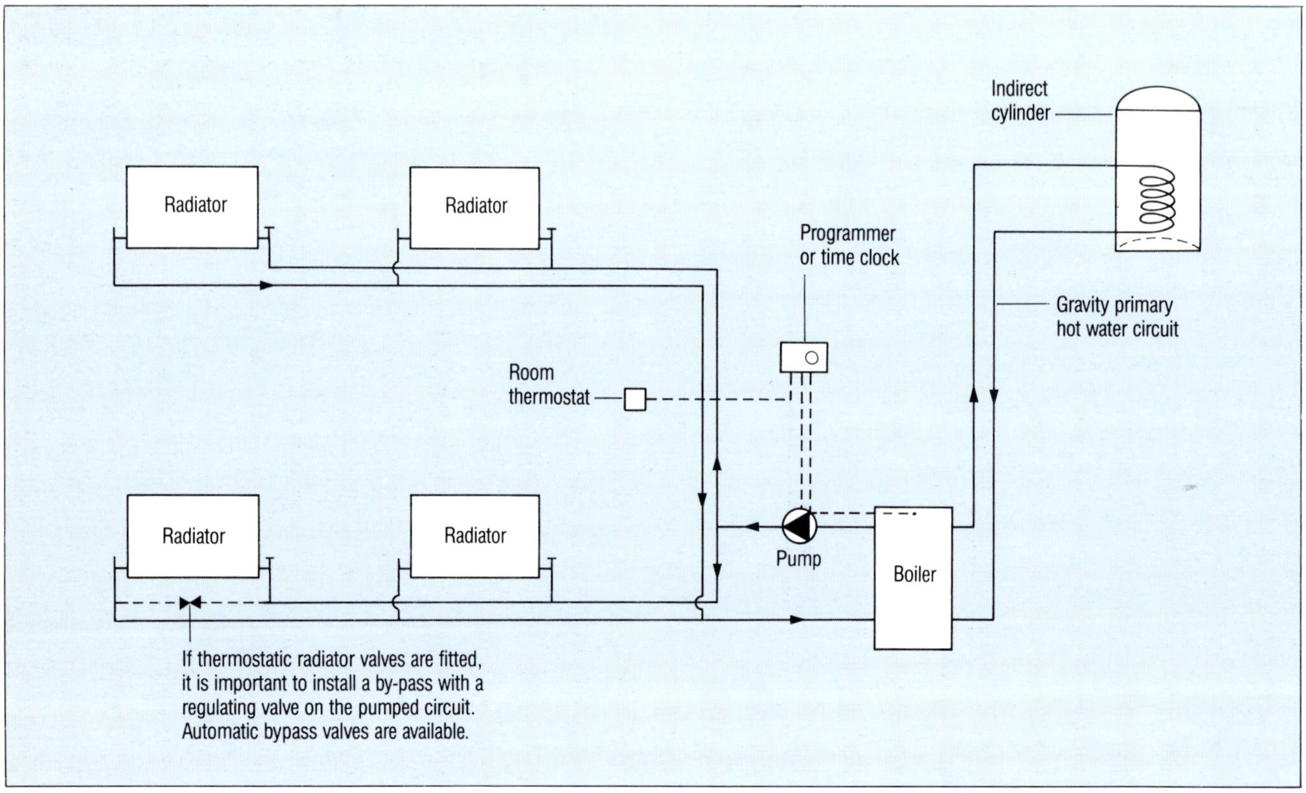

Figure B2 A two-port motorised valve has been added to the system shown in Figure B1 to give full temperature control to the domestic hot water.

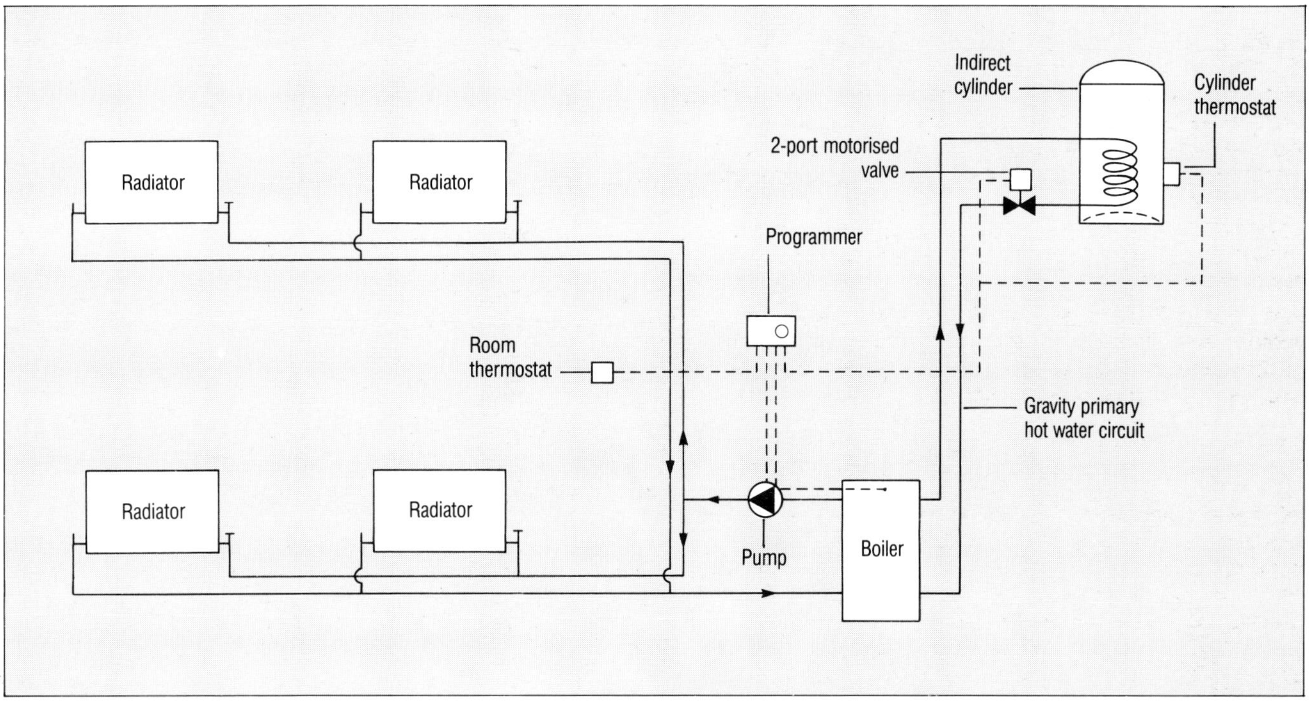

Fully Pumped System with a 3-port Control Valve

This is probably the most economical way of controlling a fully pumped system efficiently, see Figure B3. Most control manufacturers have a three-port mid-position valve which enables the heated water to be shared between the heating and hot water primary circuits enabling both circuits to be heated simultaneously. A 3-port diverter valve can also be used. This is usually cheaper than the mid-position valve and has only two positions, so that heat is either pumped to the heating or hot water circuit but not both simultaneously. The priority is usually given to the hot water circuit so that when the cylinder is being heated no heat goes to the radiators. It is possible to fit a selector switch to override this and select heating only. Thus the 3-port diverter valve has its limitations when compared with the 3-port mid-position valve. When using 3-port valves it is imperative to ensure that the heating pipework is commoned together prior to connecting to the cylinder circuit otherwise water will circulate through the heating system when only the cylinder thermostat is calling for heat.

Figure B3 Fully pumped system with 3-port valve control.

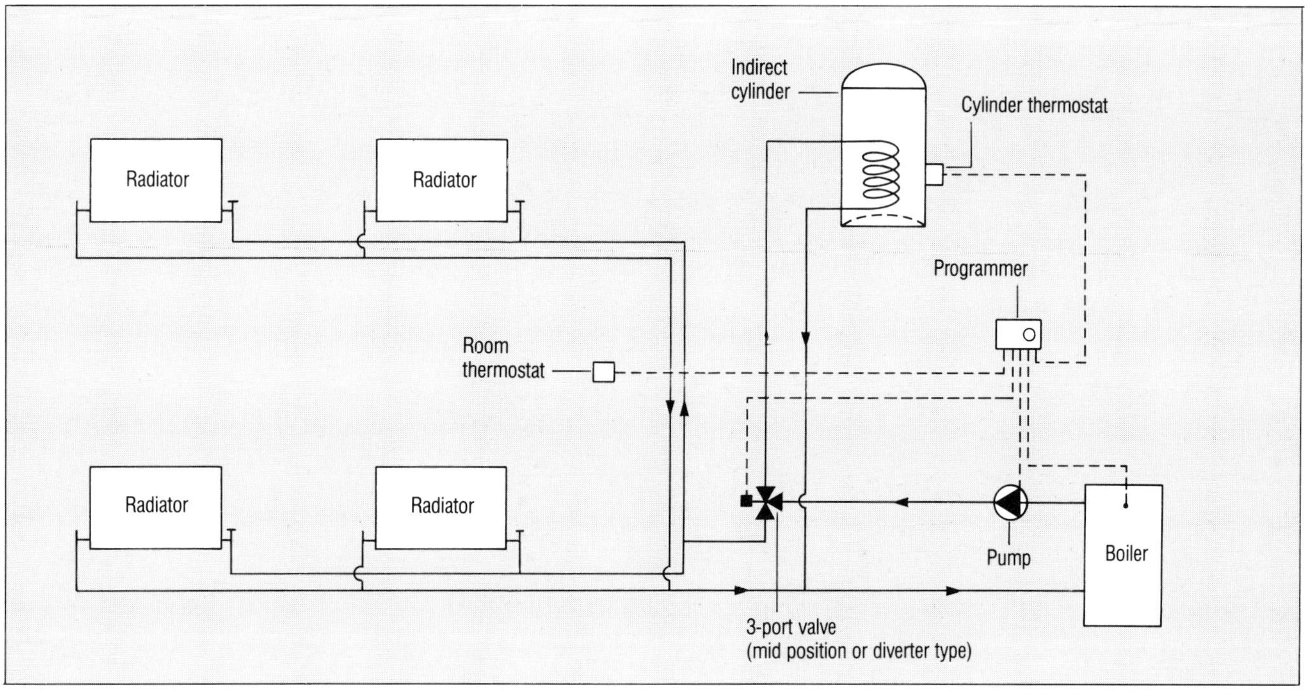

SYSTEMS OF CONTROL

Fully Pumped System with Two 2-Port Valves

A typically fully pumped system employing two 2-port valves is illustrated in Figure B4. It is generally considered better than the system using a 3-port valve shown in Figure B3 because full flow can be ensured through both circuits when required. The wiring is slightly more complex but the system does allow for more flexibility in the positioning of valves.

Zone Control

The system shown in Figure B4 can be adapted to provide zone control. This enables, say, the ground floor and first floor of a building to be controlled independently. Any number of heating zones can be controlled by fitting a separate 2-port valve controlled by its own room thermostat to each zone. It is also possible by the use of separate time controls to control times when each zone is in operation. The use of zone controls can, for instance, compensate for areas affected by solar gain or areas that are occupied only intermittently.

Figure B4 Fully pumped system with two 2-port motorised valves.
It is possible to adapt this system to give separate control to any number of zones by the addition of extra two-port valves and room thermostats.

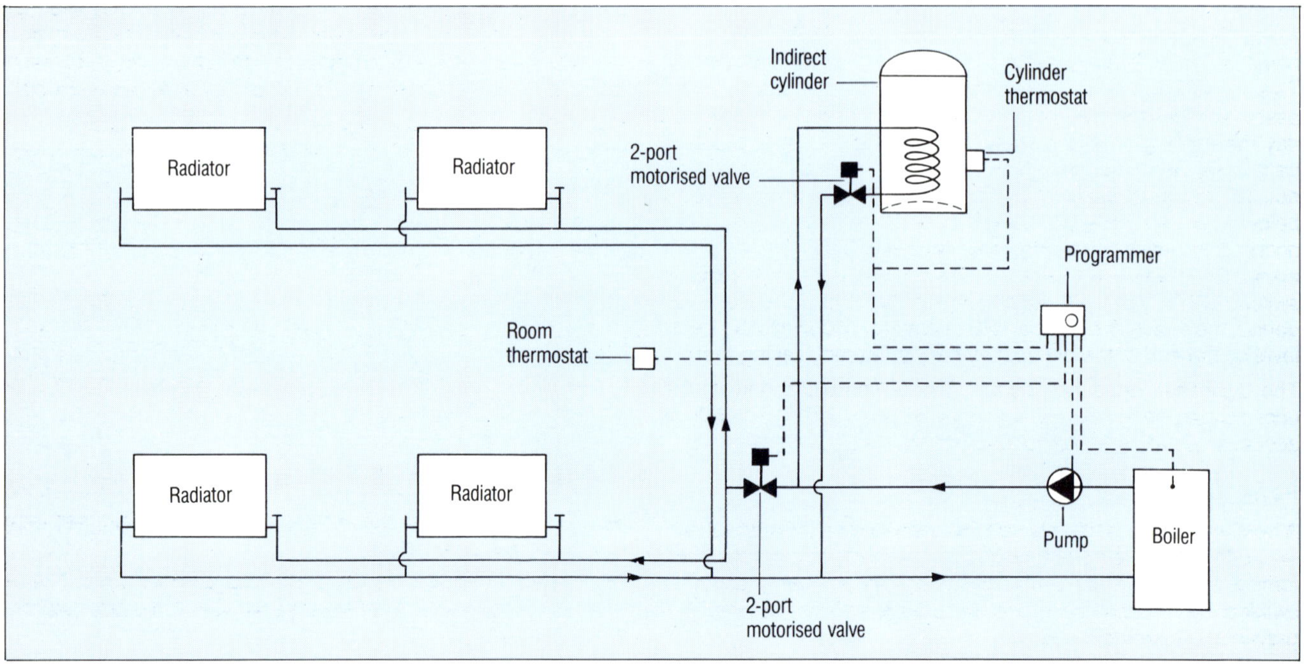

External Temperature Compensating Controls

The temperature of the water to the radiators is controlled by the use of a 3-port mixing valve activated by an external temperature sensor (see Figure B5). The temperature of the water in the system is directly related to the outside air temperature. Care should be taken when designing systems using this type of control that condensation does not take place within the boiler combustion chamber.

It is sometimes necessary to run a constant temperature circuit through the boiler using a separate pump and header pipe to ensure that the boiler return water temperature is above dew point temperature of the flue gases.

Figure B5 System with external temperature control.

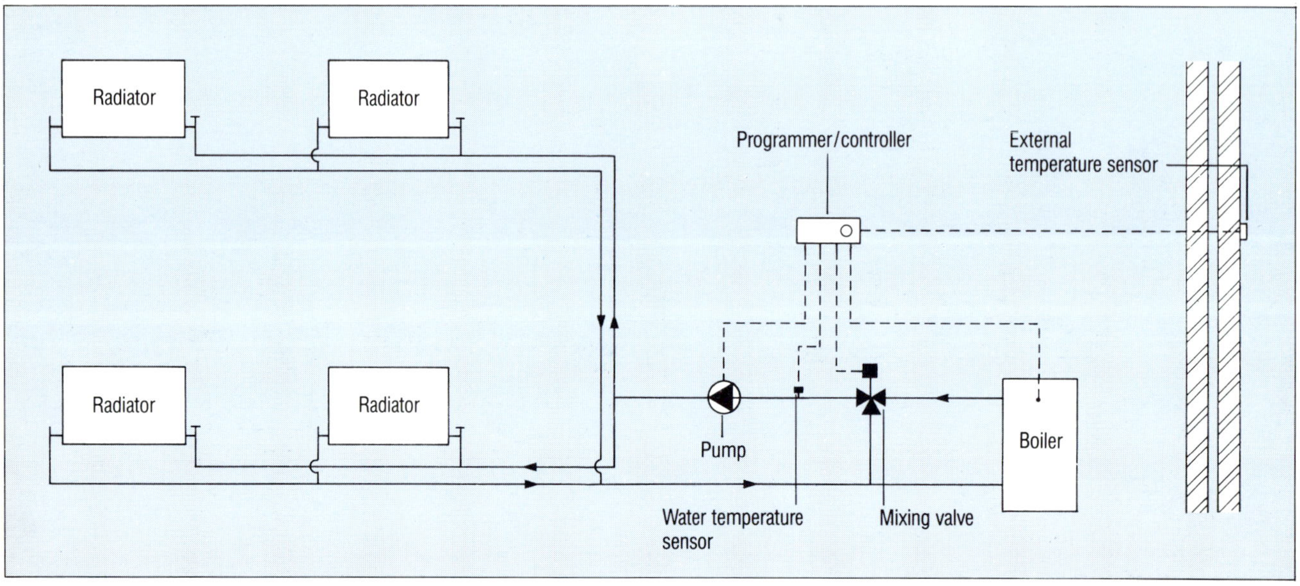

With this system the heating flow water is controlled directly by an outside air temperature detector working in conjunction with a water temperature sensor. The primary hot water circuit has been omitted from this drawing.

Optimum Start Controls Systems

These employ two sensors, one measuring outside air temperature and one measuring space temperature. The optimisers compare the two temperatures and calculate the latest possible switching-on time in order to reduce fuel consumption but achieve the desired occupancy period. In winter the optimser could optimise 'ON' the system several hours before the start of occupancy but in milder weather the optimum 'ON' time could be as little as 1 hour before the start of occupancy.

An optimiser can also calculate the optimum 'OFF' time which could be up to 2 hours before the end of the occupancy period. This is done by calculating the temperature decay curve associated with the outside and space temperatures, e.g. on a cold day the heating could be kept 'ON' until the end of occupancy because if the heating was switched OFF the temperature decay due to heat loss would result in the space temperature falling below a pre-set level, this level could be 2 or 3°C below the occupancy temperature. However, on a warm day the heat loss from a building would be less. Therefore, even if the heating is switched OFF up to 2 hours before the end of the occupancy period, the space temperature will decay slowly down to the lower level by the end of the occupancy period, thereby saving fuel.

The optimiser does not control the temperature during the occupancy period. It only switches the plant ON and OFF in accordance with the internal and external temperatures.

Frost Thermostats

A frost thermostat is usually wired to override the time control and space temperature control so that in the event of a low ambient temperature occurring in the building, the circulating pump is switched to ON and/or the heating boiler is switched to ON. This prevents the system freezing.

Sealed Systems

The basic difference between a sealed and an open system is that the feed and expansion cistern of a conventional system is replaced by a sealed expansion vessel. The vessel contains air or nitrogen at a pressure sufficient to balance the static pressure in the system, and separated from the system water by a diaphragm. As the system water is heated, it expands into the vessel, compressing the nitrogen/air and thereby raising its pressure and that throughout the system.

A sealed system can operate at temperatures higher than 82°C (180°F) but in such cases standard radiators are not recommended. It is recommended that convector radiators, skirting heaters or fan convectors are used with sealed systems operating at temperatures in excess of 82°C (180°F). In a sealed system, temperature and pressure control are achieved by a high temperature cut out and a separate pressure relief valve of a suitable type.

Only a boiler specifically intended for use with a sealed system should be employed as this will incorporate an appropriate high temperature cut out as standard. Sealed systems also require a pressure gauge, a special filling point and an air release valve at the cylinder. Correct sizing of the expansion vessel is important.

Figure B6 Three stages of operation of an expansion vessel

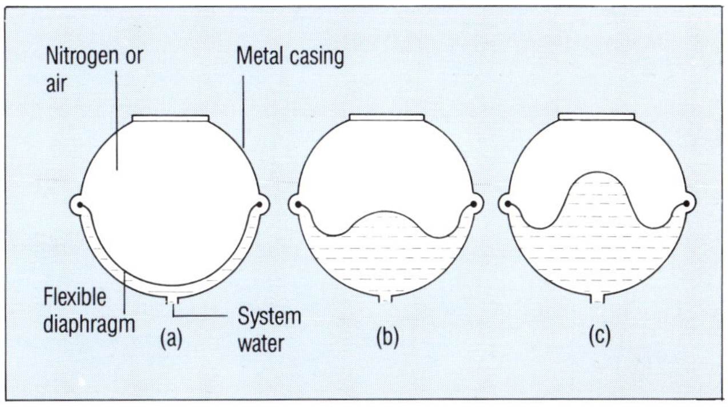

As the system water expands the nitrogen is progressively compressed from (a) to (c).

The installation should comply with the requirements of BS 5449: Part 1: 1977. The vessel itself should comply with the requirements of BS 4814: Part 1: 1976.

The diaphragm expansion vessel must have an acceptance volume sufficient to accommodate the volume change system water when heated from 10°C to 110°C. The practical acceptance volume is that which the vessel will accept when the gauge pressure developed rises to 0.35 bar (5 lb/in^2) less than the safety valve setting. Guidance on vessel sizing is given in Table B7.

Figure B7 Arrangement of Sealed System Components

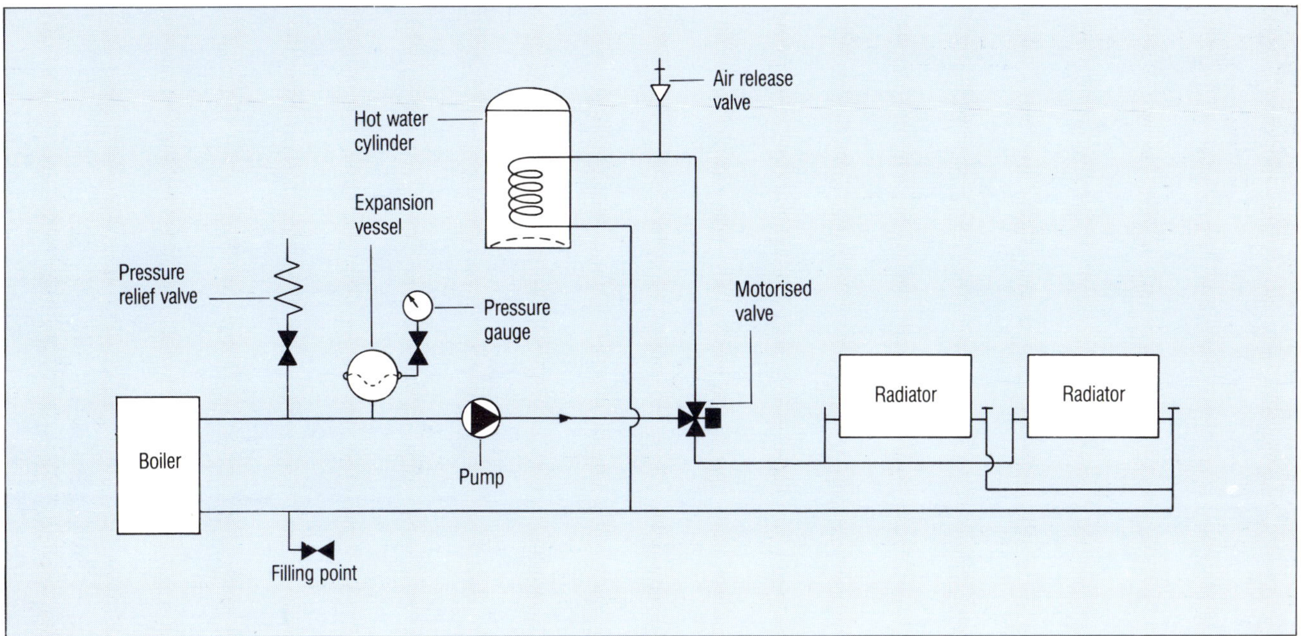

SYSTEMS OF CONTROL

Table B7 Capacities of expansion vessels

Total water content of system/litres	Safety valve setting (bar) 3.0			2.5			2.0	
	Vessel charge and initial system pressure (bar)							
	0.5	1.0	1.5	0.5	1.0	1.5	0.5	1.0
	Vessel volume							
	L	L	L	L	L	L	L	L
25	2.1	2.7	3.9	2.3	3.3	5.9	2.8	5.0
50	4.2	5.4	7.8	4.7	6.7	11.8	5.6	10.0
75	6.3	8.2	11.7	7.0	10.0	17.7	8.4	15.0
100	8.3	10.9	15.6	9.4	13.4	23.7	11.3	20.0
125	10.4	13.6	19.5	11.7	16.7	29.6	14.1	25.0
150	12.5	16.3	23.4	14.1	20.1	35.5	16.9	30.0
175	14.6	19.1	27.3	16.4	23.4	41.4	19.7	35.0
200	16.7	21.8	31.2	18.8	26.8	47.4	22.6	40.0
225	18.7	24.5	35.1	21.1	30.1	53.3	25.4	45.0
250	20.8	27.2	39.0	23.5	33.5	59.2	28.2	50.0
275	22.9	30.0	42.9	25.8	36.8	65.1	31.0	55.0
300	25.0	32.7	46.8	28.2	40.2	71.1	33.9	60.0
325	27.0	35.7	50.7	30.5	43.5	77.0	36.7	65.0
350	29.1	38.1	54.6	32.9	46.9	82.9	39.5	70.0
375	31.2	40.9	58.5	35.2	50.2	88.8	42.3	75.0
400	33.3	43.6	62.4	37.6	53.6	94.8	45.2	80.0
425	35.4	46.3	66.3	39.9	56.9	100.7	48.0	85.0
450	37.5	49.0	70.2	42.3	60.3	106.6	50.8	90.0
475	39.6	51.8	74.1	44.6	63.6	112.5	53.6	95.0
500	41.6	54.5	78.0	47.0	67.0	118.5	56.5	100.0
	For system volumes other than those given above, multiply the system volume by the factor across							
	0.0833	0.109	0.156	0.094	0.134	0.237	0.113	0.2

The system may be filled direct from the main supply using a removable flexible hose and back siphonage prevention devices in accordance with water byelaws, as shown in Figure B8.

Figure B8 Connection direct from water supply pipe in accordance with Model Byelaw 14

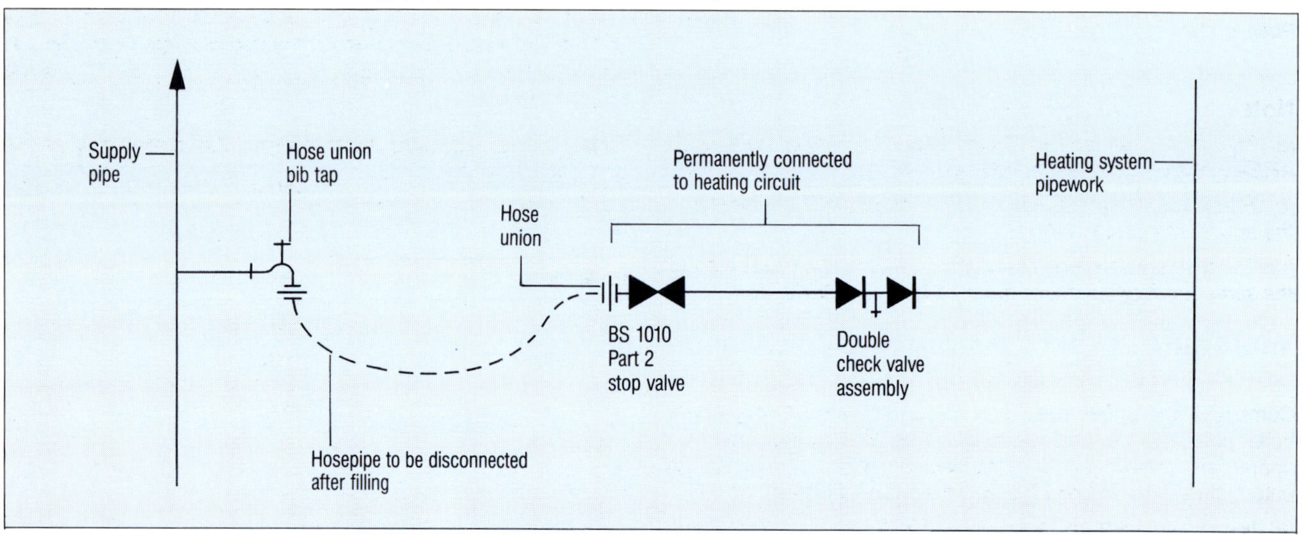

Experience with sealed systems shows that make-up may be needed in the first few months of operation. In commissioning a sealed system, it is generally advisable to maintain a low pressure and high temperature for a period of a few hours to assist in venting. When the system cools down, the pressure should be adjusted to a value equal to the static head plus a margin of 0.5 bar. The final pressure at full temperature and with all the circuits and heat emitters working should be less than 3.0 bar. Radiator valves used in sealed systems must comply with BS 2767: 1972 to reduce the possibility of leaking glands.

Boiler Air Supply

All conventionally flued boilers require a supply of air for combustion and a permanent opening should be provided for air supply into the room where a boiler is located. The openings should be direct to outside air or to an adjacent room which itself has a permanent opening to the outside air.

Solid fuel boilers with a rated output up to 45 kW should have a permanent air entry opening or openings with a total free area equal to at least the combined areas of the primary and secondary air inlets to the appliance.

Oil burning boilers with a rated output up to 45 kW require a permanent air entry opening with a total free area of at least 550 mm^2/kW of appliance rated output.

If the appliance is fitted with a draught stabilizer, or if one is fitted to the flue pipe or chimney in the same room as the appliance, then an additional permanent air entry opening should be provided with a total free area of at least 550 mm^2/kW of appliance rated output.

In the case of a gas fired open-flued appliance with a rated output up to 60 kW, the room or space in which it is located must have provision for permanent ventilation to outside air of at least 450 mm^2/kW of rated input over 7 kW.

Ventilation of Boiler Compartments

Where a boiler is in a compartment or confined space, in addition to air for combustion, provision must be made for ventilation in order to prevent overheating. Ventilation should be provided by means of permanent openings at high and low level, both communicating with the same room or on the same outside wall. Minimum free areas are given in Table B8.

Table B8 Air opening sizes

Boiler type	Opening position	Opening to room mm^2/kW rated output	Opening to outside
Open flued	High level Low level	1100 1650	550 1100
Room sealed	High level Low level	1100 1100	550 550

Notes

(a) The figures quoted refer to the MINIMUM acceptable FREE AREA when grilles are fitted to the openings.

(b) Range rated or modulating appliances should be assessed at the upper limit of their rating.

(c) The high and low level openings must both communicate with the same room or space, or must both be to outside air.

Ventilation of Living Rooms Fitted with a Combined Gas Fire and Gas Fired Back Boiler

Combined appliances (gas fires combined with gas fired back boiler units) require purpose provided ventilation, the size of the opening to outside air being based on 560 mm^2 per 1000 W output of the back boiler portion. This opening may be either:

(a) directly into the room or space containing the heater, or

Table B9 Minimum free area of ventilation to rooms where heating appliances are fitted

Location area	Type of appliance	Minimum free area of vent to outside air mm^2
Kitchen ventilation	Any flued central heating appliance or boiler	560 mm^2 per 1000 W output
Hall or passageway ventilation	Central heating appliance or boiler	560 mm^2 per 1000 W output
Basement ventilation	Flued central heating appliance or boiler	560 mm^2 per 1000 W output

(b) into an adjacent room or space which has an internal purpose provided opening of the same size to the room containing the appliance.

In rooms of 'average construction', adventitious openings into the living room are more than adequate to supply combustion air to fire and even in rooms of 'tight construction' (solid floors with tightly fitting or weatherstripped doors and windows). However it is necessary to provide a purpose-designed air vent related in size to the output of the boiler, whether or not the room is of average construction. The reason for the latter requirement is that weatherstripping could be carried out subsequent to installation in an 'average construction' type room.

Fuel Storage

For solid fuel storage a fuel store which is weatherproof with easy access for delivery and removal of fuel should be provided, capable of containing at least 6 weeks supply of fuel, but not less than 1 tonne. The fuel delivery opening should be not more than 1.4 m above ground level and not less than 600 mm.

Table B10 Solid Fuel Storage Capacities

Boiler output	Type of fuel: Dense		Type of fuel: Coke Type	
rating kW	Capacity Tonne	Floor area m^2	Capacity Tonne	Floor area m^2
Up to 15	1	1	1	1.5
15–25	1.5	1.5	1.5	2.25
25–35	2	2	2	3.0
35–45	3	3	3	4.5

The floor areas given in Table B10 above assume a stacking height of 1.4 m above floor level and are based on a specific volume of dense fuel of 1.4 m^3/tonne, and a specific volume of coke type fuel of 2.2 m^3/tonne.

Table B11 Recommended minimum capacity of oil storage tanks

Rating of Boiler kW	Capacity litres
Up to 15	1250
15–25	1360
25–35	1820
35–45	3000

ELECTRIC CENTRAL HEATING STORAGE BOILERS

Introduction

Wet central heating systems are highly favoured by many householders. The boiler market is a considerable one, both in new homes and for replacements in existing systems. Electric storage boilers were therefore developed to provide an electric alternative to fossil fuel boilers. The boilers are of two types, a dry core boiler and a hot water storage boiler.

Because of the structure of electricity tarrifs, the design and selection of electric boilers differ from fossil fuel boilers.

Economy 7 Tariff

The Economy 7 tarriff was introduced in 1978. It is a time-of- day tariff in which electricity, for whatever purpose it is used, is charged at a much lower rate during 7 hours at night. The night rate periods vary slightly depending upon area but are generally within the period from midnight to 08.00 (GMT).

Once a customer has applied for the tariff, the timeswitch and the dual rate meter are installed. There is no charge for installing these, but the quarterly standing charge of Economy 7 is a little higher than that of the standing domestic tariff. Electric boilers are specifically designed to take maximum advantage of the low rate on the Economy 7 tariff.

The Dry Core Boiler

A dry core boiler operates on the storage principle. Heat is stored during the night period when electricity is much cheaper. Heat can then be drawn off as required. Essentially, the unit comprises a core of special heat retaining bricks in a well insulated case. The bricks are heated overnight by electric elements passing through them. When required, heat is extracted by operating a fan to drive air round a closed loop within the bricks and over a heat exchanger through which water is passed. The heated water serves a radiator system in the normal way.

Should the stored heat not be sufficient on a particularly cold day, half the elements can be re-energised to top up as required.

The main components of a dry core boiler are shown in Figure B9.

Figure B9 The dry core boiler

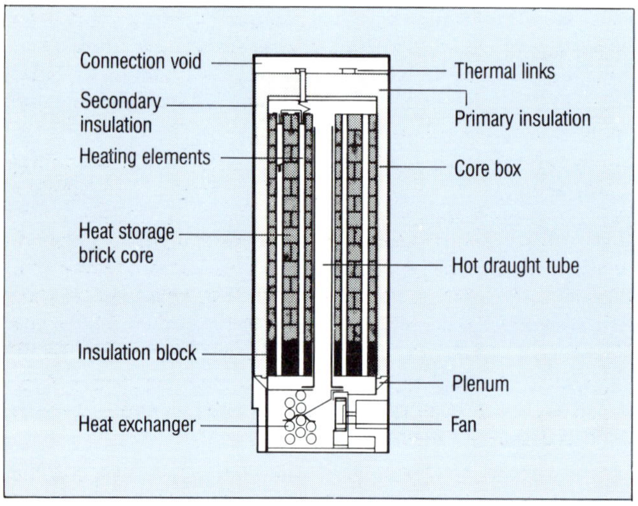

The interlocking bricks forming the core are built-up on an insulation block and surrounded by a nickel alloy core box. Two layers of Microtherm insulation panels form the primary insulation, and there is additional mineral wool insulation within the outer case. The case is attractively finished in white enamel and the boiler may be conveniently sited in a kitchen or elsewhere in a house or bungalow. The requirements for the base on which it stands must, however, be closely followed.

At maximum charge input setting, the average rate of heat emission from the boiler case measured over 24 hours is typically about 800 watts for a 19.2kW rated model. The boiler should therefore be sited where the case emission will usefully contribute to the heating of the house, but does not exceed the heat requirement of the room in which it stands. It is not recommended that the boiler is sited in a garage or outbuilding, as the case emission will not then represent useful heat.

The boiler has connections for flow, return and vent pipes. There is also an integral circulating pump and pressure relief valve, and the circuit cold feed is connected to the radiator return pipe.

Design Considerations

Before dealing with the details of the design and installation, it is important to establish a logical procedure that will ensure that all necessary factors have been considered. These steps are briefly as follows:

(a) assess the existing installation;

(b) consider improvements in thermal insulation of the building;

(c) design the incorporation of the dry core boiler into the existing system;

(d) decide on how the domestic hot water should be modified;

(e) install the dry core boiler system and Economy 7 water heating system;

(f) commission the system.

Assessing the Existing System

Electrical Supply

A basic question may arise as to the suitability of the local electricity network to meet the extra electrical load required. The installer must always check with the local Electricity Board. They will let the installer know the amount of extra load that can be accepted. Having checked this particular aspect, the remaining assessment of the installation can take place.

The total element loading of two common dry core boilers are 19.2kW and 12kW, divided into two equal banks of 9.6kW and 6kW respectively. The boiler may be operated on a single phase supply, or alternatively a three-phase version is available. With the latter, the other loads in the house are connected to the third phase to give balance.

All electrical wiring must be in accordance with the current edition of the IEE Regulations for electrical installations.

Heat Loss Calculations

The first step is to calculate the heat loss of each room for design day conditions. Approximations based on cubic capacity or the floor area of the dwellings are inadequate.

At this stage, the opportunity should be taken to consider improved insulation. If the dwelling has suitable cavity walls, these should be filled with an insulant appropriate for the area. Roof insulation should be at least 100mm thick although it is necessary to ensure that the roof space is adequately ventilated. External doors should be weatherstripped.

The dry core boiler will supply water to the radiators at temperatures similar to those used with a fossil fuel boiler, eg, a mean radiator temperature of 75°C. At full charge, the temperature can be higher. Often, it will be found that an existing fossil fuel system has generously sized radiators, and the addition of insulation to a house will increase margins. A check should however be made on the radiator sizes in each room and alterations made if the heat loss calculation shows it to be necessary.

Dry Core Boiler Capacity

A dry core unit of 19.2kW rating has a heat storage capacity of 100kWh. The house heat loss at the upper limit of its application depends on the quantity of day energy consumption over a season that is considered acceptable.

It is recommended that a maximum of approximately 5% day energy is adopted. It must be stressed that this is a seasonal value and in practice, on very cold days, 20-25% day energy may be taken whilst there will be many milder days when the store alone is sufficient. The maximum house heat loss which should therefore be considered is 10kW.

The 12kW model has a storage capacity of 60kWh and is therefore suitable for a house of up to 6kW heat loss. This sizing:

(a) allows for normal heat distribution losses assuming that pipework under the floor or in the roof space is insulated with at least 25mm pipe insulation;

(b) assumes operation of the heating system during the last 2 hours of the Economy 7 night period to ensure that most of the preheating is carried out at the cheap night rate. The night elements are sized to cope with this and to heat the store up in approximately 5 hours;

(c) assumes that the boiler is sited such that all heat emission from the case is usefully employed.

Existing Pipework

While most existing radiator systems which have been professionally installed may be expected to allow straightforward conversion, it should not be assumed that good practice has always been adhered to. The system should be examined and any departure corrected when making the new installation. It is important to ensure a sufficient water flow rate through the heat exchanger. An existing single pipe system may require modification if the required flow rate is to be achieved, particularly if 15mm copper pipe has been used.

System Controls – User

The dry core boiler is supplied complete with a control unit. This is connected to the boiler with a special multicore cable, and may be mounted as convenient within the radius of the cable length supplied, but within the same room. All the user controls are on the control unit except for the room thermostat.

The user has the following controls accessible:

(a) On-off Switch

This controls the main 24 hour supply to the boiler, (both 24 hour and Economy 7 pilot supplies must be isolated at the intake position before working on the control box or boiler).

(b) Programmer

The heating programme may be set using any sequence of the 1/4 hour tappets.

(c) Charge Control

This varies the core temperature at which the overnight charge is terminated, enabling the quantity of heat stored to be varied from about 30kWh at minimum setting to 100kWh at maximum for the larger unit. Day top-up only operates with the charge control set at 'boost'.

(d) Radiator Temperature

This adjusts the setting of the return water temperature control. Under normal conditions of use, when the room thermostat is calling for heat, the boiler fan will cycle under this control.

(e) Programme Selector

The heating may be operated according to the programmer, held-off or switched to continuous operation.

Control Sequence

The control unit contains heavy duty relays controlling power to the element banks, and a printed circuit board carrying temperature sensing and control electronics. The control inputs are the room thermostat, water temperature sensor, four core-temperature thermocouples and a 'night signal' from the Economy 7 meter system. When the night signal is on, both element banks are connected and the core temperature rises to the charge setting.

If the programmer and room thermostat demand heat, the pump is switched on, followed, subject to the water temperature controls, by the fan. The central heating water temperature rises until it reaches the set point of the return temperature. In addition, there are a high limit auto-reset and an over temperature trip sensing the flow water temperature.

When the night signal is off, the heating elements cannot be energised unless the core is discharged to a fixed low temperature – the 'boost' level. If then, the core temperature control is set at maximum, a 'boost' light comes on. Element banks are switched alternately in a timed sequence to maintain heat output. If the user prevents the 'boost' from operating by turning down the core temperature control setting, the fan is finally stopped at a very low core temperature, to prevent unnecessary operation.

Installation – Siting the Boiler

The dimensions of the 19.2kW boiler are 1664mm high, 610mm wide and 596mm deep, plus access requirements for installation and maintenance. A ceiling height of 2286mm is preferable for easy installation of heating elements through the top of the unit. However, it is possible to install elements under lower ceiling heights (possibly as low as 1900mm) provided there is access at the sides. Installers should contact the manufacturers for further details.

The 12kW boiler has the same plan dimensions as the largest unit but is only 1317mm high. The preferred ceiling height for installing the elements is 1780mm although 1575mm is possible if there is access at the sides (again, consult the manufacturers).

The 19.2kW unit weighs 825kg and has a 6mm steel baseplate to spread the load and giving a floor loading of 2393kg/m^2. The 12kW unit weighs 605kg. It is essential that the boiler is installed on a base of sufficient structural strength. Several floor types will be encountered. The traditional concrete slab on the ground floor is suitable without special support. The precast concrete beam and block floor is unsuitable unless arrangements can be made to strengthen the floor at the design stage of the building. The traditional wooden suspended floor needs a separate supporting foundation to be built. Reinforced concrete suspended floors are suitable providing the boiler is sited against a perimeter wall which supports the suspended slab.

It is important to ensure a perfectly level base, since no levelling adjustment is possible and a small error may be very noticeable, particularly alongside other units. The unit may be installed free-standing in a kitchen or utility room with large clearances all round. More normally, it will be back against a wall or in a corner of a room. In these cases, it should be positioned with the front face 600mm from the wall, with a minimum of 10mm clearance to the side wall. A skirting board at the side may touch the outer casing.

Cupboards may be fitted over the boiler after installation. They should be at least 75mm above the boiler and they should be easy to remove for boiler maintenance.

If the unit is to be enclosed on both sides with combustible materials such as wood, chipboard and plastic, suitable clearances must be allowed to avoid excessive temperatures in the unlikely event of a fault condition. Further guidance may be obtained by writing to the manufacturers giving details of the proposed installation.

When fitting the unit into a boiler cupboard, allow 50mm clearance at the sides and at the front. Fit ventilation grilles measuring at

ELECTRIC CENTRAL HEATING STORAGE BOILERS

least 225mm × 150mm at the top and bottom of the cupboard. Alternatively, a louvered door may be fitted. If the cupboard is outside the living area of the house but adjacent to it, it may be possible to fit the ventilators so that air is drawn from and returned to the house, thus making some use of the case emission.

Plumbing Considerations

There are a number of essential items to be remembered when installing a dry core boiler onto an existing system.

(a) The domestic hot water arrangement should be entirely segregated from the heating system and a suitable arrangement taking advantage of Economy 7 tariff installed. This arrangement will be more efficient and cheaper to run than if retained as part of the central heating system being heated by the dry core boiler.

(b) When using thermostatic radiator valves, care should be taken to ensure that the minimum system flow rate indicated in 'commissioning' is achieved. Although it is difficult to be hard and fast this will generally mean that at least 3 radiators need to be left without thermostatic valves.

(c) The integral circulating pump is in the flow pipe from the boiler.

Converting an Existing System

The typical existing system, and how this is generally converted is shown in Figures B10 and B11. For simplicity, the dry core boiler is shown replacing a boiler in the same position. The modification is relatively simple requiring the following main steps.

(a) The domestic hot water is segregated from the main space heating system. This can be achieved very simply by inserting a stop valve in the run of the pipe between the indirect coil feed connection to the cylinder and the system vent. The better alternative would be to disconnect the feed and return from the indirect coil and blank-off the open ends.

The existing pipe which served as both indirect coil feed and system vent in the traditional boiler system is used for the latter purpose only with the dry core boiler. The immersion heater arrangement and control for domestic hot water should then be installed in accordance with the section on water heating.

Figure B10 Typical existing system

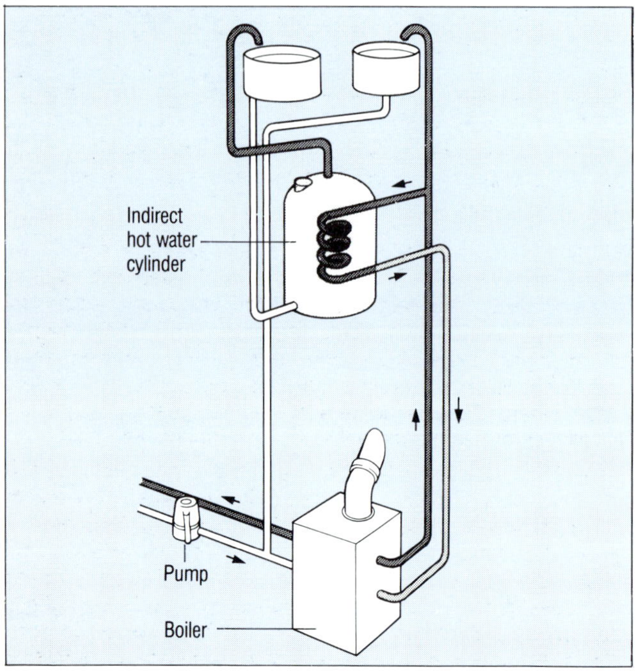

Figure B11 System conversion with a dry core boiler and Economy 7 water heating

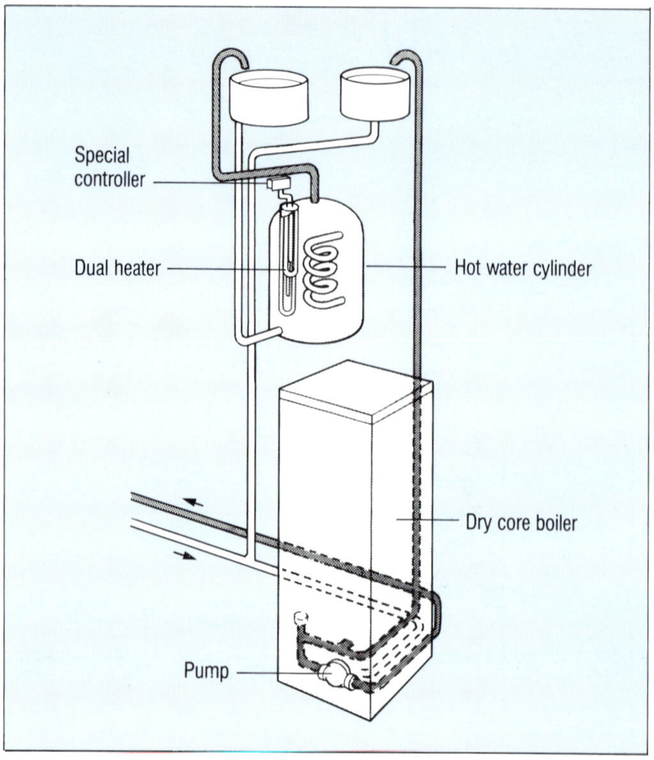

(b) The pump is located in the system flow. This is good practice since most of the system will be under positive pressure. This will reduce the chance of air getting into the system and will eliminate localised boiling that can sometimes occur at very high points in the system in the unlikely event of a series of control malfunctions. (It has to be remembered that pipework passing through the roof space will not be much below the level of expansion).

A typical system schematic is shown in Figure B12.

The dry core boiler is delivered to site in component form in labelled packages. The manufacturer's installation instructions give detailed information on the components and their assembly. The person assembling the unit should have had workshop experience in this, and should follow the written instructions closely.

Commissioning

In addition to the normal commissioning procedures to be carried out on the feed and distribution system, the following control checks should be carried out:

(1) Programme switch; on/cold/programme working.
(2) Fan inhibited at low temperature.
(3) Pump over-run working.
(4) Boost facility working.
(5) Boost cycling correctly.
(6) Room thermostat switching.
(7) Night charge operating.

After the initial night charge has been taken, a further check should be made to confirm:

(1) Fan operating.
(2) Water temperature control working.
(3) Boost inhibited when core is hot.
(4) Water temperature rise less than 30°C.

Note

The pump is set to position No 2 at the factory and it should achieve 0.23l/s in most systems, at which flow rate, the heat

Figure B12 Typical system schematic

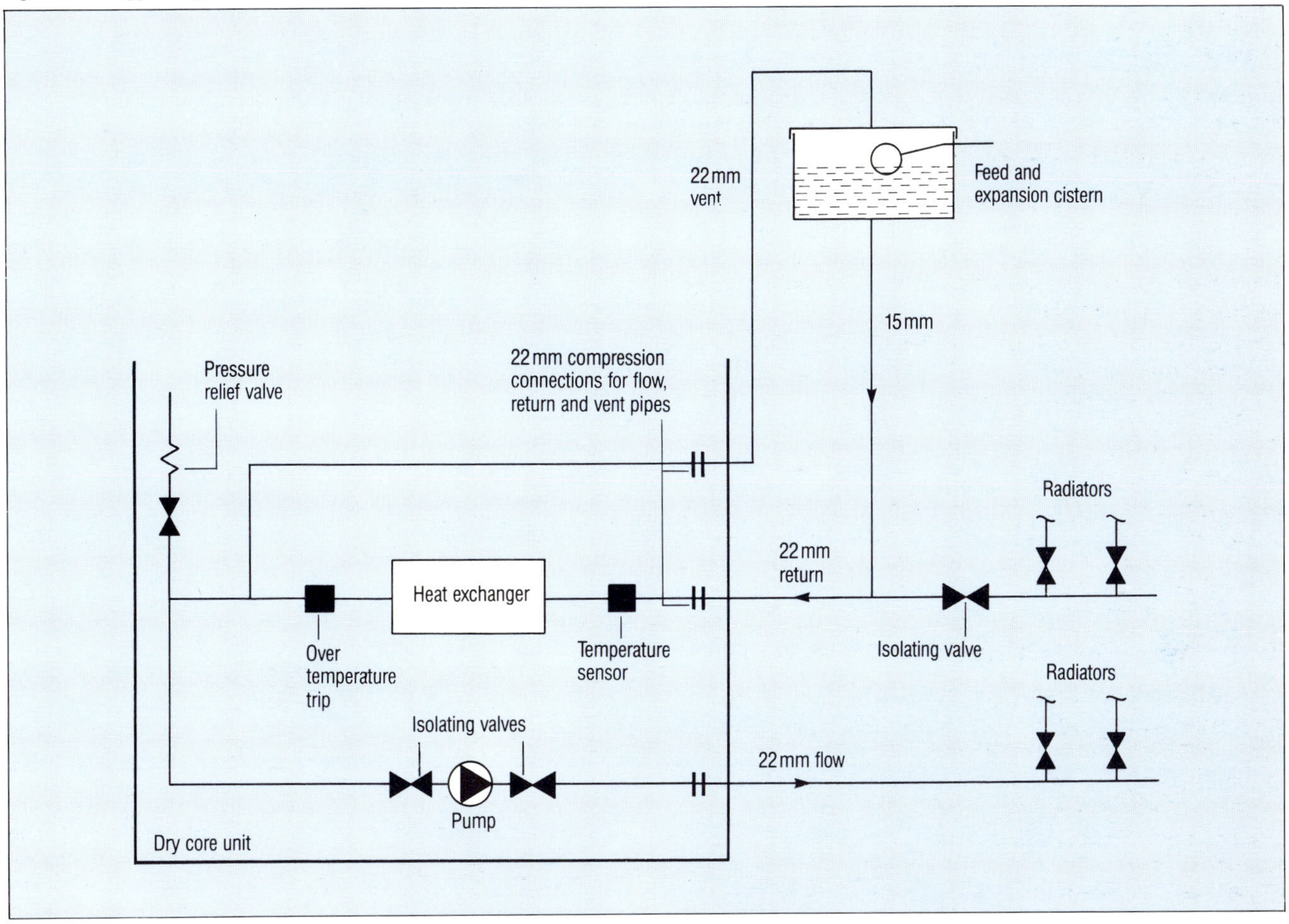

exchanger pressure drop will not exceed 12kN/m². The temperature rise would then be approximately 20°C at maximum charge. If the system pressure drop is so high that the flow is less than 0.15l/s, the temperature rise will exceed 30°C and there is a possibility of the over temperature trip operating.

If the flow rate is too low, then it will be necessary to open some of the lockshield valves in the system. If they are already fully open, then the pump setting should be raised.

The Water Storage Boiler

An alternative to the dry core boiler is the water storage boiler. An advantage of the water storage boiler is the greater flexibilty in matching the storage capacity to the heat requirement of the house. Generally however, the physical size of the water storage boiler will be larger than the dry core boiler.

In order to determine the capacity of a water storage boiler, a number of factors have to be taken into consideration. The factors relate to the calculation of the design day heat requirement of the house. The factors include:

(a) Design heat loss.
(b) Allowance for design day heat gain, electrical, metabolic and solar.
(c) Allowance for intermittency.
(d) Allowance for variation in infiltration rate.
(e) Allowance for direct acting supplementary heating.

To quantify the factors for various house types, a combination of theoretical and empirical data has been formulated by the Electricity Supply Industry. For full design guidance on the water storage boiler, the Local Electricity Board should be approached.

DOMESTIC ELECTRIC WATER HEATING

Electricity offers a wide range of options for heating water from small point-of-use units, such as undersink and oversink heaters, to large central storage cylinders heated by immersion heaters. In most cases, the point-of-use heaters are made as an all-in package requiring fairly simple connection to the electrical and mains water supplies.

An effective and economical central storage arrangement generally requires a little more design work from the plumbing engineer. The main part of this section therefore concentrates on the principles and practical aspects of designing the central storage system and only briefly considers point-of-use heaters.

Central Storage System

An electric immersion heater can be used to heat a central storage cylinder, whether or not the cylinder is also heated by an indirect coil fed from a fossil-fuelled boiler. If fitted with an indirect coil, the immersion heater may be used simply as back-up in the event of any problems or for summer use when the central heating boiler is switched off. Alternatively, it can be used for supplying all the hot water where there is no existing central heating or to ease installation problems where there is a central heating boiler.

Heating the cylinder electrically saves the need to install an indirect coiled cylinder and the secondary pipework between the boiler and cylinder, plus any associated control valves, etc. It is generally recommended that an electric central storage system is operated on the Economy 7 tariff for electricity. Economy 7 offers the householder seven hours of cheaper electricity every night. By designing the system to make the most of this cheaper night-time electricity, an electrically heated system is generally competitive

DOMESTIC ELECTRIC WATER HEATING

on running costs with any other fuel, in addition to the possible savings in installation costs. Achieving this is the general principle underlying the following design criteria.

Economy 7 Tariff

There are two main domestic tariffs for electricity, the standard domestic tariff and Economy 7. The standard tariff has a quarterly standing charge plus a single rate for each unit of electricity used. The Economy 7 tariff has a quarterly standing charge and two unit rates, one for units used during seventeen hours of daytime and a special low rate for units used during seven hours at night. The actual prices vary slightly between Area Electricity Boards but typical costs are shown in Figure B13. The standing charge of Economy 7 is slightly higher than the standard tariff and the day rate of Economy 7 is generally slightly higher than the standard unit rate (except in SEEBOARD, East Midlands and Eastern Electricity Boards where these unit rates are the same). These are offset by making the most of the seven hours of cheap night-time electricity, charged at less than half the standard unit rate.

Figure B13 Domestic electricity tariffs

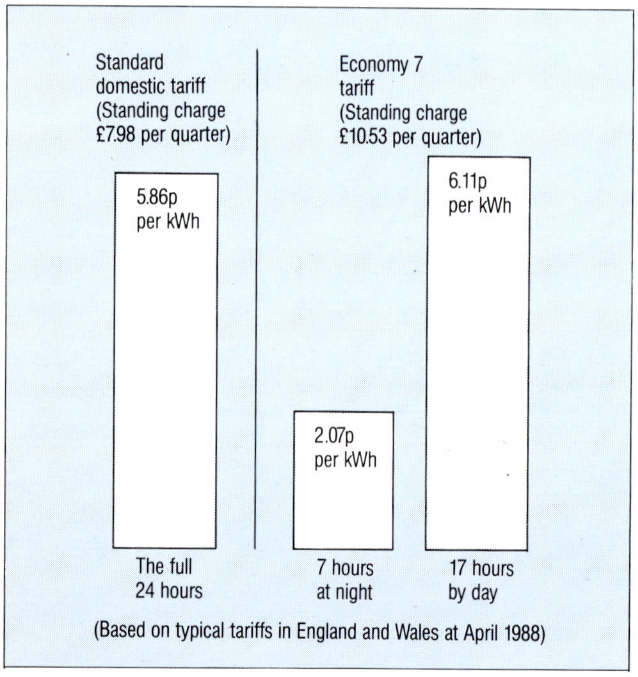

The Economy 7 tariff applies to the whole house supply, not just the heating and hot water. Any electricity used during the seven night hours, such as by the fridge and freezer or cooking breakfast, is charged at the cheap night rate.

To transfer to the Economy 7 tariff, the householder fills in an application form for the local Electricity Board. The Board then changes the householder's electricity meter for one with two sets of recording dials, one to record the day units and the other to record the night units. There is no charge made to the householder for this change of tariff.

Design Principles

For most fossil-fuelled systems, the storage cylinder can be heated at any time that the boiler is operating for the same cost. With an Economy 7 system, the aim is to rely as much as possible on heating the cylinder overnight, when the fuel is cheaper, and to minimise the use of day units for top-up quantities of water. The key therefore is to ensure a sufficient store of water is heated by the end of the night period to meet most of the household's requirements during the following day. The features which help in achieving this are as follows:

(i) A high standard of cylinder insulation, to keep the water heated overnight hot until used.

(ii) Separate immersion heaters for day and night operation. The night element heats as much water as possible at the cheaper rate, the day element heats a smaller quantity at the day rate if extra hot water is required.

(iii) Sufficient water storage capacity to suit the needs of the household.

The following sections expand these principles together with some other aspects which contribute to good system design.

Cylinder Insulation

The preferred levels of insulation for any hot water cylinder are:

(1) A lagging jacket complying with BS 5615: 1985
or
25–30 mm thick factory applied insulation

(2) 50–60 mm thick factory applied insulation.

In general, a well-fitting 80 mm thick jacket will comply with British Standard BS 5615. Where copper cylinders are being supplied with factory applied insulation, the usual thickness is about 25 mm. This has approximately equivalent performance to the 80 mm jacket, but is likely to perform better in practice since it is less prone to sagging or being moved by the householder.

The best option however is an Economy 7 cylinder complying with the Electricity Industry's cylinder specification. This specification incorporates a number of features giving high performance operation on Economy 7 including a limit on standing heat loss, generally equivalent to a 50–60 mm thickness of factory insulation.

Cylinder Size

The most commonly used existing copper cylinder is the British Standard size 7 (120 litres, 900 mm × 450 mm), followed by size 8 (144 litres, 1050 mm × 450 mm). Existing cylinders in these sizes can generally be easily converted to Economy 7 operation. However, the preferred option for new installations or where the cylinder is being replaced anyway is to install size 9E (210 litres, 1500 mm × 450 mm) size. This gives the householder a larger store of hot water heated at the cheap night rate, thereby reducing the need to top-up using day-rate electricity. In two or three person households, the night store will often meet all the requirements for the following day. (See also section on Economy 7 cylinders.)

Immersion Heater Arrangements

Figure B14 shows three arrangements of immersion heaters in cylinders. The top-entry arrangement with just a single straight immersion heater (Figure B14a) is the most common existing arrangement. Graph B3 shows the quantity of hot water given by a 27 inch straight immersion heater with an 18 inch rod thermostat in a BS 7 cylinder. This only gives about 70 litres of water above 40°C from the 120 litre cylinder. A shorter immersion heater or a shorter thermostat would further reduce the quantity of water heated.

A more suitable top-entry arrangement for Economy 7 operation is the dual element shown in Figure B14(b). The long element has the loading concentrated at the lower end of the immersion heater by manufacturing the required length from a longer element bent back and by leaving about 300 mm at the top end as an unheated cold tail. This arrangement heats the cylinder more effectively; using the long element at night gives a store of about 105 litres of water above 40°C at the start of the day (see Graph B3).

The short element of the dual is used during the day if extra hot water is needed, and heats only the top 55 litres or so, sufficient for a bath. By heating a smaller quantity the short element produces

Figure B14 Immersion heater arrangements

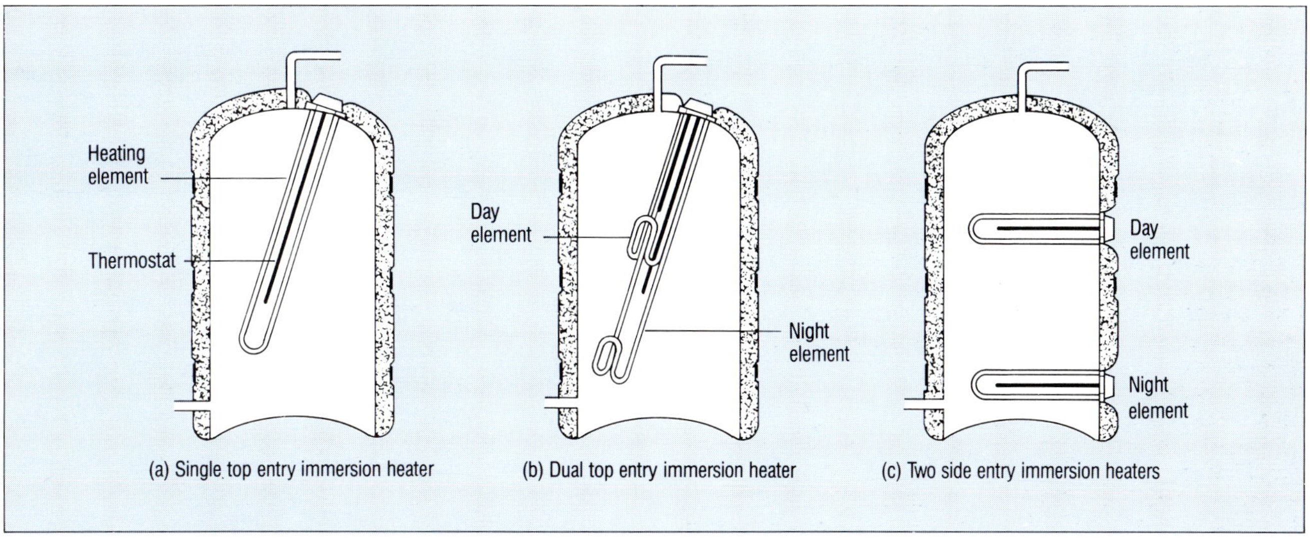

Graph B3 Profile of hot water for a size 7 (120 litre) cylinder

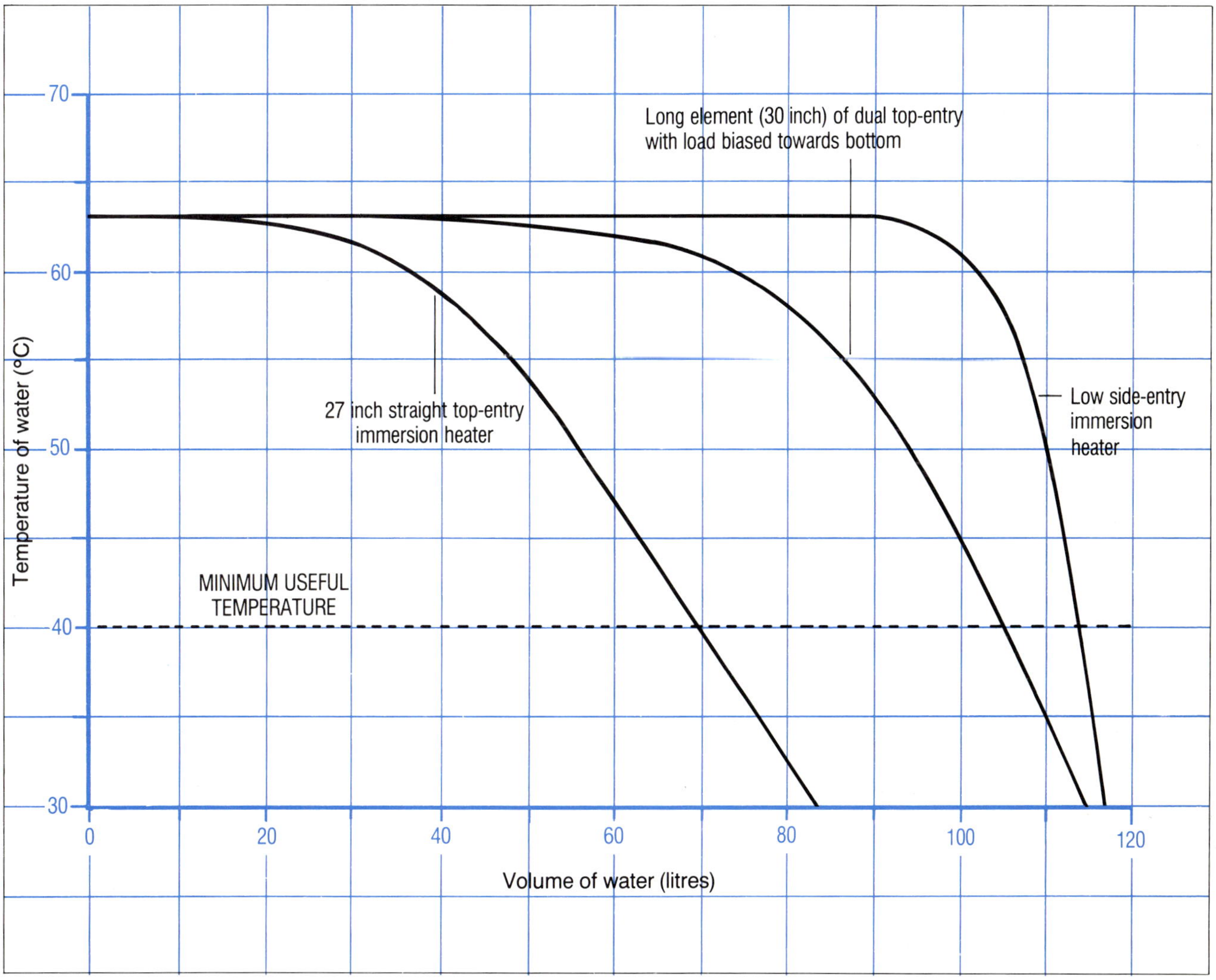

the extra hot water more quickly, and also saves on the use of day rate electricity at the expense of the following night's charge, compared to heating the whole tank again. The dual immersion heater is an effective Economy 7 conversion for cylinders with an existing top-entry boss. The thermostat controlling both elements of a dual immersion heater are normally set at about 60°C.

The optimum arrangement for Economy 7 use is a cylinder with two side entry immersion heaters as Figure B14(c). Using the lower element at night provides a store of about 113 litres above 40°C from the 120 litre size 7 cylinder (see Graph B3). In the larger size 9E, this arrangement gives about 195 litres. The high side entry element provides day top-up in the same way as the short element of the dual.

For the two side entry arrangement the thermostat controlling the

DOMESTIC ELECTRIC WATER HEATING

night (lower) element is normally set at 60–65°C, and that for the day (upper) element at 50–55°C.

Immersion Heaters

Most immersion heaters are of 3kW rating. Although the cheaper Economy 7 electricity is available for seven hours at night, a hot water cylinder generally needs less than this. A size 7 (120 litre) cylinder heats up from cold to 65°C in about 2 hours; the size 9E (210 litre) cylinder takes about 4 hours. A day boost of one hour duration supplies about 50 litres of fully heated water in a suitable arrangement.

As long as the type of immersion heater is suitable for the area, it will be long-lasting and require minimal maintenance. In hard water areas, the build-up of scale can lead to the element overheating and eventual failure. Scale builds more quickly at temperatures in excess of 60°C and thermostat settings should not generally exceed this unless using immersion heaters with a metal sheath which has low adhesive properties (eg titanium).

Naturally occurring soft water is often acidic and aggressive to common metals. Temperatures in excess of 60°C should not generally be used unless immersion heaters with resistant metal sheaths are used. Three main types of sheath are available:

(1) Tinned copper is the standard material and least resistant to attack.

(2) Incoloy 825 is a commonly used alloy sheath which offers greater resistance to attack and this should be successful in most situations.

(3) For areas of extremely aggressive water, a titanium sheath offers maximum corrosion resistance.

Economy 7 Systems

The preceding principles of good Economy 7 design have been incorporated into a range of products to assist designers and installers in providing an efficient and effective system for the end user. The Electricity Supply Industry has written specifications covering cylinders and combination units, immersion heaters, and controllers and products complying with these are available from many of the main manufacturers. These products cover most options in the conversion of existing systems to Economy 7 operation, and the installation of new or replacement systems.

The two main ways of providing an Economy 7 System are as follows:

(1) adding a dual immersion heater, 80mm insulating jacket and Economy 7 controller to a standard suitable storage cylinder (Figure 15a);

(2) installing an Economy 7 controller together with a new high-performance Economy 7 cylinder or combination unit, preferably of 210 litre capacity but 144 litre and 120 litre sizes are available for situations where space is restricted (Figure 15b).

The Economy 7 Controller

Economy 7 water heating controllers are produced by a number of manufacturers but all models operate in a similar way. The controller can operate an arrangement of two immersion heaters, giving automatic operation of one immersion heater during the cheap night period and allowing use of the other heater during the day via a customer operated run-back or electronic timer. The controller is fed from a single continuous electrical supply such as is normally installed in most cylinder cupboards and therefore offers a readily packaged and installed control system for Economy 7 operation.

Figure B16 shows a schematic diagram of the controller's wiring and a typical view of its front panel. The main components are the fixed timer arrangement, and automatic changeover switch and a boost timer. When fitted, the installer simply sets the main timer dial to the correct time of day. This will then automatically operate the night element within the cheap night period of the Economy 7 tariff, the on and off times having been factory set within the controller.

As stated earlier, even the 210 litre cylinder needs only four hours charge to heat up from cold. To reduce the cylinder standing losses, therefore, the controllers are normally set to switch on at about 2.00 to 2.30 am, rather than right at the start of the Economy 7 night period.

The automatic changeover switch ensures that the electricity circuit feeds the boost element during the day but only when the customer operates the boost timer. The boost timer is either run-back or electronic timer which automatically switches off again after the set period, usually up to one hour.

Figure B15 Recommended Economy 7 systems

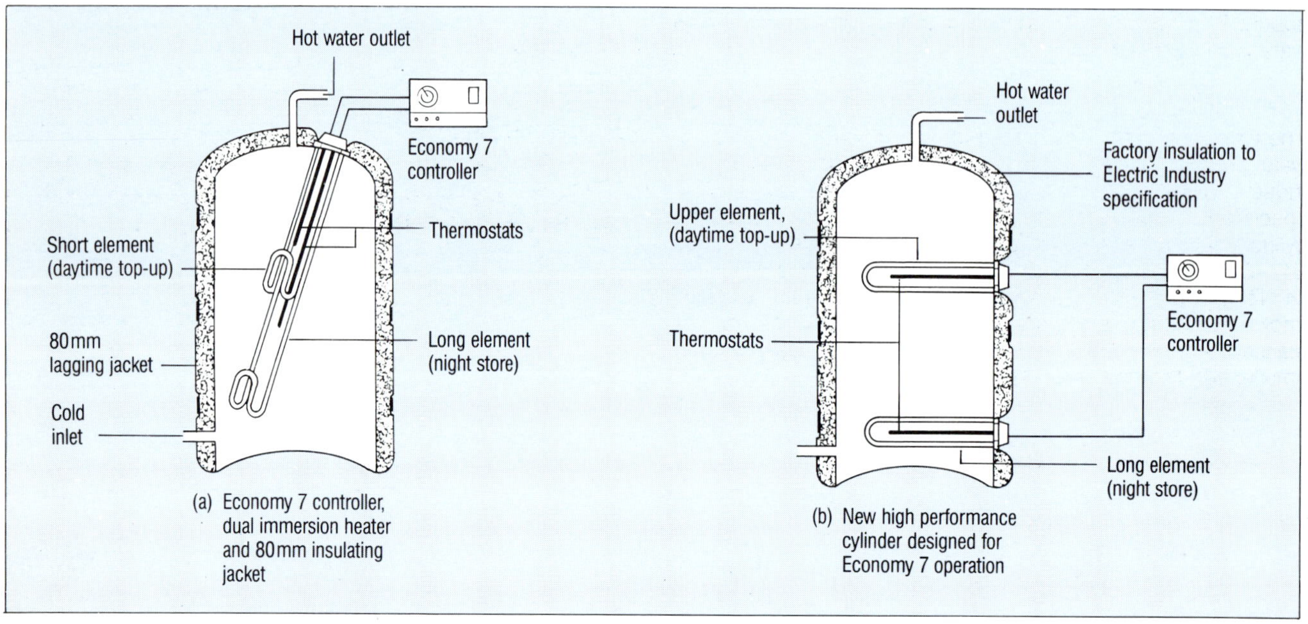

Figure B16 Economy 7 Water Heating Controller

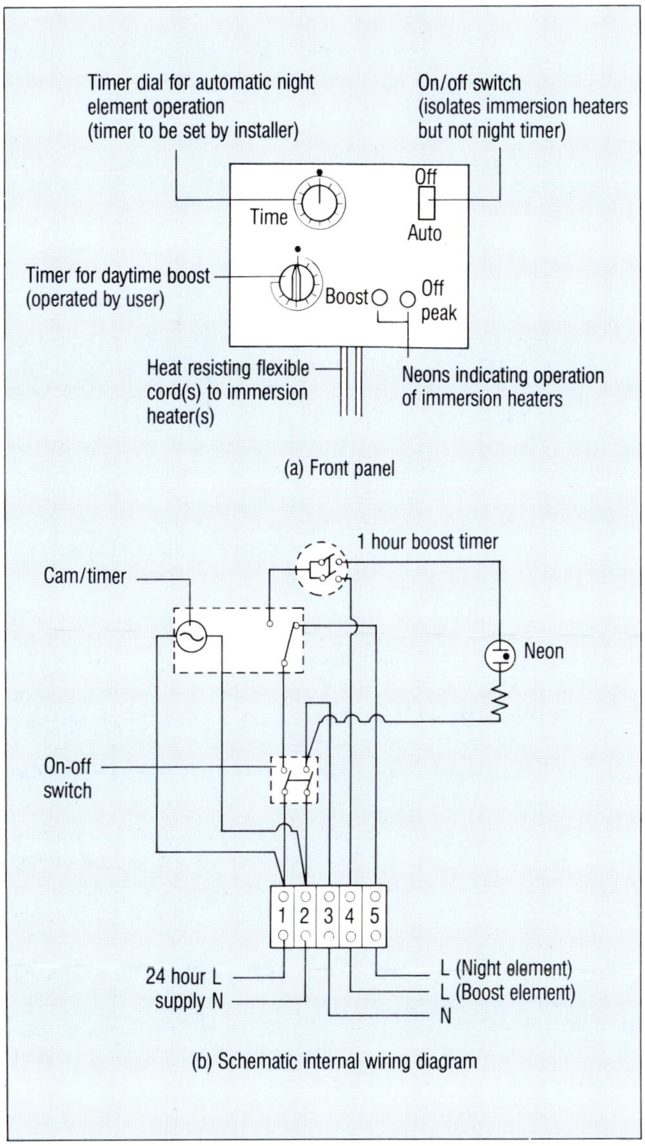

Most models have a quartz timer with a rechargeable battery reserve so that, once set, they can be expected to keep time without frequent inspection, even in the event of a break in power supply. The Ecomony 7 controller normally replaces the local water heating switch.

Economy 7 Cylinders and Combination Units

The various elements that contribute to a good Economy 7 hot water design have been incorporated in an Electricity Supply Industry specification for cylinders and combination units. The specification includes both direct and indirect cylinders and cylindrical or rectangular combination units (see Figures B17–B20). Three sizes cover the range of applications, 120 litres, 144 litres and 210 litres, or slightly less for the indirect cylinders, known as the small, medium and standard sizes respectively. The standard 210 litre size is recommended for most cases but the smaller sizes can be used where space is limited.

Figure B17 Electricity Council specification for direct cylinder

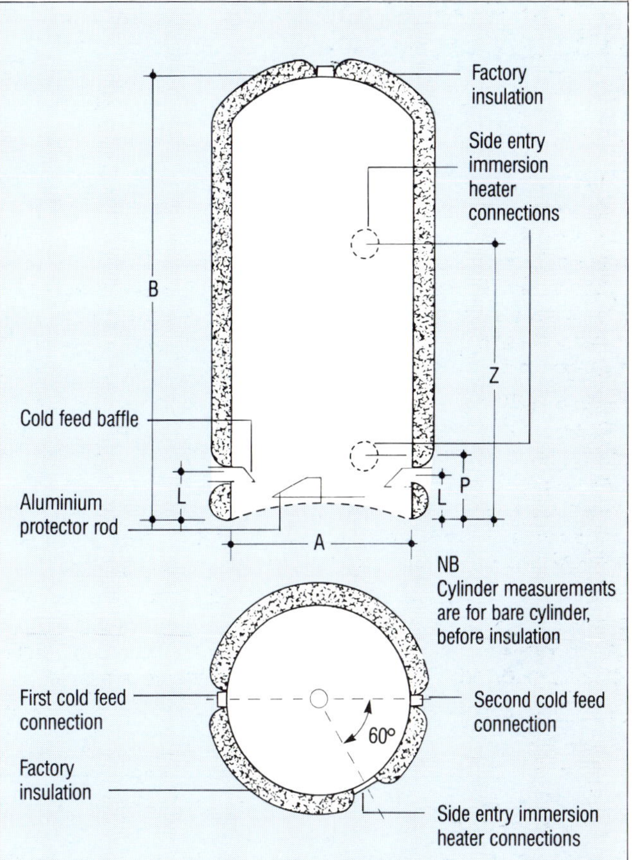

	Size 7	Size 8	Size 9E
Capacity	120	144	210 litres
Dimensions:			
A	450	450	450 mm
B	900	1050	1500 mm
L	100	100	100 mm
P	150	150	150 mm
Z	500	650	1100 mm
Internal thread size:			
Screwed connections	G1	G1	G1 ins
Immersion heater connections	G2¼	G2¼	G2¼ ins

Refer to BS 699 for grades and thicknesses of copper and other details of cylinder construction.

DOMESTIC ELECTRIC WATER HEATING

Figure B18 Electricity Council specification for indirect cylinder

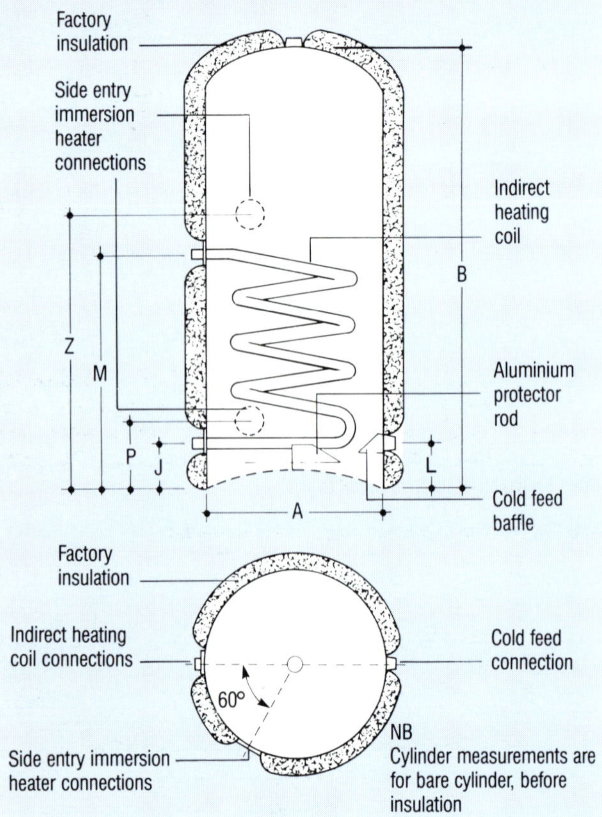

	Size 7	Size 8	Size 9E
Capacity	117	140	206 litres
Dimensions:			
A	450	450	450 mm
B	900	1050	1500 mm
L	100	100	100 mm
P	150	150	150 mm
Z	600	650	1100 mm
J	100	100	100 mm
M	475	545	695 mm
Internal thread size:			
Screwed connections	G1	G1	G1 ins
Immersion heater connections	G2¼	G2¼	G2¼ ins
External thread size:			
Heating coil connections	G1B	G1B	G1B ins

Refer to BS 1566 for grades and thicknesses of copper and other details of coil and cylinder construction.

Figure B19 Economy 7 cylindrical combination units

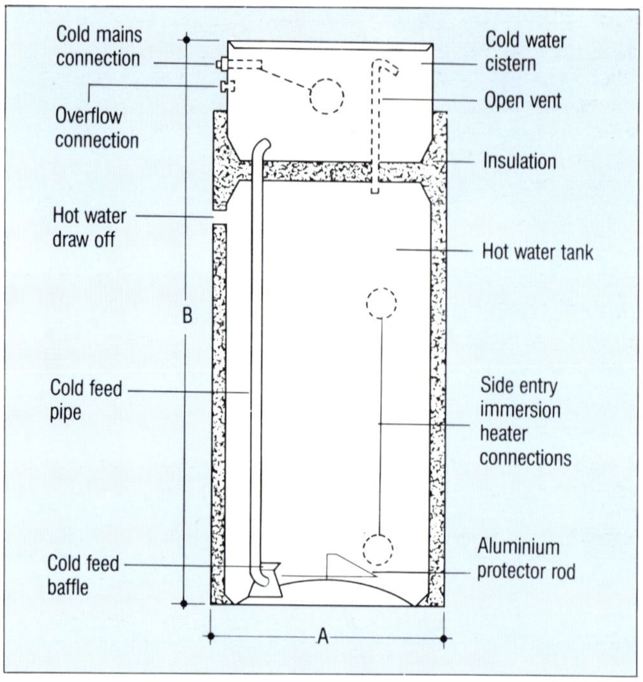

Figure B20 Economy 7 rectangular combination units

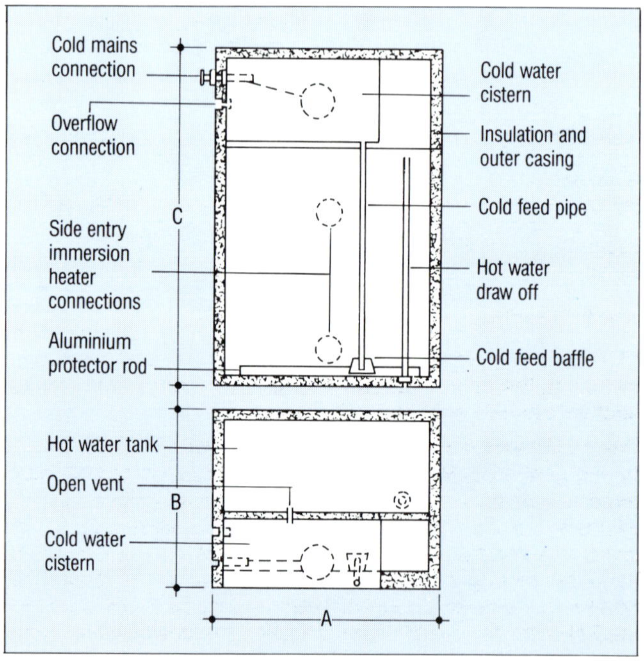

All units are kitemarked in line with the appropriate British Standards where applicable but with the following improvements:

(i) They are factory insulated so that the 24 hour standing heat loss at 45°C difference between the hot stored water and the ambient temperatures does not exceed 2.7kWh for the 210 litre size, 2.5kWh for the 144 litre size and 2.4kWh for the 120 litre size (in effect, 50-60 mm thick factory applied insulation compared to the more usual 25 mm).

(ii) The cold feed entry(s) is fitted with a baffle to minimise mixing of hot and cold water.

(iii) The cylinder is fitted with an aluminium protector rod to minimise corrosion.

(iv) The cylinder is fitted with two side entry immersion heater bosses in specified positions.

Maxistore 7 Immersion Heaters

The Electricity Supply Industry has also devised a specification for side-entry and dual top-entry immersion heaters designed particularly for Economy 7 operation and called Maxistore immersion heaters. This calls for the immersion heaters to be BEAB approved for electrical safety in accordance with BS 3456 Part 2.21 and be fitted with thermostats complying with BS 3955 and also incorporates the following features:

(a) a low watts density element for quiet operation, particularly important for the overnight elements;

(b) an appropriate length of thermostat and an element design to maximise the overnight store;

(c) an Incoloy 825 sheath to increase the element's resistance to corrosion;

(d) performance requirements for the accuracy of the thermostats.

The full details of the requirements are given in the specification but the main points are summarised below in Table B12.

Table B12 Specification for immersion heater

		Dual immersion heater	
	Side entry heater	Long element	Short element
Immersed length	14 in	27 in or 30 in made from 36 in hairpin bent back or 36 in made from 42 in hairpin bent back	16 in made from 21 in hairpin bent back
Thermostat length	11 in	18 in	11 in
Maximum watts density	8.5 W/cm	10 W/cm*	10 W/cm
Electrical loading		3 kW	
Sheath material		Incoloy 825	

* The long element on all dual heaters shall have a 300 mm long cold tail at the head end. The watts density applies only to the length of the element below this.

These side-entry immersion heaters are recommended for use in the high and low positions of all Economy 7 cylinders. The dual immersion heaters are for use in converting existing sizes 7 and 8 cylinders. The 36 inch long element is recommended for use in size 8 cylinder, although there may sometimes be problems in fitting this size eg if it is fouled by an existing indirect coil in the cylinder. In such cases the 30 inch version should be used. For the size 7 cylinder, the 30 inch element is preferred but the 27 inch can be used where there are difficulties.

Point of Use Water Heating

In many cases, the most practicable way to provide hot water is with a point-of-use water heater, located close to the basin or sink that it serves. It is particularly suitable where a long pipe run from a central hot water cylinder would otherwise be required or where there is no central storage at all. Point-of-use heating saves the costs of installing lengthy piping and avoids waste of heat from deadlegs.

Small capacity water storage units (less than 20 litres) and instantaneous heaters should normally be fitted directly to the cold water mains. In a number of instances these appliances may also be cistern fed providing there is sufficient operating head. Installers should refer to individual manufacturers' specifications.

Figure B21 Instantaneous water heater

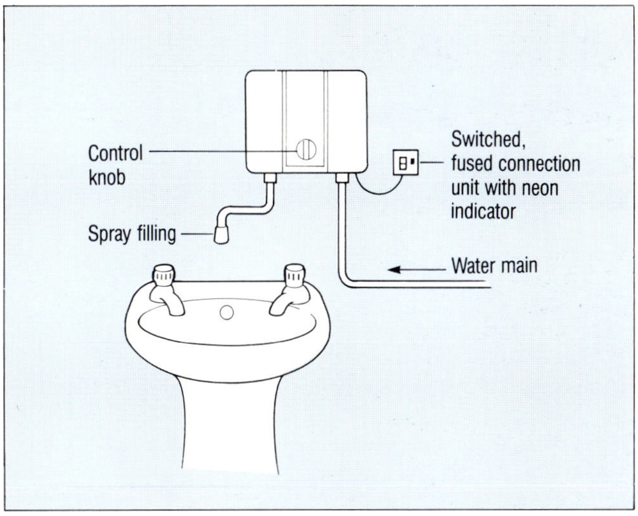

In the home the normal loading is about 3 kW. This gives a flow rate of about 1.5 litres per minute at 40°C making it suitable for a wash hand basin. Such heaters are normally fitted with a swivel outlet and a spray head. The main control knob regulates the flow of water to give temperature choice.

Figure B22 Undersink and oversink water heaters

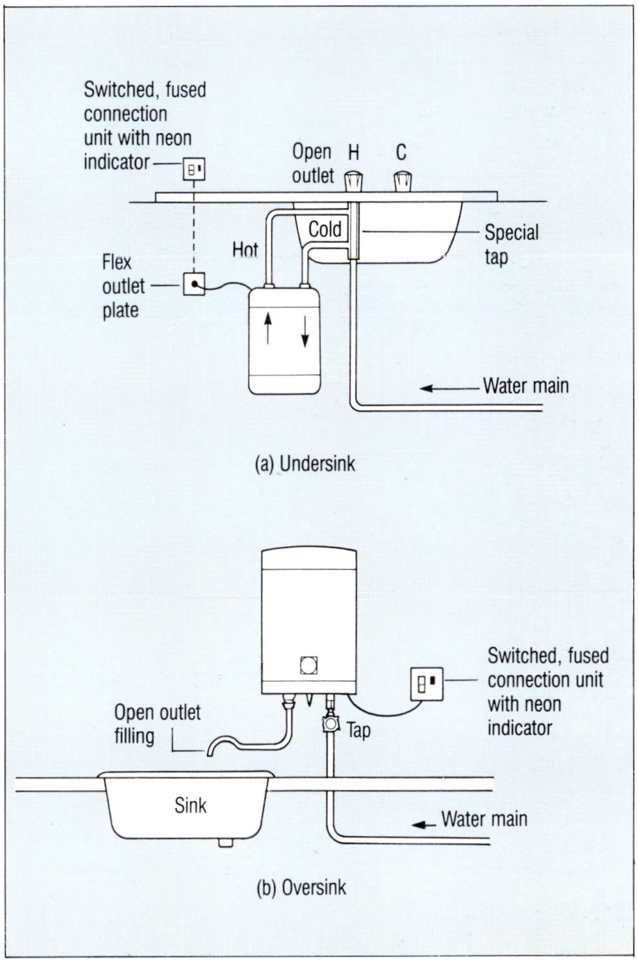

These are small storage units providing up to about 14 litres of stored hot water. Electrical rating is normally 3 kW. These are generally connected to the mains water supply and therefore have an open outlet arrangement. In the case of undersink heaters, this generally requires the installation of a special open outlet hot tap.

DOMESTIC ELECTRIC WATER HEATING

Figure B23 Wall kettle

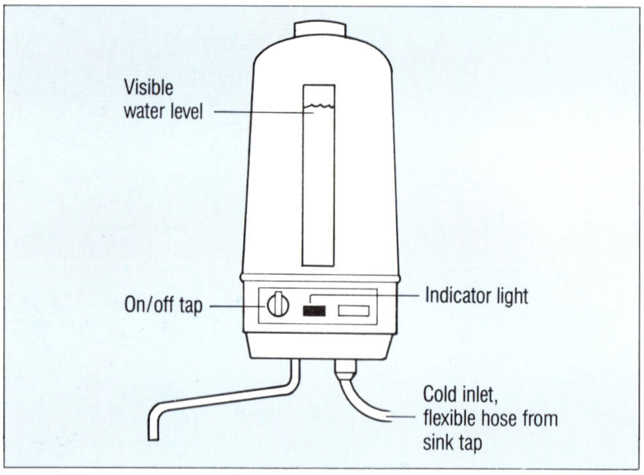

These may be permanently plumbed in but more commonly are provided with a flexible hose for filling them from a nearby cold tap. With up to about 5 litres storage capacity the kettles can heat exactly the amount of water required up to boiling point if necessary. Models are generally of 3kW rating and provided with integral thermostat control.

Electrical Installation

It is not the intention of this section to deal with the fundamentals of electrical circuit design but to guide the designer on typical circuit arrangements that are common for electric water heating.

Electric immersion heaters, whether in central storage tanks, or point-of-use water heaters are generally rated at 3kW which equates to 12.5Amps. This means that each heater should preferably be supplied via its own independent circuit direct from the point where the electricity supply comes into the house ie the meter position. It is common however for 3kW point-of-use heaters to be fed from the local ring circuit. Where a central storage tank is fitted with the recommended dual or twin side entry heater arrangement it is possible to feed the two elements from just one circuit providing an Economy 7 water heating controller is installed.

The controller is fitted with a changeover arrangement to ensure that both elements cannot be operated simultaneously which avoids the possibility of overloading the circuit.

Figure B24 shows a typical arrangement for twin side entry heaters, and Figure B25 shows a typical arrangement for a dual top entry heater.

Figure B24 Wiring installation for twin side entry arrangement

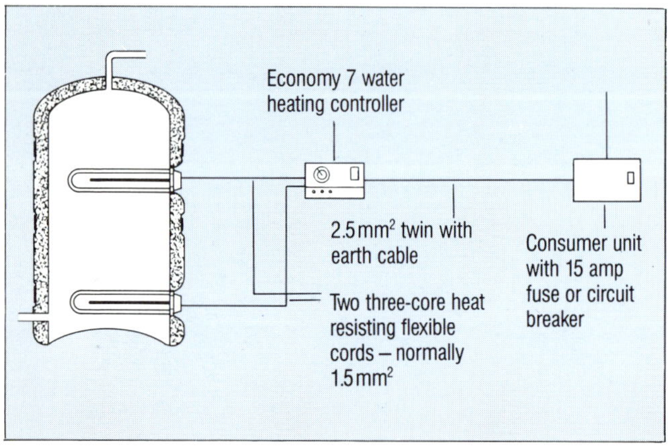

Figure B25 Wiring installation for dual element arrangement

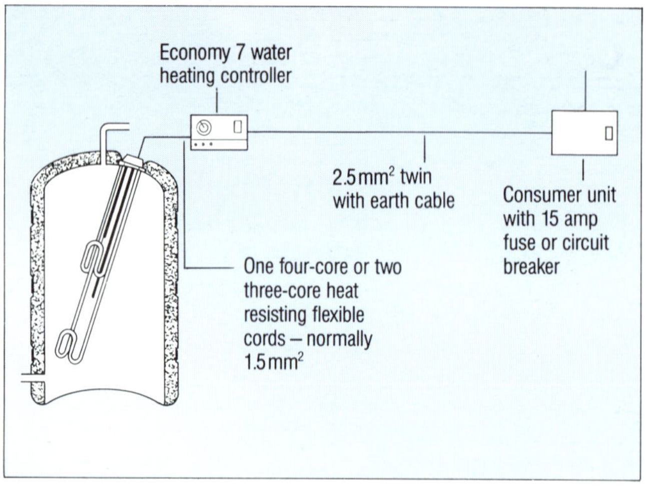

Note that in both cases:

(a) The system does not require an 'off-peak' circuit, a simple unrestricted 24 hour supply from the normal switch fuse or consumer unit is adequate.

(b) The Economy 7 controller has its own built in isolator or switch so a separate switch is not required.

(c) The size of the wiring cable from the fuse unit to the controller is normally 2.5mm² twin with earth and the circuit length can be up to 25 metres with all types of protective devices (eg fuses) and comply with the IEE wiring regulations.

(d) All normal wiring practice must be in strict accordance with the latest edition of the IEE wiring regulations.

REFERENCES

(1) Technical Information DOM 9: Design of Economy 7 Water Heating
(Electricity Council, EC 4964).

This gives a fuller account of the principles of design Economy 7 hot water systems, including comparative fuel running costs from the Energy Efficiency Office Monergy Guides which show Economy 7 to be competitive with all other fuels.

(2) Products for Use with Economy 7 Electric Water Heating (Electricity Council, EC 4806).

This lists products which are recommended as particularly suitable for Economy 7 hot water systems.

(3) Complete Guide for Specifiers: Economy 7 Water Heating (Electricity Council, EC 4669).

This lists the various combinations of water heating packages, system by system, with recommendations on the use of each one.

(4) Economy 7 Electric Water Heating in the Home: Installer Pocket Guide
(Electricity Council, EC4668).

This is a simple practical guide to Economy 7 water heating, aimed specifically at the plumber/installer.

(5) Water Supply Byelaws Guide, ISBN 0 946653 30 5; WRc Water Byelaws Advisory Service, 660 Ajax Avenue, Slough, Berkshire, SL1 4BG.

BRITISH STANDARDS AND CODES OF PRACTICE

BS 699: 1984	Specification for copper direct cylinders for domestic purposes.
BS 853: 1981 (1982)	Specification for calorifier and storage vessels for central heating and hot water supply.
BS 1387: 1985	Specification for screwed and socketed steel tubes and tubulars and for plain end steel tubes suitable for welding or for screwing to BS 21 pipe threads.
BS 1566:	Copper indirect cylinders for domestic purposes.
BS 1566: Part 1: 1984	Specification for double feed indirect cylinders.
BS 1566: Part 2: 1984	Specification for single feed indirect cylinders.
BS 2767: 1972 (1977)	Specification for valves and unions for hot water radiators.
BS 2871: Part 1: 1971 (1976)	Specification for copper tubes for water, gas and sanitation.
BS 3198: 1981 (1984)	Specification for copper hot water storage combination units for domestic purposes.
BS 3456:	Specification for household and similar electrical appliances; electric immersion heaters.
BS 3456: Part 101: 1987	General requirements.
BS 3456: Part 1: 1969 (1981)	General requirements.
BS 3456: Part 2:	Particular requirements.
BS 3456: Part 3:	Complete particular specifications.
BS 3456: Part 3: Section 3:8 1979	Food waste disposal unit.
BS 3456: Part 3: Section 3:9 1979 (1986)	Stationary instantaneous water heaters.
BS 3456: Section 102.5: 1988	Dishwashers.
BS 3456: Section 102.7: 1988	Washing machines.
BS 3456: Section 102.15: 1988	Appliances for heating liquids.
BS 3456: Section 2.7: 1970 (1986)	Stationary non-instantaneous water heaters.
BS 3456: Section 2.21: 1972 (1986)	Electric immersion heaters.
BS 3955: 1986	Specification for electrical controls for household and similar general purposes.
BS 4814: 1976	Specification for expansion vessels using an internal diaphragm, for sealed hot water heating systems.
BS 5449: Part 1: 1977	Code of Practice for central heating for domestic purposes.
BS 5615: 1985	Specification for insulating jackets for domestic hot water heating systems.

Note

(1) Year after the Standard's number indicates latest date of issue.

(2) Year in brackets gives date of latest amendment at the time of publication of this Guide.

SECTION C
Sanitary Plumbing and Drainage

DESIGN OF SANITARY PIPEWORK SYSTEMS

A sanitary pipework system should contain the minimum amount of pipework necessary to carry away discharges from sanitary appliances without risk to health and prevent air from the drain or sewer entering the building under all circumstances.

Discharge Pipe Systems
Ventilated System

A ventilated system is used in situations where there are large numbers of sanitary appliances in ranges or where they are widely dispersed, and close proximity of stack(s) to appliances is impracticable. Trap seals are safeguarded by extending discharge and ventilating stacks to atmosphere and, with the exception of ranges of WC's, providing individual branch ventilating pipes (see Figure C1).

Ventilated Stack System

A ventilated stack system is used in situations where close grouping of appliances makes it practicable to provide branch discharge pipes without the need for branch ventilating pipes. Trap seals are safeguarded by extending the stack(s) to atmosphere and by cross-connecting the ventilating stack to the discharge stack (see Figure C2).

Figure C1 Ventilated system (diagrammatic)

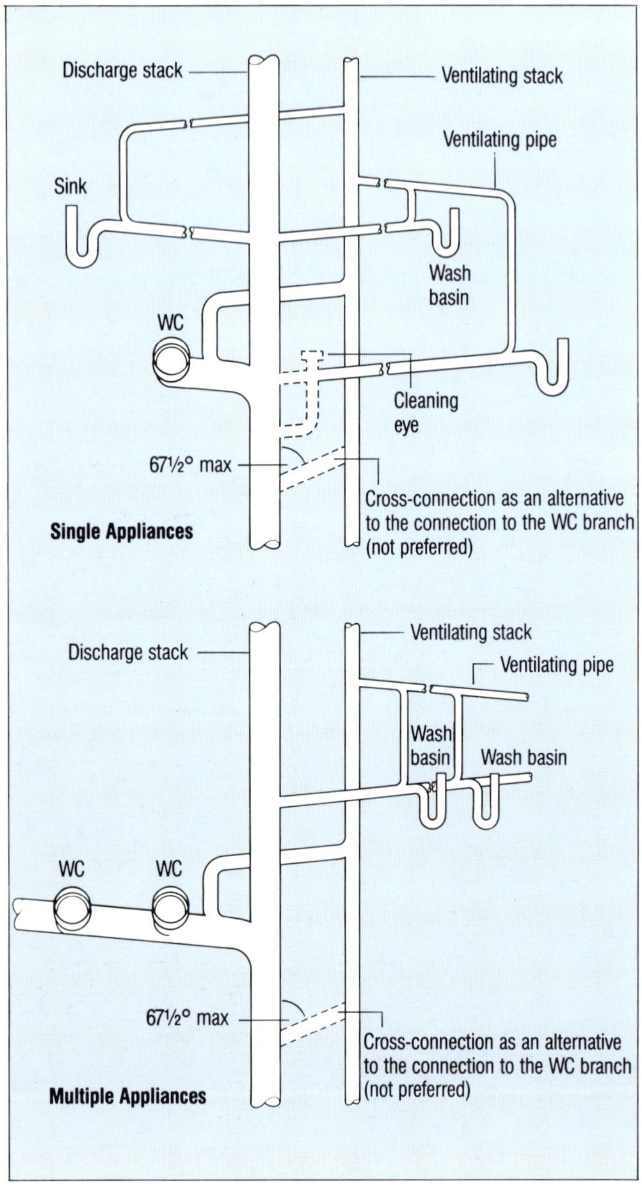

Figure C2 Ventilated stack system (diagrammatic)

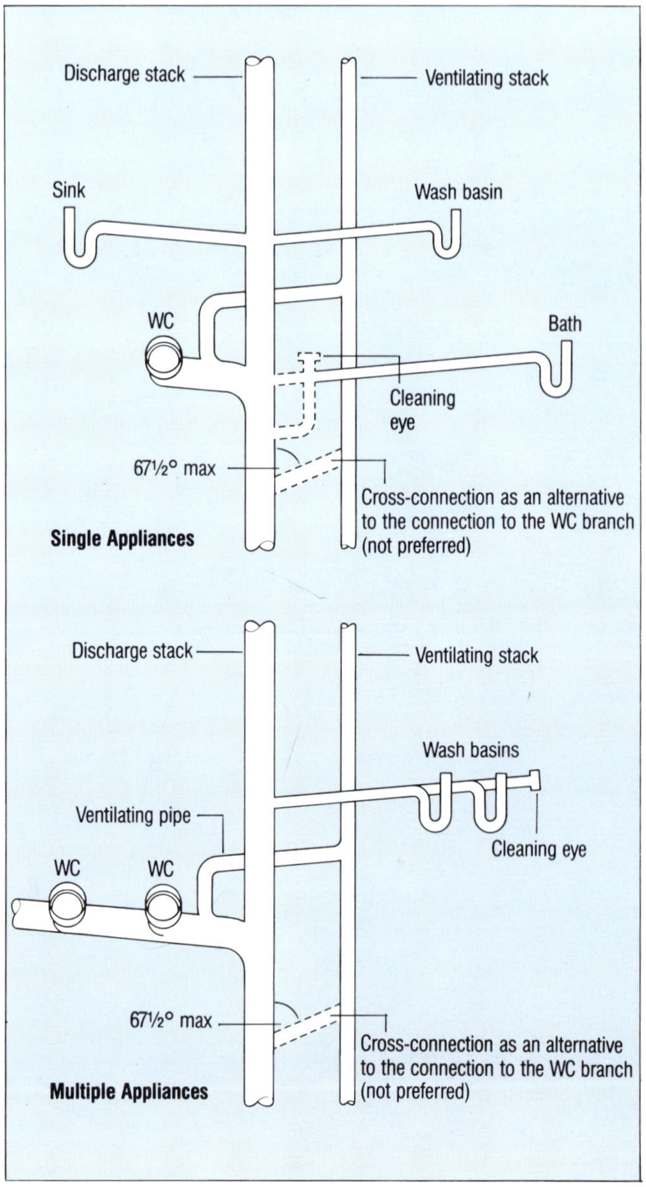

Single Stack System

A single stack system is used in situations as described for the ventilated stack system but only where the diameter of the discharge stack is large enough to limit pressure fluctuations without the need for a ventilating stack (see Figure C3).

Modified Single Stack System

A modified single stack system, providing ventilating pipework extended to atmosphere or connected to a ventilating stack, may be used where the disposition of some appliances on a branch discharge pipe could cause loss of their trap seals. The ventilating stack need not be connected directly to the discharge stack and can be smaller in diameter than that required for a ventilated stack system (see Figure C4)

Stub Stacks

For single storey buildings and ground floor appliances of taller buildings where positive pressures may cause problems if connected to the main discharge stack, a short 100mm discharge stack with the top closed, preferably with an access fitting, may be used to connect one of each of the following sanitary appliances i.e. bath, shower, wash basin, sink, washing machine and WC directly to drain providing the depth to the drain invert level is not greater than 1.5m from the crown of the WC trap and 2m to the centre of the topmost connection to the stub stack.

Drains receiving a number of stub stack connections should be adequately ventilated at the upstream end.

Figure C3 Single stack system (diagrammatic)

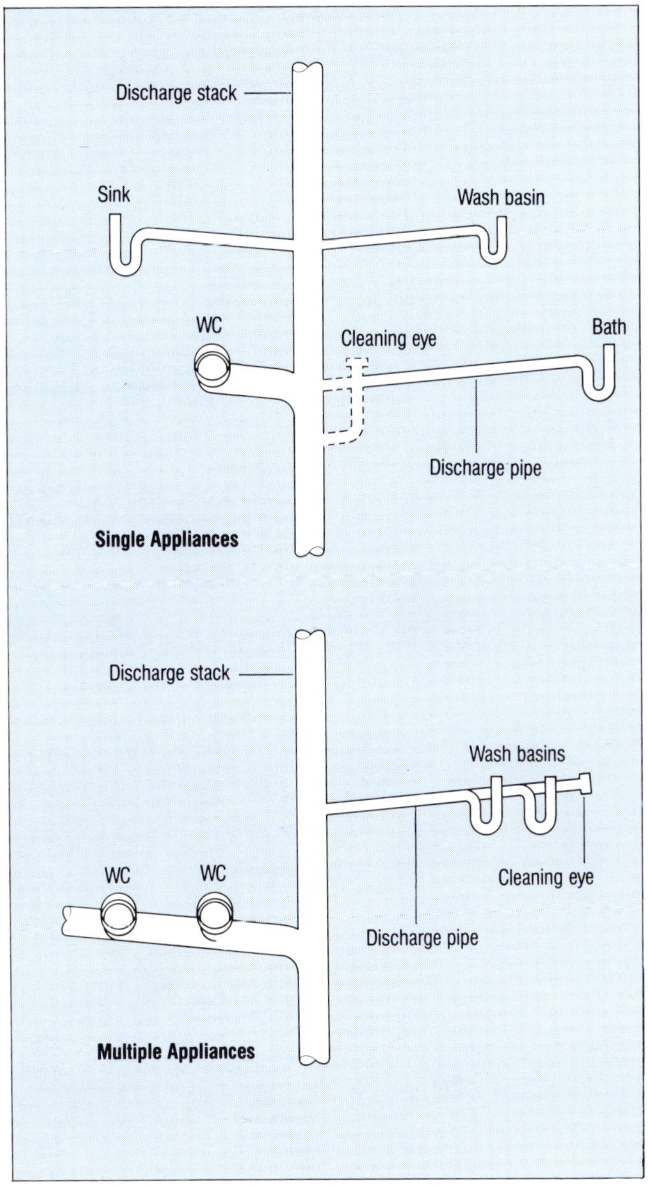

Figure C4 Modified single stack system (diagrammatic)

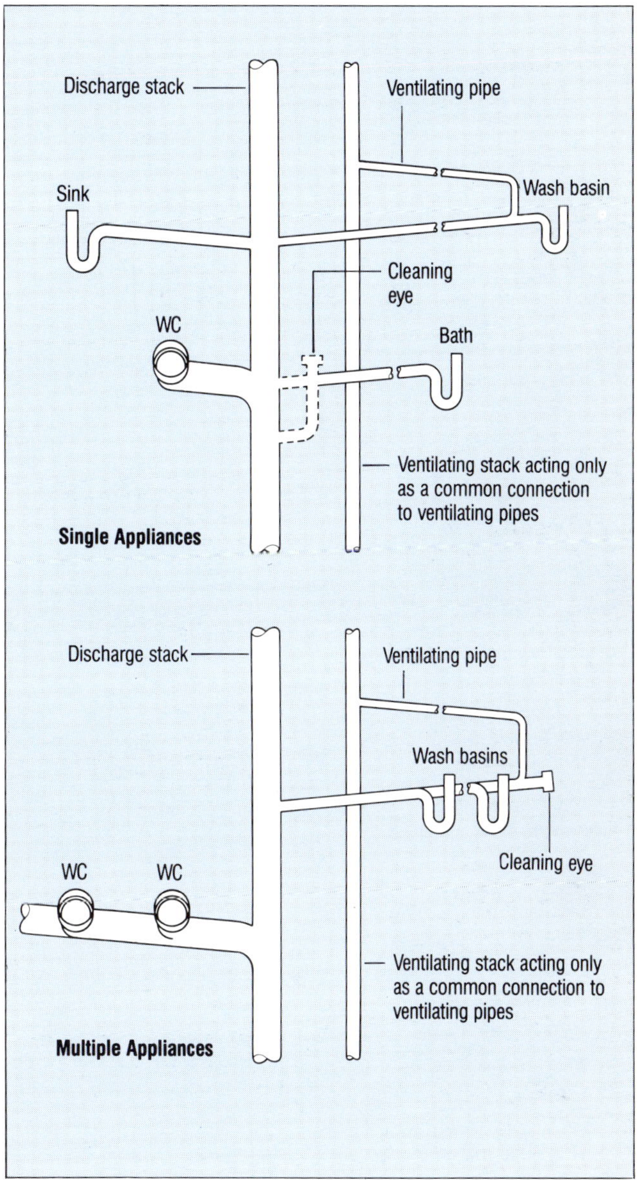

DESIGN OF SANITARY PIPEWORK SYSTEMS

Waste Traps and Discharge Pipes from Sanitary Appliances

Every sanitary appliance should be fitted with a trap either as an integral part of the appliance or attached to and immediately beneath its outlet. Two or three bowl sinks may however be fitted with manifold wastes to permit only one trap. The length of pipe from each outlet to the trap should be kept to a minimum and preferably not exceed 300mm in length. All traps should be accessible and provided with adequate means of cleaning and for this purpose there is an advantage in providing traps which are capable of being readily removed or dismantled. The minimum internal diameters of traps to various appliances are set out in Table C1.

Table C1 Minimum internal diameter of traps

Type of appliance	Minimum internal diameter mm	Minimum depth of seal mm
Domestic Appliances		
Wash basin	32	75
Bidet	32	75
Sink	40	75
Bath	40	50
Shower tray	40	50
Washing machine	40	75
Waste disposal unit (domestic; tubular trap essential)	40	75
Sanitary towel macerator	40	75
Non-domestic Appliances		
Drinking fountain	32	50
Bar well	32	75
Hotel and canteen sink	40	75
Urinal (bowl)	40	75
Urinal (stall) 1 to 6*	65	50
Waste food disposal unit (industrial; tubular trap essential)	50	75

Note

**Where more than six stalls in one range, more than one outlet should be provided.*

It is essential that traps to waste food disposal units, whether for domestic or commercial use, be of the tubular type.

Depth of Trap Seals

Traps of WC's, baths, showers and drinking fountains should have a minimum water seal of 50mm. Traps of other appliances should have a minimum water seal of 75mm for sizes up to and including 50mm diameter and 50mm water seal for sizes over 50mm diameter.

Sanitary appliances with large flat self draining surfaces e.g baths and sinks, when installed at ground floor level may be fitted with a trap having a 38mm seal where the branch discharge pipe terminates at a gully.

Discharge Pipes from Sanitary Appliances

The internal diameter of a discharge pipe from a sanitary appliance should normally be that of the trap to which it is connected and in no case less. The internal diameter of a discharge pipe from a WC should not be less than the internal diameter of the WC pan outlet to which it is connected with a minimum internal diameter of 75mm.

The fall of discharge pipes should be adequate to drain the pipe efficiently.

Waste Disposal Units

The discharge pipe from food waste disposal units should not be less than 7½° (135mm/m) to the horizontal and connect via a tubular trap, directly to a drain without an intervening gully trap or directly to a main discharge pipe or stack, preferably one taking WC discharges to help prevention of blockages.

To avoid hot grease being carried into discharge pipes and drains where it might solidify and cause blockages, food waste disposal units should always be connected to the cold water supply.

Sanitary Towel Macerators

A tubular trap should always be fitted in an easily accessible location and the discharge pipe kept as short as practicable connecting directly to a drain without an intervening gully trap or directly to a main discharge pipe or stack. The gradient of the discharge pipe should not be less than 3° (54mm/m) to the horizontal although steeper gradients are advisable and any bends should be of large radius. Other appliances connected to the discharge pipe upstream of the macerator would be an advantage to assist discharge of the waste material.

Connections from Washing Machines

The discharge pipe from automatic clothes washing machines and dishwashing machines should be connected as shown in Figures C5 and C6. In Figure C5, blockage in the branch or trap will cause overflow through the air gap. To overcome this problem the arrangement in Figure C6 may be used in dwellings not exceeding four storeys.

Figure C5 Connection of washing appliances without venting

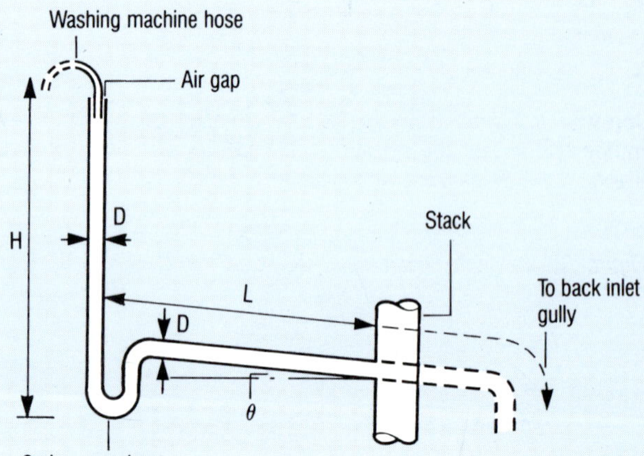

D = 40mm
θ = 1° to 2½° (18.45mm/m)
L = <3m
H = 600 to 900mm (depends on washing machine design)
Trap = 75mm seal, 40mm dia.

Figure C6 Connection of washing appliance with venting

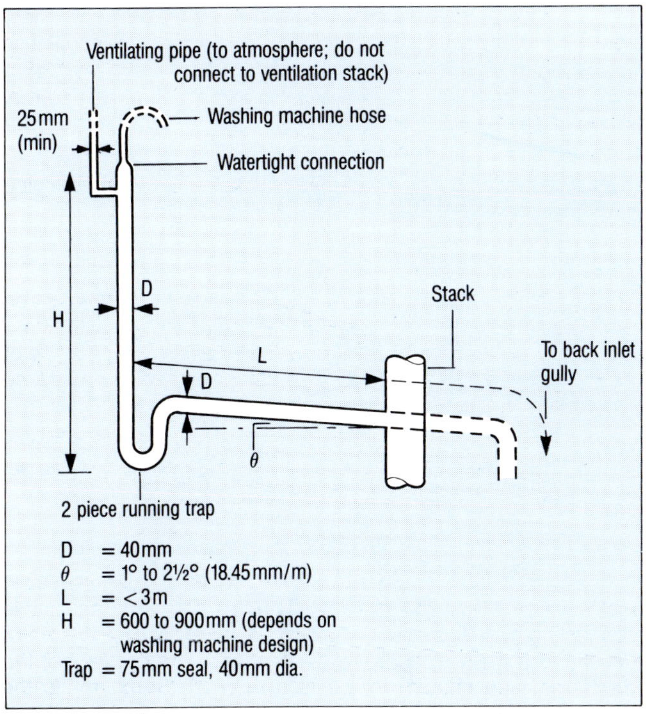

Stack Offsets

There should be no offsets in stacks below the topmost appliance unless venting is provided to relieve positive pressure. Offsets above the topmost appliance are of no consequence. Where offsets in the wet part of the stack are unavoidable, the following may be used as a guide:

(1) Unless branch discharge pipes are ventilated, they are not permitted to be connected between the ventilating pipe connections.

(2) Bends at the change of direction should be large radius (200mm minimum)

(3) Vent pipe diameter to be half of stack diameter.

(4) No offset venting is required in lightly loaded systems of up to 3 storeys in height (e.g. low rise houses).

(5) Offsets above highest branch connections do not require venting.

Prevention of Cross Flow

Opposed small diameter branch discharge pipes without swept entries should be arranged so that risk of the discharge from one branch into the other is avoided. (see Figures C7 and C8).

Small Branch Connections

Figure C7 Restricted connection area on stack

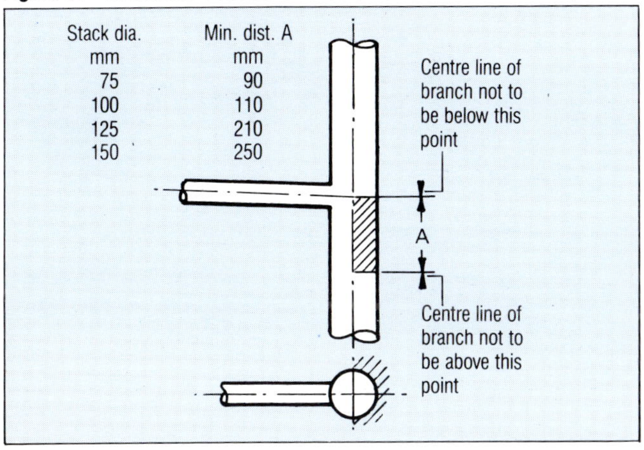

Figure C8 Examples of permitted connections

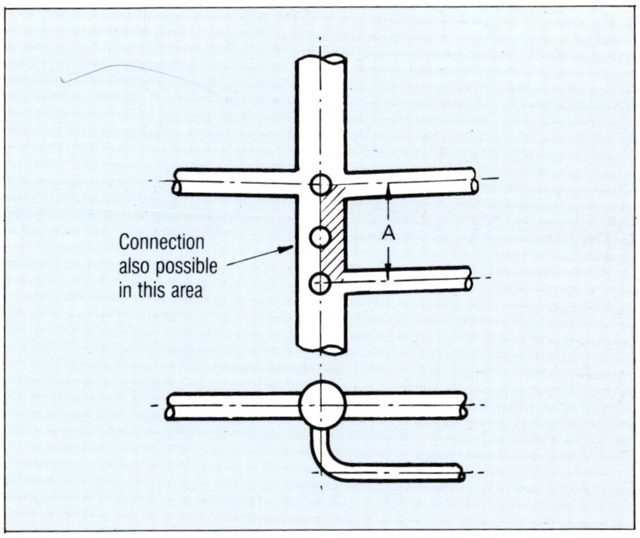

Figure C9 Restricted connection area on stack

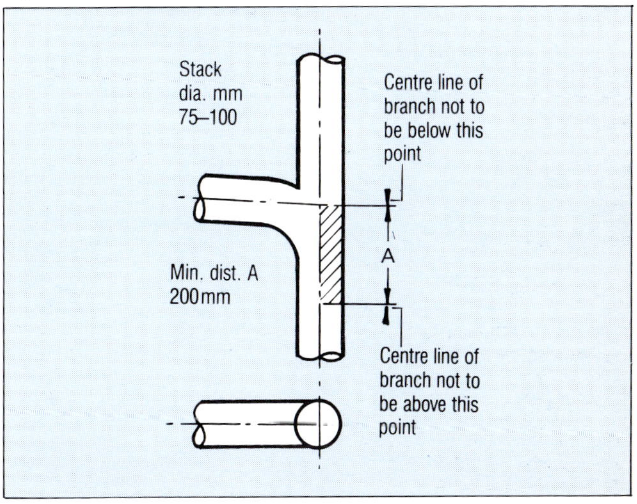

Large Branch Connections

To prevent the discharge from a large branch (eg. WC) backing up a smaller diameter (eg. bath) branch, the latter should not be connected to the vertical stack within a distance of 200mm below where the centre line on the large branch meets the centre line of the stack (see Figure C9). Where this cannot be achieved, the position of the large branch on the stack may be altered by the use of an 'S' trap WC pan or alternatively the bath waste may be connected to a 50mm parallel branch entering the stack below the danger zone (see Figure C10).

'Collar Boss' System

Modified sanitary pipework systems have been developed based on the 'Collar Boss'. Its design was prompted by restrictions imposed on the connection between the bath discharge pipe and vertical stack. The 'Collar Boss' allows the pipework from the bath to meet the stack above floor level at any point around its circumference. The 'Collar Boss'and 'Vented' systems enable the re-introduction of ventilating pipework. By determining the minimum practical size of ventilating pipe and combined branch discharge pipe, a convenient, efficient and economic assembly can be produced. It also eases some of the restrictions that govern the position of sanitary appliances.

DESIGN OF SANITARY PIPEWORK SYSTEMS

Figure C10 Example of permitted connections

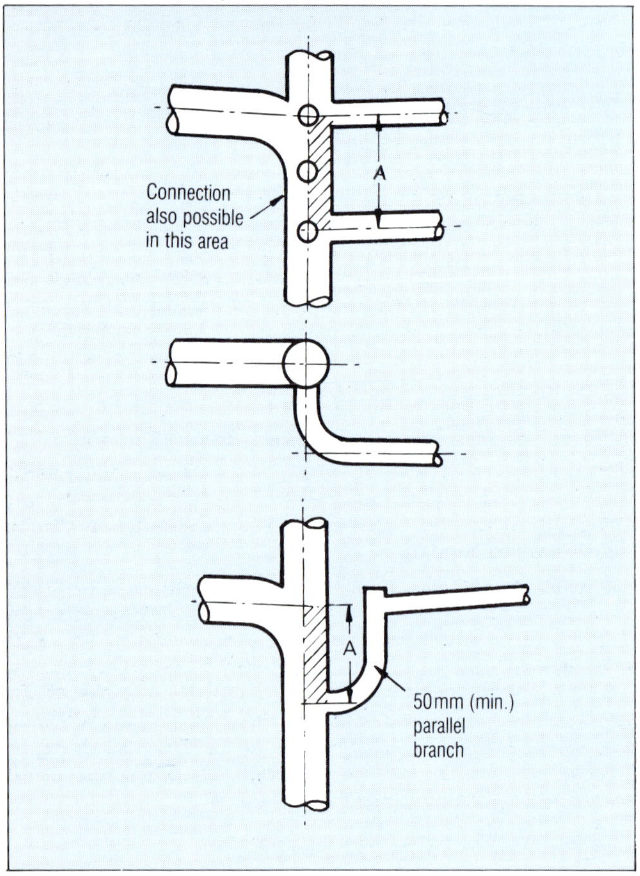

Figure C11 'Collar Boss' fitting

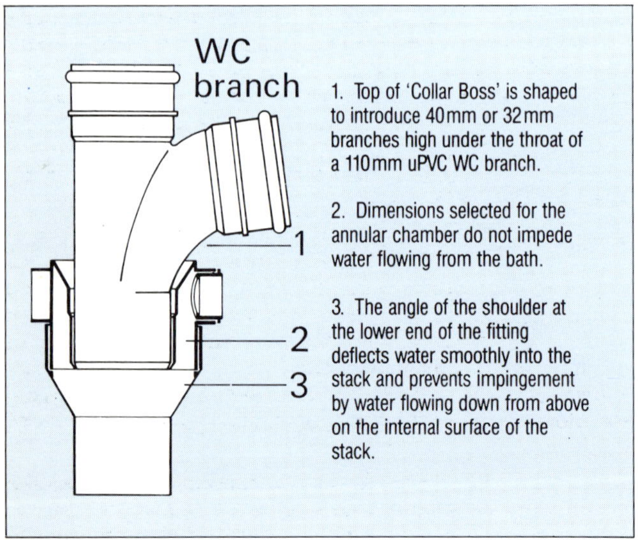

Design of Single Branches and Fittings

On unventilated branch discharge pipes to single waste appliances, precautions must be taken to prevent self-siphonage.

For washbasins the risk of self-siphonage is much more closely dependent on the design of the branch connection to the stack than it is for other appliances. 75mm seal 'P'traps must be used and any bends on the pipe should not be less than 75mm radius to the centre line. For a BS 1188 or similar wash basin, a 32mm diameter waste pipe should be limited in length to 1.7m and the maximum gradient should be determined from Figure C12.

Branch discharge pipes longer than the recommended maximum should be ventilated or a suitable resealing trap should be fitted. (Resealing traps may tend to be noisy in use). Alternatively, the diameter of the branch discharge pipe may be increased one size immediately after the trap, with a concentric reducer fitted so that its invert aligns with the trap invert. This permits the length to be increased to 3m maximum with a gradient between 1° and 2½°.

The length and slope of bath and sink discharge pipes are not so critical as the trailing discharge normally refills the trap. As a guide, 40mm diameter discharge pipes should be limited to 3m in length and the fall should not exceed 5°. It is advisable to use 'P' traps on sink wastes but bath wastes may have either 'S' or 'P' traps. Longer branch pipes may be used if suitable resealing traps are fitted, but long waste pipes tend to become blocked with sediment and are often very noisy in action.

WC branch inlets to the stack should be swept in the direction of flow with a radius at the invert of not less than 50mm and lengths should not exceed 6m.

Design of branches and fittings for commercial and public buildings should cover individual or ranges of appliances. It is necessary to limit the pressure fluctuations to ±375 N/m^2 (38mm WG) (a negative pressure corresponding to about 25mm loss of water seal from a washdown WC). Where venting is recommended, the vent pipe may be run separately or connected to a vent stack.

Vent connections should be above the spill-over level and installed with a continuous fall back to appliances. Branch discharge pipes serving ranges of WC's do not usually run full, therefore branch venting may only be necessary where bends are introduced. The occurrence of full bore flow depends upon factors such as the length and slope of branch discharge pipes and the form of connections.

Figure C12 Length and fall of basin waste. Design curve for 32mm wastes and 75mm seal 'P' traps connected to single wash basins to BS 1188 or similar

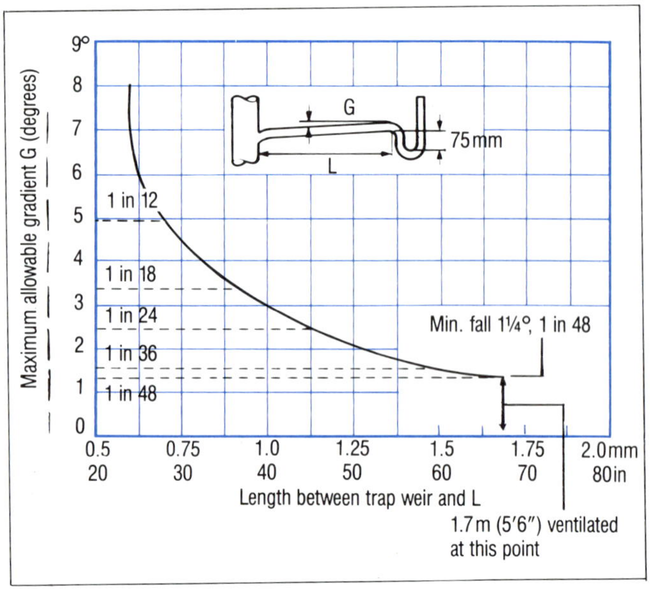

Air Admittance Valves (AAV's)

In recent years AAV's have become available which relieve negative pressures in the discharge system but prevent foul air entering the building if positive pressures occur. Only AAV's that have a current British Board of Agrément (BBA) certificate should be used. It is important that the limitations of use described in the appropriate BBA certificate should be observed. In addition, care should be taken in siting AAV's as some designs may be prone to tampering and/or problems of freezing which may cause malfunction. Preferably, AAV's should not be sited in easily accessible locations particularly those that are made for installation by 'push-fit'.

An AAV incorporating a feature to disperse condensation is shown in Figure C13.

Figure C13 Air Admittance Valve

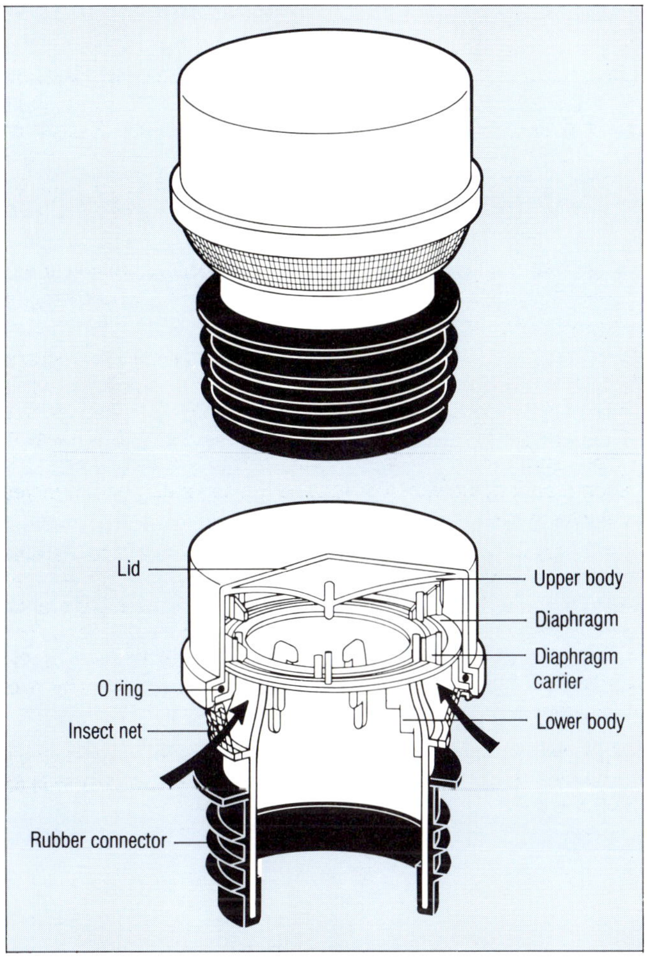

Ventilating Pipes and Stacks

The purpose of ventilating pipes is to maintain equilibrium of pressure within a system and thus prevent the destruction of trap seals by siphonage or compression. They should be installed with a continuous back fall into the discharge pipe in order to prevent condensation forming a waterlock and interfering with the free movement of air. A branch ventilating pipe should be connected to a branch discharge pipe not greater than 300mm from the crown of the trap.

Ventilating pipes to branches from individual sanitary appliances can be 25mm, but if lengths exceed 15m or contain more than 5 bends, 32mm pipe should be used. Where the connection of the ventilating pipe is liable to submergence or repeated splashing, the diameter should be enlarged (50mm min. for WC's) to above the spill-over level of the appliance.

Connections between branch ventilating pipes and any vertical stack should be above the spill-over level of the highest appliance served.

Alternative methods of connecting between discharge and ventilating stacks are shown in Figures C1 and C2. The diameters of such connections should equal the diameter of the ventilating stack. Where the ventilating stack is less than 50mm in diameter, the connection from a large (75mm min.) branch discharge pipe is preferred to the alternative cross connection.

Where cross connections are used, the slope upwards from the discharge stack to prevent cross flow should not exceed an angle of 67½°.

The lowest end of a ventilating stack should normally be connected to the discharge stack at or below the lowest branch connection; the upper end may be connected to the discharge stack above the spill-over level of the highest appliance connected or pass through the roof to atmosphere.

Termination of Discharge and Ventilating Stacks

The outlet of discharge and ventilating stacks to atmosphere should be positioned so that foul air does not cause a nuisance. This is generally achieved if the stack terminates not less than 900mm above the head of any window or other opening in a building within a horizontal distance of 3m. To reduce problems of pressure fluctuation in stacks caused by wind effect, it is advisable to position outlets away from parapets and corners of buildings. Consideration should also be given to stack outlets in relation to air intakes to mechanical ventilation systems.

Bends and Branches at Base of Stacks

The base of discharge stacks should be designed to prevent excessive positive pressure. This may be achieved by using a bend with a minimum centre line radius of 200mm or preferably two 45° radius bends.

Generally for systems up to five storeys, the distance between the centre line of the lowest branch connection and the invert of the drain at the foot of the stack should be at least 750mm. For single dwellings up to three storeys high 450mm is adequate. For multi-storey systems between five and 20 storeys, it is preferable for ground floor appliances to be connected directly to the horizontal drainage system. For buildings greater than 20 storeys, it may be necessary for both ground and first floor appliances to be disconnected from the main discharge stacks.

Admission of Rainwater into Discharge Stacks

Although contrary to Section 40 of the Public Health Act 1936, in some localities with combined drainage systems, discharge stacks are permitted to receive rainwater from roofs. As pressure fluctuations may occur in long stacks and flooding is always a possibility if a blockage occurs in the stack or drainage system, it is recommended that the practice is limited to roofs not exceeding 40m^2 per stack and to buildings not exceeding 10 storeys in height. The approval of the Local Authority should be sought before adopting this practice.

Capacities of Stacks

Owing to the terminal velocities which are attained in vertical pipes, there is a practical limit to the capacity of any given stack. In order to reduce hydraulic and pneumatic disturbances to a minimum, most authorities recommend as an upper limit a stack loading of about a quarter full. This degree of loading provides room for an air core to flow freely down the stack during fluctuating discharges.

The maximum flow possible at any given stack loading depends also upon the frictional resistance of the inner surface of the stack. Smooth pipes of copper and plastics allow slightly higher terminal velocities to be reached than in rougher pipes such as cast iron. This difference in the final velocities reached is not, however, substantial enough to affect materially the discharging capacity of the stacks.

Terminal velocity is generally in the range of 3 to 5m/s per second depending upon the size of stack, quantity of flow and discharges from branches. Terminal length is the distance that a discharge must fall to reach the constant terminal velocity and is generally achieved within approximately two floors.

DESIGN OF SANITARY PIPEWORK SYSTEMS

From the hydraulic conditions given previously and from experimental deductions, formulae have been derived to enable the flow rates in tall stacks to be computed. Work carried out at the Building Research Station and the USA National Bureau of Standards has confirmed that a formula of the type

$$q = Kd^{8/3}$$

agrees closely with the observed flows in tall cast iron stacks.

When the pipe is loaded at one quarter full, $K = 3.2 \times 10^{-5}$

For other flow loads the value of K is as follows:

1/3 full $K = 5.2 \times 10^{-5}$

1/5 full $K = 2.1 \times 10^{-5}$

1/6 full $K = 1.3 \times 10^{-5}$

where q = discharge capacity in litres per second
d = diameter of stack in mm

Table C2 Maximum capacities of stacks flowing quarter full

Stack diameter mm	Stack capacity litres/sec	Discharge units
50	1.10	10*
65	2.10	60*
75	3.20	200§
90	5.20	350
100	6.90	750
125	12.50	2500
150	20.30	5500

** No WC's*
§ Not more than one siphonic WC

Estimating Stack Loading

Most methods of estimating the probable usage of appliances and the flow loads on discharge pipes and systems of underground drainage are, in principle, methods of applying the Binomial Probability Distribution Law to practical conditions.

The Probability Graph C1 shows the probable usage, based on a mathematical certainty of 99% success, of various numbers of sanitary appliances for the given conditions or value of 'p',

where $p = \frac{t}{T}$

t = time in seconds appliance is discharging
T = time in seconds between successive usage of appliance

As an example of the application of the probability graph to a practical case, if 100 appliances each taking 10 seconds to discharge their contents are used at 500 second intervals, the value of:

$$p = \frac{t}{T} = \frac{10}{500} = 0.02$$

Reading from Graph C1, six appliances only out of the 100 installed can be expected to be found discharging simultaneously. If in addition to the above, we know that the maximum flow rate of each appliance is 2 litres/sec., then the probable peak flow load to be designed for is 6 x 2 = 12 litres/sec.

Discharge Units

In practice most systems incorporate many different kinds of sanitary appliances which may, nevertheless, finally be connected to the same discharge pipe or drain. Under these conditions it is more convenient, and indeed essential, to use a system of related unit values which have been determined mathematically to make it easier to design for these mixed systems. These unit values, which are assigned to each appliance are called discharge units (See Table C3).

For sanitary appliances with low discharge rates eg. spray taps, the rate of flow should be added to the flow ascertained by converting discharge units (see Graph C2).

Branch discharge pipes are limited to flows of half capacity and the maximum number of discharge units permitted is given in Table C4.

Graph C1 Probability Graph

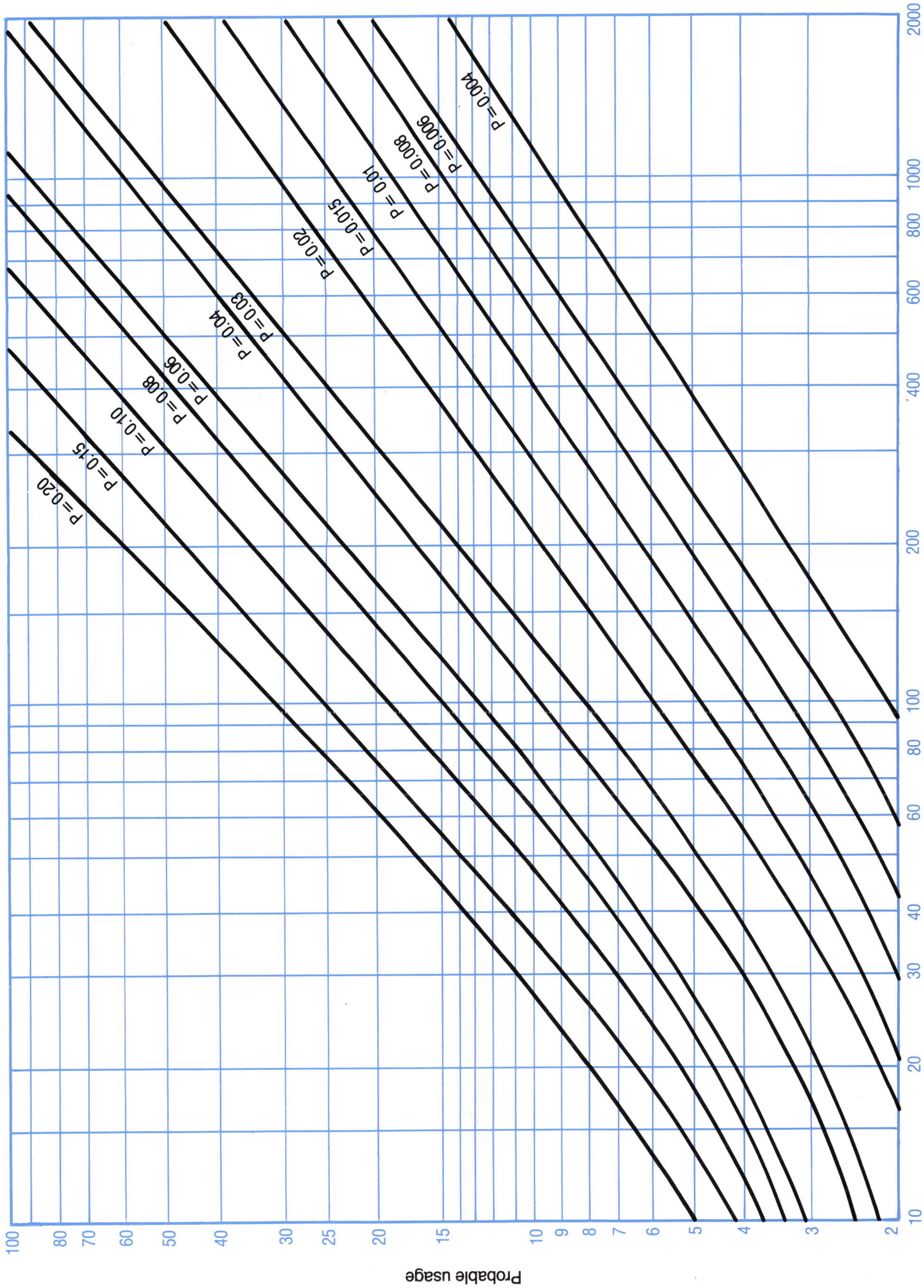

DESIGN OF SANITARY PIPEWORK SYSTEMS

Table C3 Discharge Units

Appliance	Usage T Secs	Discharge Units
WC 3 litre matched low level flush	1200	2
	600	4
	300	8
WC 4 litre matched low level flush	1200	3
	600	5
	300	11
WC 6 litre matched low level flush	1200	4
	600	8
	300	16
WC 7 litre traditional low level flush	1200	5
	600	10
	300	20
WC 9 litre traditional low level flush	1200	6
	600	12
	300	24
WC 9 litre traditional high level flush	1200	7
	600	14
	300	28
Basin 6 litre × 32mm waste	1200	1
	600	3
	300	6
Basin spray tap	0.06 l/s	use flow rate
Sink 15 litre × 40mm (domestic)	1200	3
Sink and macerator (domestic)	1200	4
Sink 60 litre × 40mm (domestic)	600	12
Sink 60 litre × 50mm (domestic)	600	15
Shower	0.1–0.45 l/sec	use flow rate
Bath 80 litre × 40mm (domestic)	4500	7
Bath 80 litre × 40mm (hotel)	1800	18
Bath 120 litre × 50mm (hospital)	1800	30
Dwelling Total	WC DU value	Plus 7
Clothes washer 180 litre	15000	4
Dishwasher (domestic)	15000	3
Urinal stall or bowl	600	0.3
Towel macerator 40mm	0.12 l/s	use flow rate

A 'use interval' (frequency of use) of 1200 corresponds to domestic use; 600 to commercial use; 300 to congested use, such as in public toilets, schools and factories. The more economic low volume WC's allow the total value of discharge units to be reduced, i.e. WC DU value plus 7.

6 and 7 litre flush WC's have been approved for use in the UK after 1st January 1991, but no information is currently available for a flush of less than 6 litres. The low volume WC flush DU's are based on empirical proportion.

Table C4 Maximum number of discharge units allowed on branch discharge pipes

Size mm	Discharge units – Gradient ½° (9mm/m)	1¼° (22mm/m)	2½° (45mm/m)
32	–	1	1
40	–	2	8
50	–	10	26
65	–	35	95
75	–	100	230
90	120	230	460
100	230	430	1050
125	780	1500	3000
150	2000	3500	7500

Note

(1) Discharge pipes sized in this way give the minimum size necessary to carry the expected flow load.

(2) Ventilating pipes may be required and sizes are shown in Table C5.

Graph C2 indicates the approximate flow in litres/second against discharge units. The three graph lines provide for attenuation of the flow.

Line 'A' – aproximate flow from one stack as BS 5572: 1978.

Line 'B' – approximate flow in a branch drain from one stack as:

$$l/s = \left(\frac{DU}{60}\right)^{0.666}$$

Line 'C' – approximate flow in main drains from multiple stacks, branch drains and buildings, as:

$$l/s = \left(\frac{DU}{60}\right)^{0.666} \times 0.7$$

For more detailed information about discharge stacks, see BS 5572: 1978.

The discharge unit method cannot give guidance on venting requirements as it is entirely based on hydraulic loadings and not on pressures developed within the stack. Table C5 giving approximate ventilating pipe and stack sizes should therefore be used with this data. This is usually a safe assumption but can lead, in some cases, to oversized ventilating systems.

Table C5 General guide for the sizes of ventilating pipes and stacks

Size of branch discharge pipe or discharge stack D	Size of branch ventilating pipe or stack
Smaller than 75mm	⅔ D (25mm min.)
75mm and above	½ D

Sizing Procedure for Discharge Unit Method

Discharge unit values are given in Table C3. For other appliances, take the discharge unit value as that given in Table C3 for an appliance with the same trap diameter. The discharge unit values of all appliances contributing to flow in a pipe or stack should be added and the appropriate size selected from Tables C2 and C4.

When stacks additionally carry continuous flows from other

Graph C2 Discharge units and flow rates

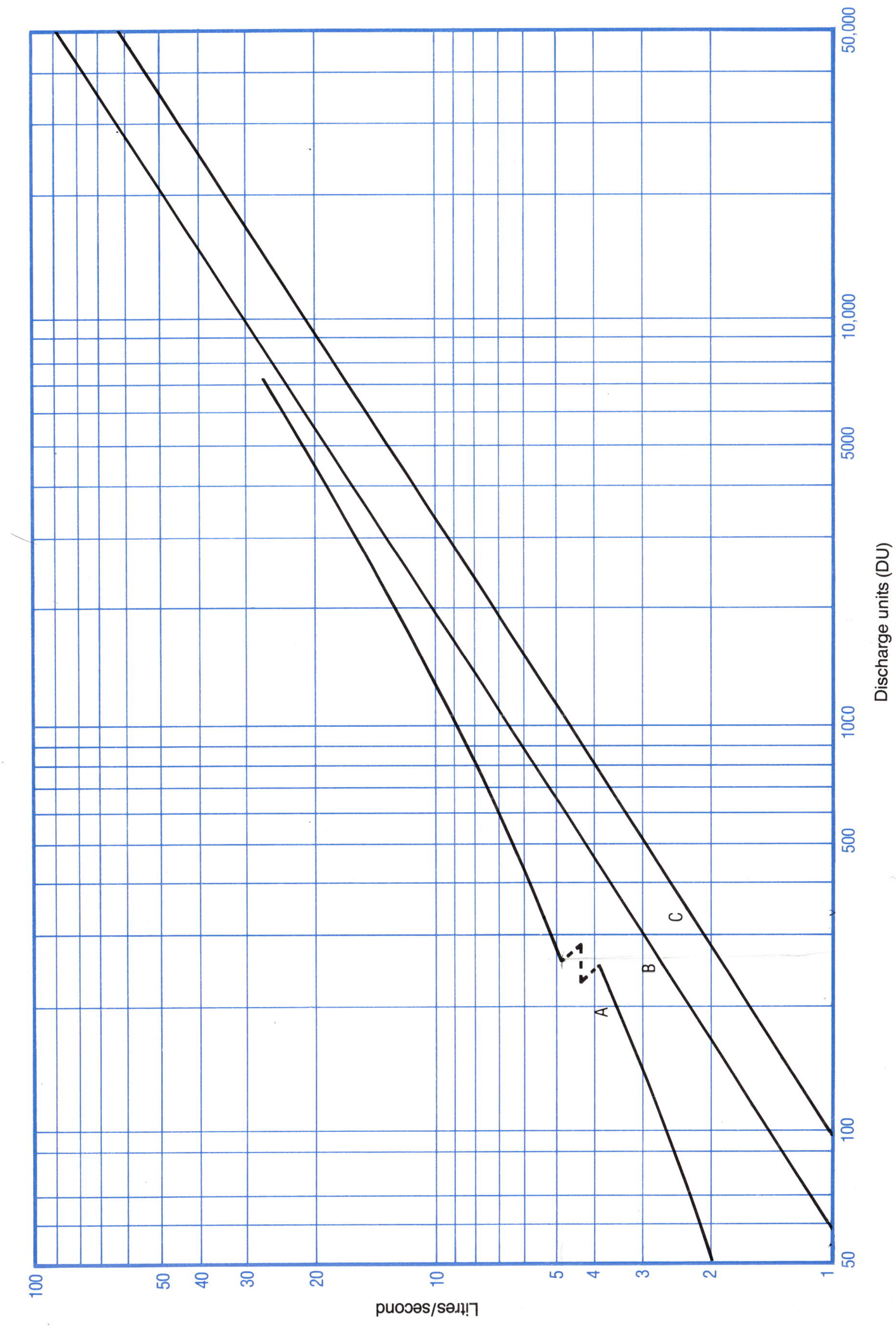

DESIGN OF SANITARY PIPEWORK SYSTEMS

sources (e.g. cooling water from equipment), the discharge unit value for the appliances should be converted into value of flow using Graph C2. The two flows can then be added and an appropriate stack size selected from Table C2.

Example 1

Determine the peak flow rate and diameter of a discharge stack required to carry the discharge from 44 flats.

Each flat has 1 WC (9 litre), 1 wash basin, 1 bath, 1 sink and 1 washing machine.

44 (6 + 7) = 44 × 13
Total discharge = 572 units

From Graph C2 it can be seen that 572 discharge units equate to a flow of approximately 6.66 litres/second.

A 100mm stack can carry 7.1 litres/second and this size will be satisfactory.

Example 2

Determine the peak flow rate and diameter of discharge stack to receive the discharge from 20 WC's (7 litre), 20 wash basins and 5 sanitary towel macerators from female toilets in an office block.

20 WC's × 10 = 200
20 WB's × 3 = 60
Total discharge units = 260

The approximate flow rate of 260 discharge units = 4.9 l/s

5 sanitary towel macerators = 5 × 0.12 = 0.6 l/s
therefore, total flow = 5.5 l/s

A 100mm stack can carry 7.1 l/s and this size will be satisfactory.

Example 3

Determine the peak flow rate and diameter of discharge stack from 10 WC's, 15 urinals and 15 wash basins in a factory.

10 WC's (9 litre) × 24 = 240
15 urinals × 0.3 = 4.5
15 wash basins × 6 = 90
Total discharge units = 334.5

The flow rate for 334.5 discharge units is approximately 5.2 l/s.

A 100mm stack can carry 7.1 l/s and this size will be satisfactory.

Refer to Table C5 to determine the size of ventilating stack if required. In each of the 3 examples, a 100mm discharge stack will require a ventilating stack of 50mm (½ D).

A convenient method of determining branch discharge pipe and discharge stack sizes for single stack (or simplified discharge) systems is given in Tables C6 and C8. Table C7 gives additional information for office and public buildings to enable discharge stacks serving WC's, wash basins and urinals to be sized from Table C6. The four combinations of appliances are based on the provision of sanitary appliances given in Table C11 and these are equated to the hydraulic equivalents of WC/wash basin combinations.

Example 4

Determine the diameter of discharge pipe required for a single stack system serving a block of flats 8 storeys high with 2 domestic appliance groups per floor. What is the maximum length permitted for the branch discharge pipe from the WC?

From Table C6 it can be seen that a 100mm stack can serve up to 10 floors each with 2 domestic appliance groups.

From Table C8 the maximum length given for a branch discharge pipe from a WC is 6m providing any bends are of easy radius and the fall is not less than 1°.

Example 5

Determine the diameter of discharge pipe required for a single stack system serving an 8 storey office block with 4 WC's, 2 urinals and 5 wash basins per floor.

From Table C7 the number of sanitary appliances per floor equates to 5 WC's and 5 wash basins. From Table C6 it will be seen that a 100mm stack serving 5 + 5 is only capable of serving 4 floors, therefore, a 150mm stack will be required.

A more comprehensive table, giving in addition, ventilating stack sizes required if conditions for single stack arrangements are not suitable, can be found in BS 5572 Table 6.

Table C6 Discharge stack sizing (a)

Stack size mm	Use: Description	Use: Interval between use mins	Number of floors of appliances	Max number of appliances/floors
75	Domestic	20	Up to 2	1 Domestic Group*
100	Domestic	20	Up to 10	2 Domestic Groups*
	Commercial	10	Up to 8	2 WCs + 2 wash basins
			Up to 4	5 WCs + 5 wash basins
	Congested	5	Up to 8	1 WC + 1 wash basin
			Up to 4	3 WCs + 3 wash basins
150	Domestic	20	Up to 30	2 Domestic Groups*
	Commercial	10	Up to 24	3 WCs + 3 wash basins
			Up to 8	5 WCs + 5 wash basins
	Congested	5	Up to 24	2 WCs + 2 wash basins
			Up to 16	3 WCs + 3 wash basins
			Up to 8	3 WCs + 5 wash basins

** 1 domestic appliance group is 1 WC, 1 sink, 1 wash basin, 1 bath (and/or shower) and 1 washing machine.*

Note

For non-swept WC branch connections, with 2 groups of domestic appliances per floor;
A 100mm stack can be used for systems up to 4 floors in height.
A 150mm stack can be used for systems up to 15 floors in height.

Branch to Stack Connections

The shape of the WC branch connection to the stack is most important as it influences induced siphonage in the stack. The recommendations in Table C6 assume that WC branch connections are swept in the direction of flow. Smaller diameter branch connections are not so critical and a swept entry is unnecessary.

Table C7 Discharge stack sizing (b)

WC		Urinal		Wash basin		WC		Wash basin
2	+	1	+	2	Equivalent	2	+	2
2	+	2	+	3	to:	3	+	3
3	+	3	+	4		4	+	4
4	+	4	+	5		5	+	5

Table C8 Design of branch discharge pipes

Appliance	Number of appliances	Trap size mm and type	Branch pipe: Max length m	Branch pipe: Size mm	Bends in branch pipe	Design precautions: Principal action to be guarded against	Design precautions: Trap design	Design precautions: Branch gradients	Design precautions: Other requirements
Wash basin	Single	32 P	1.7	32	Not more than 2 (min radius 75 mm to centre line)	Self-siphonage	Tubular	See Fig. C 12	Recommendations for BS 1188 wash basins with overflow. Some bottle traps will be suitable
	Single	32 P	3.0	40	Not more than 2 (min radius 75 mm to centre line)	Self-siphonage and excessive deposition	Tubular or bottle	1° to 2½° (18 to 45 mm/m)	Connect 40 mm branch to trap with 32 mm tail pipe, 50 mm long
	Single	32 P or S	3.0	32	Any bends to have min radius of 75 mm to centre line	Self-siphonage and excessive deposition	Resealing	1° to 2½° (18 to 45 mm/m)	See Code of Practice for vented arrangements
	Range of up to 4	32 P	3 Main 0.74 Branch	50 Main 32 Branch	None	Self and induced siphonage and excessive deposition	Tubular or bottle	1° to 2½° (18 to 45 mm/m)	Swept entry fittings at connections between 32 mm and 50 mm branches
	More than 4	32 P	10 Main 1 Branch	50 Main 32 Branch	Any bends to have min radius of 75 mm to centre line	Self and induced siphonage and excessive deposition	Resealing	1° to 2½° (18 to 45 mm/m)	Swept entry fittings at connections between 32 mm and 50 mm branches. See also Code of Practice for vented arrangements.
	More than 2	32 S	10 Main 1 Branch	50 Main 32 Branch	Any bends to have min radius of 75 mm to centre line	Self and induced siphonage and excessive deposition	Resealing	1° to 2½° (18 to 45 mm/m)	Swept entry fittings at connections between 32 mm and 50 mm branches. See also Code of Practice for vented arrangements
Bath, Sink	Single	40 P or S	3	40	Any bends to have min radius of 75 mm to centre line	Self-siphonage and excessive deposition	Tubular or bottle	1° to 5° (18 to 90 mm/m)	Longer branches possible but noisier discharge and/or blockage may result. S-trap arrangement will also give noisier discharge
Shower	Single	40 P or S	3	40	Any bends to have min radius of 75 mm to centre line	Self-siphonage and excessive deposition	Tubular or bottle	1° to 5° (18 to 90 mm/m)	
Washing Machine	Single	40 P or running trap	3	40	Any bends to have min radius of 75 mm to centre line	Siphonage from machine during refill	Tubular or bottle	1° to 2½° (18 to 45 mm/m)	40 mm standpipe, 600-900 mm high (depending on machine) with air gap at hose opening
WC	Single	75 to 100 P or S	6	100	Avoid knuckle bends	Excessive deposition	—	1° min (18 mm/m)	Branch pipe size of 75 mm can be used with siphonic WCs
	Range of up to 8	75 to 100 P or S	15	100	Avoid knuckle bends	Excessive deposition and induced siphonage	—	½° to 5° (9 to 90 mm/m)	Swept entry fittings needed between WC branches and main branch
Urinal	Single Bowl	40 P or S	3	40	Any bends to have min radius of 75 mm to centre line	Excessive deposition	Tubular or bottle	1° to 5° (18 to 90 mm/m)	Access for cleaning important
	Range of up to 5 bowls	40 P or S	As short as possible	50 Main 40 Branch	Any bends to have min radius of 75 mm to centre line	Excessive deposition and induced siphonage	Tubular or bottle	1° to 5° (18 to mm/m)	Access for cleaning important
	Range of stalls	65 or 75 P or S	As short as possible	65 or 75	Large radius bends	Excessive deposition	Tubular	1° to 5° (18 to 90 mm/m)	Access for cleaning important

KITCHEN AND LABORATORY DRAINAGE
KITCHEN DRAINAGE

Materials

Only smooth bore materials should be used for any system of drainage pipework but particular care is required when selecting materials for drainage from restaurant kitchens. It is essential that pipework should be robust and, in appropriate situations, suitable to receive high temperature discharges.

Backflow Problems

Backflow can occur and manifest itself in the kitchen system if a blockage occurs downstream at a point where the kitchen system joins the main system in a large complex or if effluents pass down stacks from floors above and through the kitchen system. It is recommended therefore that kitchen drainage systems should not be closely linked with those of other parts of the building complex and that separate drainage stacks be provided where kitchens are located other than on ground floors.

Grease Traps

Grease traps should not be used except where there are sensitive points in a drainage system eg. pumping stations, pipeline restrictions and sewage treatment works. Where used they should be designed and located to promote cooling, coagulation and retention of the grease within the trap and sized to achieve maximum efficiency.

The temperature and velocity of flow of the waste water should allow the grease to separate and collect on the surface of the water in the trap reservoir. Consideration should also be given to the general nature of the waste matter to be discharged since the reduced flow velocity through the trap will allow solid waste matter in suspension to settle and collect in the trap reservoir.

Provision should be made to facilitate the hygienic removal and disposal of the grease. Provision should also be made for the trap to be completely emptied and cleaned periodically to limit the build-up of settled matter which inhibits the effectiveness of the grease trap.

Ideally, grease traps should be located externally and siting of grease traps within kitchens should be avoided if at all possible. In no circumstances should they be sited in food storage or preparation areas.

Pumped Installations

Where kitchen drainage has to be pumped, it is essential that the pumps and sump are regularly maintained. If an effective grease trap is not installed prior to the sump, provide a chemical drip feed of a grease dissolving type into the sump to prevent grease coagulating.

Similarly, coating of the electrodes or controls with silicone will inhibit adhesion of grease to their surface.

Drainage (Above Ground)

Kitchens are of necessity, designed to ensure a natural flow of work and seldom permit the grouping of the appliances to give the best conditions of drainage. As it is of primary importance that there should be no loss of water seal in traps on kitchen appliances, an adequate ventilated system of drainage is necessary.

Branch Discharge Pipes

In view of the high proportion of solid matter and grease from kitchen appliances, 'flat' gradients should be avoided. Branch discharge pipes from macerators should be as short as practicable and connected directly to a main discharge pipe or stack. The gradient should be at least 7½° (135 mm/m) to the horizontal although steeper gradients are advisable, and any bends should be of a large radius.

It is an advantage if other appliances can be connected to the discharge pipe upstream of the macerator connection, to assist with the discharge of the waste material. The discharge pipe should connect directly to a drain without an intervening gully trap.

Traps

Traps up to and including 50 mm diameter should be tubular with a 75 mm seal and preferably of a two piece type for easy access.

Access

Access points should be above spill-over level of the appliances at risk and located in positions permitting effective use without nuisance.

Floor Drainage

Floor channels, open gullies and gratings in kitchens, food preparation and wash-up rooms harbour dirt and grease and if not properly fitted can be hazardous to pedestrian traffic. This form of drainage is unhygienic and should be avoided. However, an open top gully with sediment bucket should be provided for discharge from mobile sinks where these are to be used in vegetable preparation areas.

Specific Requirements

(1) Drain-off valves on food containers should be of the full way plug-cock type with quick release bodies for easy cleaning. These valves must not be connected to a common waste pipe or drain without an intervening air break.

(2) Sinks and washing up machines must be individually trapped and connected directly to the building drainage/plumbing system.

(3) Vegetable paring machines should be fitted with a waste dilution unit and the waste should be trapped and connected directly to the building drainage/plumbing system.

(4) The pipes from appliances which discharge waste water containing heavy concentrations of solid matter, e.g. vegetable paring machines and waste disposal macerators, should not be connected to the head of long runs of horizontal waste pipes. They should be connected as close as is practicable to the main drain or discharge stack so as to gain the maximum flushing advantage from appliances with high waste water discharge rates.

(5) Where practicable, items of kitchen equipment such as steaming ovens, bains marie, boilers and cafe sets should discharge over a drip tray or a fixed tundish having a trapped waste pipe connected to the discharge system. In situations where this is impracticable, the waste should discharge over a suitable container which can be emptied as required.

(6) Boiling pans should be drained separately over removable tundishes into trapped gullies. The trapped gully should be fitted with a solid hinged flap set flush with the floor and kept closed when not in use.

LABORATORY DRAINAGE

General

At the outset of any design of a laboratory waste system it is imperative to ascertain and receive written confirmation from the client as to the type and probable quantites of effluents that will be discharged. It is also important to know whether any allowance for future alterations or extensions is to be made.

Distribution

There is divided opinion of the relative advantages and disadvantages of either horizontal or vertical distribution but the latter is to be preferred for discharge systems and will generally permit a large degree of planning flexibility.

Vertical distribution is ideal for laboratory premises with repetitive planning on successive floors and also when dealing with alterations, as their effect is invariably confined to the laboratory concerned.

Replacement

Due to the nature of discharges from laboratories, despite the fact that the best material for the job should be selected, the life expectancy may well be less than for ordinary discharge systems. The design must therefore take into consideration ease of replacement.

Dilution

The greater the dilution of laboratory waste obtained, the better, so as to minimise possible damage to final drains and sewers. Dilution may be achieved by individual dilution recovery traps (catchpots), by similar fittings serving groups of appliances or by dilution chambers constructed below ground level as part of the drainage system.

Incompatible Wastes

The designer should always bear in mind that chemical wastes may be incompatible e.g. chemicals each producing few problems on their own, but which, if combined in a common branch discharge pipe, may have a tendency to gel or produce an undesirable chemical effect and therefore will need to be kept separate until sufficient dilution has occurred in the system.

As a general rule it is preferable for each laboratory to have its own connection(s) to the main stack(s), and not to combine wastes by continuous branch discharge pipes passing through a number of laboratories. Obviously this will be easier where vertical distribution has been adopted. For most practical purposes, once the wastes discharge into a main stack, sufficient dilution will occur, but each instance must be considered on its own merits.

Radioactive Waste

Providing certain quantities are not exceeded (these vary for hospitals and other establishments) and discharges are to a drainage system normally used for the discharge of foul water or trade effluent, then radioactive waste may be exempt from the Radioactive Substances Act 1960.

For non-exempted discharges, where Department of the Environment (Radio Active Waste Administration)) permit discharges to a drain, they will give specific instructions on necessary installations and maintenance requirements. For exempted discharges, no special precautions relating to materials are required but good design practice and sound workmanship are essential.

Sanitary/laboratory appliances should be easy to clean. Discharge pipes must be sealed throughout their run to the sewer which should be as short as possible. The routeing of pipes must take into account areas which might be put in hazard by leakage, and should be freely accessible for inspection and repair. A secondary system of containment may be considered necessary.

Radioactive pipe runs must be labelled at points of access.

The designer should consult the user about the pattern of usage and monitoring procedures. Any small or intermittent radioactive discharges should be well flushed out to main drainage, by other flows and not allowed to dry out along the pipe line. The character of any added discharge should however be taken into account, particularly to avoid the risk of blockage; discharges carrying a high proportion of paper or macerated materials should be kept separate. If a holding tank is installed for monitoring purposes, matter likely to turn septic must be excluded.

MATERIALS FOR ABOVE GROUND DRAINAGE SYSTEMS

General

For a well designed above ground drainage system it is essential to have considerable knowledge of the materials that are available and their fixing and jointing techniques.

Types of Materials

(a)	cast iron	(BS 416)
(b)	copper	(BS 2871)
(c)	galvanised steel	(BS 3868)
(d)	ABS (Acrylonitrile butadiene styrene)	(BS 5255)
(e)	HDPE (High density polyethylene)	(BS 5255)
(f)	MUPVC (Modified unplasticized polyvinyl chloride)	(BS 5255)
(g)	UPVC (Unplasticized polyvinyl chloride)	(BS 4514)
(h)	PP (Polypropylene)	(BS 5254 and BS 5255)
(j)	borosilicate glass (primarily for laboratory waste)	(BS 5895)
(k)	heavy duty polyethylene (primarily for laboratory use)	
(l)	heavy duty polypropylene (primarily for laboratory use)	

Selection

When deciding which material and method of jointing to use the following factors require consideration:

(a) type of effluent to be discharged (resistance to internal and external corrosion);

(b) durability of material;

(c) strength (e.g. impact or deformation resistance which may affect handling, storage, installation and operation);

(d) temperature limitation (consideration should be given to the location of the pipework e.g whether close to or in ducts, false ceilings etc. incorporating heating or hot water services, as well as the temperature of discharges the pipework is likely to receive);

(e) fire resistance, e.g. flammability, flame spread, stability, toxic fumes and gases, smoke density;

(f) flexibility of system (flexibility may be required for buildings that house equipment which, due to modern technology, soon becomes outdated and needs replacement, or where buildings house different clients for various purposes);

(g) weight of material and ease of installation (cutting, jointing, bending and fixing);

(h) physical properties (coefficient of linear expansion and if applicable Vicat softening point);

(j) comparative cost (material including supports and/or labour cost);

(k) space requirement (for material and supports and for the installer to make joints, inspect and test etc.);

(l) availability (delivery periods; also some industries are rationalising and reducing the variety of fittings, ensure latest trade literature is available).

INSPECTION AND TESTING OF DISCHARGE PIPES

Work should be inspected and tested during installation, care being taken that all work which is to be concealed is tested before it is finally enclosed. Final tests should be applied on completion of the installation both for soundness and performance. Normally, the air test is used for soundness, but if the water test is applied, it should be used only up to the level of the lowest sanitary appliance connected to the system, and then only in new systems.

When testing old systems, it may be necessary to limit the pressure applied because of shallow trap seals; the water test should not be used. Any defects revealed by the test should be made good and the test repeated until a satisfactory result is obtained.

Reference should be made to Local Authority and other enforcing authority requirements, particularly where pipework passes through areas where blockages and leaks cannot be detected. In general, sufficient access should be provided to enable complete systems to be tested.

Access points should be carefully sited to allow the entry of cleaning and testing equipment and consideration also be given to adjacent services. Traps and joints that are easily disconnected can be an advantage so additional access is required only under exceptional circumstances.

The discharge from urinals can give rise to heavy deposits, especially in hard water areas. Regular maintenance is therefore required and access should be provided so that all parts of the stack, branch, discharge pipe and trap can be readily cleaned. Where the vertical discharge pipe has a long connection to a manhole, access should be provided at ground floor near the foot of the stack.

In multi-storey domestic buildings, access should be provided at 3 storey intervals or less. In public and commercial buildings and more complex drainage systems, access should be provided at each floor level.

The discharge from appliances produces pressure fluctuations and the system must be designed to retain adequate water seal in all traps under excessive working conditions. Pressure effects that occur may be due to self siphonage or positive pressure. It may be necessary to use a ventilating pipe to limit the pressure fluctuations in the system within the acceptable limits.

The effect of the flow of water from an appliance into a branch discharge pipe must be considered taking account of:

(a) The design of the appliance (funnel shaped appliances increase the possibility of self siphonage).

(b) The length, slope and diameter of the branch discharge pipe.

Seal losses produced by effluent flow down the discharge stack depend on:

(a) The flow load which in turn depends on the number of appliances connected and frequency of use.

(b) The diameter of the discharge stack.

Air Test

An air test should apply a pressure equal to 3.8 mbar (38 mm) (5.0 mbar (50 mm) Scotland) water gauge and should remain constant for a period of not less than three minutes (five minutes Scotland). The water seals of all sanitary appliances which are installed should be fully charged and a test plug inserted into open ends of the pipework to be tested, each plug being sealed with a small quantity of water. One testing plug should be fitted with a tee-piece, with a cock on each branch, one branch being connected by a flexible tube to a manometer.

To apply the test, air or smoke is introduced into the system through the other branch of the tee-piece until the desired pressure is shown on the manometer scale. Alternatively, the pressure may be applied by passing a flexible tube from a tee-piece attached to a manometer through the water seal of the trap of a sanitary appliance, the test then being carried out as previously described.

Defects revealed by an air test may be located by the following:

(a) A smoke producing machine may be used which will introduce smoke under pressure into the defective pipework. Leakage can be observed as the smoke escapes.

(b) Soap solution can be applied to the pipes and joints, under test, leakage can be detected by the formation of bubbles.

Performance of Testing Systems

In addition to a test for air or water-tightness, every discharge pipe installation should be tested for stability of the trap seals on the system. When subjected to the appropriate discharge tests, every trap must retain not less than 25 mm of water seal. Each test should be repeated three times, traps being recharged before each test and the maximum loss of seal in any one test should be taken as the significant result.

The number of appliances to be discharged simultaneously for the test depends upon the number of appliances installed and the use and occupancy of the building. Suitable figures are given in Table C9 for dwellings and in Table C10 for buildings such as office blocks and schools. For other types of buildings, e.g. hotels and hospitals, the service conditions lie somewhere between these two. The number of appliances to be discharged simultaneously to simulate these conditions should be estimated from the figures given in Tables C9 and C10 and the expected service conditions.

Dwellings

To test for the effect of self siphonage, waste appliances should be filled to over flowing level and discharged in the normal way. The seal remaining in the trap should be measured when the discharge is finished.

To test for the effects of probable maximum simultaneous discharges of sanitary appliances, the number of appliances to be discharged together is given in Table C9. For the purpose of this test, baths are ignored as their use is spread over a period and they do not normally add materially to the peak flow.

Where a stack services baths only, the number to be discharged simultaneously in a test should be the same as for sinks. The worst conditions occur when appliances on the upper floors are discharged. A reasonable test therefore is to discharge up to one WC, one basin and one sink from the top floor of the building with any other appliances to be discharged on the floor immediately below.

Table C9 Number of appliances to be discharged simultaneously for testing stability of trap seals; Dwellings

Number of appliances of each kind on the stack	Number of appliances to be discharged simultaneously		
	WC 9 litres	Wash basin	Kitchen sink
1– 9	1	1	1
10–24	1	1	2
25–35	1	2	3
36–50	2	2	3
51–65	2	2	4

Commercial or Public Buildings (eg. office blocks, schools)

Waste appliances should be tested for self siphonage as for dwellings. For ranges of appliances, combinations of appliances as set out in Table C10 should be discharged, waste appliances being filled to overflow level. The worst conditions are likely to occur when the appliances at the top end of the branch are discharged. For ranges of appliances connected to a common branch in a building where very congested periods of use may be expected, e.g. schools, the figures given in columns 2 and 3 of Table C10 should be doubled.

To test the stability of trap seals when water flows down the stack, the appropriate combination of appliances from Table C10 should be discharged simultaneously with all traps fully charged. The seal remaining in the traps should be measured when the discharge has ended.

Table C10 ***Number of appliances to be discharged simultaneously for testing stability of trap seals; Commercial or Public Buildings***

Number of appliances of each kind on the stack	Number of appliances to be discharged simultaneously	
	WC 9 litres	Wash basin
1– 9	1	1
10– 18	1	2
19– 26	2	2
27– 50	2	3
51– 78	3	4
79–100	3	5

Note

The flow load from urinals, spray tap basins and showers is usually small in most mixed systems, hence these appliances need not normally be discharged.

VACUUM SYSTEM

A vacuum sanitary pipework system is based on the principle of using air, instead of water, to convey foul water away from water closets. The water from water closets is termed 'black water' and from waste appliances as 'grey water'. Vacuum flushing of WC contents requires only one litre of water and pipelines need not be laid to the gradients necessary for gravity waterborne installations. The vacuum WC suite, although similar in appearance, differs in construction from conventional types. The bowl contains no trap but has special accommodation for an airtight outlet valve and flushing apparatus. Branches from WC's, stacks and main sanitary discharge pipework need only be 40–50mm diameter.

The main stack, however, would generally be enlarged to the next diameter to ensure velocity of flow does not exceed the suggested limit of 10 m/s during peak loading. Pipe dimensions for larger installations would be calculated on the quantities of air required for removal of foul water. Branches from sinks etc. can be 32 or 35 mm with floor gully outlet of 40 mm. The special floor gully contains no water seal but incorporates an odour seal in the form of a diaphragm pressure valve. Special pressure valves, similar in principle to the floor gully, are used for connection to other appliances.

METHODS OF WASTE COLLECTION

(a) Refuse chutes provide an economic solution in multi-storey and similar type buildings. Adequate precautions must be taken to prevent entry of rodents into the bin storage area and also to minimise the entry of flies. There is much to recommend the installation of waste disposal units below each kitchen sink in dwellings.

(b) Kitchen waste disposal units are a means of disposal of wet refuse and for commercial and industrial use, it is advisable to consult specialist suppliers.

(c) Incineration of domestic refuse can be achieved by open fires but where this is not possible, the use of small or larger units should be considered.

(d) Sanitary towel disposal can be carried out quite cheaply and quickly by incineration, maceration or by collecting bins for storing and bulk disposal.

(e) Owing to its varied nature, hospital refuse should be disposed of by maceration (near as possible to the disposal source) or by incineration.

(f) Compression systems are available which reduce the volume from between 1/3rd to 1/12th.

(g) Composting refuse in agricultural areas is considered the most economic method of disposal. Refuse collection and disposal methods used should be determined in consultation with the Local Authority.

SANITARY ACCOMMODATION

(a) The planning in most instances is performed by the Architect, who should be advised of special requirements and where an alteration to the layout, within structural limitations, can affect economies and performance.

(b) Hand drying facilities can either be (i) disposable paper, (ii) roller towel, (iii) warm air electric machines.

(c) Soap solution dispensers, fed from local reservoirs provide a ready alternative to the provision of bar soap.

(d) The use of the spray mixer tap has led to an improvement in the wash hand basin. Hand rinse basins have found a ready use in general toilet accommodation. Operation other than by hand has led to removal of taps completely from basins and this can lead to a reduction of cross-infection.

Spray mixer taps produce savings in water consumption, economies in pipework and fuel costs. However, in some hard water areas the spray nozzles rapidly become blocked with lime scale and in some soft water areas, gelling of soap residues occurs in branch discharge pipes.

The additional maintenance costs may outweigh any savings made in water and fuel costs.

(e) The use of wall hung or concealed outlet type WC's makes for easier cleaning.

(f) Automatic water flow for hand washing, flushing etc. can be achieved (i) hydraulically (ii) pneumatically (iii) electrically.

Reference should also be made to BS 6465: Part 1: 1984.

SANITARY ACCOMMODATION

Scale of Provision of Sanitary Appliances

Table C11 Office buildings and shops (minimum requirements)

Appliances	Accommodation other than principals, etc.	
	For male staff	For female staff
*WCs (no urinals provided)	1 for 1 to 15 persons 2 for 16 to 30 persons 3 for 31 to 50 persons 4 for 51 to 75 persons 5 for 76 to 100 persons For over 100 persons, plus 1 for every additional 25 persons or part thereof	1 for 1 to 15 persons 2 for 16 to 30 persons 3 for 31 to 50 persons 4 for 51 to 75 persons 5 for 76 to 100 persons For over 100 persons, plus 1 for every additional 25 persons or part thereof
*WCs (urinals provided)	1 for 1 to 20 persons 2 for 21 to 45 persons 3 for 46 to 75 persons 4 for 76 to 100 persons For over 100 persons, plus 1 for every additional 25 persons or part thereof, but 1 in 4 of the additional fitments may be a urinal	
*Urinals 1 stall or 600mm of space	0 for 1 to 15 persons 1 for 16 to 30 persons 2 for 31 to 60 persons 3 for 61 to 90 persons 4 for 91 to 100 persons For over 100 persons, additional provision determined by the number of WCs (see previous item)	
§Wash basins, trough or washing fountain	1 for 1 to 15 persons 2 for 16 to 30 persons 3 for 31 to 50 persons 4 for 51 to 75 persons 5 for 76 to 100 persons For over 100 persons, plus 1 for every additional 25 persons or part thereof	1 for 1 to 15 persons 2 for 16 to 30 persons 3 for 31 to 50 persons 4 for 51 to 75 persons 5 for 76 to 100 persons For over 100 persons, plus 1 for every additional 25 persons or part thereof
Cleaners' sink	At least 1 per floor, preferably in or adjacent to a sanitary apartment	
Bins, incinerators or macerator units for sanitary dressings disposals		1 in every sanitary accommodation regularly used by females

Refer also to Para. 7.5 of BS 6465: Part 1: 1984.
**Sanitary Convenience Regulations 1964 S.I. No. 966.*
§Washing Facilities Regulations 1964 S.I. No. 965.

Notes

1. The scale of provision shown in this table should be increased by 1, and in male toilets a single bowl urinal should be provided, if the sanitary conveniences serve a staff of more than 10 persons and are also used by members of the public visiting the premises.

2. Where either:
(a) the number of people employed does not regularly exceed 5 at one time (whether or not both men and women are employed), or
(b) each of the regular employees normally works in the premises for only 2 hours or less daily, only one WC and one wash basin need be provided for use by all staff.

3. Where trough or washing fountains are provided instead of basins, 600mm length of trough, if rectangular, or 600mm of the circumference of a washing fountain or circular or oval trough counts as one wash basin for the purposes of the scale of provision.

4. Attention is drawn to the Food and Drugs Act 1955, the Food Hygiene (Markets, Stalls and Delivery Vehicles) Regulations 1966, the Food Hygiene (General) Regulations 1970 and BS 5071, Food and Drink Vending Machines.

5. Attention is also drawn to the Food Hygiene Code of Practice No. 7, Hygiene in the Operation of Coin Operated Food Vending Machines issued jointly by the Department of Health and Social Security and the Ministry of Agriculture, Fisheries and Food.

6. This table is based on the requirements of the Offices, Shops and Railways Premises Act 1963.

Table C12 Dwellings (minimum requirements)

Type of dwelling	Appliances	Dwellings suitable for up to: 2 persons	3 persons	4 persons	5 persons	6 persons and above
On one level, e.g. bungalows and flats	WC	1*	1*	1§	1§	2+
	Bath	1	1	1	1	1
	Wash basin	1	1	1	1	1
		and in addition 1 in every separate WC compartment which does not adjoin a bathroom				
	Sink and drainer	1	1	1	1	1
On two or more levels, e.g. houses or maisonettes	WC	1*	1*	1§	2+	2+
	Bath	1	1	1	1	1
	Wash basin	1	1	1	1	1
		and in addition 1 in every separate WC compartment which does not adjoin a bathroom				
	Sink and drainer	1	1	1	1	1

Refer also to BS 6465: Part 1: 1984; Para. 7.1.
**May be in the bathroom.*
§Recommended to be in a separate compartment.
+Of which, one may be in a bathroom.

Notes

1. In the private sector, a shower is equal to a bath (Public Health Act 1961, Section 33).

2. In Scotland, a wash hand basin is required in or adjacent to every WC compartment (The Building Standards (Scotland) Regulations 1981 Q7 (6)).

3. In Scotland, a shower may be provided instead of a bath in all dwellings both private and public (The Building Standards (Scotland) Regulations 1981 Q7 (1)).

4. This table is based on Circular 36/37 of the Ministry of Housing and Local Government and the Scottish Housing Handbook Bulletin No. 1, 'Metric space standards'. Attention is directed to the Building Standards (Scotland) Regulations, 1981 where these apply.

Table C13 Accommodation for elderly people (minimum requirements)

Type of accommodation	Appliances		Remarks
Self contained for 1 or 2 old people, or grouped apartments for 2 less active old people	WC	1	WC may be in bathroom
	Bath Wash basin	1 1	Bathroom within dwelling
	Sink and drainer	1	
Grouped apartments for single less active old person	WC	1	
	Wash basin	1	In WC compartment
	Sink and drainer	1	
	Bath:	not less than 1 to four flatlets	Some may be sitz baths
Common room for self contained or grouped apartments*	WC	1	Minimum number required
	Wash basin	1	In or adjoining the WC compartment
Tea pantry or kitchen*	Sink and drainer	1	Adjacent to common room
Laundry room for grouped apartment schemes*	Sink Washing machine Spin drier	1 1 1	

Refer also to BS 6465: Part 1: 1984; Para. 7.2.
**Where communal facilities are provided, these additional provisions should be made.*

Note

This table is based on Circular 82/69 of the Ministry of Housing and Local Government and Circular 84/69 of the Welsh Office, 'Housing Standards and Costs – Accommodation Specially Designed for Old People'. In Scotland the requirements are different; reference should be made to the Scottish Housing Handbook, Part 5 'Housing for the Elderly' and Part 6 'Housing for the Disabled'.

SANITARY ACCOMMODATION

Table C14 Residential homes for elderly people (minimum requirements)

Appliances	Residents' accommodation	Staff accommodation/activities[§]
Baths	1 per 15 persons*	
WCs	1 per 4 persons	At least 2 for non-residential staff
Wash basins	1 to each bedsitting room	Medical room, hairdressing, chiropodist, non-residential staff toilets, kitchen area
Sinks		General purpose room(s), kitchen areas
Cleaners' sink		Cleaners' room
Hand/bedpan sluice		Laundry area

Refer also to BS 6465: Part 1: 1984; Para. 7.3.
**Sitz baths with hand showers (not fixed overhead) and/or shower units suitable for use by residents in wheelchairs or sani-chairs, may be suitable alternatives.*
§Where residential accommodation is provided for staff, sanitary appliances should be in accordance with Table C 12

Note

Attention is drawn to the Department of Health and Society Security Local Authority Building Note No. 2 (revised 1973), 'Residential Accommodation for Elderly People', and the requirements of the National Assistance (Conduct of Homes Regulations 1962.

Table C15 Factories (minimum requirements)

Appliances	For male personnel	For female personnel
*WCs	1 per 25 persons, or part thereof. For over 100 persons, and if 'sufficient' urinals also provided, add 1 WC per 40 persons	1 per 25 persons, or part thereof
*Urinals	No specific interpretation of 'sufficient'; for guidance see Table C11	
Wash basins, troughs or washing fountains[§+]	1 per 20 persons, for clean processes 1 per 10 persons, for dirty processes 1 per 5 persons, for injurious processes	1 per 20 persons, for clean processes 1 per 10 persons, for dirty processes 1 per 5 persons, for injurious processes
Baths or showers	As required for particular trades or occupations	

See also to BS 6465: Part 1: 1984; Para. 7.6.
**Attention is drawn to the Factories Act 1961, Sanitary Accommodation Regulations, S.R. & O. 1938 No. 611, The Sanitary Accommodation (Amendment) Regulations 1974 No. 426, and other regulations applying to specific trades or occupations which may require more extensive provisions.*
§Based on H.M. Inspectors of Factories recommendations where no order specifying minimum standards is in force.

+Where troughs or washing fountains are provided instead of basins, 600 mm of length of trough, if rectangular, or 600 mm of the circumference of a washing fountain or circular or oval trough counts as one wash basin for the purposes of the scale of provision.

Note

Attention is drawn to the necessity to provide facilities for the disposal of sanitary dressings.

Table C16 Schools and higher educational establishments (minimum requirements)

Appliances	Special schools	Primary schools	Secondary schools
All 'fittings', i.e. WCs + urinals	$^1/_{10}$ of the number of pupils	Aggregate of $^1/_{10}$ of the number of pupils under 5 and $^1/_{20}$ of the number of others. Not less than 4	$^1/_{20}$ of the number of pupils. Not less than 4
	Rounded up to nearest whole even number	Rounded up to nearest whole even number	Rounded up to nearest whole even number
WCs only	Girls: all fittings	Girls: all fittings	Girls: all fittings
Urinals	Boys: not more than ⅔ fittings may be urinals	Boys: not more than ⅔ of fittings may be urinals	Boys: not more than ⅔ of fittings may be urinals
Wash basins	As primary or secondary schools	Not less than the number of fittings	1 in each washroom, at least 2 basins per 3 sanitary fittings
Showers	*	*	Sufficient for physical education
Wash basin (medical inspection)	At least 1, accessible to WC	At least 1, accessible to WC	At least 1, accessible to WC
Cleaners'/slop sink	At least one per floor	At least one per floor	At least one per floor

Nursery schools and play schools

WCs	1 per 10 pupils (not less than 4)
Wash basins	1 per WC
Sinks	1 per 40 pupils

Boarding schools§

WCs	1 per 5 boarding pupils
Wash basins	1 per 3 pupils for first 60 boarding pupils 1 per 4 pupils for next 40 boarding pupils 1 per 5 pupils for every additional 5 boarding pupils
Baths	1 per 10 boarding pupils
Showers	May be provided as alternative to not more than ¾ of the minimum number of baths

See also BS 6465: Part 1: 1984; Para. 7.7.
**Whilst not required by statute, it is suggested that sufficient showers should be provided for physical education.*
§Where sanitary accommodation for day pupils is accessible to, and suitable for the needs of, boarders, these requirements may be reduced to such an extent as may be approved in each case.

Notes

1. Some teaching and other staff may be subject to Regulations under the Ofices, Shops and Railway Premises Act, but for all members the Regulations should be used as a guide; see Table C11.

2. Attention is drawn to the necessity to provide facilities for the disposal of sanitary dressings.

3. For educational establishments not mentioned above reference should be made to the Department of Education and Science.

4. Requirements in Scotland are based upon the School Premises (General Requirements and Standards) (Scotland) Regulations 1967 and 1973.

5. Requirements in Northern Ireland are based upon the statutory requirements of the Department of Education (Northern Ireland).

6. This table is based on the Education (School Premises) Regulations, 1981 (S.I. 909).

SANITARY ACCOMMODATION

Table C17 Cinemas, concert halls, theatres and similar buildings used for public entertainment* (minimum requirements)

Appliances	For male public§	For female public§
WCs	In theatres, concert halls and similar premises: Minimum 1 for up to 250 males plus 1 for every additional 500 males or part thereof	In theatres, concert halls and similar premises: Minimum 2 for for up to 50 females, 3 for 51 to 100 females, plus 1 for every additional 40 females or part thereof
	In cinemas: Minimum 1 for up to 250 males plus 1 for every additional 500 males or part thereof	In cinemas: Minimum 2 for up to 75 females, 3 for 76 to 150 females, plus 1 for every additional 80 females or part thereof
Urinals	In theatres, concert halls and similar premises: Minimum 2 for up to 100 males plus 1 for every additional 80 males or part thereof	
	In cinemas: Minimum 2 for up to 200 males plus 1 for every additional 100 males or part thereof	
Wash basins	1 per WC and, in addition, 1 per 5 urinals or part thereof	1, plus 1 per 2 WCs
Cleaners' sinks	Adequate provision should be made for cleaning facilities, including at least one cleaners' sink	

See also BS 6465: Part 1: 1984; Para. 7.8.
**The scale of provision relates to appliances per cloakroom serving a given number of persons.*
§Where a more reliable estimate is not available it may be assumed that the total audience will consist of equal numbers of males and females.

Notes

1. *Attention is drawn to the necessity to provide facilities for the disposal of sanitary dressings.*
2. *Some staff may be subject to the Regulations made under the Office, Shops and Railway Premises Act; for other members, the Regulations should be used as a guide; see Table C11.*
3. *The sanitary accommodation provided for performers will depend upon individual circumstances, but guidance may be obtained from the Regulations made under the Offices, Shops and Railway Premises Act; see Table C11.*

Table C18 Hotels (minimum requirements)

Appliances	For residential public and staff	For public rooms	
		males	females
WCs	1 per 9 persons, omitting occupants of rooms with WCs 'en suite'	1 per 100 persons up to 400 persons. For over 400 persons, add at the rate of 1 per 250 persons or part thereof	2 per 100 persons up to 200 persons. For over 200 persons, add at the rate of 1 per 100 persons or part thereof
Urinals		1 per 50 persons	
Wash basins	1 per bedroom and at least 1 per bathroom	1 per WC	1 per WC
Bathrooms	1 per 9 persons, omitting occupants of rooms with baths 'en suite'		
Cleaners' sinks	1 per 30 bedrooms, minimum 1 floor		

See also BS 6465: Part 1: 1984; Para. 7.9.

Notes

1. *Some non-residential staff may be subject to the Regulations under the Offices, Shops and Railway Premises Act, but for all members the Regulations should be used as a guide; see Table C11.*
2. *Attention is drawn to the necessity to provide facilities for the disposal of sanitary dressings.*

Table C19 Restaurants and canteens (minimum requirements)

Appliances	For male public*	For female public*
WCs	1 per 100 persons up to 400 persons. For over 400 persons, add at the rate of 1 per 250 persons or part thereof	2 per 100 persons up to 200 persons. For over 200 persons, add at the rate of 1 per 100 persons or part thereof
Urinals	1 per 25 persons	
Wash basins	1 per WC and, in addition, 1 per 5 urinals	1 per 2 WCs

See also BS 6465: Part 1: 1984; Para. 7.10.
**It may be assumed that there will be equal numbers of males and females.*

Notes

1. Some staff may be subject to the Regulations under the Offices, Shops and Railway Premises Act, but for all members the Regulations should be used as a guide; See Table C11.

2. In canteens, the need for sanitary accommodation for persons using the canteen should be determined by the accessibility of normal workplace facilities, but at least one WC with a wash basin should be provided for each sex.

3. Attention is drawn to the necessity to provide facilities for the disposal of sanitary dressings.

4. Attention is drawn to the requirements of the Food Hygiene (General) Regulations 1970; the Food Hygiene (Scotland) Regulations 1959 to 1966.

5. This table is based on the Greater London Council, Places of Public Entertainment Technical Regulations (revised 1982). In the case of licensed buildings which are conversions, the licensing authority may permit less than these requirements, where the total number of persons does not exceed 200.

Table C20 Swimming pools (minimum requirements)

Appliances	For spectators		For bathers*	
	Males	Females	Males	Females
WCs	1 for 1 to 200 persons 2 for 201 to 500 persons 3 for 501 to 1000 persons Over 1000 persons, 3 plus 1 for every additional 500 persons or part thereof	1 for 1 to 100 persons 2 for 101 to 250 persons 3 for 251 to 500 persons Over 500 persons, 3 plus 1 for every additional 400 persons or part thereof	1 per 20 changing places	1 per 10 changing places
Urinals	1 per 50 persons		1 per 20 changing places	
Wash basins	1 per 60 persons	1 per 60 persons	1 per 15 changing places	1 per 15 changing places
Showers			1 per 8 changing places	1 per 8 changing places

See also BS 6465: Part 1: 1984; Para. 7.12.

Notes

1. Attention is drawn to the necessity to provide facilities for the disposal of sanitary dressings.

2. This table is based on Department of Education and Science, Sports Council, Technical Unit for Sport, Bulletin No. 1 – 1973 'Public Indoor Swimming Pools'.

Table C21 Places of public assembly – art galleries, libraries and museums (minimum requirements)

Fitments	For male public*	For female public*	For male staff	For female staff
WCs	1 per 100 up to 400 For over 400, add at the rate of 1 per 250 or part thereof	2 per 100 up to 200 For over 200, add at the rate of 1 per 100 or part thereof	1 for 1 to 15 persons 2 for 16 to 35 persons	1 for 1 to 12 persons 2 for 13 to 25 persons
Urinals	1 per 50 persons		Nil up to 6 persons 1 for 7 to 20 persons 2 for 21 to 45 persons	
Wash basins	1 to each WC or range of WCs		1 for 1 to 15 persons 2 for 16 to 35 persons	1 for 1 to 12 persons 2 for 13 to 25 persons
Cleaners' sinks		At least 1 per floor		

**It may be assumed that there will be equal numbers of males and females.*

SANITARY ACCOMMODATION

Table C22 New public houses (minimum requirements)

Customer accommodation	Female WCs	Males	
		WCs	Urinals
Up to 50	1	1	2
51 to 100	2	1	2
101 to 150	2	1	3
151 to 200	3	1	3
201 to 250	3	2	3
251 to 300	4	2	4
301 to 350	4	2	5
351 to 400	5	2	6
401 to 450	5	3	6
451 to 500	6	3	7
501 to 550	6	3	7
551 to 600	7	3	8
601 to 650	7	3	9
651 to 700	8	4	9
701 to 750	8	4	10
Wash basins	1 per 2 WCs	1 per WC, plus 1 per 5 urinals or part thereof	

See also BS 6465: Part 1: 1984; Para. 7.11.

Notes

1. *Accommodation calculated at the rate of 4 persons per 3m² of effective drinking area (EDA).*

2. *It may be assumed that the ratio of persons will be 75% male and 25% female.*

3. *Sanitary accommodation for employees should be related to the provisions of the Offices, Shops and Railway Premises Act 1963; see Table C11.*

Table C23 Marinas and yacht harbours (minimum requirements)

Sanitary appliance		Number of berths	
Type	Number	Male accommodation	Female accommodation
WCs	1	Per 50	Per 75
Wash basins	1	Per 40	Per 75
Urinals	1	Per 75	—
Showers	1	Per 80	Per 120
Sinks (deep)	1	Per block	Per block
Sanitary towel disposal units	1	—	Per block

Sanitary Accommodation for Disabled People

The Building Regulations 1985 Schedule 2 – Facilities for Disabled People (people with a physical impairment which limits their ability to walk or makes them dependent upon a wheelchair for mobility) requires facilities to be provided in:

(a) buildings which are office premises or shop premises within the meaning of Section 1 of the Offices, Shops and Railway Premises Act 1963;

(b) single storey buildings which are factories within the meaning of Section 175 of the Factories Act 1961;

(c) single storey buildings to be used for the purposes of a school or other educational establishment;

(d) other single storey buildings if they are buildings to which the public are admitted, whether on payment or otherwise.

In the case of any building to which schools and other educational establishments applies, the requirements shall be deemed to be satisfied by compliance with the minimum requirements specified in 'Design Note 18 (1984) – Access for Disabled People to Educational Buildings' published on behalf of the Secretary of State for Education and Science.

In any other case, the requirements as to the design of sanitary conveniences for use by disabled people shall be deemed to be satisfied by compliance with the design recommendations in clauses 6.2 to 8.4.4 (inclusive) of BS 5810: 1979.

For further information see Section I, 'Designing for the Disabled'

THE DESIGN OF DRAINAGE SYSTEMS

All drainage systems should be designed to be as simple and direct as possible. All vertical stacks should terminate with long radius bends, and except in special circumstances, all drains between manholes should be laid in straight lines, both in the horizontal and vertical planes. All changes in direction and gradient should be at an inspection chamber or manhole. The drain soffit should invariably be level, to facilitate the free ventilation of the system, thus minimising the problems associated with foul smells, gases and the creation of positive and negative pressures.

Branches from individual gullies or sanitary appliances may be square, but any branch that may run more than a proportional depth of 0.3 (30% full) should be connected into the main stream with a 45° baron bend or ¾ slipper bend, plus a long radius bend where necessary, to minimise turbulance. Except in special circumstances, all manhole branch inverts should be joined level with the horizontal centreline of the main stream drain.

All drainage systems should be designed to flow at a constant velocity. Where this cannot be achieved, the method of dealing with the problem will depend upon the capacity of the downstream receiving sewer under flood conditions.

Some sewers have spare capacity, in which case a gradual rise or fall in the system velocity is relatively unimportant. Some sewers surchage under storm conditions and where this situation occurs, the local authority may require the design and installation of 'Blucher' type flow restrictors and temporary storage/retention of flood waters.

Gradients and velocities should be kept low or reduced over the total length, and this will tend to increase the drain sizes, volume and transitional capacity/storage of the system.

If it is not possible or economic to achieve sufficient storage in the drain system, then suitable areas for surface flooding, (not for sewage) such as car parks or flood plains should be considered.

Wherever possible, steep drain gradients should be avoided. Most difficult site conditions can be overcome with backdrops, allowing the horizontal drain to flow at an acceptable velocity and the increased velocity down the backdrop can be attenuated at the bottom of the manhole. It is sometimes necessary to ramp drains and sewers in hilly districts and it is common practice to maintain the horizontal drain size down the slope and then continue horizontally at the bottom of the slope.

Under storm conditions, the bottom of the sloping drain becomes surcharged causing flood water to issue from low level gullies and manholes.

Here again, 'Blucher' type flow restrictors and temporary retention of flood water at the top of the ramp/s plus the sealing of low level manholes can be used to prevent or reduce the surface flooding of low lying areas, both for new and existing systems.

There is no practical restriction to a maximum velocity, but it is widely accepted that 3.6 metres / second should not be exceeded.

All drainage systems should be designed using an appropriate method. A large computer program called the 'Wallingford Procedure' consists of a series of six programs developed at the Hydraulics Research Station with assistance from the Institute of Hydrology, the Meteorological Office, the Transport and Road Research Laboratory and the National Physical Laboratory.

These programs are primarily intended for town and district council civil engineering projects. Information is available from the Hydraulics Research Station, Wallingford, Oxon, OX10 8BA. Briefly the Wallingford programs consist of:

(1) WASSP-HYD The 'Hydrograph Method', for sites of about 150ha and above, to size pipes and/or calculate discharge hydrographs for design rainfall events, which may include storm water overflows, bifurcations, holding tanks and pumping stations.

(2) WASSP-RAT The 'Modified Rational Lloyd-Davies Method' for sites up to about 150 ha with times of concentration up to about 30 minutes and outfall pipe diameters up to about one metre. To size pipes and calculate discharge, which may include storm overflows and bifurcations. This is the simplest of the three design methods in the package.

(3) WASSP-CHECK This 'Checks the Data' put into the program for inconsistences and errors in the connections between pipes etc.

(4) WASSP-SIM This can 'Simulate' increasing flow rates and allows both existing installations and new designs to be subjected to a gradually increasing rainfall intensity/surcharge, to establish the capacity of the system, with or without surface flooding. This program gives the engineer the ability to analyse the performance of the system and in particular, to identify the degree of surcharging in manholes and areas of surface flooding.

(5) WASSP-OPT To 'Optimise' the design of a drainage system for a minimum construction cost using the Modified Rational Method, but excluding overflows, bifurcations, storage tanks and pumping tanks. This method is meant to produce a cheaper design. The program may however require more care and intervention by the engineer to ensure overall co-ordination with other services, particularly when there are a large number of restrictions on the depth at which pipes may be laid.

(6) WASSP-COST To 'Calculate the Cost' of the resources used in constructing a sewer system, excluding the cost of ancillary structures. This cost may be used to compare different designs.

Sites smaller than approximately 150 hectare do not normally warrant the use of the full 'Wallingford', but the smaller site can be adequately designed by using (The Wallingford Modified Rational Lloyd-Davies Method) 'RAT'.

For normal estates and villages with times of concentration up to about 20 minutes and outfall pipe diameters up to 600mm, the long established 'Rational Lloyd-Davies Method' is adequate.

Drainage design programs for small sites now have the facility to simulate surcharge from increasing rainfall intensities and surface flooding.

Public and main trunk sewers are often designed to run full during periods of calculated peak loading, but it is recommended that estate drains and drains in the immediate vicinity of buildings should be designed to run part full.

The proportional depth of flow in drains should not exceed (d/D = 0.75). All estate drains should be designed with a small margin to allow for changes in site use/system design loading, illegal cross connections etc., plus adequate ventilation of the system. This requirement applies equally to surface water drains that may be polluted by petrol or chemical spillage.

The Rational Lloyd-Davies Method

The following criteria are suggested for the design and sizing of estate sewers with a computer program. The processing power and speed of these machines allows the most complex and tedious calculations, such as depth of flow, hydraulic mean depth, proportional area, proportional velocity, drain volumes etc to be obtained from the minmum data input. The over-riding benefit from such programs, is the ease with which changes and adjustments can be made to any part of a program and the computer will recalculate the whole system from start to finish, in a few seconds.

The following is offered as a Standard for all normal construction sites and buildings up to say 15 ha.

(i) The program should be able to identify each section of drain.

(ii) Calculated data for each section should be stored in the memory, using the section number as an identifier.

(iii) The program should be able to calculate to at least sub-branch level and size a minimum of 200 sections at each run.

(iv) The program should contain as many of the design parameters as possible, so that recourse to tables, charts and maps is reduced to a minimum.

(v) It should be based on established and proven principles of design.

(vi) It must be easy to add, delete or change data pertaining to any item or section, without having to re-enter all the data.

(vii) The program should calculate and display the maximum number of parameters for a minimum input of data.

(viii) The program should be easy to use by non-engineers (i.e. those who may enter the data). Generally the thinner the instruction manual the better the program!

(ix) The data required by the program should be no more than is required to size the drainage by hand.

(x) It should be possible to change "constants" such as roughness factors.

(xi) The program must cater for multi-branch manholes.

(xii) Once data has been entered into the program, it should be stored permanently for further use or modification.

(xiii) The calculations should proceed directly from the stored data and constants in the program, with no further intervention by the user.

(xiv) Each length of drain and row of data in the program should be calculated as the work proceeds, line by line.

Typical Computer Program

A typical print-out of an estate storm water sewer sizing program is shown in Figure C14. Note the 'variable constants' are shown in the header to the print-out. They are all set to a default value, but each constant can be altered to suit a particular site, as follows:

(a) Number of years between storms
(b) Run-off factor (impervious areas)
(c) Run-off factor (pervious areas)
(d) Time of entry (minutes)
(e) Max rainfall intensity (mm/hr)
(f) Drain roughness factor (mm)

Taking the print-out columns one by one:

(1) Input, the section number is shown in one of three columns,

THE DESIGN OF DRAINAGE SYSTEMS

indicating that the section is in the 'Main', 'Branch' or 'Sub-branch'.

(2) Input, the drain length must be given in metres.
(3) Input, the fall must be given in millimetres, OR
(4) The gradient must be shown as a ratio of 1 in XX.

Note:

The program only requires one entry (fall or gradient), and the machine calculates the other value or the program may have an adjustable default value using 'percentage gradients'.

(5) Output, the velocity in 'metres per second' is calculated by the Colebrook – White equation (with hydraulic mean depth of 1.00).
(6) Output, time of flow for each section in minutes.
(7) Output, time of concentration of rainfall in minutes.
(8) Output, rate of rainfall (mm/hr) as calculated by the Bilham formula (RRL 35).
(9) Input, impervious area as entered in the data file multiplied by the run-off factor. (b)
(10) Input, as for (9) but for pervious areas. (c)
(11) Output, total equivalent area in square metres.
(12) Output, the cumulative total area drained in hectares.
(13) Output, the flow rate in litres per second.
(14) Output, the required pipe diameter (mm).
(15) Output, the percentage depth of flow.
(16) Output, the hydraulic mean depth.
(17) Output, the proportional depth of flow as a percentage of the cross sectional area of the drain.
(18) Output, the proportional velocity i.e. the velocity as calculated by the Colebrook – White equation modified in relation to the hydraulic mean depth.
(19) Output, drain volume in (m^3) for each section.
(20) Output, total system volume in (m^3), including the current section.

Figure C14 Typical computer program

Roughness factor = 0.60 mm   Maximum depth of flow = 70 %   Minimum velocity = .6 m/s
Storm period = 2 yrs   Time of entry = 3 mins.   Runoff factor = 0.95   Permeability factor= 0.40

SECTION NO. M	B	S	LENGTH M	FALL mm	GRAD 1 in	VEL. M/S	TIME FLOW min.	TIME OF CONC. min.	RATE OF RAINFALL mm/Hr	CATCHMENT AREA IMPERV x fact	PERVIOUS x fact.	TOTAL	CUMULATIVE (Ha)	RATE FLOW l/s	PIPE DIAM mm	DEPTH FLOW <= %	HYDR. MEAN DEP.	PROP. AREA %	PROP. VEL. M/S	PIPE VOL. M3	SYSTEM VOL. M3	PIPE CAP. FULL
1			100.00	667	150	1.65	4.01	4.01	82.03	5700.0	0.0	5700.0	0.57	129.99	450	50	1.003	50.3	1.66	15.90	26.95	263.04
2			70.00	389	180	1.66	0.70	4.71	74.56	4275.0	0.0	4275.0	1.00	206.75	525	56	1.075	58.2	1.79	15.15	53.24	359.90
	3		160.00	1333	120	1.43	4.86	4.86	73.13	1995.0	0.0	1995.0	0.20	40.56	300	43	0.899	40.8	1.29	11.31	70.91	101.11
4			80.00	400	200	1.71	0.78	5.64	66.89	3420.0	0.0	3420.0	1.54	286.20	600	58	1.088	59.7	1.87	22.62	110.84	484.76
5			80.00	400	200	1.71	0.78	6.42	61.85	5320.0	0.0	5320.0	2.07	356.08	600	70	1.183	74.4	2.03	22.62	133.46	484.76
	6		50.00	333	150	1.65	3.50	3.50	88.73	6650.0	0.0	6650.0	0.67	164.03	450	60	1.112	62.8	1.84	7.95	146.94	263.04
	7		60.00	343	175	1.53	0.65	4.16	80.28	1045.0	0.0	1045.0	0.77	171.73	450	67	1.169	71.6	1.79	9.54	156.48	243.33
	8		******** DUMMY SECTION ********																			
	9		70.00	427	164	1.22	3.96	3.96	82.67	1805.0	0.0	1805.0	0.18	41.48	300	49	0.987	48.7	1.21	4.95	164.21	86.32
10			100.00	333	300	1.61	1.04	7.46	56.42	0.0	0.0	0.0	3.02	473.84	750	64	1.143	67.3	1.84	44.18	208.39	709.79
11			150.00	429	350	1.49	1.68	9.14	49.73	2090.0	0.0	2090.0	3.23	446.58	750	65	1.152	68.8	1.71	66.27	274.66	656.61

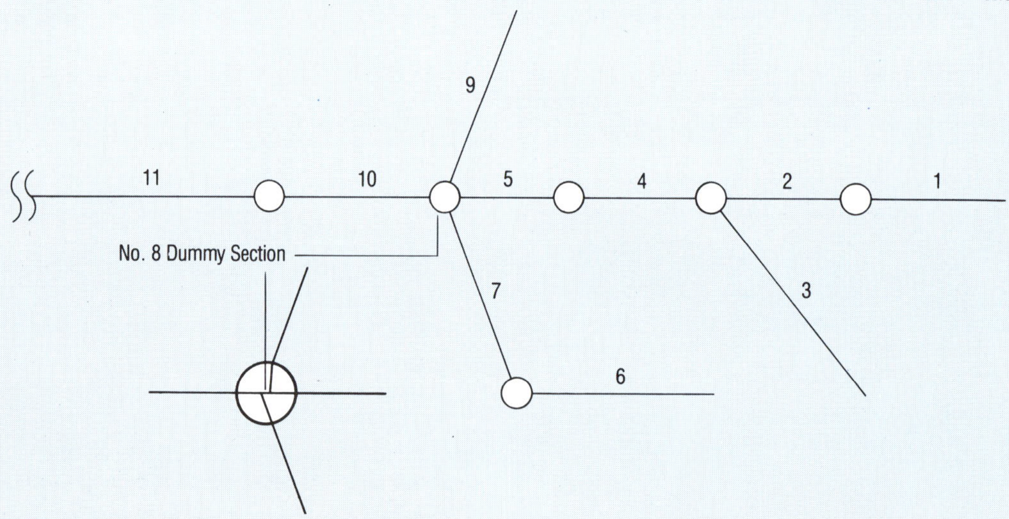

Hand Calculations

Longhand calculations for drain sizing are still an essential part of an engineers work, especially for small sites and individual buildings. Even the larger schemes that are going to be put onto computers can benefit from a few quick size checks by hand.

For those who decide to continue with hand calculations, the preceding section headed 'Typical Computer Program' may be used as a check list together with the suggested column headings on the example; items 5, 13, 14, 15, 16, 17 and 18 may be read from the Thomas Telford charts, (details shown elsewhere).

Item 7 for rainfall intensity; this may either be a blanket figure as BS 8301 for small sites with drains not exceeding 300 mm diameter, or for the larger sites, intensities as calculated by the Bilham formula (see BS 8005: Part 1, Sections 6 to 9 inclusive).

Percentage Gradients

The maximum flow rate in a drain occurs at a proportional depth of 82 to 83%. Any rise in water level above this figure will cause increased frictional resistance, a hydraulic jump and a reduced flow rate, which will tend to maintain the full surcharge condition.

The maximum depth of flow for stormwater in estate drainage generally should not exceed 75%. This will allow for changes in site development, the ingress of ground water, a margin to allow the free ventilation of the system and spare capacity for the infrequent peak storm.

Modern drain pipes, such as those of plastic, vitrified clay and spun iron are made in long straight lengths with the minimum of precision joints. All have true and smooth bores, which provide a considerable improvement by comparison to their predecessors, manufactured prior to the 1950's.

This improvement in standards and quality allow drain gradients to be reduced whilst still maintaining an adequate velocity at the calculated depth of flow.

The popular myth of a self cleansing velocity in foul drains, was arbitrarily set in the early victorian era, when the manufacturing standards and quality of materials was much lower than it is today. National standards of workmanship and supervision had not been established and much drainage work was being done for the first time, with limited and less accurate hand calculated hydraulic formula.

The discharge of waste water in branch drains is intermittent and normally starts in the form of a short wave, close to the bottom of the stack or appliance, such as from a WC. As the wave travels down the branch drain, the initial vertical discharge energy is lost and the depth of flow flattens out, this attentuation occurs because at all normal gradients, the top of the wave travels faster than the water in contact with the invert, thus levelling out the flow. Unless there is a continuous discharge of water to maintain the hydraulic depth of flow, it is not possible to achieve a constant velocity. As the numbers of discharges from branches and buildings increase, so does the diversity of use. This, together with an extended time of travel, creates the average and peak dry weather flows found in the main public sewers.

A government research publication indicates that a 100 mm foul drain with a gradient of 1 : 80 is satisfactory for flows of not less than one litre/second, (such as will commonly occur) provided at least one WC is connected; that is with a calculated velocity of 550 mm/sec.

Drainage systems should be designed to provide constant or gradually changing velocities at the design proportional depth of flow. With the exception of backdrops and ramps, all designs should avoid marked changes in velocity from one section of drain to another.

It has been found that the Rational Application of Percentage Gradients, will allow savings to be made in trench excavations, shuttering and strutting, time and cost, whilst providing almost constant velocities in main drains under design conditions. Branch drains carrying low volume intermittent flows from individual fittings and stacks should be laid to steeper gradients than main drains that carry the discharge from multiple stacks and buildings. The reasoning behind this approach is extended as follows:

(1) Main surface water drains may be laid at a gradient equal to their diameter in millimeters, e.g. 100% gradient = 300 mm dia. drain laid at a gradient of 1:300.

(2) Alternatively, main foul drains may be laid at a gradient equal to 80% of their diameter, e.g. 300 mm dia. drain laid at a gradient of 1:240.

(3) Branch drains with low flow rates may be laid at 50% of their respective main drain gradients as follows:

	Surface water drains		Foul drains	
Diameter mm	Main drain 100% gradient	Branch drain 50% gradient	Main drain 80% gradient	Branch drain 40% gradient
75	1: 75	1: 35	1: 60	1: 30
100	1:100	1: 50	1: 80	1: 40
150	1:150	1: 75	1:120	1: 60
225	1:225	1:110	1:180	1: 90
300	1:300	1:150	1:240	1:120

The above schedule is an example of what may be applied to a relatively flat site; other percentage values may be usefully applied to suit different site surface slopes.

Discharge Units

Research studies have shown that water volumes and frequency of use, vary among households in relation to factors such as household size, social and economic class and availability of appliances and fittings, i.e. in households of similar size, the frequency of toilet flushing is higher in houses with two toilets than it is in houses with only one.

The social and economic status of a household is relative to the use of water, i.e., small families of people in professional employment use more water per head per day and large families of unskilled people use the least.

It will be seen from Table C24 that there is a definite trend in the use of water across the social and economic spectrum and therefore discharge units should not be applied as a blanket figure without the application of 'engineering judgement'.

Table C24 Average water consumption for varying households

Tenure	Litres/person/day
Owner occupied	117
Private rented	102
Other rented	103
Local Authority	97
Household size	
1	126
2	131
3	117
4	106
5	101
6	90
7 +	75
Social group	
1 Professional	133
2 Intermediate	122
3 Skilled non-manual	115
4 Skilled manual	107
5 Partly skilled	105
6 Unskilled	101

Wherever possible, all foul drainage calculations should be cross checked against information from a second source, e.g. the average Dry Weather Flow for domestic foul water discharge, per head of population is 130 to 150 litres per 24 hours and this includes the ingress of ground water and a percentage of surface water cross connections. The peak DWF is normally twice the average.

In the past, the peak discharge period occured early morning, now the trend is for the peak flow to occur when the cheap electric tariff starts, when it is more economic to run clothes and dishwashing machines etc.

The Model Water Byelaws

The July 1986 Model Water Byelaws make provision for water savings in the UK particularly with WC's and urinals.

A number of other countries, such as the USA, Germany, Japan, Mexico, Africa, Asia and the Scandinavian countries have for

THE DESIGN OF DRAINAGE SYSTEMS

some years, been using 3 & 4 litre low volume designs for matched WC's and cisterns, now the UK is moving towards 6 and 7.5 litre flushes and metred water supplies, where, what is consumed is directly paid for.

The general trend in the volume of water used, both in the UK and abroad has been downwards and it is now more important not to oversize foul drains.

The new less wasteful low volume flush designs will make it possible to utilize the capacity of existing drainage systems more efficiently and the demand for new sewage treatment works will be reduced. The overall effect should make substantial savings to water supply, treatment and disposal.

Graph C2, Sanitary pipework (see page 87) indicates the approximate flow in litres/second against discharge units. The three graph lines provide for attenuation of the flow.

Line 'A' = aproximate flow from one stack as BS 5572: 1978.

Line 'B' = approximate flow in a branch drain from one stack as:

$$\text{l/s} = \left(\frac{DU}{60}\right)^{0.666}$$

Line 'C' = approximate flow in main drains from multiple stacks, branch drains and buildings, as:

$$\text{l/s} = \left(\frac{DU}{60}\right)^{0.666} \times 0.7$$

It will be seen from the graph that Line B (flow in a branch drain) gives a lower flow rate than Line A and that Line C (flow in a main drain) gives a lower flow rate than in Line B.

The three categories A,B & C for one value of discharge unit provides the attenuation of flow from the sanitary fitting to the main drain and thus minimises the oversizing of branch and main drains.

For more detailed information about discharge stacks, see Sanitary Pipework (page 78).

Gravity Flow Pipe Design Charts

Charts for the design of circular pipes have been available from various sources for many years and have been used, with varying degrees of success, to design pipes running part full under gravity. Recently, new charts (see Graphs C3 to C7) have been developed which overcome the difficulties encountered with the older style charts, making it possible to read velocity and flow rate directly from a single chart, for all depths of flow.

Theory

Each chart has been prepared for a particular combination of pipe diameter (D) and roughness (K) and is composed of two families of curves (S and Q) superimposed on a pair of orthogonal axes. The intersection of these curves represents the relationship between flow rate (Q), pipe gradient (S), proportional depth of flow (d/D) and flow velocity.

The Colebrook-White equation has been selected to represent gradient (s) curves because of its sound basis in both theory and experimentation.

$$V = -2\sqrt{(2gs\psi D)}\,\log\left[\frac{K_s}{3.7\psi D} + \frac{2.51\nu}{\psi D\sqrt{(2gs\,\psi D)}}\right]$$

where ν = kinematic viscosity and g = gravitational acceleration

In this form of the equation, D has been modified by a correction factor (ψ) to account for depths of flow less than full. This factor is only dependent on the geometry of flow, and is defined in terms of the angle (θ) subtended at the pipe centre by the free water surface.

$$\psi = \frac{\theta - \sin\theta}{\theta} \text{ where } \theta = 2\cos^{-1}\left(1 - \frac{2d}{D}\right)$$

The continuity equation forms the basis of the flow (Q–) curves which, when modified for gravity flow, can be written as:

$$V = \frac{8Q}{\psi\theta D^2}$$

Example

It is estimated that a new housing development will produce daily peak foul sewage flow of 52 l/sec. Determine the pipe size and minimum gradient required to carry this flow whilst ensuring that a velocity of 0.76 m/sec is achieved and that a proportional depth of 0.75 is not exceeded. Site conditions dictate that the maximum grade available is 1:40 and the ground surface falls at 1:240. Confirm that a satisfactory velocity is also achieved before completion of the development when the flow is 10 l/sec.

Assume K_s = 1.5 mm.

Step (1) Select a design chart by choosing:

(a) A pipe roughness (eg. K_s = 1 5 mm) relevant to this application.

(b) A pipe diameter (eg. D = 250 mm) which can carry the design flow (Q). Q must lie within the range of flows on the right hand side of the chart.

Step (2) Sketch on the limiting factors:

(a) Minimum self-cleansing velocity (V) eg. 0.76 m/sec.

(b) Maximum proportional depth of flow (d/D) eg. 0.75.

(c) Maximum available fall (eg. 1:40).

(d) Minimum available fall (eg. 1:240).

This will define the allowable gradient 'window' (shown shaded in Graph C7) within which all combinations of flow and gradient satisfy the constraints listed above.

Step (3) Locate the actual design flow Q-curve (e.g. 52 l/sec) and verify that it passes through the 'window'. If it does not, change the pipe diameter and go to Step (1).

Step (4) Read along the Q-curve and:

(a) Note the point where it leaves the 'window'.

(b) Locate the gradient S-curve which intersects it at this point (eg. 1:110). This value represents the minimum gradient at which the pipe can be laid whilst satisfying the velocity and proportional depth constraints given in Step (2).

Also read-off for this flow and grade:

(c) Proportional depth (eg. 0.75).

(d) Velocity (eg. 1.32 m/sec).

Step (5) For other flows in a pipe laid at the gradient found in Step 4(b) (eg. 1:110),

(a) Locate and read along the new Q-curve (eg. 10 l/sec) to the intersection with the chosen S-curve.

Read off;

(b) Proportional depth (eg. 0.28).

(c) Velocity (eg. 0.89).

Note

It is recommended that a value of K_s = 0.6 is used when designing stormwater drains and K_s = 1.5 is used for foul drains.

For a fuller account of the theory, more examples and a range of forty-four different two colour charts, the reader is referred to the reference below. Graphs C3 to C7 have been reproduced by permission of the authors and the publishers; Butler D & Pinkerton B.R.C. 'Gravity flow pipe design charts', Thomas Telford Limited 1987.

Graph C3 100 mm diameter pipe – gravity flow ($K_s = 0.6$)

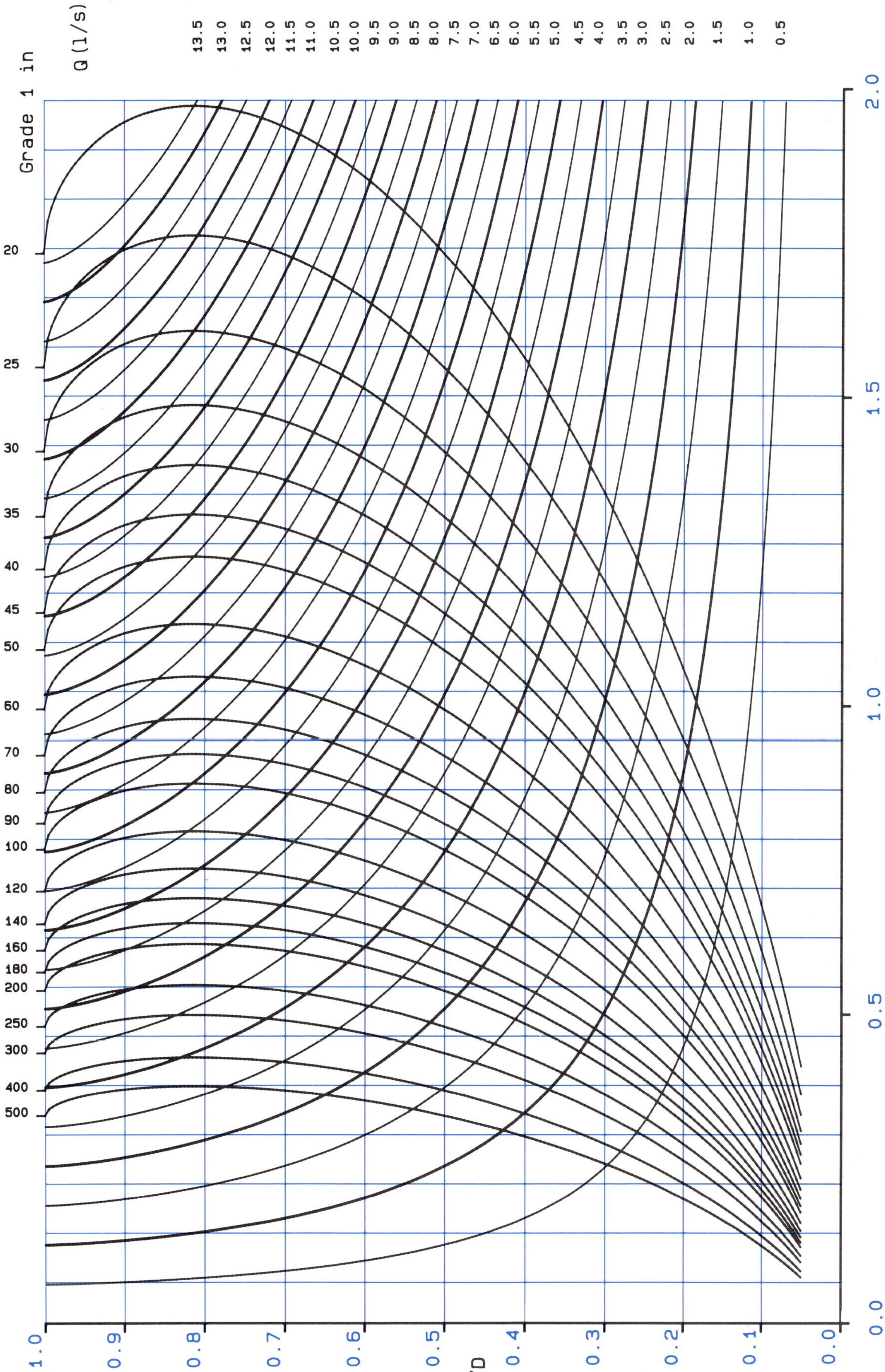

Graph C4 100 mm diameter pipe – gravity flow (K_s = 1.5)

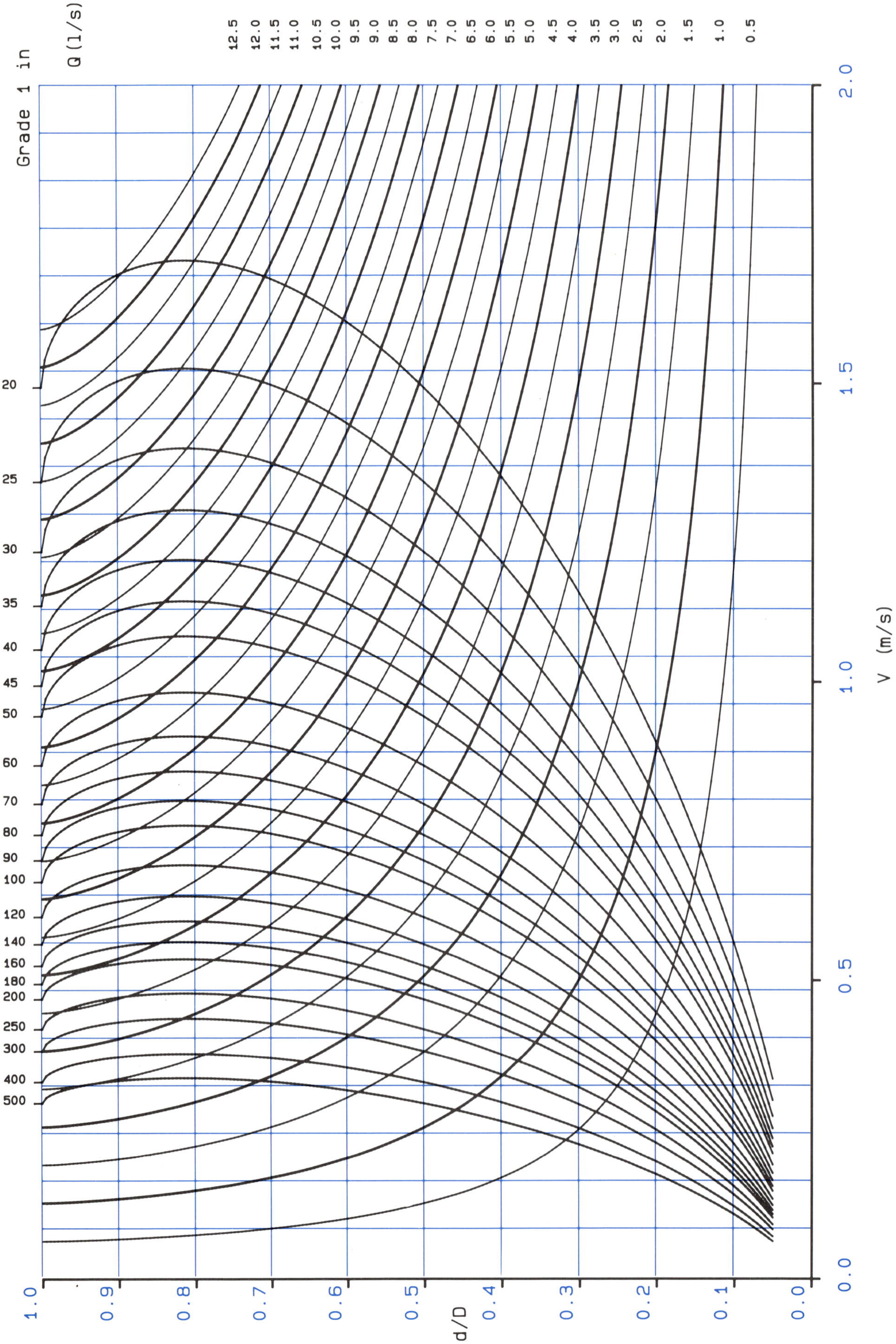

Graph C5 150 mm diameter pipe – gravity flow ($K_s = 0.6$)

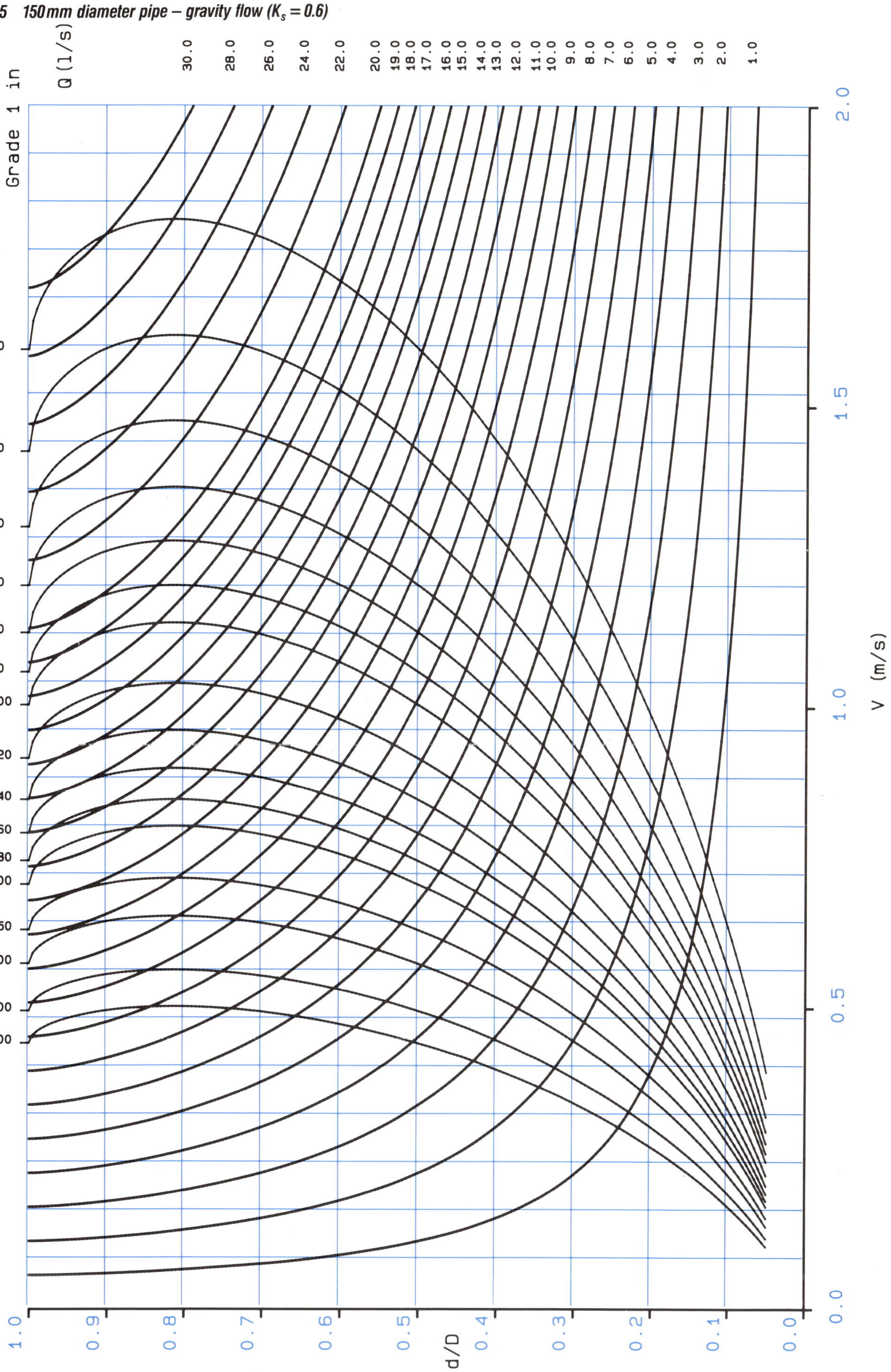

Graph C6 150 mm diameter pipe – gravity flow (K_s = 1.5)

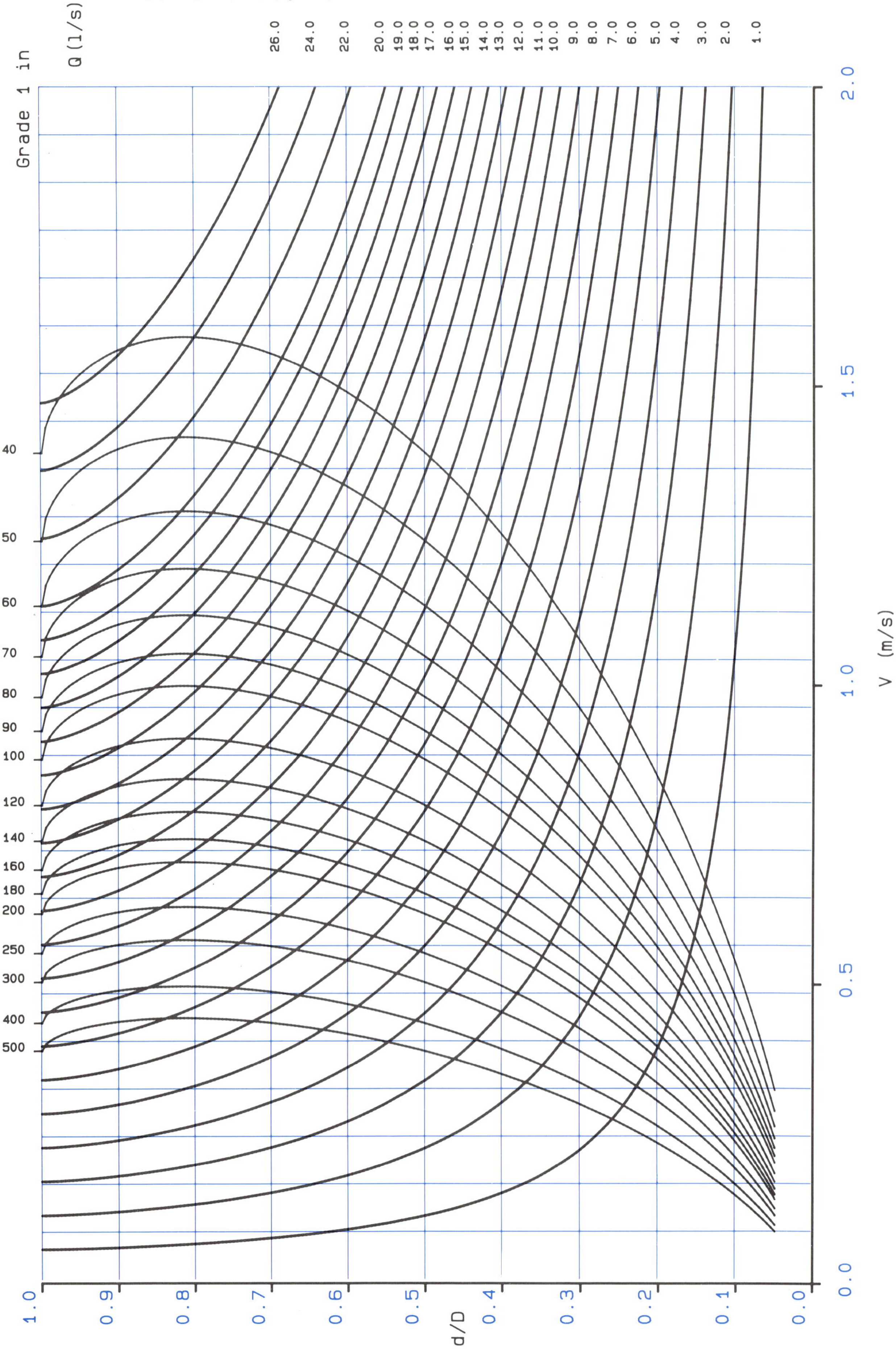

***Graph C7** 250 mm diameter pipe – gravity flow (K_s = 1.5)*

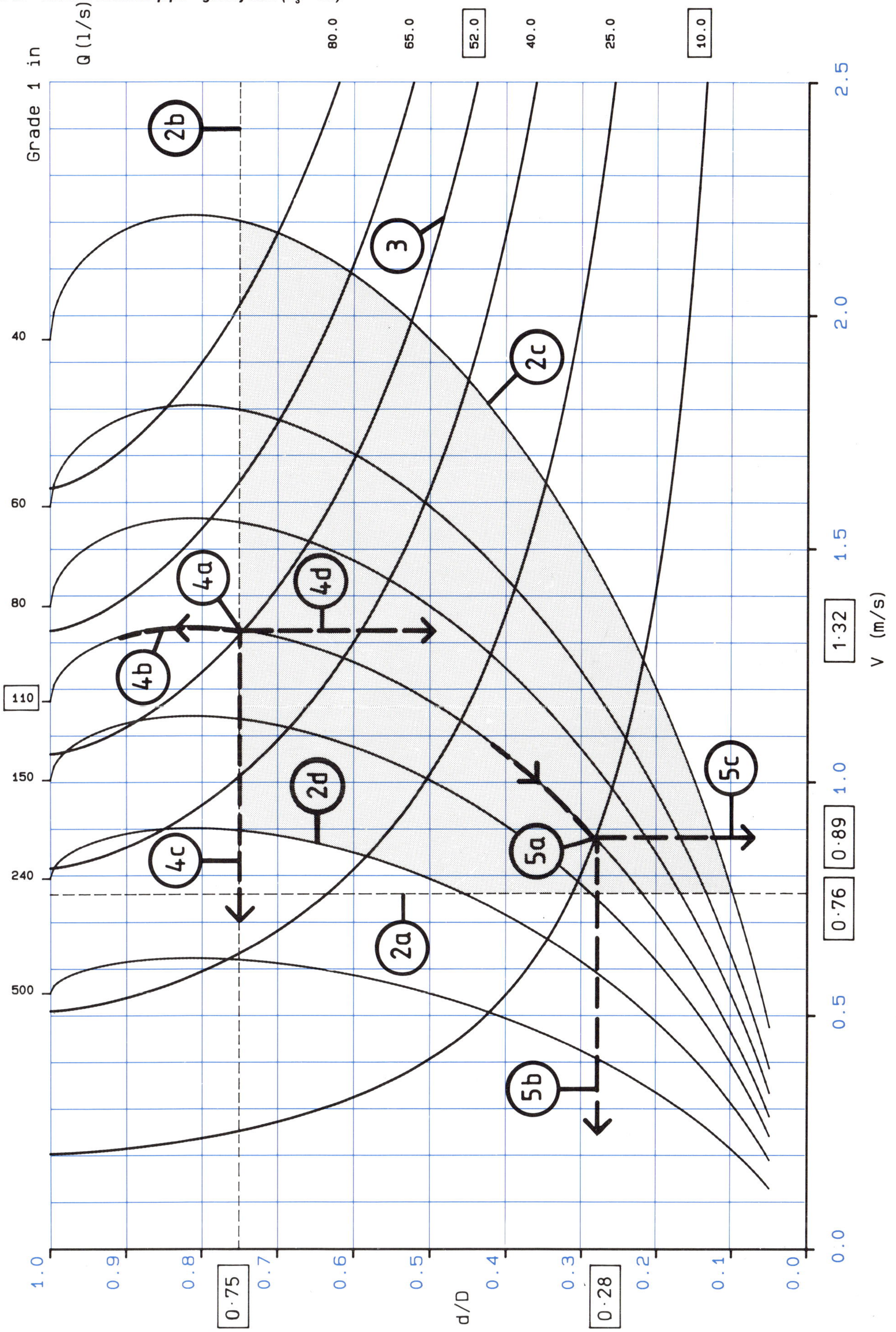

SEWAGE PUMPING SYSTEMS

Areas of flat and low land such as in Norfolk, Suffolk, old gravel pits and marshy ground that have been landfilled with refuse etc., present particular problems for development schemes that require modern drainage facilities.

Surface water can usually be dealt with by the careful use of open waterways and pumps, such as are used in the Norfolk Broads, but foul drainage problems cannot be solved quite so simply.

Traditional systems of gravity drains and their gradients quickly get down to ground water and uneconomic depths in large areas of virtually level ground, and low duty electrically powered sewage pumps are not normally considered suitable for multiple pump systems feeding into one extended common main.

Two solutions are available, the first, a proprietary system of suction drainage marketed under licence from the USA by Cowells Sewerage Systems Limited of Norwich, Norfolk and used in the UK since about 1980.

The design principle is as follows, one or a group of dwellings are provided with a traditional domestic gravity drainage system that discharges to a conveniently located collection chamber. The chamber is fabricated from glass reinforced plastic, approximately 1200 mm deep × 750 mm diameter and this acts as a collecting sump and storage chamber in the event of an emergency. The collection chamber is fitted with a sewerage suction pipe and an air operated water level sensing pipe, both connected to an interface (vacuum control) valve in a second GRP chamber similar to before but shallower.

When the waste water rises in the chamber, it causes an increase in the air pressure in the level sensing pipe when approximately 50 litres has been collected. The rising air pressure opens the interface valve, drawing the sewage into the system. The valve remains open for a few seconds after the evacuation of the waste and a volume of air is also drawn into the system and this plug of sewage and air travels along the main at 4.5 to 5.5 m/sec, as long as the differential remains.

There are reports of systems consisting of up to fifty eight interface valves on approximately 4,500 metres of polyethylene main laid in 90, 110 and 160 mm diameters. The pipework is rated at 6 bar and laid in a series of gradients (at 0.2%) and lifts (rising ramps). The main is buried between 1 and 1.5 metres below ground and valves fitted at intervals along the main to allow for maintenance and the addition of more interface valves at a later date.

The sewage is drawn through the main by differential pressure into a steel receiving vessel of approximately 9 m capacity, and a negative pressure of 0.5 bar is generated within the vessel by two vacuum pumps. All the above is contained in a small pump house which may be located in a discreet position away from public view.

When the sewage reaches a preset level in the receiving vessel, sewage discharge pumps cut in and evacuate the vessel via a traditional pump and delivery main to a public sewer or sewage treatment works. As the interface vacuum valves cycle and use up the available vacuum in the system, a vacuum pump commences operation to return the system to its negative operating pressure.

The second solution consists of sewage ejector stations driven by compressed air, such as supplied by Adams Hydraulics Limited, Peaseholm Green, York. The ejectors are manufactured in a variety of sizes ranging from 45 to 900 litres capacity and may be used in any combination of sizes to suit each situation. The system is especially suitable for modern industrial parks that include a phased construction of business units and restaurant facilities and can be designed to suit just a few users or many thousands.

Sewage ejectors have been in use for over one hundred years either as individual stations or on extended systems with multiple stations. Plastics sewage mains have been used since the early 1970s.

The design principle is that one or a group of buildings is drained by traditional gravity drains to an underground ejector station, which may take a variety of forms to suit each situation, ranging from one ejector pot in a factory built packaged ejector station to a pair of ejectors in a builders' work chamber.

Where there are two ejector pots in one station, they would normally operate as duty and standby, so that should the duty pot fail, the standby pot would automatically take over the duty. Maintenance personnel would normally change the duty pots over three or four times a year to maintain an even workload on both pots. This can be monitored by fitting mechanical trip counters to the air valve mechanisms, enabling the discharge from each station to be calculated and recorded in a maintenance/running cost log book.

As the ejector pot fills with sewage, it actuates an air valve which drives the contents of the pot and some air into the main, the period of discharge typically being 15 to 30 seconds. The remaining compressed air is ventilated to atmosphere which in turn allows the pot to refill.

Systems consisting of up to thirty ejector stations have been recorded and approximately 3000 metre of polypropylene main laid in 100, 150 and 200 mm diameters.

The pipework is rated at 6 bar and laid flat in the ground below the frost line. Valves should be fitted at intervals along the main to allow for maintenance and the addition of more ejector stations at a later date. The main should rise to a point where it can discharge into a sewer or treatment works.

The compressed air main follows the route of the sewage main around the site in the same trench and provides the ejector stations with typically 1 to 2 bar of working pressure. The compressors may be accommodated in a suitable remote housing and contain an air receiver and air compressors with after coolers if the air is to be distributed in plastic mains, plus duty and standby pressure reducing valves to provide the minimum working pressure for the system.

Most plastics are sensitive to temperature and pressure and if hot water from commercial kitchens etc is to be discharged in any volume, then design changes should be made to reduce the temperature, reduce the working pressure and/or change or upgrade the pipe strength and material.

Both systems have ejector pots or interface valves that operate either on demand, in isolation or together in a random number. Positive and negative air pressures are such a variable source of energy that they can be dormant over prolonged periods, waiting for the next discharge cycle to occur. At random moments of peak loading, the system will work at a continuously variable reduced state until it returns to normal.

Ventilating of Pumping Chambers

Care must be taken in the design of pumping chambers etc. to ensure adequate ventilation in order to prevent the hazards of the following gases:

Methane

Methane is a high explosive gas which can be produced in both clean water and foul drainage systems from the anaerobic digestion of organic matter. It can enter and accumulate in drains and chambers with potentially disastrous consequences. Methane is soluble in water, and increasingly so above ambient pressure. It has a specific gravity of 0.56 and therefore is lighter than air, hence the recommendation that all drainage systems should be ventilated. The flammable limits of methane are 5% to 15% volume in air and between these limits an explosion can occur.

Poisonous Gases

As well as the risk of explosive gases, drains have the additional risk of the presence of poisonous gases. Industrial effluents and their admixture can cause the creation of poisonous or explosive gases, or there could be high levels of carbon dioxide (e.g. 1.53 heavy) which could displace the oxygen in the atmosphere.

Oxygen is normally present in air at approximately 21% by volume. If this level is reduced to less than 16%, then this can itself constitute a dangerous atmosphere.

Hydrogen Sulphide

In domestic sewage, the most common poisonous gas is hydrogen sulphide. It is produced in acid conditions from organic and inorganic compounds containing sulphur and can be recognised by its smell of 'rotten eggs'. In a humid atmosphere, it is corrosive and high temperatures provide ideal conditions for its manufacture. The specific gravity of hydrogen sulphide is 1.19 and concentrations of 700 ppm can cause a loss of conciousness and death in only a few minutes. A person can smell the gas when it is present in small amounts, down to 0.2 ppm and the maximum level of safe, long term (8 hrs) exposure to the gas is thought to be 10 ppm.

Drainage Definitions

Proportional Area

The area of flow expressed as a percentage of the cross sectional area of the pipe.

Proportional Hydraulic Mean Depth

The cross sectional area of flow divided by the wetted perimeter of the pipe.

Note that this figure enables an allowance to be made for the increased resistance to flow in drains running partly full. Drains running full bore have a proportional HMD of 1; drains running half full also have a proportional HMD of 1.

If proportional area is used and the wetted perimeter is expressed as a percentage of the total, then:

$$\text{Proportional HMD} = \frac{\text{Proportional area}}{\text{Proportional wetted perimeter}}$$

If absolute measurements are used, e.g. metres squared, then in order to arrive at unity for full-bore flow, a correction factor must be used:

$$\text{Proportional HMD} = \frac{\text{Area of flow} \times 4}{\text{Wetted perimeter} \times D}$$

where: Area of flow is in square metres
Wetted perimeter is in metres
D = Diameter of pipe in metres

Multiplying the HMD as calculated (by either method), by D/4 where it is in metres, will give the actual HMD.

Proportional Velocity

The variation of the actual velocity of flow in a drain from the calculated full-bore velocity. An approximation of the proportional velocity may be calculated by multiplying the calculated full bore velocity by proportional $HMD^{0.666}$

Proportional Flow

See proportional area.

Proportional Depth

The depth of flow in a drain expressed as a percentage of the diameter.

Hydraulic Gradient

The fall in a drain per unit length. (More accurately, the difference in height of water levels per unit length).

When sizing foul drainage, the calculation of flow (in litres/sec) from the usage probabilty and the estimated flow rate is generally performed by using a graph. However, in a computer-aided design program, this is not convenient, and it is better for the program to calculate the flow rate directly from a formula.

Poisson's distribution is one such formula:

$$P(y) = \frac{e^{-np}(np)^y}{y!}$$

where:

$e = 2.71828$
n = total number of appliances
p = sum of the probabilities/number of appliances
y = an integer variable
$y!$ = y factorial

The formula is used in an iterative manner, i.e. a repetitive loop. The limit of probability, P(y) is less than 0.99, then y is incremented by one and the calculation repeated, adding the previous value of (y) to the new answer. This process is repeated until (y) is greater than 0.99. The actual (probable) flow rate, in litres/sec, is obtained by multiplying (y−1) by the total flow rate for all appliances, divided by the total number of appliances.

If the total number of appliances is greater than 500, then this figure is divided by 5, and the final flow rate calculated by:

$$\frac{(y-1) \times (3.2 + (y-1/50)) \times \text{total flow rate for all appliances}}{\text{total number of appliances}}$$

Experience has shown that this gives a close match to published data.

Drainage Formula

Where:

A = cross sectional area in m^2
C = Chezy coefficient
D = pipe diameter in metres
f = friction coefficient
g = acceleration due to gravity
HMD = hydraulic mean depth
n = Manning friction coefficient
Q = volume of flow in m^3
s = gradient (slope)
ν = kinematic viscosity
V = velocity in m/sec

(1) Colebrook-White Equation

This equation, developed about 50 years ago, is the preferred method of calculating the velocity of flow in pipes running full. The equation is the most reliable for modelling pipe flow, and can cater for a wide range of pipe diameters and roughness.

$$V = -2\sqrt{(2gs\,D)}\,\log\left[\frac{K}{3.7\,D} + \frac{2.51\nu}{D\sqrt{(2gs\,D)}}\right]$$

(2) Darcy-Weisback Formula

$$V = \sqrt{\frac{8.g.HMD.s}{f}}$$

The friction coefficient is dimensionless, and varies with the diameter of the pipe, the flow velocity and the kinematic viscosity.

(3) Chezy Formula

This formula is generally written as:

$$V = C\sqrt{HMD \times s}$$

The value of the dimensionless coefficient C may be calculated by:

$C = K.HMD^{-0.1666}$ where K = a coefficient related to roughness.

A general value for K is 84, and if the pipes are sized as running full, then the Chezy coefficient C is equal to 84.

SEWAGE PUMPING SYSTEMS

(4) Manning's Formula

This is an exponential form of the basic Chezy formula where:

$$V = \frac{HMD^{0.666} \times s^{0.5}}{n}$$

The friction coefficient 'n' varies with material and diameter. Generally, the larger the diameter the rougher the bore, the greater is 'n':

Diameter	n
80–120mm	0.014
121–175mm	0.015
175+	0.016

Smooth bore pipes have a lower value which rises with the roughness of the pipe:

Type of pipe	n
New smooth drain	0.010
Cast iron pipes	0.011
Old sewer	0.012

The figures for roughness should be adjusted with the values for diameter.

(5) Crimp & Bruges Formula

This is similar to the Manning formula, but with a constant value for 'n' (0.012). The formula may be rewritten:

$$V = 84 \times HMD^{0.666} \times s^{0.5}$$

(6) Continuity Equation

$$Q = V \times A$$

(7) By combining the Continuity Equation and the Manning Formula, the following is obtained:

$$Q = \frac{A \times HMD^{0.666} \times s^{0.5}}{n}$$

Drainage Charts for Obtaining the HMD

The proportional area of flow should first be calculated by dividing the calculated flow rate by the full-bore flow rate, and multiplying by 100. Look up the proportional area in Table C25 and read off the proportional depth and the Hydraulic Mean Depth.

Designing Pipework Systems to Resist Seismic Disturbances

Seismic disturbances occur regularly in many countries around the world, but fortunately not very often in the UK. However, small scale disturbances have been recorded in this country and under certain circumstances precautionary measures are necesary to resist failure of the pipework systems.

Table C25. Proportional areas and depths at HMDs

Proportional area %	Proportional depth flow	Proportional HMD	Proportional area %	Proportional depth flow	Proportional HMD	Proportional area %	Proportional depth flow	Proportional HMD
0.003	0.001	0.002	22.146	0.273	0.633	83.018	0.774	1.213
0.021	0.002	0.007	24.737	0.296	0.676	85.173	0.794	1.216
0.071	0.006	0.015	27.447	0.319	0.719	87.178	0.814	1.217
0.168	0.010	0.026	30.264	0.342	0.761	89.026	0.833	1.216
0.327	0.016	0.041	33.176	0.366	0.802	90.717	0.851	1.213
0.563	0.022	0.059	36.170	0.390	0.842	92.249	0.869	1.208
0.888	0.030	0.080	39.232	0.415	0.880	93.622	0.885	1.201
1.315	0.039	0.103	42.347	0.440	0.918	94.839	0.901	1.192
1.857	0.050	0.130	45.500	0.465	0.953	95.904	0.915	1.182
2.523	0.061	0.159	48.676	0.490	0.987	96.821	0.928	1.170
3.323	0.074	0.190	51.859	0.515	1.018	97.598	0.941	1.157
4.265	0.087	0.223	55.032	0.540	1.048	98.243	0.952	1.143
5.355	0.102	0.259	58.180	0.564	1.075	98.764	0.962	1.128
6.598	0.118	0.296	61.287	0.589	1.100	99.174	0.971	1.113
7.998	0.134	0.335	64.339	0.614	1.123	99.483	0.979	1.097
9.556	0.152	0.375	67.320	0.638	1.143	99.704	0.985	1.080
11.274	0.170	0.417	70.217	0.662	1.161	99.852	0.991	1.063
13.149	0.189	0.459	73.017	0.685	1.176	99.940	0.995	1.047
15.179	0.209	0.502	75.707	0.708	1.189	99.984	0.998	1.030
17.359	0.230	0.545	78.277	0.731	1.200	99.998	1.000	1.013
19.684	0.251	0.589	80.717	0.752	1.208	100.00	1.000	1.000

The effect of such a disturbance on normal domestic plumbing is likely to be of negligible consequence and at worst, assuming the building is still sound, can be put right relatively easily.

There are however, certain industries where the effluent is either toxic or highly dangerous and any discharge to atmosphere could have disastrous results. Substances containing radioactive materials in the discharge from hospitals, research laboratories, universities and the nuclear industries require special precautions to ensure the integrity of the discharge pipework system. The conveyance of bacteria contaminated effluents, e.g. from research buildings, laboratories etc, before sterilization, also comes into this category and the discharge pipework requires similar protection.

Although this section refers to seismic shocks, other closely related disturbances can occur from subsidence in areas where mine workings are in operation or have been abandoned and consideration should be given to ensure as much flexibility as possible in all pipework systems, both above and below ground level.

Precautionary Considerations

Where it is known that the effluent is not toxic or radioactive, sufficient precautions may be assumed to have been taken by the use of flexible joints on pipework and carefully engineered supports to allow movement between the ground and/or the building fabric and the pipework system.

If the effluent is known to be toxic and/or radioactive, then the precautions in the pipework become more stringent. Flexibility is still required in the pipework system, but will have to be provided without the use of the normal commercial 'O' ring joint. In fact, the whole of the discharge pipework would be better without any joints and providing that it is compatible with the chemical condition of the effluent to be discharged, all-welded stainless steel tube is often the best solution. This material is also ideal where steam sterilization of the pipework system is necessary before allowing access to the system. Borosilicate glass has been used for low level radioactive discharge pipework providing adequate secondary containment is provided.

System Integrity

The degree of toxicity and/or radio activity will determine the amount and type of secondary containment necessary and this may vary from simple co-axial pipework (see Figure C15), to more complex thick concrete secondary containment ducts (see Figure C16), the wall thickness being compatible with the possible levels of radiation. Certain systems may require the use of co-axial pipework and concrete shielding. Any secondary containment and/or shielding must also be designed to preserve flexibility using slip joints for ducting, but joined with flexible sealant to preserve the integrity of containment.

Pipework Suspension

All suspension systems must be designed to allow for differential movements. Co-axial pipework must provide for both the inner and outer pipe to move within specific limits totally independent of each other along their axis and the outer pipe shall be of sufficient diameter to allow some transverse movement but without the system becoming unstable.

Pipework supported from the structure or contained within metal or concrete ducts will probably require to be supported using spring loaded hangers (see Figures C17 & C18), tensioned to allow the main supporting member, be it duct or part of the fabric, to move within limited tolerances and not impose any undue stresses on the pipework system.

It is not possible to define here all the requirements of a design to resist seismic disturbances where toxic and/or radioactive substances are conveyed by pipeline. The nuclear and chemical industries normally lay down guide lines and methods of protection required for each type of effluent to be conveyed.

The purpose of this section therefore is not to be specific but to make the designer aware of the necessity of maintaining the integrity of the discharge pipework system under conditions which may damage the system and allow toxic and/or radioactive substances to be leaked to atmosphere.

Figure C15 Co-axial piping

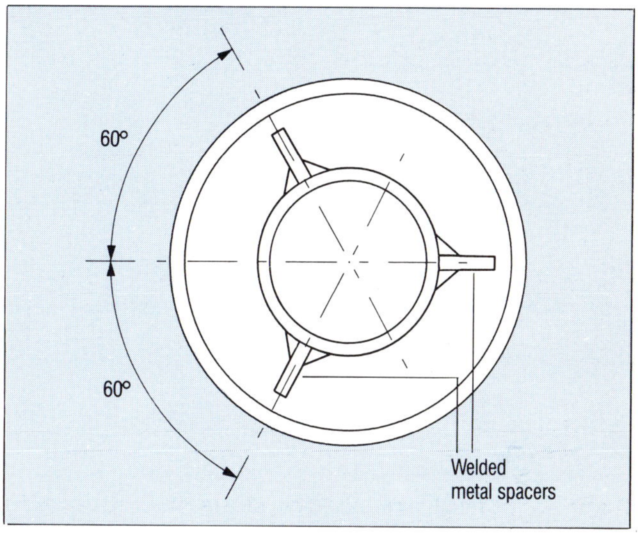

Figure C16 Concrete duct (radioactive protection)

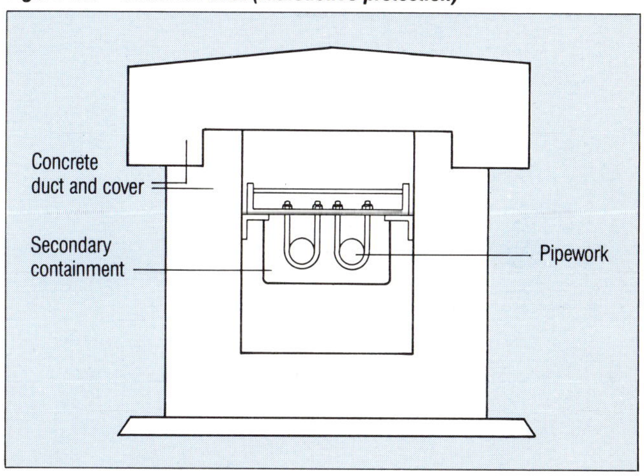

Figure C17 Spring hammer

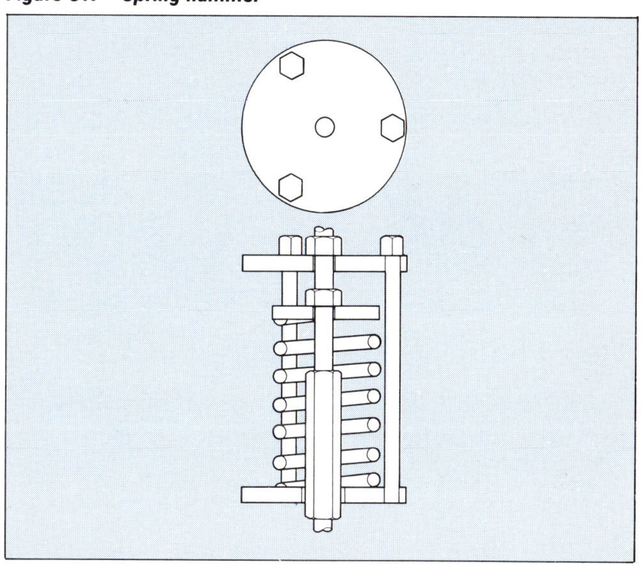

SEWAGE PUMPING SYSTEMS

Figure C18 Sway brace

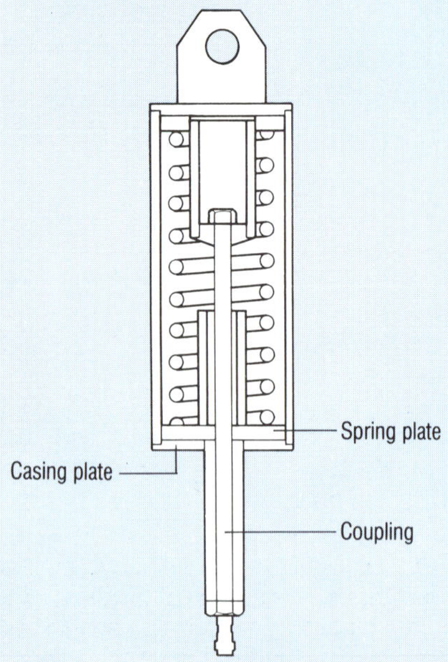

BELOW GROUND DRAIN RENOVATION

Introduction

Before replacing a faulty drain, the possibility of renovation should be considered. The benefits of the latter over the former can be assessed both in terms of construction cost and disruption.

The advantages of renovation over renewal are much greater for drains laid under buildings or hardstanding areas or for drains that require deep excavations. However, with shallow drains in soft ground or paving where excavations will not cause undue disruption, then the advantage of renovation will be greatly reduced.

Careful consideration should be given to the type and extent of renovation, bearing in mind that the life expectancy of most types of renovation are much less than that of new drains.

Points to examine when considering renovation or renewal include:

(1) What is the condition of the drains and could they be successfully renovated?

(2) What is the use of the building the drain serves and would renewal cause undue disruption?

(3) What is the life expectancy of the building the drain serves?

(4) What is the position and route of the existing drain?

(5) Is the existing drain correctly sized?

(6) Are there any particular constraints regarding access or costs?

In order to ascertain the condition of an existing drain a detailed physical and internal closed circuit television survey (CCTV) should be carried out to confirm the following:

(1) Size, material falls and route of drain.

(2) Size, construction and condition of connecting manholes and chambers.

(3) Flow characteristics and condition of drains and component fittings.

(4) Load on drain from foul and surface water connections.

(5) Types of effluent being discharged into the drainage system.

With this information the most suitable method of renewal or renovation can be assessed and a detailed specification produced.

Drainage Renovation

Constant maintenance of drains and sewers has always been necessary. Modern technology has opened up new opportunities in this field, offering a wide variety of techniques and materials for renovation work.

Sophisticated equipment has been adapted to investigate the internal condition of pipes, together with machinery and chemicals designed and produced to help rectify some of the problems that exist.

This section describes in general terms, the renovation of drains and small sewers and the factors to be considered before deciding whether a drain or sewer is suitable for renovation and which of the methods available would be best suited for the particular job.

Contracts in different areas may be treated differently due to financial constraints and a cautious approach by local authorities and engineers or because of the conditions of the ground or buildings.

Renovation or Renewal?

Opinions whether to renovate or renew drains are divided. Some authorities insist that if a drain no longer holds a reasonable air test it should be renewed. Others may not be so concerned if a drain leaks or infiltrates until damage occurs to adjacent property.

Infiltration of ground water into foul drains and sewers greatly increases the work of the pumps at the sewage works. There may be as many disadvantages as there are advantages to renovation but the main influences may be the saving of time, money and no excavation.

Even the uncertainty that a new drain would be installed correctly may influence final judgement. For example, when a building or housing estate is to be converted or modernised it may be justified to install new underground gas and water services when the age of these service pipes are approaching 20 or 30 years from the end of their life expectancy.

If there are financial considerations, renewing the drains at the same time may not be considered possible. An alternative solution would be to renovate existing drains to a good standard, using one of the methods described.

Other items for consideration include:

Item	Remarks
Clayware pipes	After correct renovation, could last indefinitely.
Cast iron pipes	May need descaling and inspection chambers replaced.
Concrete pipes	Check for serious deterioration.
Brickwork	Missing bricks, re-pointing, new invert, insert new liner.
Age of pipework	Not as important as the type of materials used.

Item	Remarks
General condition	This is more important than the one or two bad areas which would probably be dug up and relayed.
Areas with severe damage	Excavate and renew.
Drain oversized	Could be relined or new invert channel fitted.
Undersized	For small diameter drains consider using a pipe burster to pull through larger diameter pipe.
Depth of pipe	Underpinning foundations could be required if pipe was renewed.
Pipes under buildings	Considerable saving in time and money if able to renovate.
Insufficient access for maintenance	Install extra rodding eyes and manholes to comply with BS 8301
Capacity for future development	May not be valid.
Allowance for movement	Check video for circumferential cracks.
Displaced joints with flow	Normally acceptable if not severe.
Displaced joints against flow	Could be a problem if severe.
Longitudinal fractures	Pipe has lost much of its through pipe strength, possibly by pressure from above.
History of blockages	Reason for blockages should feature in video survey, do not discount serious misuse and excessive build up of fats.
Slack gradient	May not be a problem.
Back fall	Excavate and relay to gradient as required.
Lateral branches on run between manholes	Could make some systems of renovation difficult
Life expectancy of buildings	The renovated drains should outlive the building by a good margin.
General improvement to manholes	Renew step irons, increase size of manhole and cover, improve benching, all to comply with BS 8301.

The standard of a completely renovated drainage system must comply with the National Building Regulations 1985 stipulating that drains are to work efficiently without nuisance or hazard to the public or surrounding property, discharge its contents and withstand an appropriate air or water test as laid down in BS 8301.

Cast Iron Drainage

Even after 40 years, cast iron drains may be as sound structually as the day they were installed but they are subject to a build-up of scale where thickness and composition varies from one situation to another. The consistency of the scale can produce a hard rough surface which in time could reduce the bore sufficiently to cause regular blockages. This problem may be solved by cutting away the scale by high pressure water from a specially designed descaling jetting head (see jetting).

Advice must be obtained from jetting manufacturers before using such techniques.

Cast Iron Bolted down Inspection Chambers

The cast iron chambers manufactured with spigot and socket collars for caulked lead joints were manufactured with level inverts between the branches and the main channel. In some instances, this has resulted in cross flow of effluent from one branch to another resulting in blockages.

Where possible, these should be removed and replaced with the current types manufactured with branch inverts 65 mm above the main channel; alternatively change to vitrified clay or uPVC channel within the manhole area. Check if a double sealed replacement manhole cover would be required.

Many maintenance engineers are not in favour of bolted down chamber covers due to a possibility of unbolting the cover while the drain is surcharged. However, a manhole with an open channel allows the drain to be rodded from above which has certain advantages.

Position of Access Doors at the Base of Soil Stacks

During the early stages in the erection of a building, the access doors at the base of the soil stack are installed. Often these access doors have been placed at the foot of the stack connecting directly onto the bottom bend. By the time the building has been completed the possibility of finding a removable panel to the soil stack has diminished.

Once the panel is removed to expose the access door it is likely to be either too close to the ground or obstructed by other services.

Consideration should be given to installing a new access door at a practical height of about 1 m above F.F.L.

High Pressure Jetting

Jetting machines are available in a variety of sizes each designed to provide X litres per second at a pressure of Y bar.

A unit designed to discharge a high volume of water at a moderate pressure is required for large diameter pipes whilst for small diameter pipes, a unit to discharge a low quantity of water at high pressure would be used.

There is a wide range of jetting heads, one which is heavy and flat, known as 'the plough' is designed to lie flat in the bottom of a pipe for clearing silt with its rear facing jets. Most jet heads are designed for clearing blockages and loose matter, but to descale a pipe or remove roots, a high pressure pencil cutting jet is required at right angles to the pipe.

The Combi-Jet manufactured by Harbens Systems Limited is one such jet designed for this purpose. When the jetting unit is pumping water at half pressure, this jetting head will propel itself through the drain by its rear facing jets. Turning the unit up to full pressure will activate a special valve fitted to the head which will close the rear jets in favour of two rotary cutting jets set at 90°.

These pencil cutting jets will cut through scale and roots etc, the debris is then pulled back through the drain by returning to half pressure.

Scale on salt glazed pipes is normally easier to remove than on cast iron.

BELOW GROUND DRAIN RENOVATION

Manufacturers of jetting machines in this country include:

Harbens Systems Limited Watt Road Churchfields Salisbury SP2 7UD	Neolith Pumps Limited Peel Mill Market Street Whitworth Nr Rochdale Lancs OL12 8HN

Imported machines are available from:

Wards Flexible Rod Company Limited Finchley Avenue Mildenhall Suffolk IP28 7BQ	Ridge Tool Royston Road Baldock Herts SGT 6NR

Small portable electric and petrol driven jetters suitable for basic drain cleaning are available from a number of trade outlets.

Closed Circuit TV

CCTV surveys are no longer regarded as a luxury, but an essential surveying instrument to record the internal condition of pipes. Most local authorities will not adopt a sewer unless they have a video record of its condition from an independent CCTV surveying firm.

Colour cameras have now replaced black and white, and the modern camera is no longer fitted with a tube but has been replaced with a CCD chip similar in size to a postage stamp. This has resulted in a new generation of smaller cameras reliable and robust, able to negotiate around a 90° bend. Individual house branch connections from inspection chambers can now be inspected.

Fibre optics may be of valuable use in other commercial aspects but due to the cost and delicate nature, they have so far been unsuitable for the rough treatment they might receive with drainage works.

METHODS OF PIPE RENOVATION

Cement Grouting with Pistons

This method has been used for many years, and may still be considered if a rigid system of drainage construction is acceptable.

The principle is rather primitive, two rubber pistons separated by a filling of cement grout are pulled through the drain run from one manhole to the other by the cables attached to each one. The pressure of one piston being pulled against the other squeezes the cement grout into the cracks and gaps, this procedure is repeated as necessary until the drain holds up to the required water test.

Branches to soil stacks and gullies cannot be effectively treated.

Heavy imposed loads or further ground movements may result in re-opening the cement grouted joints.

Liquid Resin Grouting

Perhaps the simplest form of sealing cracks but very effective. The resin arrives on site in a powder form, pre-blended with a catalyst in the factory and requiring only water to produce the low viscosity grout solution.

The material has been used for many years for the stabilizing of soil in deep excavations below water table level, and in tunnelling projects.

The site procedure is to plug-off the drain at the lower end and then fill from the upper end to a sufficient height to force the solution into every crack or porous joint along the run. While the resin remains in a liquid state, it will penetrate through the crack in the pipe and into the surrounding concrete or soil, thereby grouting not only the drain but also the surrounding material adjacent to the area of failure.

Should there be further ground movement the elastic rubbery resin would allow slight movement of the drain without breaking its watertight seal.

Because the resin grout strengthens and stabilizes the surrounding material adjacent to the damage pipe, even 'open'· cracks can be made watertight.

This material used in its liquid form is able to treat bends and branches which would not normally be accessible by other means.

One company operating a liquid resin grouting service is Rice-Hayward Limited, Huddersfield.

Figure C19 Cement and resin grouting

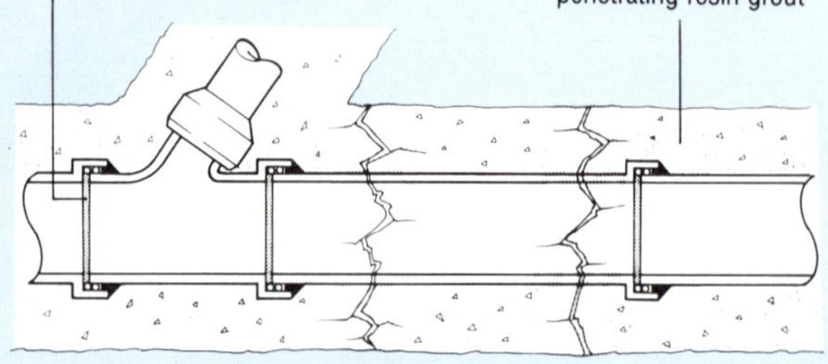

Two-part Chemical Grout

The principle of 'Superaqua' is to fill a cleaned and plugged-off length of sewer with sodium silicate solution and allow it to seep out through the leaks. The solution is quickly pumped back into the tanker, then diluted hydrochloric acid is discharged into the same sewer and allowed to stand.

A chemical reaction takes place between the two in the ground around the pipe, effectively grouting up the holes. Once the acid has been pumped back into its tanker, the sewer is flushed with water and should be leak free.

In severe cases the procedure is repeated until the sewer is sound.

Special Grouting to Individual Joints or Cracks

Unlike the two previous forms of grouting where whole areas were treated, this method restricts the grout to one area at a time.

Joints and cracks may be grouted individually by the positioning of two special plugs either side of the defected area, a special grout is then pumped between the two pistons to seal the cracks. After completion the pistons are moved on to the next defective area and the process repeated, along the line.

This process or similar is used by a number of firms including Sykes, AMK and D.J.Ryan & Sons Limited.

General Comment

The use of liquid resin grouts or liquid chemicals to seal pipes is a simple process. Lateral branches are treated at the same time as the main run unless purposely plugged-off.

Many other methods of lining the main drain leave the problem of having to deal with the lateral connection by either cutting a hole through the new liner by a remote control cutter, or digging down to excavate the connection.

Drain Liners

Drain liners may be divided into two groups:

(1) Insitu liners
(renovating an existing pipe with a new liner)
(2) Preformed pipes and units
(replacement of an existing pipe)

Insitu Liners

(1) Continuous Insitu Liners

For some years, Insitu Permaline have been installing a new hard structural liner within an existing drain. The flexible felt liner is saturated in a resin and delivered to site in a refrigerated container lorry. The felt sock liner is fed through the drain by filling it with water.

Figure C20 Insertion of liner

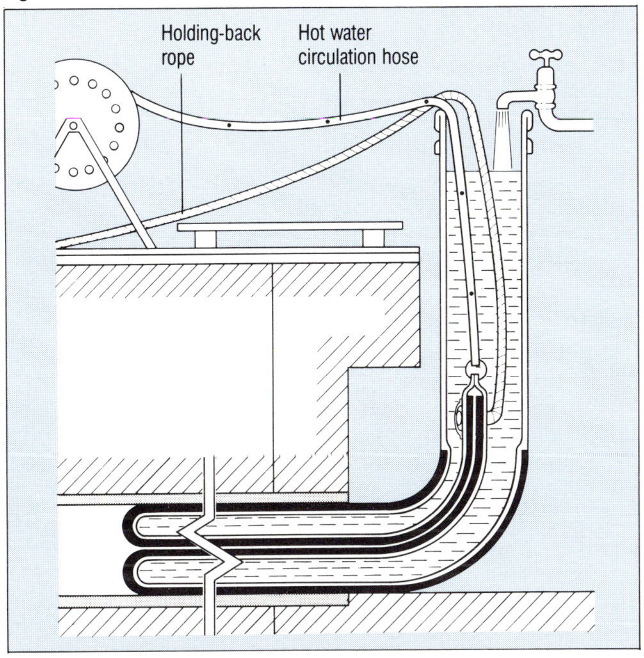

Figure C21 Curing liner

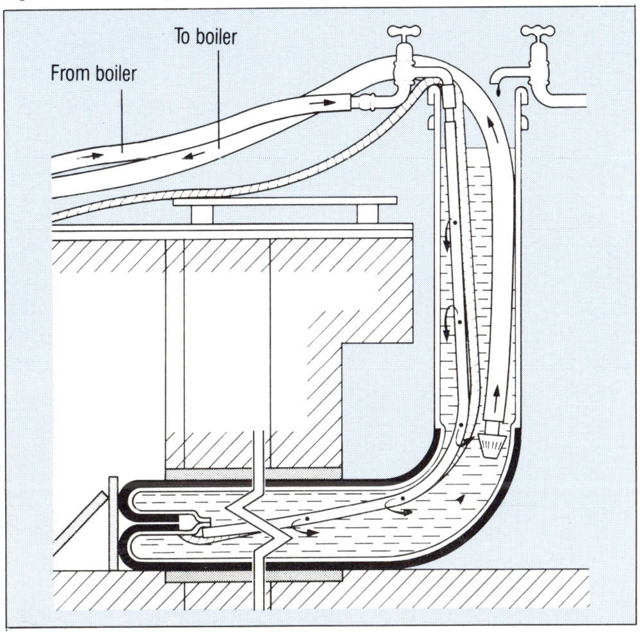

Figure C22 'Permaline' liner

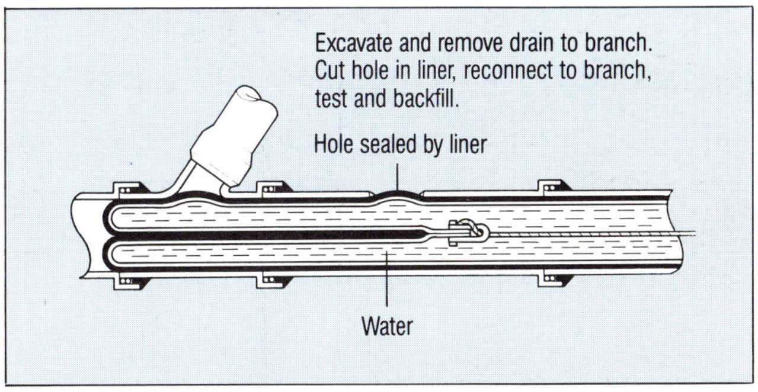

Heating the water within the lining sock for an hour or so will cure the resin. Once curing is completed both ends are cut to suit the manhole.

A similar process is carried out by Pheonix with their seamless resin tube liner, but this is fed through the pipe by compressed air and finally cured by the introduction of steam.

E.W. Avent also carry out a similar lining process with compressed air, but also have an internal tab to help pull the liner around slow bends.

(2) Sectional Insitu Liner

Figure C23 Renoline sleeve held in position by inflatable bag

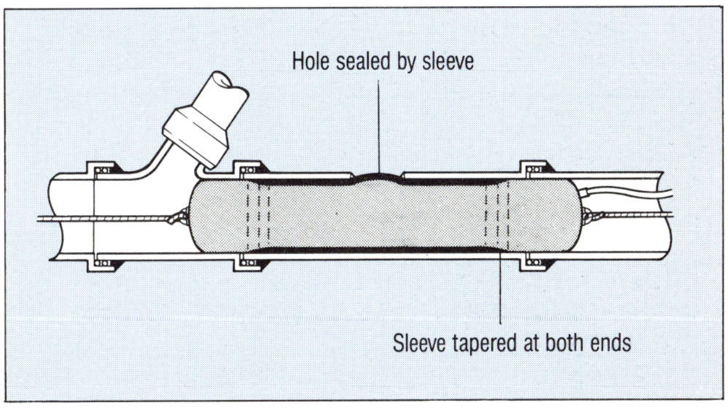

Where only a short section of pipe needs to be repaired the following method can be used.

The Renoline system is a fibre glass sleeve up to 1.2m long, impregnated on site with chemically resistant polyester resin.

The sleeve is wrapped around a special deflated air bag and then pulled into its required position, visually aided by a TV survey camera. The sleeve is then expanded tightly against the damaged pipe by simply inflating the special air bag. It is left to cure for an hour, the bag is deflated and removed leaving the drain ready for immediate use.

One advantage of this method over continuous lining is that it gives the option of treating only the damaged pipes and avoiding the entry connections from gullies etc. that would normally be sealed off in the case of a continuous liner.

uPVC Spiral Relining

The SR System

This Rice-Hayward sewer renovation system comprises the in-situ formation of a structural liner from a helically wound PVC strip. The strip, which is available in various widths and thicknesses to suit different diameters and strength requirements, is fed into a winding machine located at the bottom of a manhole or excavation.

The edges of the strip interlock to produce a positive joint between adjacent turns of the helix without the use of adhesives.

Annulus grouting is carried out with the liner full of water to resist

METHODS OF PIPE RENOVATION

grout pressures and flotation forces. Various grout formulations are available to suit site conditions.

The SR system is principally for non-man-entry sewers up to 900mm diameter and lengths in excess of 100m are said to be feasible.

Sleeved uPVC Pipes

The InterLine System

This involves the insertion of short lengths of unjointed uPVC pipes inside one or more plastic sleeves. The system provides structural strength, prevention of exfiltration or infiltration, satisfactory hydraulic characteristics and resistance to corrosion and chemical attack.

The system is designed for use on non-man-entry sewers of any shape with laterals and/or severe infiltration problems. Theoretically, there is no limit to size or length.

Further details of SR and InterLine systems may be obtained from Rice-Hayward Limited, Nortonthorpe Mills, Scissett, Huddersfield, HD8 9FB, Tel: (0484) 861916 who operate a consultancy service for any potential renovation project.

Spraying

A number of firms have been carrying out research into spraying a new surface onto an exisiting pipe, but few have been totally successful.

Avon Lippiatt Hobbs have developed their 'Spinseal resin spray pipe liner' which can negotiate pipes from 305mm to 1220mm diameter.

Preformed Pipes (new replacement pipes)

Continuous Slip Lining

A number of firms use long lengths of polyethylene insertion pipe to line sewers from manhole to manhole. This may be done size for size pipe bursting or oversized pipe bursting, to pull through a larger diameter pipe.

Sectional Slip Liner – Snap Lock

'Snap Lock' polyethylene pipe developed specifically for relining sewers from within manholes, is manufactured by Stewarts and Lloyds Plastics.

Short lengths of pipe are assembled in the drain by just snapping together, and pushing through the drain, or behind a pipe bursting machine.

Sectional Slip Liners – Demco

'Demco' integral threaded, jointed sewer lining pipes are provided in short lengths from 500mm up to 6m long. They are simply screwed together within a manhole and pushed through the drain as a slip liner. Alternatively they can be used with a pipe bursting machine.

These pipes are made of modified polypropylene for extra strength.

Sectional Slip Lining – Vitrified Clayware

Both Naylor and Hepworth have plain ended vitrified clay pipes, designed for use with micro tunnelling and jacking etc.

Sectional Slip Lining – Spun Concrete

ARC Pipes Ltd produce reinforced concrete jacking pipes from 900mm to 2100mm diameter.

Pipe Bursting

This is a method using an existing pipe to receive a new pipe of the same size or even larger. A pneumatic mole is placed into the existing pipe in order to 'burst it open' to form a hole the diameter of the pipe being pulled along, which may be larger than the existing pipe.

Rice-Hayward's 'MagnaLine' is one such proprietary upsizing and repair system. It involves the use of an hydraulic burster together with either uPVC clips or expanded metal to form a duct into which conventional pipes or a liner can be inserted. This system is more controllable than percussive moling and overcomes the soil regression problems.

Small Man Entry 'Egg and Barrel' Shaped Sewers

Very often these brick sewers are greatly oversized and in the case of the barrel shaped sewer the invert is very broad and flat. The pointing and brickwork to the invert is normally in a poor state.

To improve the sewers hydraulic gradient, a 9in. vitrified channel bedded onto the exisiting brick invert and benched both sides would provide a new clean, efficient, self cleansing sewer invert.

Where the general condition of a brick sewer is poor, glass reinforced plastic sewer liners are produced by Stanton of Nottingham for installing into the sewer in one piece sections, up to 6m long.

The annular gap between the new section and existing brickwork is pressure grouted to complete the job.

RAINWATER SYSTEMS

This section has been produced for use in conjunction with BS 6367 with the aim of guiding the designer through the relevant parts of the British Standard. It also offers charts and tables as an alternative to otherwise lengthy calculations.

The flow chart (See Figure C24) page is designed to suggest appropriate routes through both BS 6367 and this design guide section for the various situations.

Method of Design

The method of design is based on the following assumptions:

(a) The gutter is normally level.
(b) The gutter has a uniform cross-sectional shape.
(c) The outlets are large enough to allow the gutter to discharge freely.

Sources of Data

(a) Data and formulae for rainfall intensities are from BS 6367.
(b) Formulae used for production of gutter sizing nomograms are from BS 6367.
(c) Actual flow rates for grated roof outlets should be obtained from the manufacturers.
(d) Flow rates for ungrated outlets are from the formula in BS 6367.
(e) The rainwater downpipe sizing chart is based on the formula

$$q = Kd^{8/3}$$

resulting from work carried out by BRE and the US National Standards Bureau, where:

q = discharge capacity in litres/sec
d = stack diameter in mm
K (1/5 full) = 2.1×10^{-5}
K (1/4 full) = 3.2×10^{-5}
K (1/3 full) = 5.2×10^{-5}

Figure C24 Flow chart showing calculation procedure

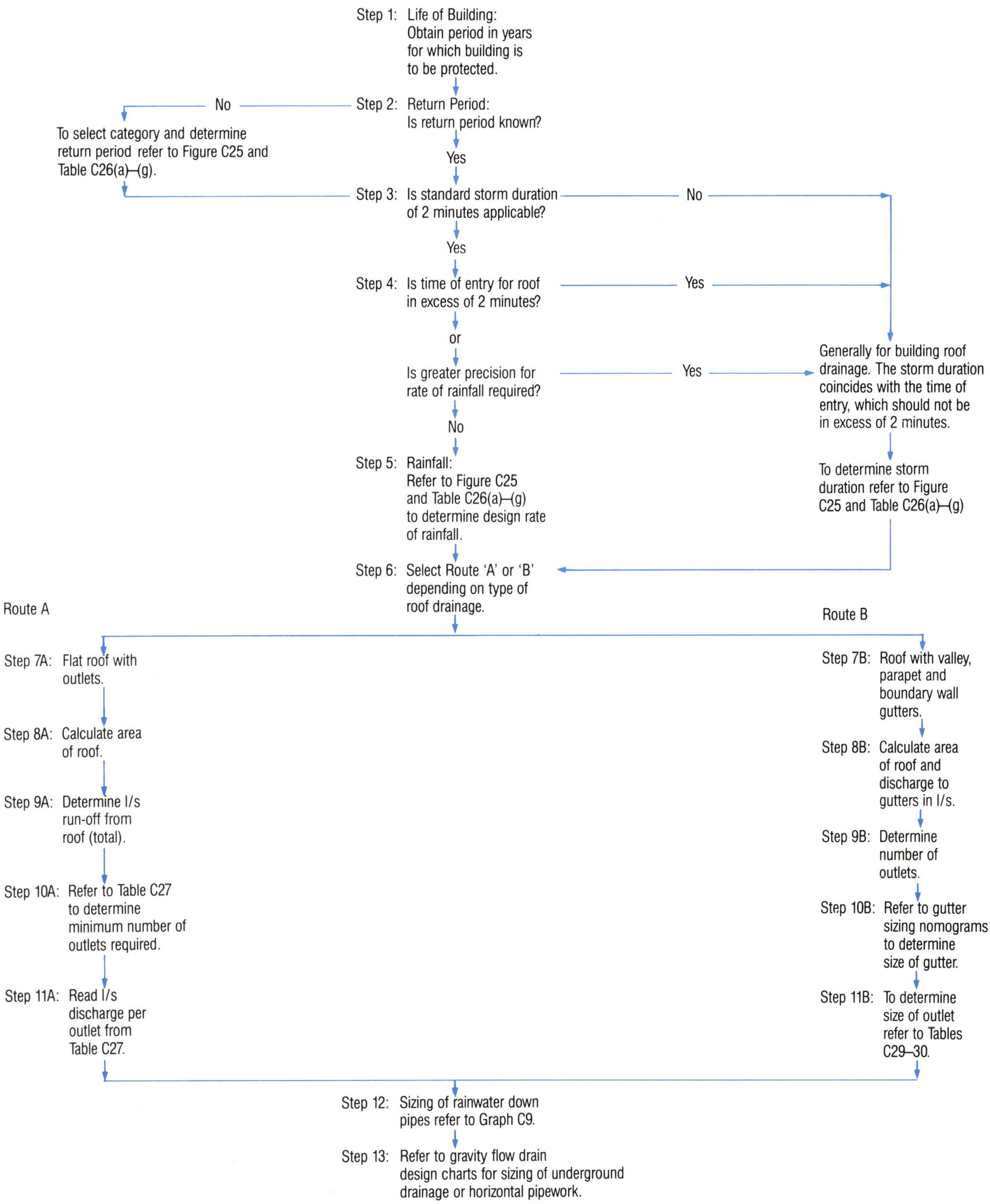

The following two examples are for information purposes only, giving a step-by-step guide through the calculation procedure.

Example 1: Route A – Flat Roof

A typical spec office block of approximate plan dimensions of 50 m × 20 m sited in Newcastle. The expected life of the building is 40 years and it has been decided to allow for a 60 years return period.

The Architect has specified a flat roof, asphalt finish with a parapet around the perimeter of the roof.

RAINWATER SYSTEMS

Step 1 Life of building 40 years
Step 2 Yes 60 years
Step 3 Yes 2 min M5
Step 4 No
Step 5 Location of building – Newcastle

Design rate of rainfall

By reference to the UK rainfall map, Figure C25, we can see that the site of our building (Newcastle) is adjacent to contour line D.

We now refer to Table C26, from which we can determine that, for a 60 year return period with a 2 minute duration rainfall, the required design intensity is 150mm per hour.

Step 6/7 In this example we have an upstand flat roof with grated rainwater outlets. Therefore we follow Route 'A' through the guide.

Step 8A Determine area of roof and flow rate in litres per second from the roof. We are assuming in this example that the area of roof to be drained is the same as the plan area, i.e. no overhangs.

Area of Roof A = length × width
Therefore A = 50m × 20m
= 1000m^2

Step 9A Total run off from roof in litres per second.

Now from Step 5 we know that the design intensity is 150mm/hr (0.15m/hr).

So flow rate = Area × intensity (in metres)
Therefore flow rate = 1000m^2 × 0.15m/hr
= 150m^3/hr or 41.66l/sec

Step 10A Select type of outlet

For installation reasons a 100mm diameter vertical grated type outlet Type 17B is selected.

Details of this outlet are to be found in Table C27.

From this Table we can determine the discharge rate in litres per second that the chosen outlet is capable of receiving at a given head. So calculating the number of outlets required to drain the roof of the building:

$$\text{Number of outlets} = \frac{\text{Total run-off in litres per second (from step 9)}}{\text{litres per second discharge per outlet}}$$

An outlet has been chosen, having a head of 30mm over the outlet. The roof is protected with an asphalt upstand which is capable of containing at least 60mm head over the outlet without risk of damage to the building.

$$\text{Therefore number of outlets} = \frac{41.66}{2.92}\ \text{l/sec}$$

Therefore 14 outlets required.

Step 11A Flow rate per outlet is 2.92 litres per second (from Table C27)

Step 12 To determine diameter of rainwater pipes. We have assumed for the purpose of this exercise that there is one outlet per rainwater downpipe.

Therefore flow down any one pipe is 2.92 litres per second.

By reference to Graph C9, it can be seen that a 100mm diameter pipe flowing 1/5 full (curve A) is capable of receiving 4.2 litres per second and is therefore satisfactory.

As the maximum flow rate of 150mm/hr (ie. the 60 year return storm) the head of water over the outlet will exceed the 30mm for a short period of time at the peak of the storm. However, the increase in head will provide an increase in velocity and the total head should not exceed 1.5 × design head of 30mm.

Step 13 Horizontal pipework should be sized using underground gravity flow charts.

When using high intensities for roof areas the flow rates should only be carried through to the branch drains immediately outside the building.

Design flow rates may be reassessed based on the sensitivity of the underground drainage system.

Note

Flow rates should be obtained from manufacturers, as these are usually higher than the BS standard types listed due to their type of design and construction.

Example 2 – Route B – Valley Gutter

Design and size roof drainage for a large warehouse with pitched roof and valley gutters.

The warehouse is to be located in Swindon and is to be designed for a 40 year building life, with a high degree of confidence that the building will be protected from rainwater for the duration of its life.

The roof construction is to be steeply pitched roof with valley gutters.

Step 1 Life span of building: 40 years

Sensitive warehouse with steeply pitched roof and valley gutter.

Step 2 Return period. This is not known and has therefore to be determined. Refer to Table C28.

Return period, T = 'C' factor × life of building
Selecting category 4 from Table C28.
'C' factor is 4.5
Life of building is: 40 years
Therefore T = 4.5 × 40 years
180 years.

Step 3 Is standard storm duration of 2 minutes applicable?

Step 4 No: storm duration (time of entry) is required to be 1 minute.

Step 5 To determine the rainfall design intensity:

Referring to Figure C25, locating the building on the map of UK, Swindon lies on contour F. This shows the amount of rainfall occuring on average once in 5 years for a duration of 2 minutes. This corresponds with Table C26 and shows that, for a storm duration of 1 minute with a return period of 180 years, the design rainfall intensity is 275mm per hour (0.275m/hr).

Step 6/7 Type of roof drainage
Steeply pitched roof with valley gutters, follow Route 'B' (Trapezoidal Gutter).

Step 8B Plan area of roof and discharge to gutter.
Plan area of roof draining into each valley gutter = 1000m^2

Flow rate in litres per second to the gutter:

= area of roof × design rainfall intensity (Step 3)

= 1000m^2 × 0.275m/hr

= 275m^3/hr

$$= \frac{275000}{3600}\ \text{l/hour}$$

= 76l/sec (run-off per second per length of gutter)

Step 9B Determine number of outlets and discharge in litres per second per outlet:

Number of outlets provisionally determined by potential stack location (one per outlet). Therefore number selected; 8.

$$\text{Therefore flow rate per outlet} = \frac{\text{total flow rate}}{\text{number of outlets}}$$

$$= \frac{76}{8} \text{ l/sec}$$

$$= 10 \text{ l/sec approximately}$$
(5 l/sec from each length of gutter).

Step 10B Determine size of gutter:
Known criteria:

1. Gutter sole to be suitable for foot traffic, therefore minimum width of sole to be 300 mm.

2. Length of gutter 7.0 metres (assuming equal outlet centres).

Now refer to respective nomogram; in this exercise we are using trapezoidal gutters with a 1.1 side slope. (Refer to page 124 for explanation of how to use sizing charts).

Reading from gutter sizing nomogram, Graph C11 where

yc = 29 mm (critical depth)
yuf = 63 mm (depth upstream of gutter – high point of water)

Freeboard must be added to yuf; this should be a minimum of 2/5ths of the maximum depth of flow (yuf) to a maximum of 75 mm.

Therefore freeboard = 25 mm (minimum)
So, total depth of gutter is, yuf + freeboard

= 63 mm + 25 mm
= 88 mm

Details of roof construction and weathering might dictate that a greater overall depth of gutter is required.

For added safety overflow pipes or weirs should be installed at either end of the gutter.

For the gutter to discharge freely, the outlet should be ungrated and of not less than three quarters of the width of the gutter. It may not be possible to select an outlet to conform with this, therefore we must alternatively construct a box receiver at the outlet. Box receivers should be the width of the gutter and the depth should not be less than H + 25 mm, where H = head of water over outlet.

Step 11B Size of rainwater outlet (and depth of box receiver). Having selected a circular ungrated outlet refer to Table C30 page 127.

Known criteria:

1. Flow rate per outlet = approximately 10 l/s

2. Head of water over outlet to be not more than yc of 29 mm (from Step 10).

Referring to Table C30, a 250 mm diameter outlet requires a head of water H of 45 mm over the outlet to receive a flow rate of 10.06 l/s. It is not possible to select a standard outlet.

A smaller outlet could be chosen by using a box receiver to give a free discharge, as the outlet would then be less than three quarters of the width of the sole of the gutter.

Selecting a 150 mm diameter outlet, the head of water over the outlet, H, is 65 mm.

65 + 25 mm = 90 mm (minimum depth of box receiver)

Plan area of Box receiver would be made up as follows:
Width = Not less than the width of flow in the gutter.
Length = the length of the box for a flow of water from one direction should not be less than (0.75 × depth of gutter). If the flow of water is from both directions, the length of the box should be (1.5 × depth of gutter) minimum.

Step 12 Sizing of rainwater downpipes.
Refer to Graph C9,

Known criterion:

1. Flow rate = 10 l/s
A 150 mm diameter rainwater downpipe flowing 1/3 full is capable of carrying 10 litres per second.

When using high intensities for roof areas the flow rates should only be carried through to the branch drains immediately outside the building.

Design flow rates should be reassessed based on the sensitivity of the underground drainage system.

Design Rate of Rainfall for Varying Storm Durations and Return Periods

Appendices A2 and A3 of BS 6367 give methods of calculating the return period or duration of varying storm intensities using the standard 2 min once in 5 year storm as a base line (2 min M5).

The following tables have been calculated using the formulae in A2 and A3, and have been drawn up to reduce calculation time.

The tables should be used as follows:

1. Read from Figure C25 the contour reference for the site location.

   The areas between contour lines on Figure C25 (which correspond with Tables C26 (a)–(g) should not be read as having a blanket coverage of rainfall, but as two extreme rates of rainfall that coincide with the contour lines, having a constant graduation from one contour to the other, e.g. Derby is situated mid-way between contours E and F, therefore a judgement must be made between Tables (a) to (g).

2. Now read off from the selected table the intensity in mm/hr against the required storm duration and return period.

RAINWATER SYSTEMS

Figure C25 Key to rainfall Tables C26(a) to (g)

Table C26(a) Rainfall amount 1.5 mm in 2 minutes, occurring on average once in 5 years (2 min M5)

Duration (mins)	Intensity (mm/hr)				
	50 mm	75 mm	100 mm	150 mm	225 mm
1	4.5	35	185	1800	—
2	5.5	55	300	3500	—
3	10	175	800	—	—
4	20	210	—	—	—
5	40	400	2000	—	—
10	300	300	—	—	—
	Return period (years)				

Table C26(b) Rainfall amount 2.0 mm in 2 minutes, occurring on average once in 5 years (2 min M5)

Duration (mins)	Intensity (mm/hr)				
	50 mm	75 mm	100 mm	150 mm	225 mm
1	1.75	7	35	325	4000
2	2	18	60	850	—
3	4.5	35	175	1800	—
4	7	45	350	3000	—
5	8	60	750	4000	—
10	45	450	2500	—	—
	Return period (years)				

Table C26(c) Rainfall amount 2.5 mm in 2 minutes, occurring on average once in 5 years (2 min M5)

Duration (mins)	Intensity (mm/hr)				
	50 mm	75 mm	100 mm	150 mm	225 mm
1	7 mths	3	8	90	1000
2	1.5	5	20	300	2500
3	2	7.5	40	500	—
4	3	15	60	800	—
5	4	20	100	1000	—
10	15	150	700	—	—
	Return period (months/years)				

Table C26(d) Rainfall amount 3.0 mm in 2 minutes, occurring on average once in 5 years (2 min M5)

Duration (mins)	Intensity (mm/hr)				
	50 mm	75 mm	100 mm	150 mm	225 mm
1	6 mths	1.75	4.5	35	400
2	7 mths	2.5	10	60	750
3	8 mths	4.5	18	190	2000
4	9 mths	7	20	350	3000
5	2 yrs	11	40	—	—
10	7 yrs	40	300	—	—
	Return period (months/years)				

Table C26(e) Rainfall amount 3.5 mm in 2 minutes, occurring on average once in 5 years (2 min M5)

Duration (mins)	Intensity (mm/hr)				
	50 mm	75 mm	100 mm	150 mm	225 mm
1	5 mths	1	2	20	180
2	6 mths	1.75	4.5	35	320
3	7 mths	2	10	60	700
4	9 mths	4	12	100	800
5	10 mths	4.5	18	185	1800
10	4 yrs	20	95	750	—
	Return period (months/years)				

Table C26(f) Rainfall amount 4 mm in 2 minutes, occurring on average once in 5 years (2 min M5)

Duration (mins)	Intensity (mm/hr)					
	50 mm	75 mm	100 mm	150 mm	225 mm	275 mm
1	4 mths	7 mths	1.5	7	50	180
2	5 mths	8 mths	2	18	110	450
3	5 mths	1.75	4	35	250	700
4	6 mths	1.75	6	40	350	1000
5	7 mths	2	7.5	50	650	—
10	2 yrs	7.5	40	300	3500	
	Return period (months/years)					

Table C26(g) Rainfall amount 4.5 mm in 2 minutes, occurring once in 5 years (2 min M5)

Duration (mins)	Intensity (mm/hr)						
	50 mm	75 mm	100 mm	150 mm	225 mm	275 mm	300 mm
1	3 mths	5 mths	9 mths	4.5	40	75	350
2	3 mths	6 mths	2	7	55	190	750
3	4 mths	1	3	18	120	350	1100
4	6 mths	1.5	4.5	20	200	450	1600
5	6 mths	2	5	38	250	650	—
10	1 yr	6	50	200	1800	1900	—
	Return period (months/years)						

Series of Tables Relating Rainfall Intensity, Duration and Return Period

NB For intensities of 225 mm per hour and above, alternative means of protecting the building other than increasing the size of rainwater pipes etc. should be considered, such as the provision of overflows or additional freeboard.

The average roof should have a duration/run off time not exceeding two minutes. Periods of greater duration shown for information and completeness only.

RAINWATER SYSTEMS

Table C27 Estimated capacities of outlet for flat roofs

Outlet type	Table in BS 416: 1973	Outlet size	Area drained at rainfall intensity of 75mm/h at depth of water above outlet: 5mm	10mm	15mm	20mm	25mm	30mm
		mm	m^2	m^2	m^2	m^2	m^2	m^2
Square flat grating	17A	100	17	49	90	135	190	250
Circular flat grating	17B	50, 65	4.5	12	23	36	51	67
	17B	75, 90, 100	9.6	27	50	77	105	140
'D'-shaped flat grating	17C	65	5.2	14	27	42	58	77
	17C	75	5.4	15	28	43	61	80
	17C	90, 100	7.6	21	39	60	85	110
Circular flat grating with horizontal pipe	17D	50	4.5	12	23	36	51	67
	17D	75, 100	7.1	20	37	57	80	105
		Outlet size	l/s	l/s	l/s	l/s	l/s	l/s
Square flat grating	17A	100	0.354	1.021	1.875	2.812	3.958	5.208
Circular flat grating	17B	50, 65	0.093	0.250	0.479	0.750	1.062	1.396
		75, 90, 100	0.200	0.562	1.040	1.604	2.187	2.920
'D'-shaped flat grating	17C	65	0.108	0.292	0.562	0.875	1.208	1.604
		75	0.112	0.312	0.583	0.896	1.270	1.667
		90, 100	0.158	0.437	0.812	1.250	1.771	2.292
Circular flat grating with horizontal pipe	17D	50	0.093	0.250	0.479	0.750	1.062	1.396
		75, 100	0.148	0.417	0.771	1.187	1.667	2.187

Procedure

1. Determine type of outlet required.
2. Refer to the Table for cast iron roof outlets as BS 416, (see also Table 7 in BS 6367).

Actual flow rates for grated roof outlets should be obtained from the manufacturers.

Rainwater downpipes should be sized separately using Graph C9 if vertical, or drain sizing data if horizontal.

Table C28 Categories of risk

Type of Roof	BS 6367 category	C Factor
Pitched roof with external eaves gutters	2	
Flat roof with outlets	2 or 3	1.5
Flat roof with sealed gutters	2 or 3	1.5
Flexible (low loading) flat roof	4	4.5
Flat roof (low loading) with gutters	4	4.5
Pitched roofs with internal valley gutters	4	4.5

The C factor is used when calculating the return period, i.e. Return period, T = C factor × life of building.

Note

For extreme cases requiring a return period of 35000* years, refer to BS 6367 Appendix A.

*This return period is an empirically derived statistical figure for the frequency of occurrence encompassing all known events (see BS 6367 Appendix D: References, Item 1)

Use of Gutter Sizing Charts

The charts provide a unique method of selecting gutter sizes in line with BS 6367.

A series of four charts cover four basic gutter profiles.

British Standard gutters are manufactured in a limited range of shapes and sizes, but with the aid of the charts it is possible to select a gutter to suit any application.

Procedure

(1) Calculate the flow of water (Q)

(2) Select the breadth (B) and length (L) to read off depth of water at point of discharge (yc) and the upstream depth (yuf) to which a suitable freeboard depth must be added, (minimum 2/5ths to a maximum of 75mm).

Example

Roof size: 20 × 19.2m = 384m^3

(Refer to Graph C8)

Rainfall Intensity = 75mm/hr
100% run off = 8l/sec

The gutter runs off full length (20m) of the building and has one outlet at each end. Half of the water will discharge in each direction; ie. 4.0l/s into 10m length of gutter. Select the appropriate nomogram for gutter profile required and range of sizes.

Starting at the bottom right-hand side of the chart select a gutter breadth (B), say 200mm, follow vertically up the nomogram to the flow rate curve of 4.00l/s. Read off and note value of yu, 60mm.

Read across to the centre axis noting the depth (yc) at the outlet, of 32 mm. Continue to the left to the curve representing the gutter length (10 m) and drop into the bottom left-hand section of the nomogram to the value of the depth (yu) of 60 m previously established.

Follow right (horizontally) to the intercept with the yuf axis and read off the value at yuf. Add freeboard to yuf.

Graph C8 Gutter sizing nomogram

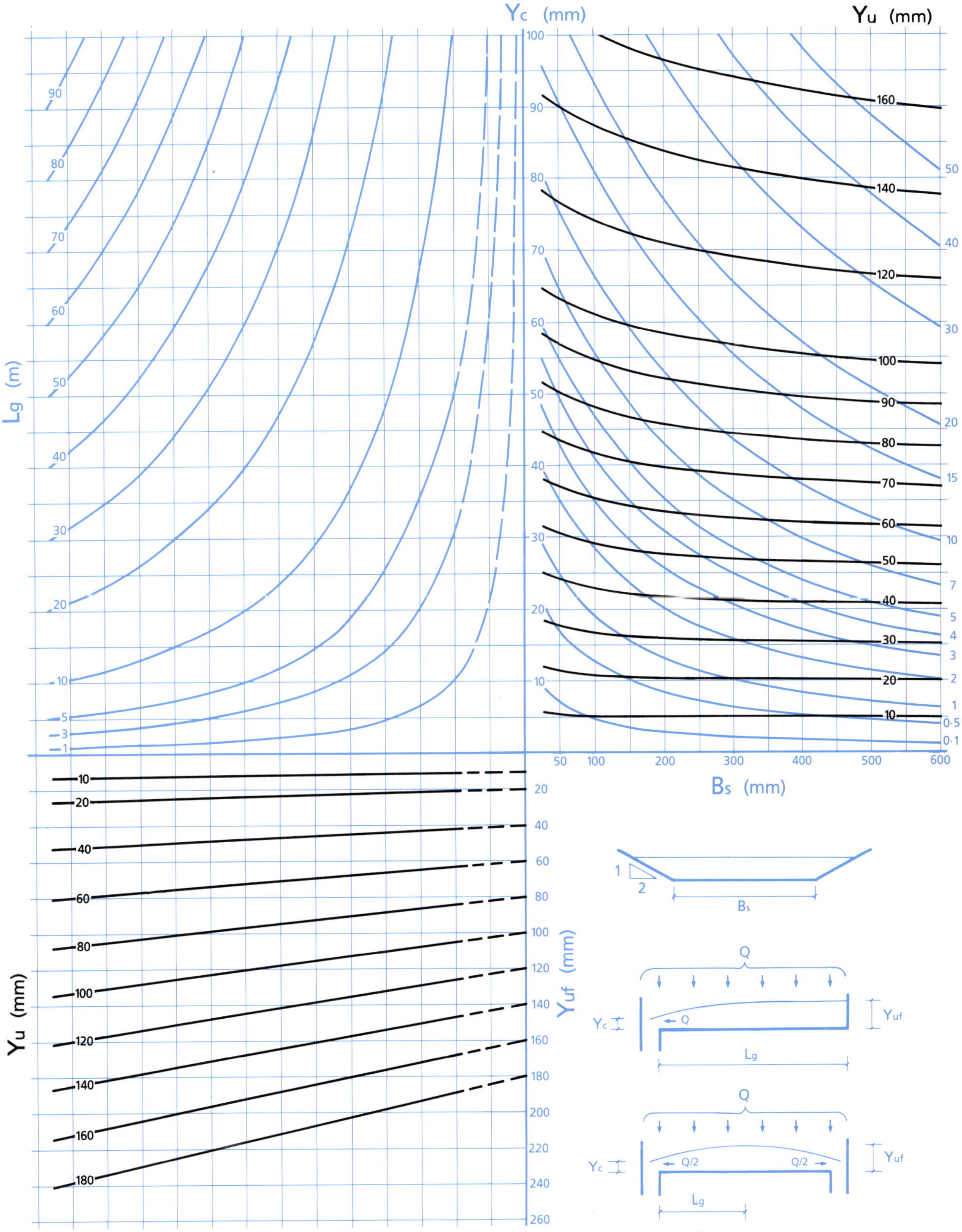

RAINWATER SYSTEMS

Gutter Outlets

(Use of Tables C29 and C30)

Having determined the depth of water at the downstream end of the gutter (yc). The depth (yc) is for a free flowing gutter and a depth not less than this must be used to size the box outlet from the Tables.

Procedure

Select Table C29 or C30 for type of outlet required, i.e. square/rectangular or circular, and read outlet size from table against depth of water over outlet.

Note

(a) It is recommended that a box receiver is constructed at the outlet to be full width of gutter, and depth no less than H + 25 mm. The gutter will not otherwise act as a free outlet, which would invalidate previous data.

(b) Diameter of outlet to be a minimum of 3/4 the width of gutter to ensure free discharge where a sump is not used.

(c) Bs (mm) = Sole width of gutter
Lg (m) = Length of gutter
Q (l/sec) = Discharge of flow in gutter
yc (mm) = Depth of flow at downstream end of gutter discharging freely
yu (mm) = Depth of flow at upstream end of gutter
yuf (mm) = Depth of flow at upstream end of gutter taking into account the affect of frictional resistance
yg (mm) = Overall depth of gutter
yg × 1.5 = Length of sump

Table C29 Flow rates, square/rectangular ungrated gutter outlets (litres per second)

Head of water over outlet (mm)	Size of rainwater outlets in mm						
H	75 × 75	100 × 75	100 × 100	150 × 100	150 × 150	200 × 150	200 × 200
5	0.14	0.16	0.19	0.23	0.28	0.33	0.37
10	0.40	0.46	0.53	0.66	0.79	0.92	1.05
15	0.73	0.85	0.97	1.21	1.45	1.69	1.94
20	1.12	1.30	1.49	1.86	2.24	2.61	2.98
25	1.56	1.82	2.08	2.60	3.13	3.65	4.17
30	2.05	2.40	2.74	3.42	4.11	4.79	5.48
35	2.59	3.02	3.45	4.31	5.18	6.04	6.90
40	2.96	3.69	4.22	5.27	6.32	7.38	8.43
45	3.14	4.19	5.03	6.29	7.55	8.80	10.06
50	3.31	4.42	5.89	7.37	8.84	10.31	11.79
55	3.48	4.64	6.18	8.50	10.20	11.90	13.60
60	3.63	4.84	6.45	9.68	11.62	13.56	15.49
65	3.78	5.04	6.72	10.08	13.10	15.28	17.47
70	3.92	5.23	6.97	10.46	14.64	17.08	19.52
75	4.06	5.41	7.22	10.83	16.24	18.94	21.65
80	4.19	5.59	7.45	11.18	16.77	20.87	23.85
85	4.32	5.76	7.68	11.52	17.29	22.86	26.12
90	4.45	5.93	7.91	11.86	17.79	23.72	28.46
95	4.57	6.09	8.12	12.18	18.28	24.37	30.86

Table C30 Flow rates, circular ungrated gutter outlets (litres per second)

Head of water over outlet (mm) H	Size of rainwater outlets in mm				
	75	100	150	200	250
5	0.11	0.15	0.22	0.30	0.37
10	0.32	0.42	0.63	0.84	1.05
15	0.58	0.77	1.16	1.55	1.94
20	0.89	1.19	1.79	2.39	2.98
25	1.25	1.67	2.50	3.33	4.17
30	1.64	2.19	3.29	4.38	5.48
35	2.07	2.76	4.14	5.52	6.90
40	2.37	3.37	5.06	6.75	8.43
45	2.52	4.02	6.04	8.05	10.06
50	2.65	4.71	7.07	9.43	11.79
55	2.78	4.94	8.16	10.88	13.60
60	2.90	5.16	9.30	12.39	15.49
65	3.02	5.37	10.48	13.97	17.47
70	3.14	5.58	11.71	15.62	19.52
75	3.25	5.77	12.99	17.32	21.65
80	3.35	5.96	13.42	19.08	23.85
85	3.46	6.15	13.83	20.90	26.12
90	3.56	6.32	14.23	22.77	28.46
95	3.66	6.50	14.62	24.69	30.86

Graph C9 Vertical rainwater pipe sizing chart

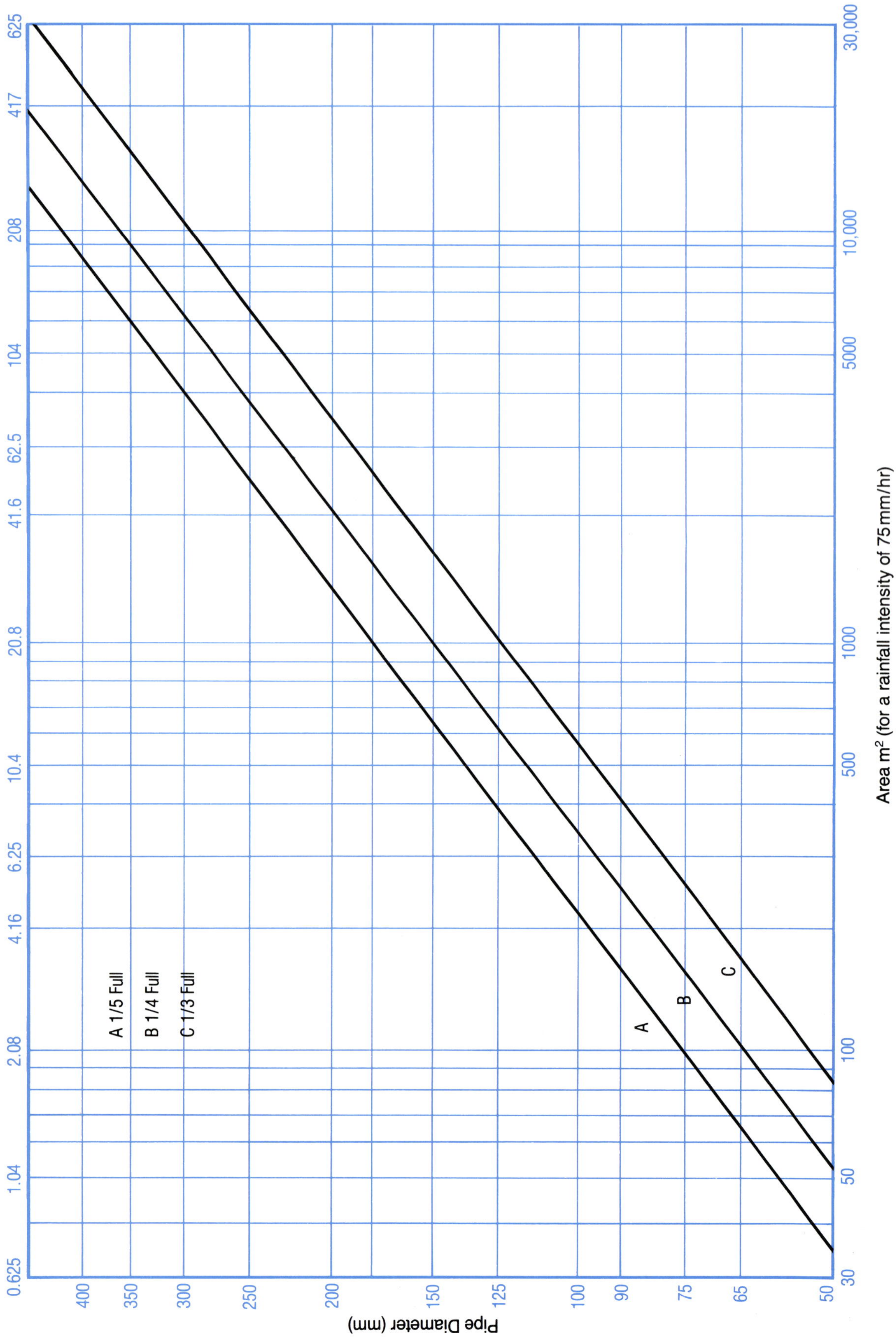

Graph C10 Rectangular Box Gutter

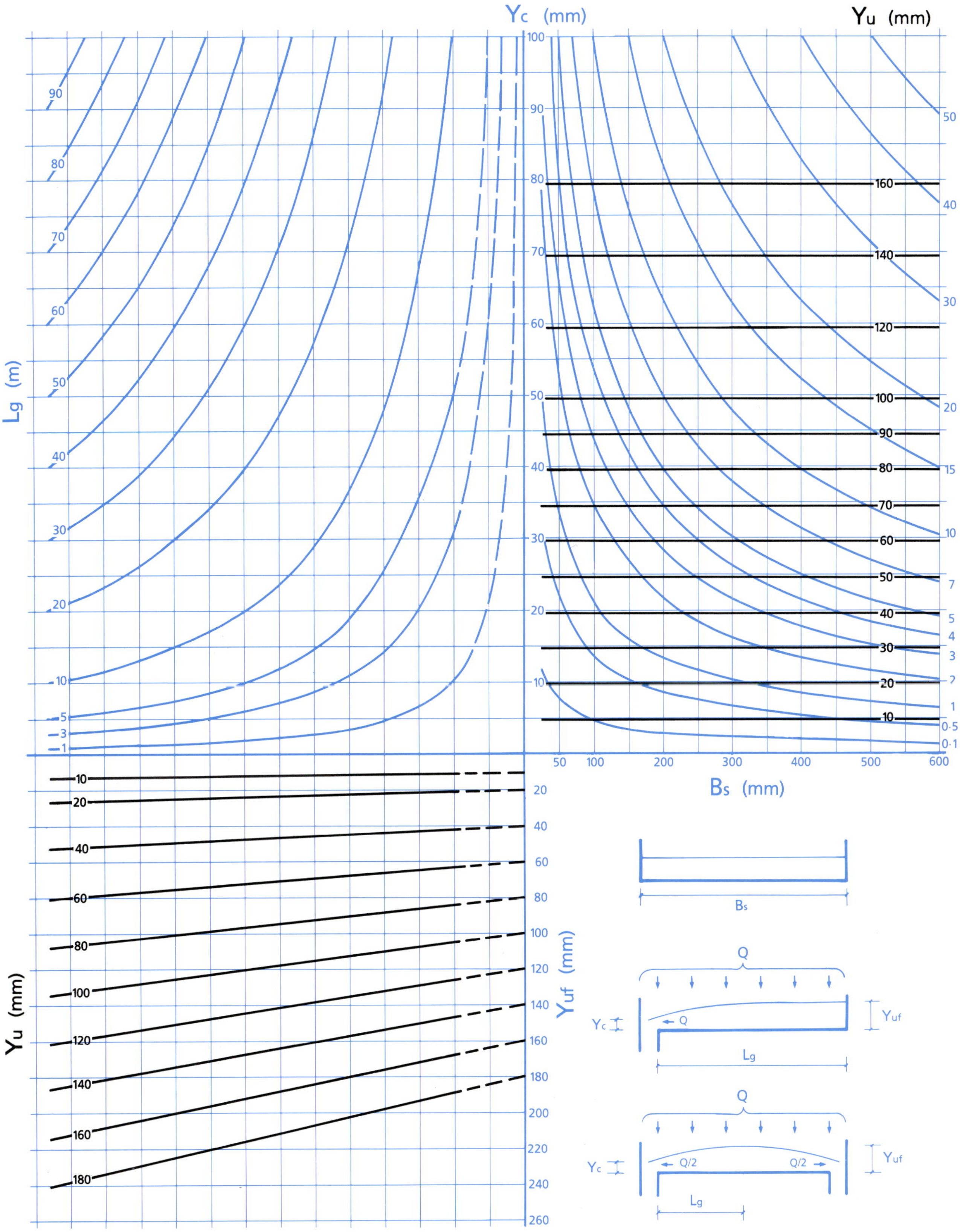

Graph C11 Trapezoidal gutter with 1 to 1 side slope

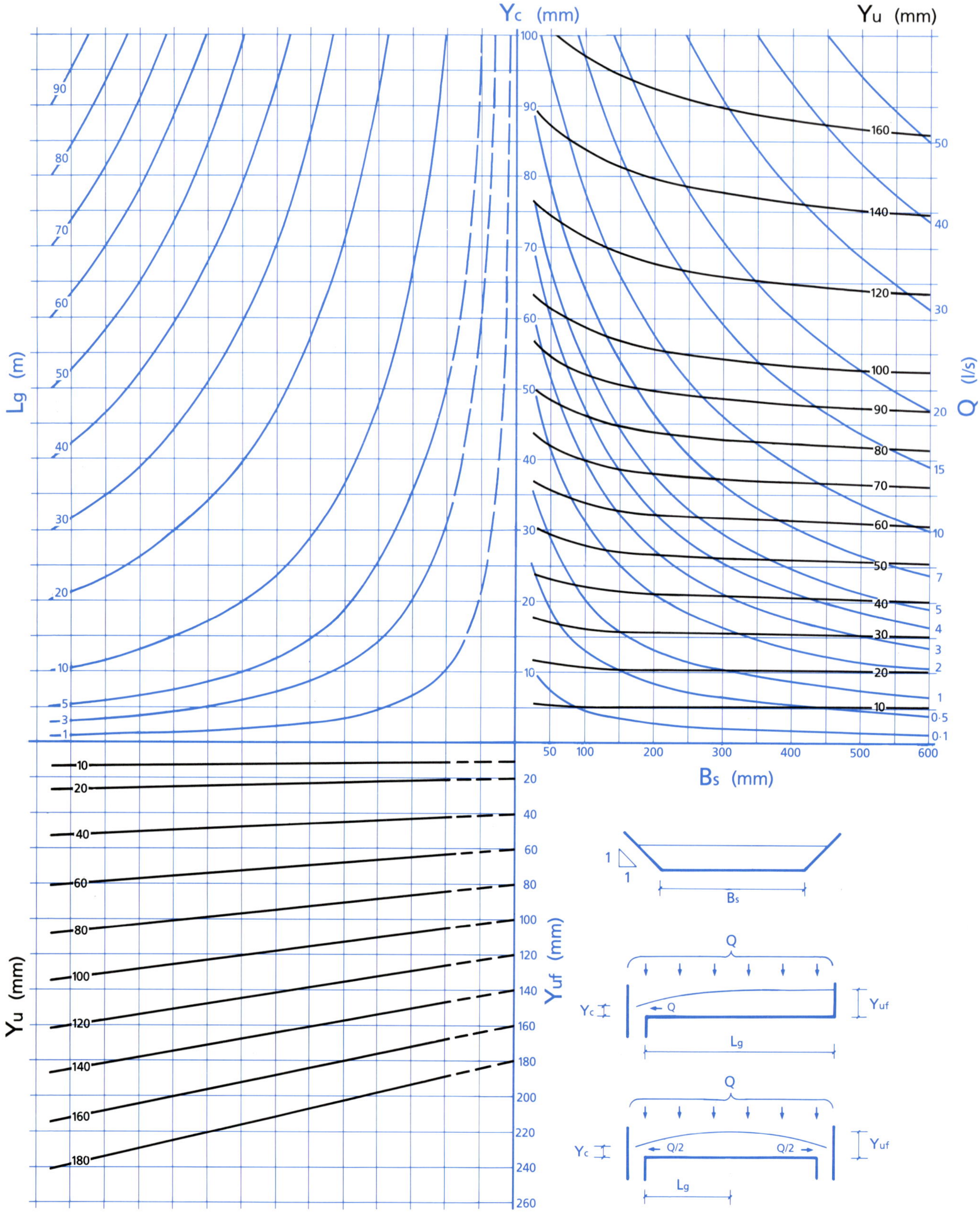

Graph C12 Trapezoidal gutter with 1 to 1.5 side slope

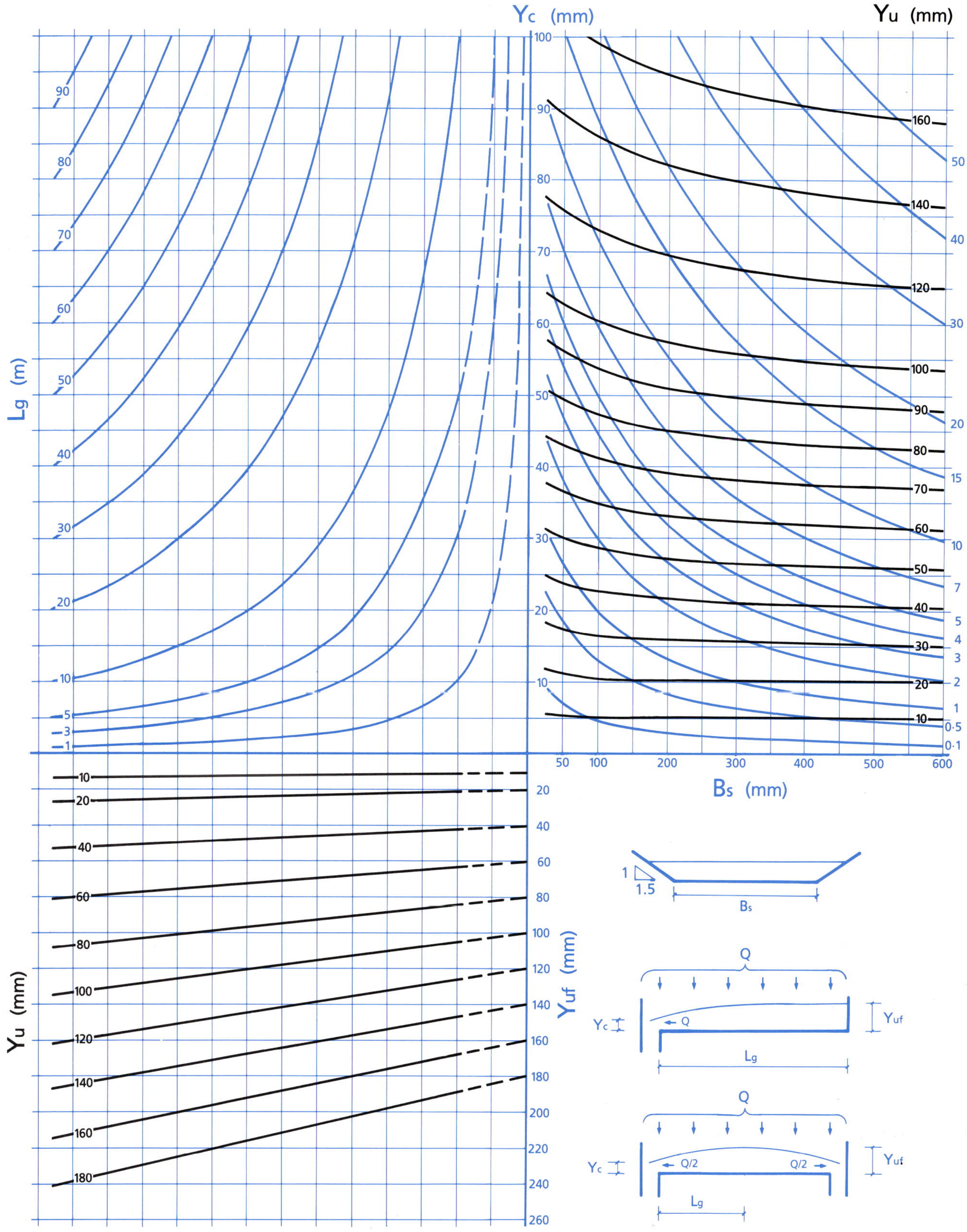

Graph C13 Trapezoidal gutter with 1 to 2 side slope

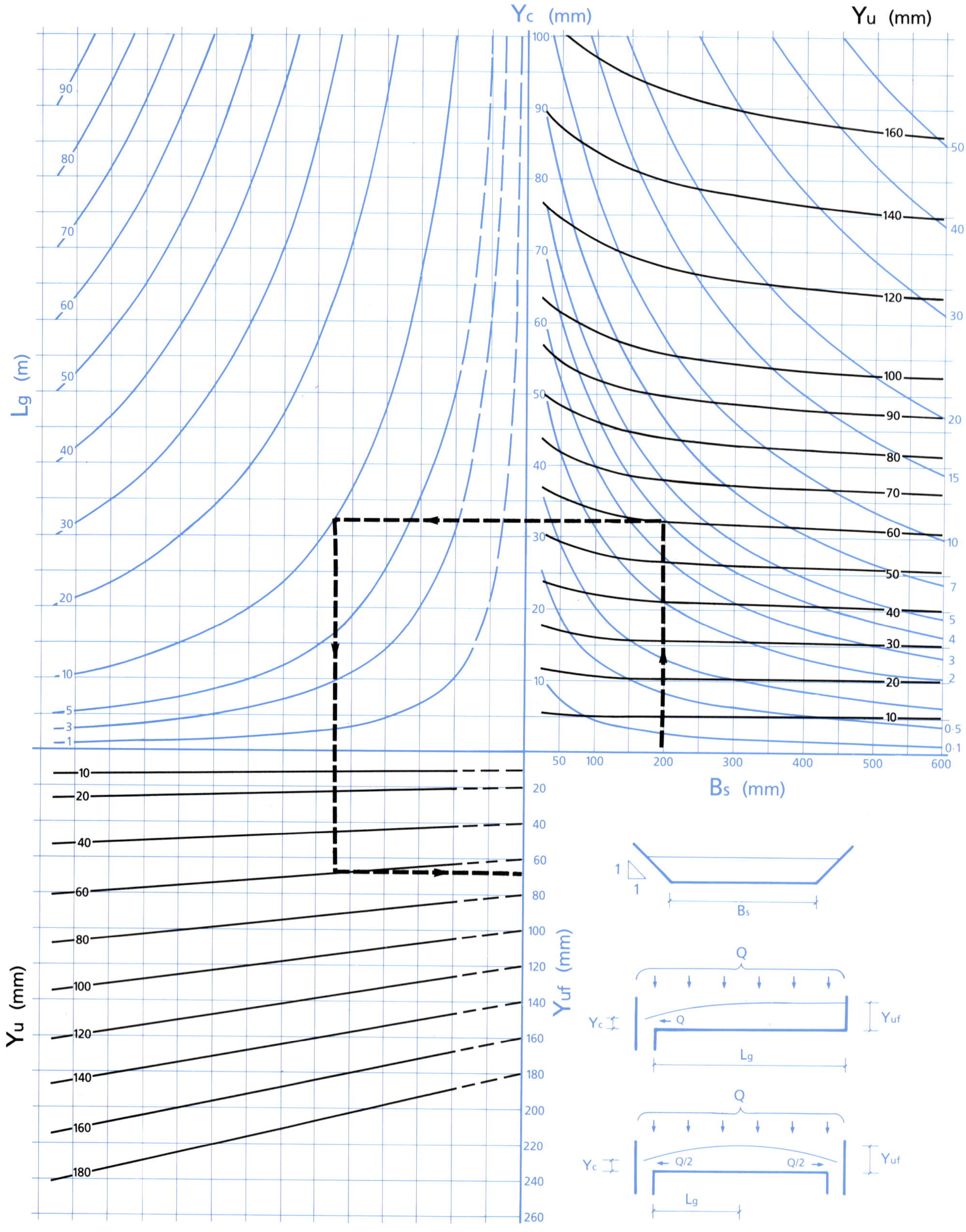

Prevention of Ice on Roofs

Snow that has built-up on roofs melts as a result of heat from two sources, sun exposure and the building's internal heating.

In either case, a potential problem can start when melted snow from roof overhangs, freezes in cold gutters and drainpipes, forming successive layers of ice. This can continue until the flow is blocked, leading to damaged gutters and drains – as well as the formation of dangerous icicles.

Water can also infiltrate the roof and building, leading to structural damage, displaced and broken roof tiles, cracked walls, burst pipes and damage to facades.

A proprietary system designed to prevent the above problems is 'IceStop', manufactured by Raychem Limited, Altrincham, Cheshire.

This comprises a self-regulating heater, developed for ice prevention on almost any type of roof. The heater adjusts its heat output to the ambient conditions at any point, the largest output is in snow and ice water. IceStop is laid along gutters and in drain pipes and does not require additional fixings.

Figure C26 'IceStop' self regulating heating system

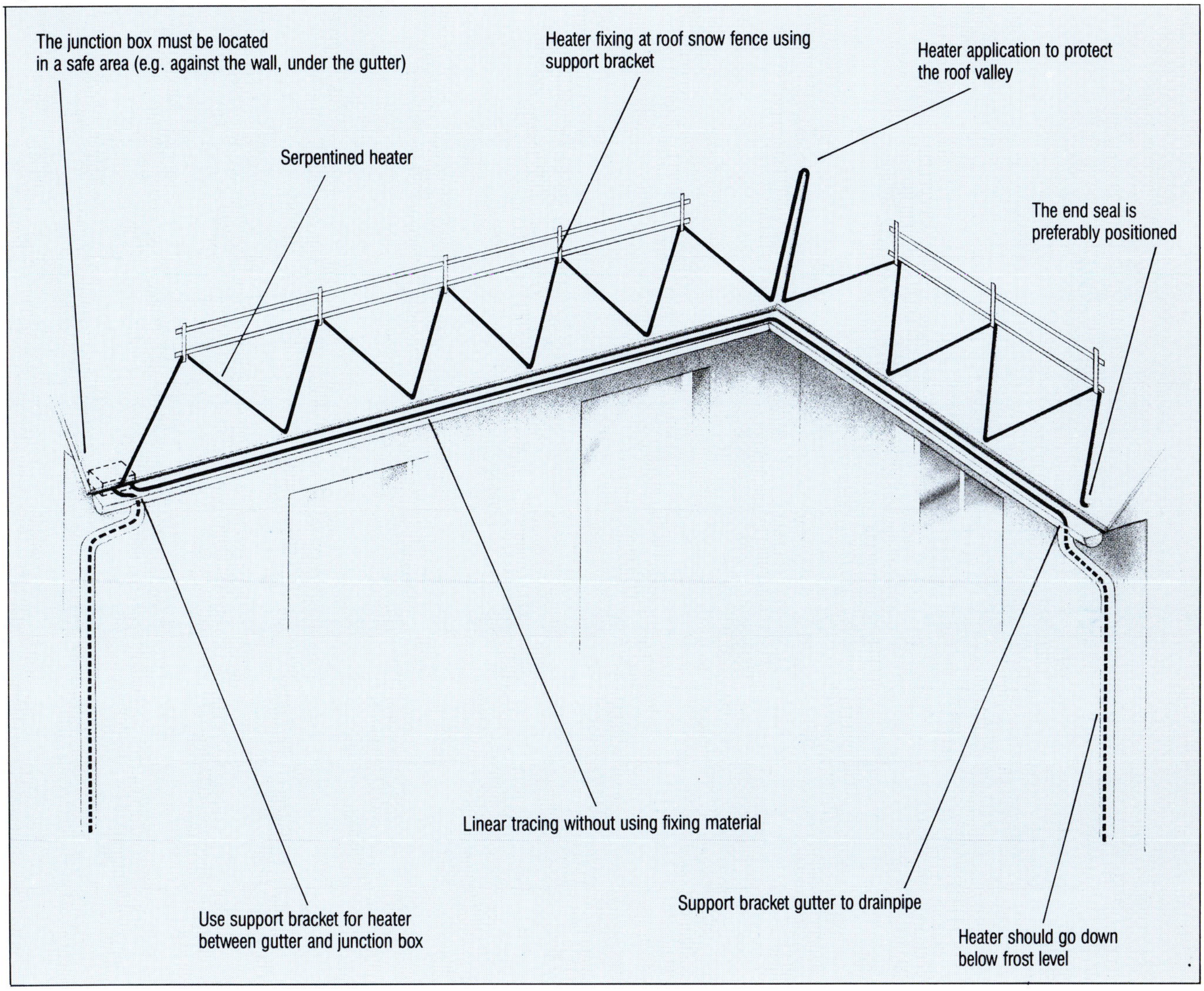

SOLID WASTE REFUSE DISPOSAL

Scope

This section deals with the on site storage and collection of refuse from dwellings and commercial premises, but does not cover hospitals, factories or buildings producing hazardous wastes, where special provisions will be required.

Refuse Collection

Refuse collection can normally be separated into two categories, namely domestic and trade.

With house refuse, the local authority is required under the Public Health Act 1936, to provide a free collection service, normally at a weekly interval.

Trade refuse may either be collected by arrangement with the local authority or by a private contractor and in both cases, a charge is made on the volume of refuse removed from the site.

Design Considerations

At the commencement of the building design process, the method of refuse collection and on site storage facility requires to be outlined and ultimately agreed with the local authority or waste removal contractors.

SOLID WASTE REFUSE DISPOSAL

It will be necessary to determine the volume of refuse that will be produced within the building and the frequency of collection from site. If on site treatment, e.g. provision of compaction equipment, is to be employed, the space requirements will need to be incorporated, together with suitable provision of access for the collection vehicle.

The volume of refuse generated in the building can be calculated by reference to Tables 1 and 2 in BS 5906: 1980. This standard also contains details of the likely composition of the refuse from various types of premises.

Site collection and storage facilities in current use are as follows:

(a) Housing

Normal method for storage is the provison of a metal dustbin or plastic sack, which is kept in a well ventilated area, with easy access for manual collection.

(b) Blocks of Flats

The provision of storage facilities for refuse collection is dependent on the number of flats and the height of the building. Up to a height of 4 storeys, consideration should be given to the installation of vertical refuse chutes on the basis of one per 60 metres of building length, in order that the occupants do not have to walk more than 30 metres to use it.

If a chute is not provided, then a communal refuse container is located at the same space intervals stated for the chute, housed in an enclosed chamber.

Flats over 4 storeys in height should be provided with a vertical refuse chute complete with ground floor chamber as detailed in BS 5906: 1980.

Consideration should also be given to the installation of sink macerator units in each flat, to dispose of organic waste. These units are not favoured by the local authorities, due to the likely increase in rodent infestation to the adjacent sewers and therefore approval of their use must be obtained.

Where chutes are installed the following points should be included:

(i) The chute must be located in an area where it will not cause a noise problem to nearby residents.

(ii) The chute must be ventilated to atmosphere.

(iii) Fire precautions need to be considered.

(iv) A hose facility complete with drainage gully should be provided in the ground floor storage chamber to enable the area to be kept clean. The water supply to the hose point should comply with the requirements of the local Water Authority, to avoid possible contamination of the supply.

(v) Hopper inlets should be housed in its own ventilated compartment constructed to provide the necessary fire rating.

(iv) Chutes should have a minimum diameter of 450mm and comply to BS 1703.

(c) Commercial Buildings

Current practice for refuse collection to commercial buildings in the United Kingdom tends to use manual means carried out by cleaning staff, with storage facilities at basement or ground level. However, there is some indication that refuse chutes are now being provided.

Where large projects involving a mixture of retail shops with flats above are involved, careful consideration on the method of transporting the refuse to the storage area is necessary.

Occupants of some commercial buildings require to separate their refuse for salvage purposes e.g. waste paper and kitchen swill for animals. In these cases, additional space must be provided within the storage area to cater for the material between collections and in the case of kitchen waste, the area may need to be cooled in order to provide suitable hygienic storage facilities.

On Site Treatment

To reduce the cost of refuse removal from site, there are various methods employed to reduce the volume and the following equipment may be used:

(1) Compaction Machines

This equipment is the most common method currently used and employs a hydraulic ram to compress the refuse into a container thereby reducing the volume. In a block of flats, a compactor can be fitted to the base of a refuse chute in the form of an automatic revolving "Carousel" unit, which loads the refuse into plastics sacks.

For commercial buildings where large volumes of refuse are generated, a static packer unit coupled to a sealed container may be used. Refuse is manually loaded into the top of the static packer and the internal ram mechanism compresses the refuse into the sealed container. When the container is full, it is uncoupled from the packer machine and removed from site on a special vehicle for emptying.

Using compaction equipment, the volume of refuse can be reduced by ratios varying from 1.5 to 15 depending on the type of equipment employed.

(2) Baling Machines

In instances where waste paper or cardboard is salvaged, the use of a baling machine is essential to reduce the bulk.

Waste material is manually fed into the charging box and is compacted by a ram mechanism into bales held in place with wire straps before removal from the machine.

(3) Shredding and Crushing Equipment

These machines are used to reduce the bulk of refuse by some 50% and are necessary equipment if the small bore airborne transporting pipeline system is used.

The refuse is placed into a feed hopper where revolving cutters shred the material into small particles.

(4) Incineration

The use of incinerators to burn refuse in a building has been reduced since the introduction of the Clean Air Acts 1956 – 1968 and are currently only used for disposal of contaminated waste from hospitals etc.

Pipeline Systems

There are available both airborne and waterborne pipeline systems for the removal of refuse to a collection point.

Smallbore airborne systems comprise of distribution pipework up to 300mm diameter, for the transportation of shredded material with the use of vacuum machines. Although this type of system has been developed and used abroad, no installation has been carried out to date in the United Kingdom.

REFERENCES

National Water Council – Technical Paper No 6 1982
American Society of Plumbing Engineers
National Plumbing Code Handbook (USA) 1984
Plumbing Code of Sewerage Regulations – Melbourne and Metropolitan Board of Works: 1984
Hydraulics Research Station, Wallingford, Oxon OX10 8BA
Crimp and Bruges; Tables and Diagrams: 1967
Road Research Laboratory: 1976 Road Note 35; A Guide for Engineers to the Design of Storm Sewer Systems
The Work of the Public Health Engineer, Escrit
Thomas Telford (Publishers): Gravity Flow Pipe Design Charts, 1987
BRE Digest 249: Sanitary Pipework: Part 2 – Design of Pipework

BRITISH STANDARDS AND CODES OF PRACTICE

BS 21: 1985	Specification for pipe threads for tubes and fittings where pressure tight joints are made on threads (metric dimensions).
BS 416: 1973	Specification for cast iron spigot and socket soil, waste and ventilation pipes (sand cast and spun) and fittings.
BS 437: 1978	Specification for cast iron spigot and socket drain pipes and fittings.
BS 864	Capillary and compression tube fittings of copper and copper alloy.
BS 864: Part 2: 1983 (1987)	Specification for capillary and compression fittings for copper tubes.
BS 1125: 1987	Specification for WC flushing cisterns (including dual flush cisterns and flush pipes).
BS 1184: 1976 (1981)	Specification. Copper and copper alloy traps.
BS 1188: 1974 (1982)	Specification for ceramic wash basins and pedestals.
BS 1189: 1986	Specification for baths made from porcelain enamelled cast iron.
BS 1206: 1974 (1976)	Specification for fireclay sinks; dimensions and workmanship.
BS 1244	Metal sinks for domestic purposes.
BS 1244: Part 2: 1982	Specification for stainless steel sink tops (excluding inset sinks).
BS 1329: 1974 (1976)	Specification for metal hand rinse basins.
BS 1390: 1972 (1983)	Specification for sheet steel baths.
BS 1710: 1984	Specification for identification of pipelines and services.
BS 1876: 1972 (1977)	Specification for automatic flushing cisterns and urinals.
BS 2494: 1986 (1987)	Specification for elastomeric joint rings for pipework and pipelines.
BS 2871: Part 1: 1971 (1976)	Copper tubes for water, gas and sanitation.
BS 3380: 1982	Specification for wastes (excluding skeleton sink wastes) and bath overflows.
BS 3868: 1973 (1980)	Specification for prefabricated drainage stack units: galvanised steel.
BS 3943: 1979 (1985)	Specification for plastics waste traps.
BS 4118: 1981	Glossary of sanitation terms. (to become BS 6110 Section 3.3)
BS 4127	Specification for light gauge stainless steel tubes.
BS 4127: Part 2: 1972 (1986)	Metric units.
BS 4305: 1972 (1978)	Specification for baths for domestic purposes made from cast acrylic sheet.
BS 4514: 1983 (1984)	Specification for unplasticized pvc soil and ventilating pipes, fittings and accessories.
BS 4880	Specification for urinals.
BS 4880: Part 1: 1973	Stainless steel slab urinals.
BS 5254: 1976 (1983)	Specification for polypropylene waste pipe and fittings (external diameter 34.6 mm, 41.0 mm and 54.1 mm)
BS 5255: 1976: 1984	Specification for plastics waste pipe and fittings.
BS 5503	Specification for vitreous china wash down WC pans with horizontal outlet.
BS 5503: Part 1: 1977	Connecting dimensions.
BS 5503: Part 2: 1977 (1980)	Materials, quality, performance and dimensions other than connecting dimensions.
BS 5504	Specification for wall hung WC pan.
BS 5504: Part 1: 1977	Wall hung WC pan with close coupled cistern. Connecting dimensions.
BS 5504: Part 2: 1977	Wall hung WC pan with independent water supply. Connecting dimensions.
BS 5504: Part 3: 1977	Wall hung WC pan. Materials, quality and functional dimensions other than connecting dimensions.
BS 5505	Specification for bidets.
BS 5505: Part 1: 1977	Pedestal bidets over rim supply only. Connecting dimensions.
BS 5505: Part 2: 1977	Wall hung bidets over rim supply only. Connecting dimensions.
BS 5505: Part 3: 1977	Vitreous china bidets over rim supply only. Quality, workmanship and functional dimensions other than connecting dimensions.
BS 5506	Specification for wash basins.

BRITISH STANDARDS AND CODES OF PRACTICE

BS 5506: Part 2: 1977	Wall hung wash basins. Connecting dimensions.
BS 5906: 1980 (1987)	Code of Practice for storage and on-site treatment for solid waste from buildings.
BS 6087: 1981	Specification for flexible joints for cast iron drain pipes and fittings (BS 437) and cast iron soil, waste and ventilating pipes and fittings (BS 416).
BS 6340	Shower units.
BS 6340: Part 1: 1983	Guide on choice of shower units and their components for use in private dwellings.
BS 6340: Part 2: 1983	Specification for the installation of shower units.
BS 6340: Part 3: 1985	Specification for prefabricated shower enclosures and shower cabinets.
BS 6340: Part 4: 1984	Specification for shower heads and related equipment.
BS 6340: Part 5: 1983 (1986)	Specification for prefabricated shower trays made from acrylic material.
BS 6340: Part 6: 1983 (1986)	Specification for prefabricated shower trays made from porcelain enamelled cast iron.
BS 6340: Part 7: 1983 (1986)	Specification for prefabricated shower trays made from vitreous enamelled sheet steel.
BS 6340: Part 8: 1985	Specification for prefabricated shower trays made from glazed ceramic.
BS 6347: 1984 (1986)	Specification for polyethylene pipes (Type 50) in metric diameters for general purposes.
BS 6367: 1983 (1984)	Code of Practice for drainage of roofs and paved areas.
BS 6455	Sanitary installations.
BS 6465: Part 1: 1984	Code of Practice for scale of provision, selection and installation of sanitary appliances.
BS 8005: Part 1 (1987)	Guide to new sewerage construction.
BS 8233: 1987	Code of Practice for sound insulation and noise reduction for buildings.
BS 8301: 1985	Code of Practice for building drainage.
CP 312	Code of Practice for plastics pipework (thermoplastics material).
CP 312: Part 1: 1973	General principles and choice of materials.
CP 413: 1973 (1980)	Code of Practice for ducts and building services.

Note

(1) Year after the Standard's number indicates the latest date of issue.

(2) Year in brackets gives the date of the latest amendment at the time of publication of this Guide.

DEFINITIONS AND DESCRIPTIONS

Absolute Pressure

The gauge pressure, plus the pressure of the atmosphere. (When a pressure is referred to without 'gauge' or 'absolute' being stated, it is normally understood to be the gauge pressure).

Atmospheric Pressure

Normal atmospheric pressure at sea level is 1.00 bar = 10.20 m head. Atmospheric pressure varies with the altitude above sea level and with the weather conditions.

Capacity

The volumetric flow rate delivered by a pump, normally expressed in cubic metres per hour (m^3/h), or litres per second (l/s).

Cavitation

Cavitation occurs when the static pressure somewhere locally within the pump falls to be below the pressure of the liquid and the word originates from the Latin word for 'hollow'.

Cavitation is a rather complex and undesirable condition that may occur in a pump installation and can usually be recognised by a metallic knocking which may vary from very mild to very severe.

In general terms and referring to centrifugal pumps, when liquid flows through the suction line and enters the eye of the impeller, an increase in velocity occurs which is accompanied by reduction in pressure. If the pressure falls below the vapour pressure corresponding to the temperature of the liquid, the liquid will vaporize and then consist of liquid plus pockets of vapour. The liquid then reaches regions of higher pressure through the impeller and the cavities of vapour collapse, giving rise to the noise described above.

Although cavitation is popularly associated with centrifugal pumps, it is a condition that can arise with any pump installation unless care is taken with pump selection and planning the installation. In general terms, the following conditions should be avoided as far as possible.

(1) Suction lift higher or suction head lower than recommended by manufacturer.

(2) Liquid temperatures higher than that for which the system was originally designed.

(3) Speeds higher than manufacturer's recommendation. For centrifugal pumps, care should also be taken that they are not operated with heads much lower than the head for peak efficiency or with capacities much higher than the capacity for peak efficiency. It is advisable to operate as near the peak efficiency as possible, particularly if there are other adverse conditions present which may tend to promote cavitation.

Efficiency

Efficiency of a pump is its ability to perform with the least energy input.

Friction Head

The head necessary to overcome resistance to motion in pipes and fittings. It is of great importance as it may exceed the static head.

Gauge Pressure

The intensity of pressure exerted by a pump, recorded by gauge, calibrated above atmospheric pressure.

Net Positive Suction Head (NPSH)

The method of specifying suction performance under critical suction conditions including altitude, high temperature and high vapour pressure and the limitations of the pump itself.

(i) NPSH available
The absolute pressure of the liquid at the pump entry, less its vapour pressure at the pumping temperature. It is an expression of the installation conditions.

(ii) NPSH required
The minimum absolute pressure necessary at the pump entry to maintain the required flow without cavitation.

Note

The NPSH required must always be less than the NPSH available.

Power

The total amount of energy required to drive a pump or a pumping set.

Power and Efficiency

$$\text{Power of a pump} = \frac{\text{work done}}{\text{time (seconds)}}$$

$$= \frac{\text{Newton} \times \text{metres}}{\text{seconds}}$$

$$= \frac{\text{force} \times \text{total head}}{\text{seconds}}$$

$$= \frac{\text{kg} \times 9.81 \times \text{head}}{\text{seconds}}$$

$$= \frac{9.81 \times \text{litres} \times \text{head}}{\text{seconds}}$$

$$\text{Pump power} = 9.81 \times \text{l/sec} \times \text{m}$$

$$\text{Watts} = \frac{\text{Nm}}{\text{s}}$$

Note

1 kg/s = 1 l/sec
1 Newton = 1 kg × m/sec^2 = force
1 Newton = kg × 9.81

The total pump head must include the measured pump head, plus all pipe and fitting resistances plus the velocity head.

Total Head = L + Lf + $\frac{1}{2}mV^2$

The efficiency of a pump is the ratio of the input power to the brake power. The brake power is the power absorbed by the pump. output

$$\text{Efficiency \%} = \frac{\text{output}}{\text{input}} \times 100$$

Input or power to be provided:

$$\text{Input in kW} = \frac{\text{output in kW}}{\text{efficiency: \%}}$$

Owing to the combined effectiveness of the pump and motor, overall efficiencies of about 50% are quite normal for coupled sets.

Self-Priming

When a pump is operating under suction lift (negative) conditions, self-priming is the characteristic which enables the pump to evacuate air from the suction line, thus creating a vacuum which allows the atmospheric pressure to push the liquid through the suction pipe into the pump.

Static Delivery Head

The vertical distance between the centre-line of the pump and the free surface of the discharged liquid.

DEFINITIONS AND DESCRIPTIONS

Static Head

The pressure exerted by a liquid at rest, to the surface, expressed in terms of vertical head of the liquid.

Suction Head (positive)

This exists if the source of liquid is above the pump, a condition generally known as 'flooded suction', and is the vertical distance from the centre-line of the pump up to the free surface of the liquid.

Suction Lift (negative)

This exists if the source of liquid is below the pump and is the vertical distance from the centre-line of the pump up to the free surface of the liquid.

When a fluid (liquid or air) is sucked into a pump, it is pushed in by the pressure of the atmosphere as a result of the pressure of the fluid in the pump being reduced. The pressure of the atmosphere and hence the depth from which a pump will 'suck', varies with the altitude above sea-level and with the state of the weather (see definition of Atmospheric Pressure).

The suction lift for which a pump is listed is, unless otherwise stated, the total suction lift of which the pump is capable as the listed speed, assuming normal temperature at normal atmospheric pressure, deductions being made for high altitude and for high temperature and vapour pressure. The suction lift of a pump is also affected by the viscosity of the liquid, but this should be covered by the makers' ratings at stated viscosities.

The effect of vapour pressure on permissible suction lift is important, particularly with volatile liquids such as gasoline, aromatic solvents and most liquids at elevated temperatures and where suction lift is necessary, full details of the installation should preferably be referred to the supplier for consideration.

Where the suction conditions are liable to be critical it will be necessary to calculate NPSH values to ensure correct pump selection.

Total Delivery Head

The static delivery head, plus the velocity head and the friction head in the inlet pipe system.

Total Head on the Pump

The static delivery head, plus the velocity head and the friction head in the entire pipe system, plus the static suction lift, or minus the static suction head (positive), as the case may be. (See Figure D1, which illustrates various head conditions on a pump).

Total Suction Head (positive)

The static suction lift, minus the velocity head and friction head in the inlet pipe system.

Total Suction Lift (negative)

The static suction lift, plus the velocity head and friction head in the suction pipe system.

Vacuum

A space in a pump system from which, some or all the air has been removed and the pressure is below atmospheric pressure.

Vacuum is normally measured in millimetres head of mercury, indicating the degree of partial vacuum in a pipe or vessel. A complete or perfect vacuum is impossible to attain.

Vapour Pressure

Is the minimum pressure necessary to prevent evaporation of a liquid at a given temperature. All liquids evaporate spontaneously unless prevented by external pressure. Vapour pressure increases rapidly by a rise in temperature. The vapour pressure of water is 0 at freezing point but reaches normal atmospheric pressure, 1.00 bar at 100°C, which is the boiling point of water sea level.

Velocity Head

This is the pressure required to set a liquid in motion and is generally of practical importance only in the case of pumps of large capacity or where the suction lift is near the limit.

Velocity head = $0.5 \times M \times V^2$

where M = kg

V = velocity in m/second

Figure D1 Pumping definitions

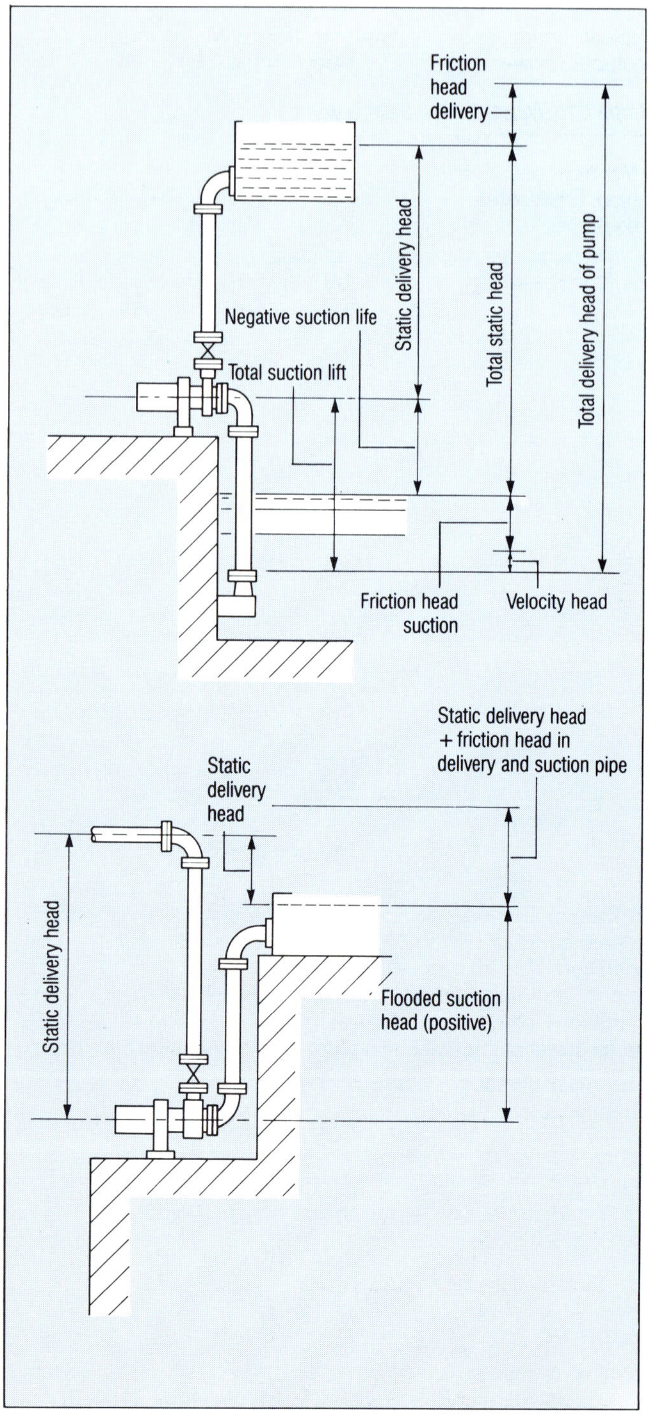

PUMP TYPES

The two main pump classifications are centrifugal and positive displacement, there being many different types within each category. However, the main characteristics referred to below can be regarded as generally applicable.

Centrifugal

A centrifugal pump in its simplest form consists of an impeller and a volute casing which has to be filled completely with liquid when the pump is in operation, the impeller 'throwing' the pump liquid to the outside of the volute, thus imparting kinetic energy. In this way a centrifugal pump is capable of generating a certain head which varies according to the pump speed and the accepted method of expressing the relationship between capacity and head is by means of a 'characteristic curve', often referred to as the Q/H curve where Q is the quantity (flow rate) and H is the head.

Typical characteristic curves for a simple centrifugal pump including power and efficiency are illustrated in Graph D1.

Graph D1 Typical characteristic curves

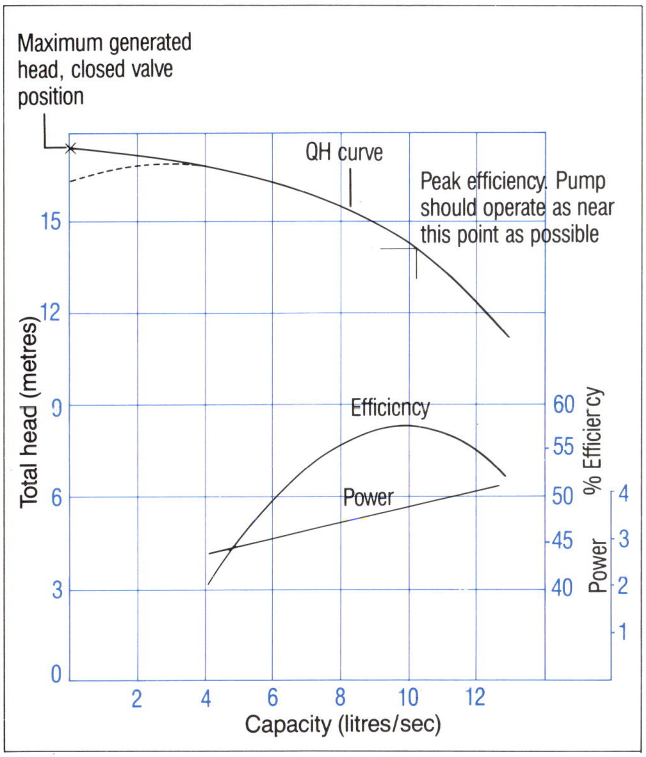

At the point where the curve intersects the head ordinate, the generated head is at a maximum and corresponds to a no-flow condition. This is often referred to as the 'closed valve pressure' and a centrifugal pump should not be left operating in this condition for any length of time as it will rapidly over-heat due to the horse power at this point being turned effectively into heat energy.

The main characteristics of centrifugal pumps can broadly be summarized as follows:

(a) Capacity varies with head (see Q/H curve).

(b) Capacity proportional to pump speed.

(c) Head proportional to the square of the pump speed.

(d) Non self-priming.

(e) Suitable for low-viscosity liquids.

Positive Displacement

Positive pumps usually consist of a casing containing gears, vanes, pistons, lobes, screws, sliding-shoes, etc. operating with minimum clearance, the liquid being positively transferred from suction to discharge port. Due to fine clearances involved, most positive pumps are self-priming and some will handle entrained gas or air.

Neglecting leakage, they deliver almost constant capacity irrespective of variations in head so that the usual Q/H curve when drawn for a typical positive-displacement pump is an almost vertical straight line. For this reason it is not usual to provide Q/H curves for positive pumps.

Their main characteristics are as follows:

(a) Capacity substantially independent of head.

(b) Capacity proportional to speed.

(c) Self-priming.

(d) Suitable for viscous liquids (reduced speeds usually necessary for high viscosities).

Applications

Pumped systems or a booster set will be required where any one or number of the following conditions prevail.

(a) Where there is no piped public water supply and the only available water is at or below ground level.

(b) Where the pressure from the public supply main is insufficient to reach the highest or most distant draw-off point for the whole or part of the day.

(c) Where the available pressure is sufficient for normal domestic and culinary purposes but is too low for other equipment or fittings to the building, eg. hose reels, fire hydrants, wash down points, showers, process equipment etc.

(d) Where the pressure is adequate for all the building's requirements but the Water Supply Undertaker is not prepared to guarantee the pressure in the foreseeable future.

SYSTEMS

Packaged Systems

Most pump manufacturers supply equipment that is prefabricated using all the components necessary to provide a system that is factory tested to a proven performance. Package systems reduce the amount of site work and ensure that the arrangement of all components can be adequately maintained.

It is usual for package units to be positioned on a prepared base; only final connections to the suction and discharge pipework and connection of the electricity supply to the pre-wired integral control panel is required on site.

Unpackaged Component Systems

It is sometimes necessary to design and install separate items of pumping equipment and controls to form a system on site. The collection of items are all supplied separately and the installer is required to connect the pumps, vessels, valves and control equipment into an operational system.

Although satisfactory systems can be arranged, there are a number of disadvantages. When properly costed, it is often found that this method of providing a pumping system is expensive and the responsibility for resolving problems is impossible to apportion between the pump manufacturer, and ancillary equipment supplier and the installer. There may also be over design of some systems due to the 'ad hoc' selection of various components.

SYSTEMS

Effects of Pump on System

The pressure developed from the pump is proportioned throughout the system as indicated on the following pressure diagrams.

Figure D2 (a) Open circuit

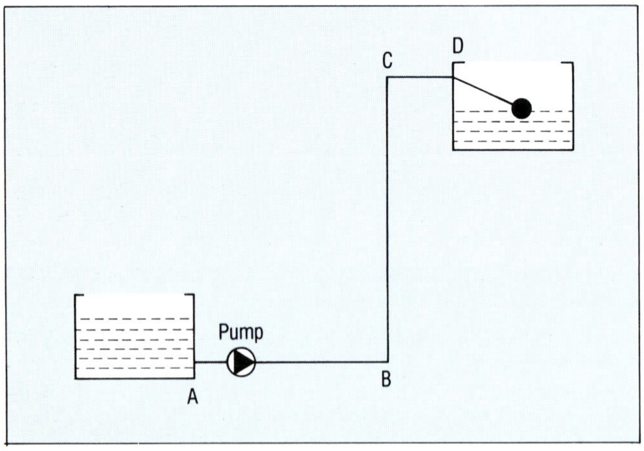

Figure D2 (b) Open circuit pressure diagram

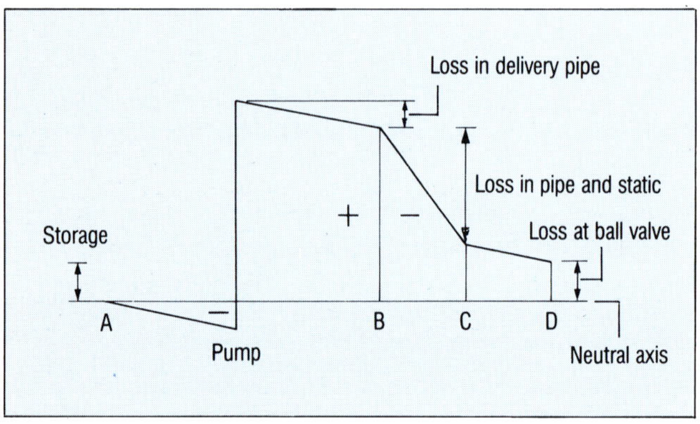

Figure D3 (a) Closed circuit

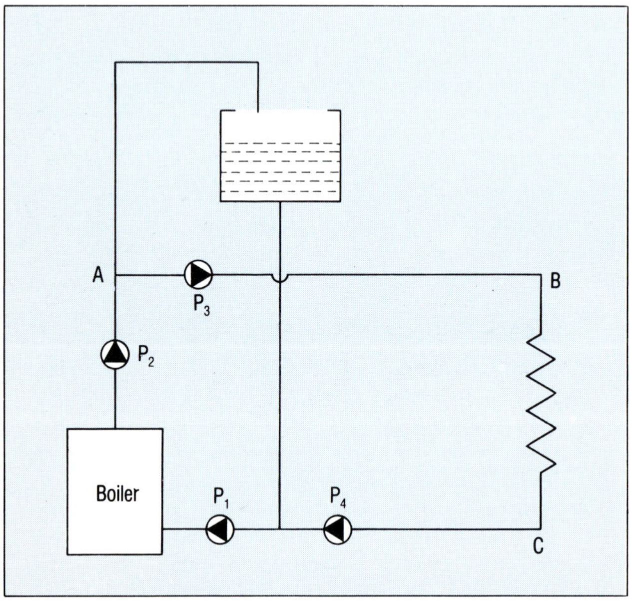

By varying the position of the pump, the pressure diagram changes accordingly.

Figure D3 (b) Closed circuit pressure diagram for pump position P1

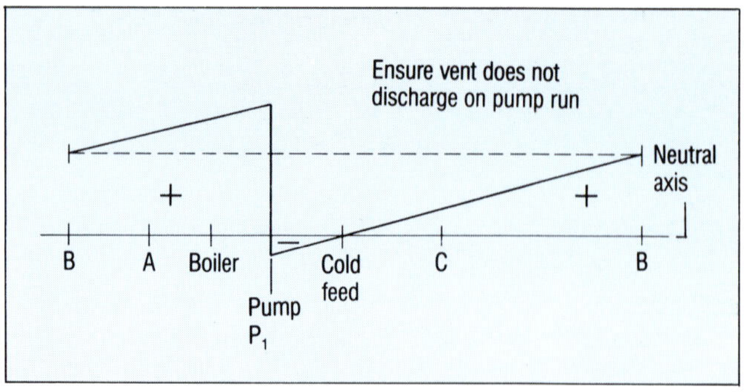

Figure D3 (c) Closed circuit pressure diagram for pump position P2

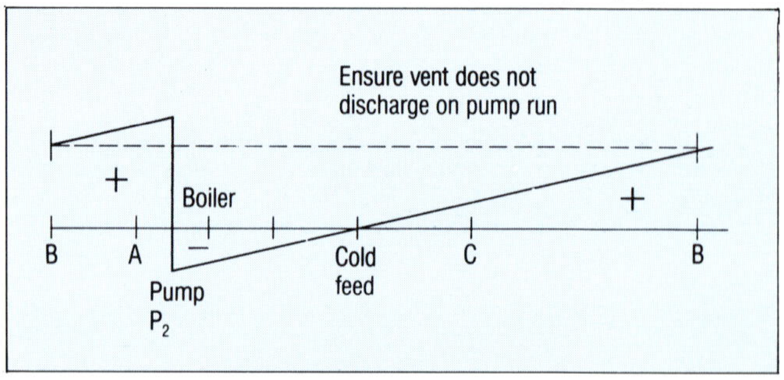

Figure D3 (d) Closed circuit pressure diagram for pump position P3

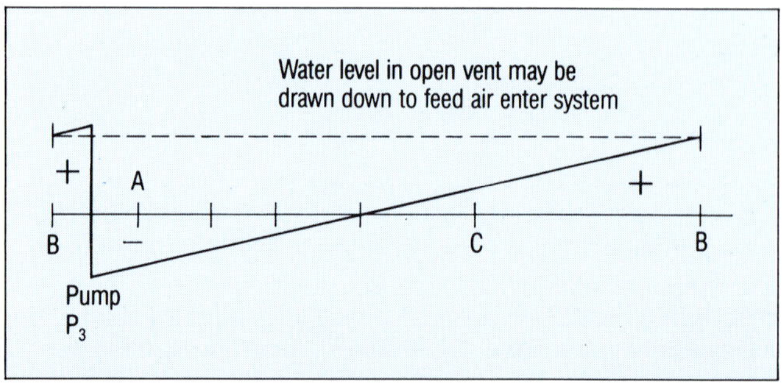

Figure D3 (e) Closed circuit pressure diagram for pump position P4

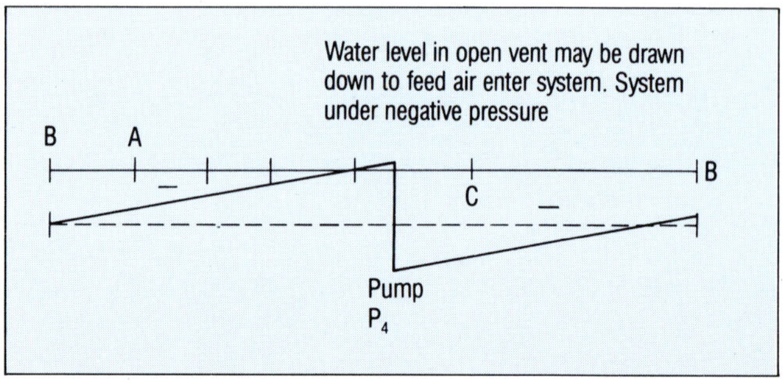

A regulating valve in a closed circuit system is frequently needed to reduce the flow back to the design condition as illustrated in Graph D2.

Graph D2 Pump graph

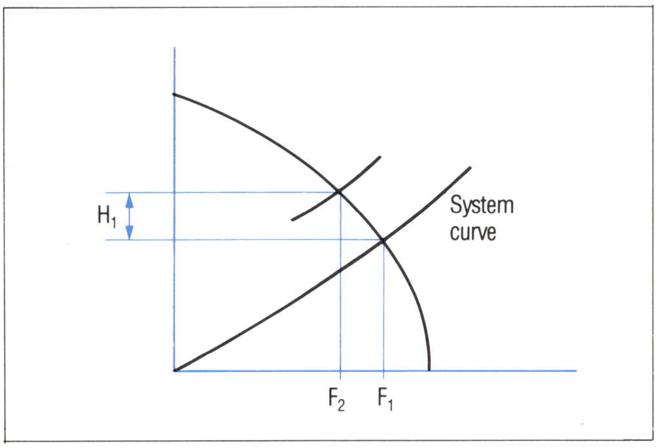

If the curve to the varying flows through the system is plotted on to the pump graph as Graph D2, it would indicate that a flow of F_1 would be achieved. However, if the design caused for a flow of F_2, additional resistance as indicated by H_1 would be required. A regulating valve should be adjusted to produce this pressure loss.

PUMP SELECTION

Between 90 to 95 per cent of the world's pumping is carried out using centrifugal pumps and wherever the conditions are suitable, a centrifugal pump is normally the simplest and most economical type available.

For installations operating under flooded suction conditions, ie. with a positive suction head, or where large volumes of water have to be moved at relatively low heads, then the centrifugal pump is the natural choice. Many special types have been developed, including multi-stage designs in which a number of impellers are assembled from a common shaft with the pumped liquid being delivered to each impeller in turn. Neglecting internal losses, the total head generated by a multi-stage pump is the sum of the heads generated by each impeller. Multi-stage pumps may be of horizontal or vertical design, the latter having the advantage of requiring appreciably less floor-space.

For water boosting services where relatively small capacities are required at fairly high heads, vertical multi-stage centrifugal pumps have been developed. These have small diameter impellers having a low tip velocity, one of the factors contributing to the quiet operation which is so often essential for this class of application, particularly in residential premises and hospitals. Where viscous liquids are to be pumped or where self-priming with good suction function performance is essential, a positive-displacement pump would be the first choice. With some applications however, it is necessary for the pump to operate for periods with a completely dry suction and one specialized type, the sliding shoe design, is capable of handling this condition for reasonable periods without harm. Such a dry suction condition could arise with recirculation systems where the suction source may be starved for periods due to cyclical or abnormal operating conditions, with sump pumps emptying engine-room bilges aboard ship and similar duties.

With the growing interest in pollution control, there is an increasing interest in pumps suitable for handling oily water without excessive emulsification so that the liquid can be handled efficiently by conventional oil/water separators and the sliding-shoe design, for example, operating at reduced speed, has proved very suitable for this application.

Centrifugal pumps are generally considered unsuitable as they churn the liquid excessively and the oil/water separators are unable to handle the resultant emulsification.

Although these notes on pump types have referred briefly to the main characteristics of centrifugal and positive displacement pumps, the past few years have seen a rapid growth in the number of specialized designs available for particular applications. It is therefore advisable for pump manufacturers to be approached during the early stages of installation layout and design for their recommendations.

NOISE

Noise can be very simply defined as unwanted sound. It is generally accepted as an irritant which can affect our mental and physical state and its effect can be particularly significant in the relatively quiet environmment of the office, domestic and hospital buildings.

Noise transmission associated with pumping systems falls into three main categories, airborne noise, water borne noise and noise transmitted through the actual structure of a building. Careful selection of equipment is most important to ensure that noise generation is kept to an absolute minimum.

Airborne noise is not normally a particularly serious problem with pumping equipment and its effects can be minimized by careful siting of the plant room in relation to working or living areas and by the use of sound insulating materials on the walls and ceiling of the plant room.

Water borne noise may be more of a problem as it carries through the pipework to any part of the building and the most effective way of dealing with it therefore is to prevent it at source as far as possible, rather than to suppress it. Particular attention should be paid to pipe sizing and layout to keep water velocities low and avoid turbulence and pipework should be securely anchored to prevent vibration. Where risers pass through floors they should not be grouted in but supported by soft packing materials and where pipework is clipped to walls, ceilings etc., pipe clips should preferably be insulated with rubber, felt or similar absorbent material. While it is generally considered that 1440 rpm pumps are inherently quieter than 2900 rpm pumps, modern vertical multi-stage pumps with small diameter impellers having a low top speed operating at 2900 rpm have proved to be extremely quiet in operation.

Noise can be transmitted to the structure of the building through the foundations of the pump and in certain instances it may be desirable to mount the pump sets on resilient mats or anti-vibration mountings in which case it is essential that flexible connections are fitted to the suction and discharge flanges of the pump set to minimize the possibility of vibrations being transmitted to the structure through the pipework and to prevent undue strain on the flanges.

These simple precautions are normally sufficient to produce satisfactory results but where exceptionally quiet operation is essential for a very specialized application it may be necessary to take additional precautions and such cases should be referred to companies who specialize in noise suppression.

PUMP INSTALLATION

It is most important that pumps are correctly installed and the following notes are intended as a general guide.

Location

The pump should be in an accessible location and there must be room for dismantling and maintenance. It is normally advisable to raise the pump above floor level so that it is more accessible and can be easily drained.

Piping

The size and suction of delivery pipes should be carefully calculated and should in any case, not be smaller than the pump connections. For long runs of piping or when handling viscous liquid they should generally be larger than the pump connection. For other than very simple runs, the actual pipe losses should be calculated so as to check the total head on the pump. It is quite common for the friction head to exceed the static head.

Piping should be accurately cut and fitted so that it can be bolted up to the pump branches without putting any strain on the pump or pipe joints. Easy bends should be used and sharp elbows and tees avoided. Particular care must be taken that all joints in the suction line are absolutely tight to avoid loss of capacity or difficulty in priming due to airleaks. In the case of non-self priming centrifugal pumps, the suction pipe should slope up towards the pump suction to avoid the possibility of trapping air in the top of the suction piping.

Suction Strainer

For most industrial applications or where solids may be present, a suction strainer should be fitted to protect the pump. The total area of the holes in the strainer should never be less than twice the cross-sectional area of the suction pipe, preferably three times as a general rule and four times or more if there is likelihood of frequent clogging.

Foot Valve

With most pumps operating under suction lift conditions a foot valve at the end of the suction pipe is desirable to keep the suction pipe full at all times and eliminate the need for priming after a shut-down. Depending upon the application, the foot valve may be combined with the suction strainer.

Discharge Non-return Valve

If there is any appreciable discharge head or length of delivery pipe, a non-return valve should be fitted, both to make it possible to open up the pump without draining the pipe and to prevent the head of water from driving the pump in reverse after stopping.

Relief Valve

Positive displacement pumps will develop excessively high pressures if run with the discharge shut-off or throttled and a spring-loaded relief valve should be fitted to the pump.

A relief valve is designed to give temporary protection against an abnormal condition and, should the valve operate, the cause should be ascertained and rectified.

Electric Motors

Squirrel-cage AC motors are the normal choice wherever possible and can usually be started direct-on-line. Under these conditions, the average motor takes four to six times full load current on starting and where local regulations or conditions prohibit direct-on-line starting, the use of slip-ring motors or other reduced-current starting methods should be considered. The type of enclosure, eg. drip proof or totally enclosed, should be considered with reference to conditions under which the pump will be working.

Switch Gear

It is not possible to protect a motor adequately by means of fuses only as a fuse that will carry the normal starting current is too heavy to give protection against ordinary overloads. Pumps should be controlled by starters with suitable overload releases, set in accordance with the manufacturer's recommendations.

BRITISH STANDARDS

BS 1394	Stationary circulation pumps for heating and hot water service systems.
BS 1394: Part 1: 1971	Glanded and glandless pumps.
BS 1394: Part 2: 1971	Domestic glandless pumps.
BS 4082	Specification for external dimensions for vertical in line centrifrugal pumps.
BS 4082: Part 1: 1969	"I" type.
BS 4082: Part 2: 1969	"U" type.
BS 5257: 1975	Specification for horizontal and end section centrifrugal pumps (16 bar).
BS 6700: 1987	Specification for design, installation, testing and maintenance of services supplying water for domestic use within buildings and their curtilages.

Note

The year given after the Standard's number indicates the latest date of issue.

NATURAL GAS

The piped gas provided for domestic purposes and for commercial and industrial utilization is currently supplied primarily from gas fields in the British sectors of the North Sea, with additional supplies from other fields, eg. Morecambe Bay.

Substitute natural gas (SNG) is manufactured as a direct substitute for natural gas and as a means of providing additional gas to meet peak loads. It can be made from a range of feed stocks in a number of different types of plant. Feed stocks commonly used are liquefied petroleum gas (LPG) and naphthas.

The properties of these gases compared with their methane content are given below.

Data	Natural gas	SNG	Methane
CV Btu/ft^2 CV MJ/m^2	1034.0 39.2	1000.0 38.0	995.0 37.0
Specific gravity	0.58	0.555	0.56
Wobb number (Btu.ft^2)	1330.0	1343.0	1134.0
Air/gas (Volume/volume)	10.5	10.6	9.6
Flame speed	35cm/sec	35cm/sec	35cm/sec
Temperature for ignition	700°C	700°C	700°C

The Flow of Gas in Pipes

The Pole formula is used in the gas industry for determining the flow rate of gas in pipes. It is a simplification of the Darcy fluid flow formula.

The friction coefficient (f) applied in the Darcy formula is taken as a constant ie. f = 0.0065 for gas pipes of the diameter used for domestic supplies.

Pole Formula (SI units)

$$Q = 0.0071\sqrt{\frac{d^5 \times h}{s \times l}}$$

where Q = flow (m^3/hr)
d = diameter of pipe (mm)
h = pressure drop (millibar)
l = length of pipe (m)
s = specific gravity of gas

Pole Formula (imperial units)

Q = flow rate (ft^3/hour)
d = diameter of pipe (inches)
h = pressure drop (inches WG)
l = length of pipe (feet)
s = specific gravity of gas

Factors Affecting Pressure Loss

By re-arranging the metric version of the Pole formula;

$$h = \frac{Q^2 \times s \times l}{d^5(0.0071)^2}$$

It can be seen that a pressure drop of h;

varies with length $= \frac{1}{l}$

varies with quantity = Q

varies with specific gravity = s

varies with $\frac{l}{d}$ (ie. inversely as diameter)

The pipe friction is assumed to be constant and the specific gravity can also be assumed constant.

Example

Find Q (m^3/h) when d = 20mm, h = 1mb, l = 9m, s = 0.58

$$Q = 0.0071\sqrt{\frac{d^5 \times h}{s \times l}}$$

$$= 0.0071\sqrt{\frac{20^5 \times 1}{0.58 \times 9}}$$

$$= 0.0071 \times 782.96$$

$$= 5.559 \text{ m}^3/\text{h}$$

Gas Flow Tables

When using Tables E1 and E2 to solve practical problems, the pressure drop allowed for must include the losses of the measured pipe run plus an addition for fittings in the line from Table E3.

In the example given, h = 1mb and l = 9m, from which pressure loss

$$\text{mb/m} = \frac{1}{9} = 0.1111 \text{ mb/m}.$$

Referring to Table E2, it will be seen that the flow rate at this pressure loss and at the pipe diameter given, d = 20mm is 5.40m^3/h. This agrees adequately with the calculated result.

NATURAL GAS

Table E1 Pipe sizing table *(Discharge in a straight horizontal copper tube (to Table X of BS 2871: Part 1: 1971) with 1.0 mbar differential pressure between the ends, for gas of relative density 0.6 (air = 1))*

Size of tube mm	Length of tube (m)									
	3	6	9	12	15	20	25	30	40	50
	Discharge (m^3/h)									
6	0.12	0.06								
8	0.52	0.26	0.17	0.13	0.10	0.07	0.06	0.05		
12	1.5	1.0	0.85	0.82	0.69	0.52	0.41	0.34	0.26	0.20
15	2.9	1.9	1.5	1.3	1.1	0.95	0.92	0.88	0.66	0.52
22	8.7	5.8	4.6	3.9	3.4	2.9	2.5	2.3	1.9	1.7
28	18	12	9.4	8.0	7.0	5.9	5.2	4.7	3.9	3.5
35	32	22	17	15	13	11	9.5	8.5	7.2	6.3
42	54	37	29	25	22	18	16	15	12	11
54	110	75	60	51	45	38	33	30	26	23
76.1	280	190	150	130	120	98	86	78	66	58
108	750	510	410	350	310	260	230	210	180	160

* 1 mbar = $10^2 N/m^2$ = 0.1 kPa.

Table E2 Pipe sizing table *(Discharge in a straight horizontal steel pipe (to Table 2, medium, of BS 1387: 1967) with 1.0 mbar differential pressure between the ends, for gas of relative density 0.6 (air = 1))*

Nominal size mm	in	Length of pipe (m)									
		3	6	9	12	15	20	25	30	40	50
		Discharge (m^3/h)									
6	(⅛)	0.29	0.14	0.09	0.07	0.05					
8	(¼)	0.80	0.53	0.49	0.36	0.29	0.22	0.17	0.14	0.11	0.08
10	(⅜)	2.1	1.4	1.1	0.93	0.81	0.70	0.69	0.57	0.43	0.34
15	(½)	4.3	2.9	2.3	2.0	1.7	1.5	1.4	1.3	1.2	1.0
20	(¾)	9.7	6.6	5.3	4.5	3.9	3.3	2.9	2.6	2.2	1.9
25	(1)	18	12	10	8.5	7.5	6.3	5.6	5.0	4.3	3.7
32	(1¼)	39	27	21	18	16	14	12	11	9.2	8.1
40	(1½)	59	40	32	27	24	21	18	16	14	12
50	(2)	110	76	61	52	46	39	34	31	26	23
65	(2½)	220	150	120	100	92	79	70	63	54	47
80	(3)	340	230	190	160	140	120	110	97	83	73
100	(4)	690	470	380	330	290	250	220	200	170	150

* 1 mbar = $10^2 N/m^2$ = 0.1 kPa.

Table E3 Pipe sizing table *(The effects of elbows, tees or bends inserted in a run of pipe (expressed as the approximate additional lengths to be allowed)*

Nominal size				Approximate additional lengths to be allowed					
Cast iron or mild steel		Stainless steel or copper		Elbows		Tees		90° bends	
mm	(in)	mm	(in)	m	(ft)	m	(ft)	m	(ft)
Up to 25	(1)	Up to 28	(1)	0.5	(2)	0.5	(2)	0.3	(1)
32 to 40	(1¼ to 1½)	35 to 42	(1¼ to 1½)	1.0	(3)	1.0	(3)	0.3	(1)
50	(2)	54	(2)	1.5	(5)	1.5	(5)	0.5	(2)
80	(3)	76.1	(3)	2.5	(8)	2.5	(8)	1.0	(3)

Table E4 Pipe supports *(Maximum interval for cast iron, mild steel and stainless steel pipes)*

Nominal size Cast iron, mild steel	Stainless steel		Interval for vertical runs		Interval for horizontal runs	
mm	mm	in	m	ft	m	ft
15	15	½	2.5	8	2.0	6
20	22	¾	3.0	10	2.5	8
25	28	1	3.0	10	2.5	8
32	35	1¼	3.0	10	2.7	9
40	42	1½	3.5	12	3.0	10
50	54	2	3.5	12	3.0	10
80	76.1	2	3.5	12	3.0	10
100	108	4	3.5	12	3.0	10

Table E5 Pipe supports *(Maximum interval for light gauge copper pipes)*

Nominal size		Interval for vertical runs		Interval for horizontal runs	
mm	in	m	ft	m	ft
Up to 15	Up to ½	2.0	6	1.2	4
22	¾	2.5	8	1.8	6
28	1	2.5	8	1.8	6
35	1¼	3.0	10	2.5	8
42	1½	3.0	10	2.5	8
54	2	3.0	10	2.7	9
—	2½	3.5	12	3.0	10
76.1	3	3.5	12	3.0	10
108	4	3.5	12	3.0	10

Meter House Dimensions

There are currently no national standards in respect of commercial/industrial gas meter houses/compartments. Any attempt to produce 'national' criteria in this Guide would inevitably attract criticism from British Gas Regions who adopt their own standards in respect of meter houses. There is also a trend towards the use of prefabricated 'skids' which include the meter and it's associated pipework together with valves, by-pass arrangements and the necessary pressure and temperature compensating equipment. Individual gas regions currently approve their own individual designs which might not be accepted elsewhere. Designers are therefore recommended to refer to the gas region in whose area the work is to be undertaken to establish accepted methods and dimensions.

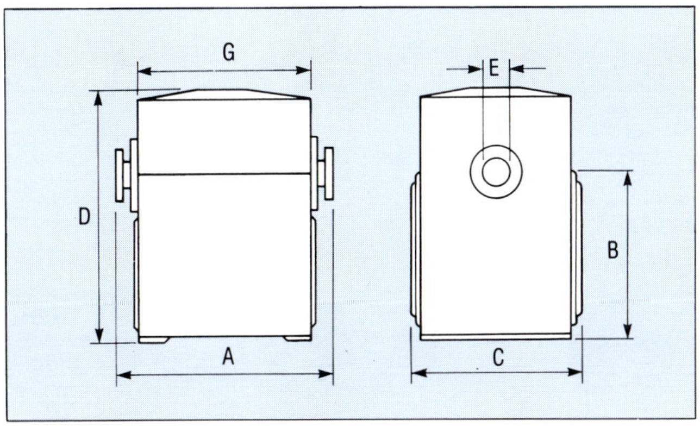

Table E6 Industrial gas meter sizes (IGE Table)

Capacity per hour		Capacity per rev.	Type of index	A	B	C	D	E nom.		G	Weight
cu. ft.	m^3	m^3		m	m	m	m	mm	in.	m	kg
1200	34	0.0566	A	0.6858	0.508	0.4128	0.7366	75	3	0.4953	50.80
1800	51	0.0944	A	0.762	0.508	0.5842	0.7938	75	3	0.5652	79.38
3000	85	0.1888	A	0.8763	0.6795	0.635	0.9462	100	4	0.6795	111.13
4500	127.5	0.2832	B	1.0668	0.8763	0.7811	1.124	100	4	0.870	175.53
6000	170	0.2832	B	1.0795	0.8763	0.7811	1.124	150	6	0.870	175.53
9000	255	0.5664	B	1.2319	1.0414	0.9017	1.4415	150	6	1.0287	323.40
12000	340	0.708	B	1.3462	1.143	1.016	1.5812	150	6	1.143	419.10
15000	425	0.708	B	1.4097	1.2192	1.016	1.6574	200	8	1.143	457.20
20000	566.5	1.416	B	1.5431	1.2446	1.1621	1.8098	200	8	1.2891	609.60
25000	708	1.416	B	1.5431	1.2446	1.1621	1.8098	200	8	1.2891	609.60
30000	850	1.70	C	1.6764	1.4288	1.3081	1.9939	250	10	1.4224	1111.25
50000	1416	3.40	C	1.9558	1.6193	1.6256	2.286	300	12	1.6574	1320.80
60000	1700	3.40	C	1.9558	1.6193	1.6256	2.286	300	12	1.6574	1320.80
70000	1982	3.40	C	1.9558	1.6193	1.6256	2.286	300	12	1.6574	1320.80

NATURAL GAS

Table E7 Typical equipment gas consumption figures

Appliance	Gas consumption m³/hr	ft³/hr	litres/sec
45 litre boiling pan	2.5	90	0.7
90 litre boiling pan	3.4	120	0.95
135 litre boiling pan	4.3	150	1.2
180 litre boiling pan	5.0	175	1.4
1200mm hot cupboard	2.7	95	0.75
1800mm hot cupboard	3.0	110	0.85
Steaming oven	2.1 to 2.9	80 to 100	0.6 to 0.8
Double steaming oven	5.75	200	1.6
2 tier roasting oven	2.9	100	0.8
Double oven range	10.0 to 12.0	350 to 400	2.75 to 3.2
Roasting oven	1.7	60	0.47
Gas cooker	4.3	150	1.2
Hot cupboard	1.0	35	0.275
Drying cupboard	0.3	10	0.08
Gas iron heater	0.3	10	0.08
Washing machine	1.1	40	0.31
Wash boiler	1.7 to 2.9	60 to 100	0.47 to 0.8
Bunsen burner	0.15	5	0.04
Bunsen burner, full on	0.6	20	0.16
Glue kettle	0.6	20	0.16
Forge	0.85	30	0.23
Brazing hearth	1.7	60	0.47
Incinerator	0.36 to 1.2	12 to 40	0.1 to 0.32

LIQUEFIED PETROLEUM GAS INSTALLATIONS

Introduction

Liquefied petroleum gas (LPG) is a generic term used to describe gases (predominantly C_3 and C_4 hydrocarbons) which exist as vapour at normal atmospheric temperature and pressure but which can be liquefied at only moderate pressure. When the pressure is released, the gas returns to its vapour state.

The two main liquefied petroleum gases in general use are commercial butane and commercial propane, both conforming with BS 4250. All persons involved with LPG installations should be familiar with the properties of commercial butane and commercial propane and the potential hazards involved, both for their own safety and that of others in the vicinity, including those involved in fire-fighting and control.

Building Control and Planning Permission

It is essential that the local Building Control and Planning Departments are consulted at an early stage in any proposal to site LPG storage vessels, either cylinders or tanks. Planning permission is required and the Planning Department will normally consult with the Environmental Health Department, Health and Safety Executive and Fire Brigade. If in doubt, ask – a telephone call to the local Planning Department could save considerable expense if the installation has to be changed later in order to comply with the Regulations.

Typical properties of commercial butane and propane gases, based on Appendix 1 in LPGITA Code of Practice No 1 are as follows:

	Commercial Butane		Commercial Propane	
Liquid Phase				
Relative density (to water) of liquid at 15.6°C	0.57		0.51	
Litres/tonne	1750		1960	
Kg/litre	0.57		0.51	
Imperial gallons/ton	390		440	
lb/imperial gallon	5.7		5.1	
Vapour Phase				
Relative density (to air) at 15.6°C and 1015.9 mbar	2.0		1.5	
Ratio of gas to liquid volume at 15.6°C and 1015.9 mbar	240		270	
Boiling point at atmospheric pressure	−2°C		−45°C	
Vapour pressure (abs) (max)	bar	psig	bar	psig
20°C	2.5	40	9.0	130
50°C	7.0	100	19.6	283
Limits of flammability at atmospheric pressure				
(% gas in air)	1.8 to 9.0		2.2 to 10.0	
Calorific values (gross)	121.8 MJ/m³		93.1 MJ/m³	
(net)	112.9 MJ/m³		86.1 MJ/m³	

General Properties

All personnel working with LPG should receive adequate initial and refresher training as appropriate. The notes in this Guide are not intended to cover all aspects of LPG installations but are a guide to the extent of information personnel engaged in small industrial, commercial and domestic installations would be expected to fully understand.

For reference and further information, the Codes and Standards listed at the end of this Section should be consulted. The values of the physical characteristics of the product have been 'rounded' to facilitate remembering; for exact values, refer to British Standards, LPGITA or the suppliers of the gas.

LPG is normally stored at ambient temperature as a liquid in steel vessels (or special lightweight alloy cyclinders for touring caravans) under pressure. There are special applications where the liquid is stored under refrigerated conditions at a lower pressure but these applications do not normally apply to the type of projects being considered here.

The liquid is colourless and is approximately half the weight of an equivalent volume of water. If LPG is spilled on water, it will float on the surface before vapourising. The liquid occupies about 1/250th of the volume needed if the product was stored as a gas. It is more practical therefore, to store and transport the product as a liquid under pressure than as a gas. However, a leakage of a small quantity of the liquid product can lead to large volumes of vapour/air mixtures and possible hazard.

LPG vapour is denser than air, commercial butane being about twice as heavy as air and commercial propane about one and a half times as heavy as air. Leakage of LPG will therefore flow to low points, for example, along the ground to the lowest level of the surroundings, into basements or into drains without water seals. If the air is still, any LPG vapour will disperse slowly and could be ignited a considerable distance from the original leakage, the flame travelling back to the source of leakage.

When LPG vapour is mixed with air in certain proportions, a lower flammable mixture is formed. Within the range of 2% (lower limit) to 10% (upper limit) of the vapour in air at atmospheric pressure, there is a risk of explosion. Outside this range, any mixture is either too weak or too rich to propagate flame but it is important to understand that over-rich mixtures can become diluted with air and becomes hazardous. At pressures higher than atmospheric, the upper limit of flammability is increased but the increase with pressure is not linear, eg. doubling the pressure does not double the upper limit of flammability.

ON NO ACCOUNT SHOULD A NAKED FLAME BE USED TO DETECT A LEAK

An explosimeter, properly calibrated for LPG, must be used for testing the concentration of LPG vapour in the air. LPG vapour is slightly anaesthetic and may cause suffocation if present in sufficiently high concentrations. Extreme caution must be taken when testing for leaks and there should always be a second person stationed outside the area to supervise. The LPG supplier should be consulted if a leak is suspected, particularly a leakage of liquid product. There is no acceptable percentage or degree of leakage with LPG.

LPG is normally odorised by the manufacturer by the addition of an odorant such as ethyl mercaptan or dimethyl sulphide which give LPG its characteristic odour and enable leaks to be detected by smell at concentrations down to 0.4% of the gas in air (ie. one-fifth of the lower limit of flammability). There are special applications where the LPG is not odorised, eg. where the odorising material is harmful to a process or does not serve any useful purpose as a warning agent. In such applications, extra precautions and safety procedures must be taken in respect of marking storage vessels, installing pipelines externally, inspections daily, provision of automatic flammable gas detectors etc.

Vessels are never completely filled with liquid, the maximum percentage fill varying between 80-87% depending on the vessel size. The space above the liquid level allows for liquid expansion (due to temperature changes) and for a supply of compressed vapour for drawing-off by the consumer. When the pressure of the vapour in the space above the liquid falls, the liquid boils to produce more vapour and restore the pressure. The boiling point of propane is around –45°C at atmospheric pressure. The liquid takes heat (latent heat of vapourisation) from the liquid itself, the metal of the vessel in contact with the liquid and from the surrounding air. This cooling effect may cause condensation and even freezing of the water vapour in the air local to a leakage. This effect may show as a 'white frost' at the point of escape and make it easier to detect leakage. Leaks can sometimes be seen as a 'shimmering' due to the refractive index of LPG.

LPG, particularly liquid product, can cause severe frost burns (for reasons similar to those outlined in the previous paragraph) if brought into contact with the skin. Goggles, gloves and protective clothing should be worn if exposure to this hazard is likely to occur.

A vessel which has contained LPG and is 'empty' may still contain LPG in vapour form. In this state, the pressure in the vessel is approximately atmospheric and if the outlet valve is left open or is leaking, air can diffuse into the vessel forming a flammable mixture and creating a risk of explosion. Alternatively, LPG may diffuse from the vessel to the atmosphere.

LPG Bulk Tank Location and Safety Distances

This section covers LPG bulk storage installations at industrial, commercial and domestic consumers' premises where the LPG is stored in a tank or tanks larger than 150 litres water capacity. For installations in refineries, bulk plants and large industrial plants, reference should be made to the publications issued by the Health and Safety Executive and the LPGITA.

Storage tanks should normally be installed above ground in the open air in a well ventilated position and should NOT be installed in basements or open pits. Tanks should not be installed one above the other. Vertical cylindrical storage tanks are commercially available and although used extensively abroad, vertical tank installations in the UK are not commonplace and are generally more expensive. However, a vertical tank may solve a problem on a restricted site but the LPG supplier should be consulted at an early stage.

Tanks may be installed UNDERGROUND but most LPG suppliers would prefer not to put tanks underground mainly because such installations have to be accessed for full internal visual examination and ultrasonic or hydrostatic testing every 5 years instead of every 10 years for an above ground tank. The cost of the associated attendances may make underground tank installations prohibitive.

Storage tanks should be sited and located in accordance with Table E8. Note that details for underground tanks have been omitted for clarity but details for these can be obtained from Table 2 in HSE Guidance Note HS(G)34 or Appendix 2 in LPGITA Code of Practice No 1.

Table E8 Minimum recommended safety distances for LPG storage vessels

Maximum water capacity of any single tank in a group	Nominal LPG capacity	Maximum total water capacity of all tanks in a group	Minimum separation distance	
			From building boundary, property line (whether built on or not) or fixed source of ignition*	Between tanks
litres	tonnes	litres	metres	metres
150 to 500	0.05 to 0.25	1500	2.5	1.0
> 500 to 2,500	0.25 to 1.10	7500	3.0	1.0
>2,500 to 9,000	1.10 to 4.00	27,000	7.5	1.0
>9,000 to 135,000	4.00 to 60.00	450,000	15.0	1.5

** A fire wall may be used under certain circumstances to reduce (to approximately half) the minimum separation distances shown above. For details, see HSE and LPGITA publications*

The number of tanks in one group should not exceed six, subject to the maximum total capacity of a group given in Table E8. If more than one group is required, then any tank in one group should be

LIQUEFIED PETROLEUM GAS INSTALLATIONS

at least 7.5 metres from any tank in another group unless a radiation wall is erected between the groups or adequate fixed water spray systems are provided.

Separation distances are intended to protect the LPG facilities from the radiation effects of fires involving other facilities as well as to minimise the risk of escaping LPG being ignited before being dispersed or diluted. The distances given are minimum recommendations and refer to the horizontal distance in plan between the nearest point on the storage tank and the nearest point of a specified feature (eg. an adjacent storage tank, building, property line etc).

Radiation walls or adequate fixed water drenching systems may be provided to enable separation distances for above ground tanks to be reduced, but specialist advice should be obtained. If separation distances are reduced, it may be necessary to provide diversion walls or kerbs (maximum height 500 mm) to ensure that the path of leaking gas from a storage site to a specified feature is not less than that shown in Table E8.

Conventional bunds, (an enclosure capable of retaining the total capacity, plus 10% margin, of all the vessels within the enclosure) around LPG storage tanks should NOT be used.

No LPG storage tank should be installed nearer than:

(a) 6 metres to the bund wall of any tank containing a flammable liquid with a flash point below 32°C.

(b) 6 metres from any tank containing a flammable liquid with a flash point between 32°C and 65°C.

(c) 3 metres from the top of the bund wall of any tank containing a flammable liquid.

LPG storage tanks should be installed well away from tanks containing liquid oxygen or other toxic or hazardous substances (eg. chlorine) – distances between 6 and 45 metres are not uncommon depending on the relative sizes of the vessels – but specialist advice should always be obtained, (see Table 4 in HSE Guidance Note HS(G)34).

No LPG storage tank should be installed within the bunded enclosure of a tank containing a flammable liquid, liquid oxygen or any other hazardous or cryogenic (producing low temperatures) substances. No LPG storage tank should be located in any bund where there is a permanent source of heat (eg. steam mains) or within the bunded enclosure of a heated storage tank (eg. fuel oil tank).

The vicinity of LPG storage tanks should be free of pits and depressions which might form gas pockets and affect the safety of the tanks. The ground beneath storage tanks should be either compacted or concreted and should be sloped to;

(a) prevent the accumulation of any liquid, including rainwater and cooling water applied under fire-fighting conditions, beneath the tanks;

(b) ensure a flow of any liquid away from tanks so that other vessels or important areas are not affected.

LPG storage vessels should not be sited in locations known to be susceptible to flooding eg. near rivers and streams which could overflow their banks during abnormal weather conditions.

To prevent tampering and possible vandalism, storage tanks should be enclosed by an industrial-type fence which is at least 1.8 metres high and at a distance of not less than 1.5 metres from the LPG tanks, unless it is a boundary fence when the distances given in Table E8 will apply.

Around the immediate vessel area, fences should have at least two non-self locking gates, not adjacent to each other and preferably at diagonally opposite corners of the fenced enclosure, opening outwards to provide easy means of exit in an emergency situation. It is preferable that the padlocks fitted to these gates should suit the master key system operated by most LPG suppliers as this can prevent problems and delays during deliveries. The gates must be unlocked when the compound is occupied.

The provision of a fence need not apply to tanks of 9,000 litres water capacity or less which are provided with a hinged, lockable cover to deny access to valves and fittings. However, should it be elected to erect a fence to prevent unauthorised interference, then the full provisions of the previous paragraph must be applied including the provision of two means of exit. Where there is surveillance at industrial premises, the site perimeter fence may suffice for security.

Where there is a possibility of mechanical damage to LPG storage and associated equipment from vehicles (eg.in goods delivery yards, near site roads etc.), suitable protection must be provided by the use of crash barriers, vehicle impact bollards or a non-continuous wall not more than 500 mm in height.

At least two NON SMOKING OR NAKED LIGHTS notices in red on white background shall be fixed to the outside of the compound surrounding wall or fence or, if these have not been provided, the notices shall be attached to the tank. The size of the lettering shall be such that notices can be clearly read at the safety distances applicable to the installation and from points of access to the storage site.

Each storage tank shall be clearly and boldly marked:

HIGHLY FLAMMABLE – BUTANE
(or PROPANE as appropriate)

Long grass, weeds and combustible material should be kept clear from an area within 3 metres of any storage tank of up to 2,250 litres of water capacity and within 6 metres of larger vessels. Certain weed killers are a potential source of fire hazard and should not be used.

An effective earthing point and/or bonding connection should be provided at the consumer's storage site for discharging static electricity from bulk tanker vehicles prior to commencing the delivery operation. Consumer vessels greater than 2,250 litres water capacity should be electrically earthed as a protection against the accumulation of static electricity. The earthing point for the bulk tanker vehicle and the earth for the storage vessel should have electrical continuity and should preferably be a common earth. The resistance to earth should not exceed 1 × 10 ohms.

Access is required for delivering and positioning the tanks. The smaller domestic tanks may be off-loaded from the delivery vehicle using the onboard vehicle crane and then man-handled into position using a tank trolley. A separate crane may be required to off-load and position larger tanks. Bulk LPG delivery tankers normally carry a 30 metre long hose which means that the tank has to be within approximately 25 metres from the road or hard standing for the tanker. Extended fill pipes are practicable but are normally used as a last resort in difficult commercial and industrial sites.

The tanker driver must be able to stand at the storage tank or filling point and be able to observe the vehicle whilst he is filling the tank. Access to the premises for a bulk tanker vehicle may be required under some site conditions. If the bulk tanker vehicle has to negotiate difficult bends or steep gradients, it is usually advisable to arange for the LPG supplier to make a 'dummy delivery' before any work commences. Suppliers may charge more for the LPG if it is necessary to send a helper with the driver for deliveries to awkward sites.

Figure E1 shows the layout of a typical small bulk vessel adjacent to a building and Figure E2, a 1 tonne bulk vessel at domestic premises.

Figure E1 Small bulk vessel adjacent to a building

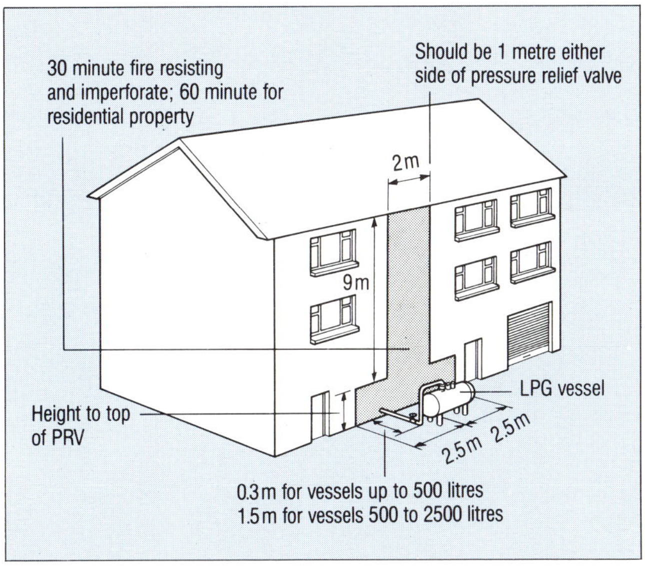

Figure E2 Small bulk vessel at domestic premises

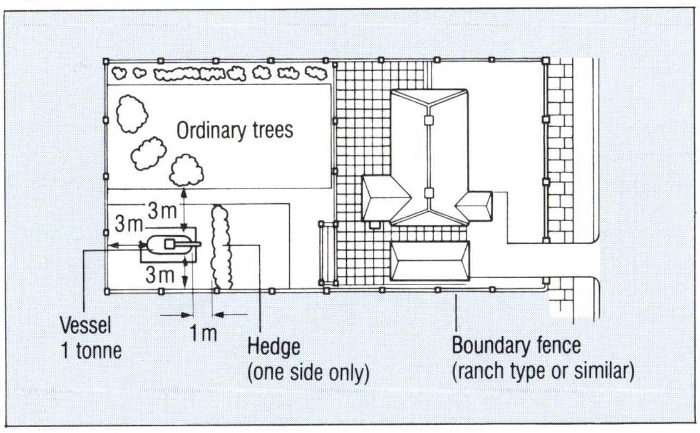

Storage Tanks and Fittings

Storage tanks are designed, fabricated and tested in accordance with British Standards, AOTC Rules and other recognised pressure vessel codes. Storage tanks are normally purchased or hired from the LPG supplier, who should assume responsibility for ensuring that the tanks, tank supports, protection against corrosion, testing and provision of tank fittings all comply with the rules and regulations.

The pressure of the vapour within a bulk storage vessel, under normal UK weather conditions, can vary between 2 and 9 bar depending on the liquid product temperature. On average, the vapour pressure is about 7 bar when liquid product temperature is 15°C. Safety pressure relief valves protect the vessel against excessive pressure due to exposure to heat in adjacent facilities.

A first stage regulator, mounted on the tank at the vapour off-take, reduces the varying high pressure to 0.75 bar. A second stage regulator reduces the intermediate pressure of 0.75 bar down to 37 mbar which is the industry standard propane operating pressure for domestic appliances.

The second stage regulator incorporates safey features to protect the appliances against excessive pressure.

Tanks are sized in accordance with three main criteria:

(a) Vapour Offtake Capacity

The tank must be able to boil-off liquid product faster than the vapour product is being drawn-off when all the gas appliances are operating at maximum capacity. This rate of gas production is known as the vapour offtake capacity of the tank and is a function, amongst other things, of the surface area of the tank; the larger the tank the larger the offtake capacity. It is normally quoted in cubic metres per hour (m^3/hr) or in kilograms per hour (kg/hr). Sometimes two figures are quoted; a figure for intermittent offtakes and a lower figure for continuous offtakes.

(b) Storage Capacity Required

To ensure continuity and security of gas supplies under adverse weather conditions, storage should be sufficient for a minimum of six weeks gas supply at maximum use. (This allows for the tank being partially full, some of the gas being used and for three weeks adverse weather conditions in the UK when it may not be possible for delivery vehicles to get to the site). The consequences of the gas supply failing during adverse weather conditions, industrial action, etc should be considered (eg. hospitals, old persons homes, etc) and additional storage provided if necessary.

(c) Location for Tank

Tanks must be located in accordance with the Codes of Practice and Regulations.

Each tank should be provided with at least one each of the following fittings suitable for LPG service over the range of pressures and temperatures appropriate to the operating conditions in service:

(a) pressure relief valve connected directly to the vapour space;

(b) drain or other means of removing the liquid contents;

(c) a fixed maximum level device and preferably also an independent contents gauge;

(d) a pressure gauge connected to the vapour space if the vessel is over 5,000 litres water capacity;

(e) a suitable earthing connection if the vessel is over 2,500 litres;

(f) a filling connection.

It should be noted that some LPG suppliers provide a combination valve for certain sizes of tanks which includes;

(a) a threaded filling connection incorporating a manual shut-off valve, a spring loaded back check valve, a relief valve to prevent pressure build up between the back check valve and the manual shut-off valve and a protective cap to prevent thread damage and ingress of foreign matter;

(b) a vapour off-take connection controlled by a diaphragm valve which can be fitted with a first stage regulator and pressure gauge;

(c) a fixed liquid level gauge incorporating a replaceable control valve.

For a detailed specification of tank fittings and piping, consult the references at the end of this section.

Figure E3 shows typical LPG storage tank dimensions. For exact dimensions and a copy of a certified drawing, consult the LPG supplier of the tank(s).

LIQUEFIED PETROLEUM GAS INSTALLATIONS

Figure E3 Storage tank dimensions

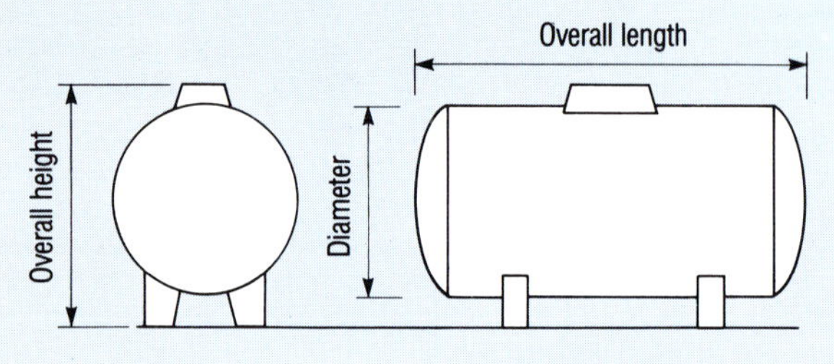

Nominal LPG capacity (kg)	Water capacity (litres)	Diameter (mm)	Overall length (mm)	Approx overall height (mm)	Continuous off-take (m^3/hr)
200	450	610	1675	900	2.26
600	1400	1000	1985	1310	5.66
1000	2250	1000	3042	1460	7.08
2000	4500	1220	4100	1685	10.19
7000	16000	1700	6782	3580*	28.30
12000	28000	2172	8600	3910*	39.62

** including platform and guard rails*

LPG Storage in Cylinders

The basic principles detailed in previous sections in respect of safety when handling and storing propane in tanks also apply to installations where the propane is stored in cylinders.

For permanently piped domestic installations with a low offtake rating, eg. single cooker, two 19kg propane cylinders with a change-over regulator should be sufficient. Larger domestic and small commercial and industrial installations, with a higher offtake rating, eg. central heating boilers and fires, bunsen burners in laboratories, etc, may require four 47kg propane cylinders. Each cylinder is fitted with a valve; the handwheel is turned anti-clockwise to open and clockwise to close. An excess pressure relief valve is also provided. The four cylinders, at the container pressure of up to 9 bar, are connected to a change-over valve by means of pigtails (see Figure E4). The change-over valve incorporates a regulator to reduce the pressure to 37mbar for the supply to the appliances. The change-over valve connects two cylinders to the pipework system and, when these two cylinders are empty, the valve automatically changes over to the two reserve cylinders and indicates that these have been brought into use.

For indoor applications, eg. portable domestic cabinet heaters where the cylinder is contained in or close to the unit, butane must be used.

Manufacturers of LPG may have different sizes and ranges of sizes of cylinders. Table E9 gives details of the popular sizes for a range of four propane cylinders and three butane cylinders:

Table E9 Cylinder sizes

Cylinder capacity (kg)	Overall height (mm)	Maximum diameter (mm)	Cylinder Offtakes Intermittent (m^3/h)	Continuous (m^3/h)
Propane				
3.9	337	248	0.28	0.17
13.0	584	318	0.57	0.28
19.0	800	318	0.85	0.42
47.0	1289	375	1.69	0.85
Butane				
4.5	337	248	0.17	0.08
7.0	495	256	0.21	0.10
15.0	584	318	0.28	0.14

Figure E4 Cylinder storage

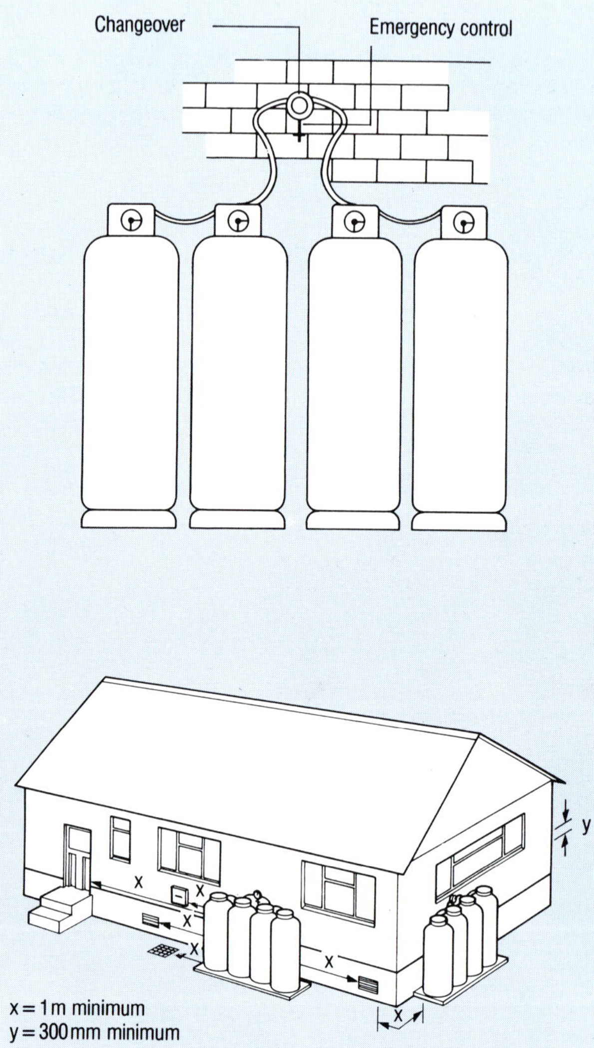

Courtesy: Calor Gas Ltd.

Pipe Sizing

The industry standard operating pressure for propane is 37 mbar and in order to obtain this pressure at the appliance, pipework must be of adequate diameter to pass the required rate of gasflow when all the appliances are operating at maximum output, without significant pressure drop due to friction losses. A pressure drop of not more than 2.5 mbar between the second stage regulator and the appliance at maximum gas flow rates has been proven to give satisfactory results.

Consideration should be given when sizing main runs of pipework to future extensions and/or appliances being added to the installation.

Table E10 gives recommended sizes for short final connections to individual domestic appliances.

Table E11 gives pipe size and flow rates in longer runs and higher rates of flow, with a 2.5 mbar pressure drop. This table should be used to determine the diameter of the main run.

Copper pipe to BS 2871: Pt 1: 1971 Type C106 Table X or Y with compression fittings to BS 864: Pt 2: 1983 and/or capillary fittings to BS 864: Pt 2: 1983 are suitable for LPG.

Galvanised screwed steel pipe, medium weight, to BS 1387: 1987 with screwed fittings to BS 143: 1968 are also suitable for LPG installations. PTFE tape or a jointing compound specifically made for LPG must only be used. General purpose jointing materials with hemp are not suitable for LPG. Existing joints on pipework being converted to LPG must be examined and if made with general purpose jointing materials and hemp, must be dismantled and remade or the pipework scrapped.

Polyethylene may be used for underground pipework but the manufacturer's instructions for jointing must be followed. Polyethylene must not be exposed to strong sunlight during transit and pending installation.

Underground pipework must be a minimum of 500 mm below finished ground level. Trenches should be dug deeper than 500 mm and all stones should be removed from the bottom of the excavation before backfilling and compacting the bottom 75 mm of the trench with sifted material. The pipework should be supported with sifted material and a warning tape should be laid 150 mm above the pipework to prevent possible damage during future excavations. All backfill should be carefully selected to remove all stones which could cause damage to the pipework due to settlement and compaction. Pipework should be protected if installed in corrosive soil conditions.

Pipework installations must be pressure tested and proved to be leak-free before connecting to the tank(s) or cylinder(s).

At the point of entry to the premises, LPG pipework must be bonded to the main earthing terminal in accordance with the IEE Regulations for Electrical Installations. The purpose of installing bonding conductors is to ensure that any metalwork within the premises, (eg. gas and water services) is at the same potential as the metalwork of the electrical installation. Bonding of gas service pipes should be made on the consumer's side of a meter, if fitted, between the meter outlet union and any branch pipework, but within 600 mm of the gas meter. The cross-sectional area of the bonding conductor is laid down in the IEE Regulations, depending on the type of installation, and this work should only be carried out by a competent electrical contractor or the Electricity Board. There is a statutory duty under Gas Safety (Installation and Use) Regulations 1984 for the installer of the LPG system to advise the occupier of the property that a equipotential bonding conductor is required.

Table E10 Guide to pipe sizes for appliances

Appliance	Metric (mm)	Imperial (in)
Central heating boilers	22	¾
Cookers, domestic	15	½
Lights	6	¼
Portable fires	6	¼
Fixed fires	10	⅜
Water heaters		
Single-point instantaneous	10	⅜
Multi-point instantaneous	15	½
Bath heaters	15	½
Optional single or multi-point	10	⅜
Sink storage	6	¼
Circulators	6	¼

Source: Calor Gas Limited

Note

All the above relate to outside diameter copper tubing with a maximum pipe run of 3 m (10 ft). Above this, the next largest pipe size should be used up to 6 m (20 ft); thereafter, pressure loss and pipe flow charts should be consulted.

Table E11 (a) Pipe sizing: BS steel pipe

Length of Tubing		8 mm (¼ in)		15 mm (½ in)		20 mm (¾ in)		25 mm (1 in)	
Metres	Feet	Btu/h	Ft³/h	Btu/h	Ft³/h	Btu/h	Ft³/h	Btu/h	Ft³/h
3	10	45	18.0	375	150	750	300	1650	660
6	20	31	12.6	250	100	525	210	1125	450
9	30	26	10.4	200	80	412	165	875	350
12	40	22	9.0	175	70	350	140	750	300
15	50	20	8.0	150	60	300	120	650	260
18	60	18	7.2	135	54	275	110	600	240
21	70	17	6.8	125	50	257	103	550	220
24	80	16	6.3	112	45	240	96	500	200

LIQUEFIED PETROLEUM GAS INSTALLATIONS

Table E11 (b) Pipe sizing: Imperial copper tube (NB)

Length of Tubing		¼ in		⅜ in		½ in		¾ in		1 in	
Metres	Feet	Btu/h	Ft^3/h	Btu/h	Ft^3/h	Btu/h	Ft^3/h	Btu/h	Ft^3/h	Btu/h	Ft^3/h
3	10	13	5.5	52	21.0	155	62	487	195	1100	440
6	20	9	3.9	35	14.0	105	42	317	127	750	300
9	30	8	3.2	28	11.5	82	33	255	102	575	230
12	40	6	2.7	25	10.0	72	29	220	88	500	200
15	50	6	2.4	22	9.0	62	25	195	78	450	180
18	60	5	2.1	20	8.3	55	22	175	70	387	155
21	70	4	1.9	19	7.6	52	21	157	63	362	145
24	80	4	1.8	17	7.0	48	19	145	58	337	135

Table E11 (c) Pipe sizing: Metric copper tube (OD)

Length of Tubing		6 mm		10 mm		15 mm		22 mm		28 mm	
Metres	Feet	Btu/h	Ft^3/h	Btu/h	Ft^3/h	Btu/h	Ft^3/h	Btu/h	Ft^3/h	Btu/h	Ft^3/h
3	10	10	4.3	77	31.0	131	52.6	707	283	1405	562
6	20	7	3.0	50	20.0	89	35.6	460	184	782	313
9	30	6	2.5	42	17.0	70	28.0	370	148	735	294
12	40	5	2.1	37	14.8	61	24.6	320	128	640	256
15	50	4	1.9	33	13.3	53	21.2	282	113	575	230
18	60	4	1.7	30	12.3	46	18.6	252	101	495	198
21	70	3	1.5	28	11.2	44	17.8	228	91	462	185
24	80	3	1.4	26	10.4	41	16.5	210	84	430	172

Table E11 (d) Pipe sizing: Polyethylene pipe

Btu/h	C.F.H.	Maximum Length 25 mm OD		Maximum Length 32 mm OD	
Heat Imput	Propane	Metres	Feet	Metres	Feet
130	52	72	236	213	698
140	56	62	204	184	603
150	60	54	178	160	525
160	64	47	154	141	462
170	68	42	138	125	410
180	72	37	121	111	364
190	76	33	108	100	328
200	80	30	98	90	295

Note: Btu/h is heat input in 000s rounded down figures based upon propane at low pressure of 37 mbar (14.8 in w.g.) and 2.5 mbar (1 in w.g.) maximum pressure drop.

COMPRESSED AIR

Compressing the Air

There are many different types of machines for compressing air-reciprocating, rotary-vane, screw and turbine compressors. This section will cover reciprocating and rotary vane compressors only; turbine types are normally used only where extremely large quantities of compressed air are needed, often at relatively low pressures, and are outside the scope of the normal industrial installation.

The reciprocating compressor may have one or several stages. The rotory vane compressor consists of a rotor, having blades free to slide in radial slots, rotating off centre in a cylindrical chamber. Rotation causes the blades to be thrown out by centrifugal force and to sweep the compression chamber. A small amount of oil is admitted to the chamber to seal and lubricate the blades and to act as an internal coolant. Again, there may be one or more stages.

The screw compressor is a rotary positve displacement machine in which two intermeshing rotors each in helical configuration, displace and compress the air. Available in lubricated and non-lubricated (oil-free) construction, the discharge air is normally free from pulsation. The machine has a high rotation speed, and is available in single or twin stages.

There is no hard and fast rule about the choice of single or multi-stage compressors. A multi-stage machine will use less power to compress a given quantity of air, the power required being appreciably less as the pressure rises. But a multi-stage compressor can be more costly to purchase and so there must be an economic balance between the initial cost and the running cost.

It is therefore usual to find that for simplicity and low initial cost, single stage compressors are used for small duties and pressures up to about 7 bar (100 psi), whereas for pressures above this and for higher duties, compressors having two or more stages are used.

There is also considerable difference in the air temperature leaving a single or two stage compressor as Table E12 will show. The sizing of compressors is outside the scope of this Guide, but there are a number of points which should not be forgotten when sizing and choosing a compressor. One of these points covers the effect of altitude on the volumetric efficiency of the compressor, as shown in Table E13. Other aspects which should be considered include the following.

(1) Future expansion requirements.
(2) Maximum and minimum pressures required in the system.
(3) Type of cooling required.
(4) Type of compressor.
(5) Running cost.
(6) Initial cost.
(7) Space.
(8) Type of control to meet anticipated plant requirement.
(9) Protection devices.

Table E12 Final temperature (0°C) of adiabatic compression from free air at 1.013 bar at 20°C (SI metric units)

Gauge pressure (bar)	Single stage (°C)	Two stage (°C)
3	164	85
4	192	97
5	218	106
6	240	116
8	278	129
10	310	141
14	365	160

Table E13 Effect of altitude on compressor volumetric efficiency (SI metric units)

Altitude (m)	Barometer pressure (mbar)	Percentage relative volumetric efficiency compared with sea level (4 bar)	(7 bar)
Sea level	1013	100.0	100.0
500	945	98.7	97.7
1000	894	97.0	95.2
1500	840	95.5	92.7
2000	780	93.9	88.0
2500	737	92.1	87.0

Compressor Cooling

Because of the temperature rise which takes place when air is compressed, some form of cooling is required so that the temperature is not too high for satisfactory lubrication and to avoid excessively high thermal stresses in the machine structure. Cooling may be either by air or water.

Nowadays, air cooled compressors may have capacities up to 350 dm^3/s (750 cfm), or be rated on continuous duty up to 14 bar (200 psi). Air cooled cylinders are finned and additional cooling is provided by arranging for the flywheel or a fan to direct a stream of air onto the cylinder. Such compressors should not be run in a confined space, otherwise the high ambient temperature will prevent adequate air cooling.

A common method of compressor cooling is of course, to provide a water jacket. There are a number of ways in which such a jacket could be supplied with cooling water. It should however be remembered that, although the colder the water the more effective the inter or after cooler, cold water fed to the compressor jackets can be harmful. This is because it can cause water vapour in the compressed air to condense to the detriment of cylinder lubrication and also lead to possible corrosion.

Thermo-syphon Circulation

Thermo-syphon circulation is satisfactory for small single-stage compressors and relies on convection to circulate the water which is heated by the compressor. The water circulates from the compressor jacket to a holding tank where the heat is lost. It is essential that the flow and return pipes have a fall from the tank to the compressor to ensure good circulation. Even a horizontal pipe will reduce the flow rate and may induce air locks. Preferably, the tank should be placed in the open-air and the top should be open to provide the maximum cooling effect; however, adequate protection of the tank from birds etc must be ensured.

A tank having a large water surface area loses heat more quickly than a tall narrow tank. The drawback of such a tank is that it is liable to freeze if the compressor is shut down in cold weather. A stop valve should therefore be fitted in the cold water make-up line and on the tank outlet so that the compressor cooling jacket can be drained. To avoid draining an overnight shutdown, it is a good idea to fit a small electric immersion heater in the tank.

Pump Assisted Circulation

For larger single stage compressors, thermo-syphon circulation is too slow to dissipate the heat and a circulating pump must be installed to increase water velocity. The required water tank capacity should be discussed with the compressor manufacturer, but where information is not available, Table E14 can be used as a rough guide for compressors running at up to 7 bar (100 psi).

COMPRESSED AIR

Table E14 Cooling tank capacities

Compressor capacity (dm^3/s free air)	Tank capacity (litres)	Compressor capacity (cfm free air)	Tank capacity (gallons)
10	170	25	40
25	370	50	80
50	700	100	150
70	1020	150	225
100	1600	200	360
140	2200	300	480
200	3000	450	700
280	3800	600	850
350	4500	800	1000

Closed Cycle Cooling

It is usually better, particularly with larger compressors, to operate them on a closed circuit, the heat being dissipated either through a cooling tower or through a mechanical cooler.

A further advantage is that a closed circuit can eliminate jacket scaling, particularly if the water is treated. The use of the closed circuit does not mean that temperature control is unimportant. With the closed circuit, it is usually preferable to use a 'three-way temperature control'. It is also important to remember that in severe winter conditions, it is not uncommon for cooling towers to freeze solid. Should this be the case, the compressor must be shut down.

Freezing of the sump in the cooling tower may be prevented by fitting a heating coil. To prevent the tower becoming a solid mass of ice, the line from the diversion control valve should be so valued that it can return the water from the compressor direct to the sump instead of to the top of the tower in such low temperature conditions. Under these circumstances, sufficient cooling can usually be maintained due to heat losses from the sump itself.

A mechanical cooler, where the cascade of water is cooled by forced or induced air draught, is much smaller than the cooling tower and the elements are almost totally enclosed offering less risk of freezing when shut down. However, any ice which does form will usually melt within seconds of the compressor starting up.

Rotary Compressors

Where the compressor is of the rotary-vane or oil lubricated screw type it should be noted that oil is usually injected into the compression chamber to form a seal between the blades and the casing and to act as an internal coolant. The oil is removed from the air by a separator at the discharge and is then passed back to the sump by way of a water cooled heat exchanger.

Cooling the Air

The whole purpose of the compressed air installation is to deliver air to the point of use in the best possible condition – clean, dry and with the minimun loss of pressure. If it fails on any one of these counts, then there is likely to be increased wear on tools, poor performance particularly of items such as paint spray equipment, and the operating costs will inevitably be higher than they ought to be.

Removing Moisture

Atmospheric air always contains a proportion of water vapour, the amount depending on the relative humidity. In Britain, this may be between 50 and 75% (RH), this being highest in foggy or rainy weather or if the inlet of the compressor is adjacent or over a pond, stream or other damp area. It is important to note this latter point when installing compressors.

The amount of water which can be held by a given volume of air will depend on its temperature. The moisture carrying capacity of air increases with a rise in temperature and conversely, and at this stage, possibly of greater importance, it decreases with a fall in temperature (see Graph E1). Its moisture carrying capacity also falls as the pressure is increased. So, when 'free air' containing water vapour under average conditions enters the compressor, two things will happen. Its ability to hold the water will decrease as the air is compressed to a smaller volume but will increase because of the higher temperature resulting from the compression. Under average conditions, the air will leave the compressor, just able to carry its initial water content.

It will follow that any cooling which then takes place must cause the air to shed its excess water vapour by condensing and this can be accelerated by introducing artificial cooling devices such as intercoolers and aftercoolers.

Although it is customary to lag steam mains to retain heat, it is a potentially bad practice to lag the compressed air main between the compressor and the first major cooling plant (ie. aftercooler or receiver). If the pipework is lagged, the high discharge temperature of the compressed air may be sufficient to spontaneously ignite the deposits of oil, dirt, scale etc. commonly found in this first section of pipe. Once the air has cooled to near ambient temperature, this danger will not arise. An intercooler is fitted between the stages of a multi-stage compressor; its purpose is to cool the air between the stages and in cooling the air, it also serves the very useful purpose of condensing out the surplus water vapour which, if allowed to pass to the next compression stage, could condense on the cylinder wall with resultant damage to the compressor.

It is essential that the water is drained away from the intercooler and this can best be done automatically using one of the range of compressed air traps. An aftercooler should be fitted immediately following the compressor so as to remove as much water as possible before the air reaches the receiver.

The water must be drained from the bottom of the aftercooler, and this is best done automatically. Manual drains will work only if they are attended to regularly. Rarely, if ever, is this possible, and an automatic drain trap is the best way of ensuring that the system operates properly.

The most efficient aftercoolers are usually water cooled and the lower the air temperature they can produce, the better. However, there is a point of maximum efficiency so, for reasons of economy, where mains water is used, it is well worth fitting a temperature control to the water outlet to keep consumption within reasonable bounds.

For those areas where cooling water is either not available or is too expensive, the air blast aftercooler becomes the first choice. Ambient air is blown by an electric motor and fan over a bank of finned tubes, through which the compressed air flows. Although the compressed air discharge temperature is likely to be on average, approximately 6°C (10°F) higher than for a water cooled aftercooler, the unit will still require automatic drainage by a trap.

Graph E1 Moisture content of air (100% RH)

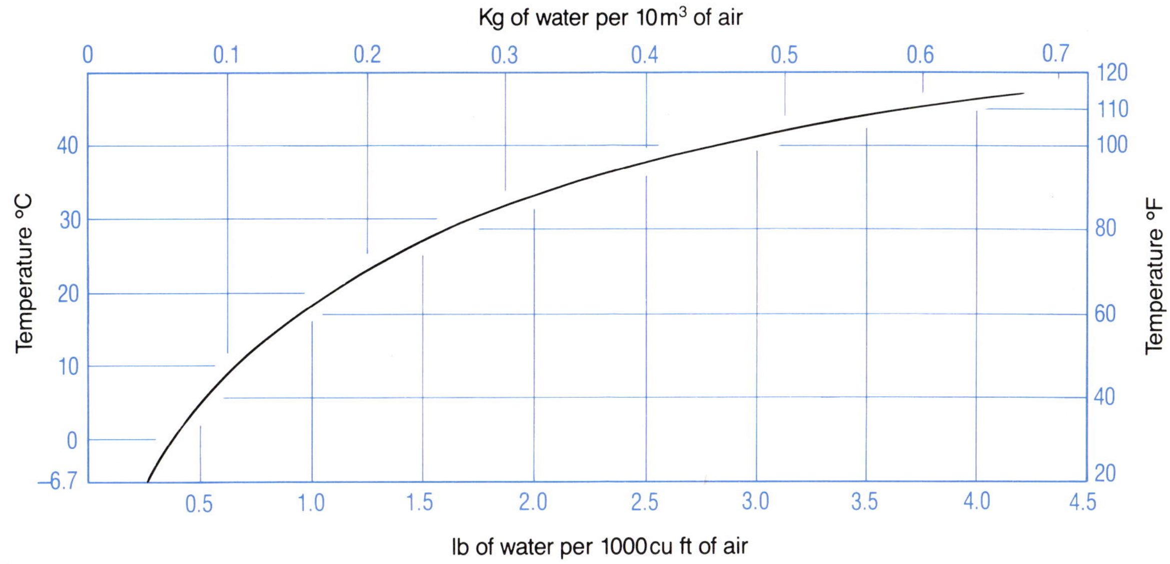

Example
(SI metric units)

How much moisture will separate out from air if the compressor inlet conditions are 20°C and 70% relative humidity; the compressor delivers 1 m^3/sec of free air compressed to 7 bar to the system at 25°C?

Compressor takes in 1m^3/sec

From Graph E1, water taken in will be:

$$\frac{0.18 \times 70}{10 \times 100} = 0.0126\,\text{kg/s}$$

Compression ratio at 7 bar = 7.91 (Table E15)

Next, we must find the volume of air after compression. Since its volume is proportional to the absolute temperature and to

$$\frac{1}{\text{compression ratio}} \times 1\,\text{m}^3$$

will occupy:

$$\frac{1}{7.91} \times \frac{(273 + 25)}{(273 + 20)} = 0.128\,\text{m}^3$$

From the graph, 10 m^3 of air at 25°C can carry 0.24 kg of water

Therefore, 0.128 m^3 can carry:

$$0.128 \times \frac{0.24}{10} = 0.00307\,\text{kg}$$

Therefore, the amount of water which will separate out is:

$0.0126 - 0.00307 = 0.00953$ kg/s

Receivers

Important as aftercoolers are, it would be very unusual for all the water vapour in the air to condense at this point. Further cooling almost always takes place in the receiver as well as from the distribution system. The water vapour, (and oil mist, if the compressor is of the lubricated type) condenses in the receiver and collects at the bottom. On those installations where the compression plant is small, an aftercooler may not be fitted, thus making the receiver the point at which most condensed liquid will be found. If these liquids are allowed to build up, carry- over into the mains system is likely. There is also the possibility of corrosion of the receiver itself.

It is therefore, important to ensure that these collected liquids and solids (atmospheric dust, pipe scale, carbon, rust etc.) are automatically removed as they collect. As the trap and its protective strainer will have to handle varying proportions of water, oil, emulsion, dirt, etc., regular cleaning is essential.

If excessive amounts of oil are being carried over from the compressor, it generally indicates that maintenance of the compressor is required. If a manual drain cock is fitted to the receiver a short distance above the drain trap outlet, oil and scum floating on the water surface (which might foul up the trap) can be periodically drained off.

Apart from the receiver's ability to cool the air and hence deposit liquid (that is why it is better to site the receiver where the ambient temperature is low), it performs two other functions. For some applications, it is important that the pressure pulses produced by a reciprocating compressor be eliminated as far as possible. The receiver therefore acts as a pulsation damper. The receiver also acts as a power storage vessel, allowing intermittent high demands for compressed air to be met from a smaller compression set.

Being a pressure vessel and thus subject to regular inspection, a receiver is fitted with inspection covers and manholes. These also allow any solid contaminent build-up to be removed. To comply with Factory and Safety Acts, a receiver must be fitted with an adequately sized safety valve and generally a pressure gauge is also fitted.

On small horizontal receivers generally supplied with the smaller industrial or garage type compressor, automatic drainage may be more difficult. The drain point is often in the centre of the dished end of the receiver or on the top. In each case, an internal dip pipe is fitted to allow the air pressure to displace the collected liquid when the manual drain is opened. An automatic trap can be used.

It is also worthwhile considering the capacity of the receiver. This is usually sized on the actual output in 1 minute from the compressor, but where consumption is high and fairly constant, the air is in the receiver for too short a time to cool down very much. Where this is so, the storage capacity is obviously low and it is better to size the receiver on plant consumption rather than on compression output.

One compressor manufacturer recommends the following as a guide to receiver size:

COMPRESSED AIR

$$\text{Receiver capacity } (m^3) = \frac{m^3 \text{ of free air required}}{\text{Allowable pressure drop (bar)}}$$

and also;

$$\text{Receiver capacity (cu ft)} = \frac{\text{cu ft of free air required} \times 14.7\text{psi}}{\text{Allowable pressure drop (psi)}}$$

Example (1)

A machine requires $3m^3$; available pressure is 7bar and the minimum suitable pressure is 5.5bar.

By using the formula to determine the receiver capacity;

$$\frac{3}{1.5} = 2m^3$$

Example (2)

A machine requires 100cu ft; available pressure is 100psi and the minimum suitable pressure is 80psi.

By using the formula to determine the receiver capacity;

$$\frac{100 \times 14.7}{20} = 73.5\text{cu ft}$$

Layout

Although in an ideal system, all cooling and condensing should be carried out before the air leaves the receiver, this is not very often achieved in practice. It is in fact impossible where aftercoolers are not fitted. The whole of the compressed air mains therefore become additional cooling surfaces, the amount of condensing which takes place depending on the efficiency of moisture extraction before the air leaves the receiver and the temperature in the mains system itself.

It is useless to provide a compressed air gun to blow out particles of swarf after a machining operation if, every time the operator used the gun, he squirts water all over the finished job. Equally, it can be very expensive to pass this water through compressed air operated tools.

Care must therefore be taken in the layout of the mains so that adequate fall is given to proper drainage points.

The general layout of the building will dictate the best positions for drain points but in general, the main should be given a fall of not less than 1 m in 100m (1 inch in 10ft) in the direction of air flow and the distance between draining points should not exceed 30m (100ft).

It is a good idea to form a distribution system as a ring main to help reduce pressure losses. It also makes the alteration of extension of the existing system easier. Drainage points should be provided by using equal tees and it assists in the separation of the water if these are arranged to change the direction of the flow as shown in Figure E5. Whenever a branch line is taken off the main, it should leave the top of it, so that any water in the main doesn't fall straight into the plant and the bottom of the falling pipe should be drained as in Figure E5.

Separators

Whilst automatic drain traps will effectively deal with water which has collected at the bottom of the main or in a receiver of some kind, they can do nothing for the mist of water droplets which may be suspended in the air. For most everyday applications, much of this water can be removed by fitting a separator in the distribution mains as Figure E5.

When a separator is fitted in a ring main system, install it to allow for the normal direction of flow.

Dryers

There are applications where the air must not only be clean, but have a reduced dew point. This may call for more sophisticated and expensive methods to lower the dew point of the compressed air. There are three common systems used for this purpose.

(a) Adsorption Dryers These consist of two pressure vessels filled with water adsorbing chemical. Wet compressed air is passed through one chamber until the chemical is saturated. Whilst this first chamber is being regenerated by heat and/or a purge of the ultra dry air, the second chamber is adsorbing moisture. An automatic control system alternates the chambers, one operation, one regeneration.

(b) Absorption Dryers These consist of a container of chemical through which compressed air passes. The chemical absorbs the water vapour, forming a solution which drains to the bottom of the container. This solution has to be discharged periodically by a drain trap, and the level of the desiccant then requires topping up. A domestic salt cellar is typical of this type.

(c) Refrigerant or Chiller Dryers These units are heat exchangers which will cool the air down to a theoretical dew point of 1 to 3°C (34 to 37°F) and thus precipitate out the moisture. The system is a straight mechanical refrigeration unit with one extra facility included. This is a second heat exchanger whereby the outgoing cold, dry air is used to pre-cool the incoming compressed air supply. In doing so, the outgoing cold air is warmed up to around ambient temperature.

All these units incur running costs of a kind, whether it be compressed air for purging, steam or electrical power to reactivate chemicals, or the replacement of desiccant.

Main Pressure Reduction

It is often necessary to reduce the mains pressure when supplying groups of plant or complete workshops. This requires a pressure reducing valve having quite a large capacity and very good flow characteristics. This unit is somewhat different from the small regulators used for individual items of plant.

Sizing Compressed Air Mains

The compressed air mains are the all-important link between the compressor and the point of usage. It is thoroughly bad to install mains which are too small and cause high pressure drop. If for example, a compressor has to work at 8bar (120psi) to cater for pressure drop conditions whereas 7bar (100psi) would normally meet the case, it calls for an additional power input of as much as 10%.

Mains which are too small also cause high velocity, making it difficult to separate the water from the air because much of the condensed vapour running as a stream of water along the bottom of the pipe, will be whipped up by and carried along with the fast moving air stream. Whilst a watchful eye must be kept on the pressure drop, it is common practice to size compressed air mains on velocity and a reasonable figure for all practical problems is 6–9m/s (20–30ft/sec) which is sufficiently low to prevent excessive pressure drop on most systems and will allow moisture to precipitate out without re-entrainment.

Many compressed air systems are working inefficiently because the demand has outgrown the supply – new pneumatic plant has been added from time to time without addition to the compressor plant or mains.

In designing a new plant, some thought might be given to possible future demands and allowances made in the mains sizes. Sizing by velocity presents an easy form of determining pipe size for a given duty, but it must be remembered that the duty of a compressor and the demand of the equipment is usually expressed in dm^3/s of free air and that when compressed, the volume will be less.

Table E15 shows that the ratio of compression and the actual volume occupied at any given pressure can be found by dividing the volume of free air by the ratio of compression.

Example

(SI metric units)

At a gauge pressure of 8 bar, Table E15 shows the ratio of compression as 8.9 so if we have 190 dm³ of free air compressed to 8 bar, it will occupy a space of:

$$\frac{190}{8.9} = 21.35\,dm^3$$

By adding the equivalent lengths to the actual length of pipe, the loss in each section of a system can be easily found by reference to the Table E16.

Table E15 Ratio of compression (metric units)

Gauge pressure (bar)	0.5	1	2	3	4	5	6	7	8	10	12	14	18
Ratio of compression	1.5	1.99	2.97	3.96	4.95	5.94	6.92	7.91	8.9	10.87	12.85	14.82	18.77

Table E16 Resistance of pipe fittings (equivalent length in m)

	Nominal pipe size (mm)									
Type of fitting	15	20	25	32	40	50	65	75	100	125
Elbow	0.26	0.37	0.49	0.67	0.76	1.07	1.37	1.83	2.44	3.2
90° Bend (long)	0.15	0.18	0.24	0.38	0.46	0.61	0.76	0.91	1.2	1.52
Return bend	0.46	0.61	0.76	1.07	1.2	1.68	1.98	2.6	3.66	4.88
Globe valve	0.76	1.07	1.37	1.98	2.44	3.36	3.96	5.18	7.32	9.45
Gate valve	0.107	0.14	0.18	0.27	0.32	0.40	0.49	0.64	0.91	1.20
Run of standard tee	0.12	0.18	0.24	0.38	0.40	0.52	0.67	0.85	1.2	1.52
Through side outlet of tee	0.52	0.70	0.91	1.37	1.58	2.14	2.74	3.66	4.88	6.40

Table E17 Formula for converting volume of compressed air to volume of free air

Air tools are usually rated in c.f.m. of free air. Where ratings of other air equipment are not given in terms of free air consumption, the following formulae may be used to convert

$$Q = Q_1\,\frac{P_1 + 14.7}{14.7}$$

Q = Cubic ft of free air
Q_1 = Cubic ft compressed air
P_1 = Compressed air pressure p.s.i.g.

$$q = q_1\,\frac{P + 1.033}{1.033}$$

q = Litres free air
q_1 = Litres compressed air
P = Compressed air pressure in kg/cm²

$$Q = \frac{Q_1(P + 101325)}{101325}$$

Q = Cubic metres free air
Q_1 = Cubic metres compressed air
P = Compressed air pressure in Newtons/metre²

Weight and Volume of Pure Air Pure air at 32°F (0°C) and 14.7 p.s.i. (101.325 kN/m²) absolute (atmospheric pressure).
Weight: 0.08073 lb/cu. ft. (1.29 g/litre). Volume: 12.387 cu. ft/lb. (772 litres/kg).

Table E18 Relative discharging capacities of steel tubes to BS 1387

Nominal dia. of larger pipe in.	mm	Nominal diameter of smaller pipe in in. and mm													
in		⅛	¼	⅜	½	¾	1	1¼	1½	2	2½	3	4	5	6
mm		3	8	10	15	20	25	32	40	50	65	75	100	125	150
		Approximate number of small pipe flows served by larger pipe													
⅛	6	1	–	–	–	–	–	–	–	–	–	–	–	–	–
¼	8	2.1	1	–	–	–	–	–	–	–	–	–	–	–	–
⅜	10	4.5	2.1	1	–	–	–	–	–	–	–	–	–	–	–
½	15	8	3.8	1.8	1	–	–	–	–	–	–	–	–	–	–
¾	20	15	8	3.6	2	1	–	–	–	–	–	–	–	–	–
1	25	30	15	6.6	3.7	1.8	1	–	–	–	–	–	–	–	–
1¼	32	60	25	13	7	3.6	2	1	–	–	–	–	–	–	–
1½	40	90	40	20	10	5.5	2.9	1.5	1	–	–	–	–	–	–
2	50	165	75	35	20	10	5.5	2.7	1.9	1	–	–	–	–	–
2½	65	255	120	55	30	16	8	4.3	2.9	1.6	1	–	–	–	–
3	75	440	210	100	55	27	15	7	5	2.7	1.7	1	–	–	–
4	100	870	400	190	100	55	30	15	10	5.3	3.4	2	1	–	–
5	125	1500	720	330	180	90	50	25	17	9	6	3.5	1.8	1	–
6	150	2400	1130	530	300	150	80	40	28	15	9	5.5	2.8	1.6	1

COMPRESSED AIR

Table E19 Discharge of air through orifices

Gauge Pressure bar	Discharge of free air in litres/sec for various orifice diameters in mm						
	0.5	1	2	3	4	10	12.5
0.5	0.06	0.22	0.92	2.1	5.7	22.8	35.5
1.0	0.08	0.33	1.33	3.0	8.4	33.6	52.6
2.5	0.14	0.58	2.33	5.5	14.6	58.6	91.4
5.0	0.25	0.97	3.92	8.8	24.4	97.5	152.0
7.0	0.33	1.31	5.19	11.6	32.5	129.0	202.0

Graph E2 Compresed air sizing nomogram

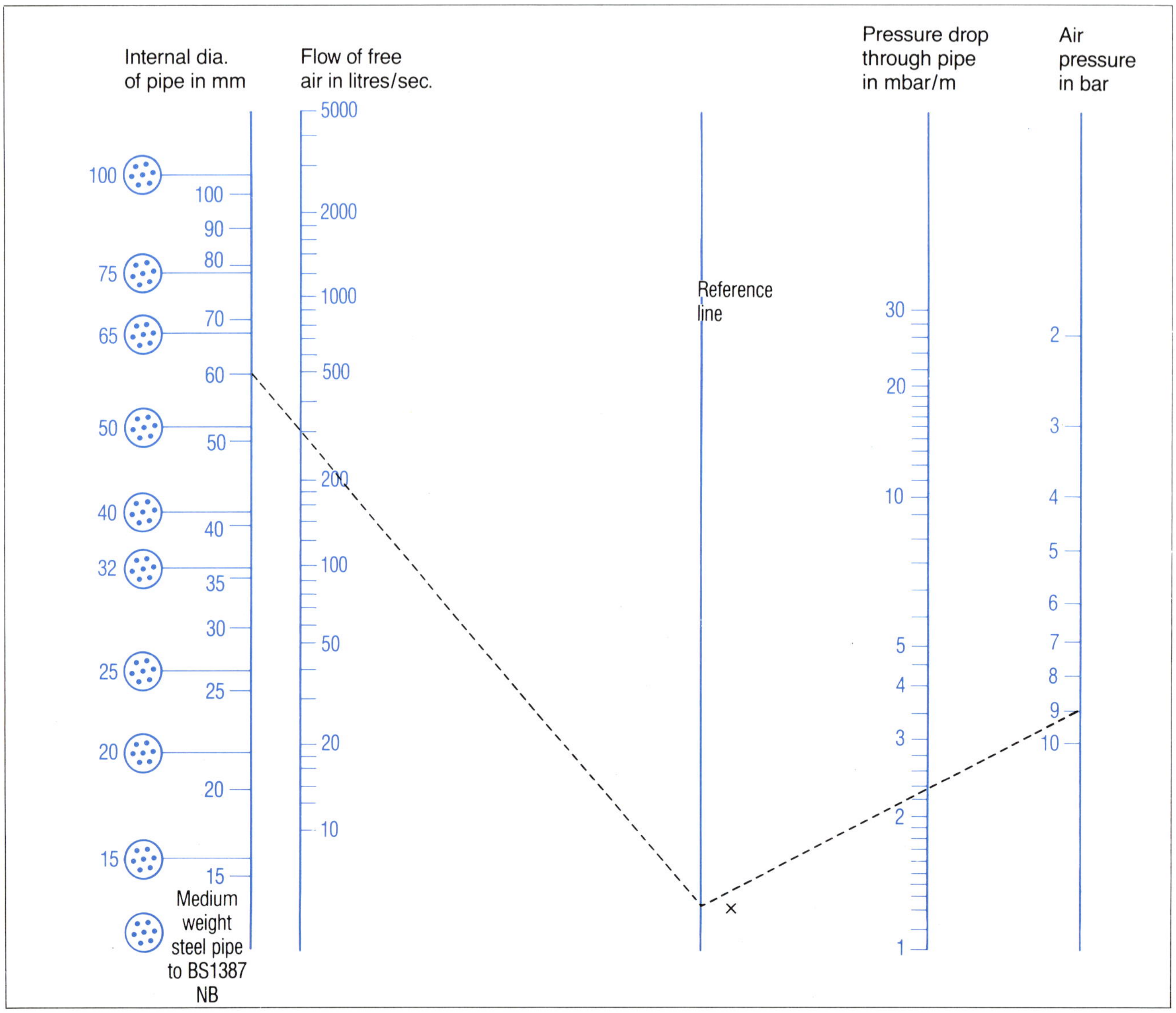

The nomogram shown in Graph E2 gives a ready means for determining pressure drops through pipes often found in industry. It is based on the following formula which can also be used for pipe sizes outside those shown in the table.

L = length of pipe (ft)
R = ratio of compression at the beginning of the pipe
d = internal pipe diameter (in)

$$\text{Pressure drop (bar)} = \frac{KLQ^2}{R \times d^{5.3}}$$

Where K = 800
L = length of pipe (m)
Q = volume of free air passing through the pipe (l/sec)
R = ratio of compression
d = internal pipe diameter (mm)

$$\text{Pressure drop (psi)} = \frac{LV^2}{R \times d^{5.3} \times 35{,}122}$$

Where V = volume of free air passing through the pipe (cu ft/min)

Example

Determine the size of pipe needed to pass 300 litres/sec free air with pressure drop of not more than 300 mbar in 125 m of pipe run, air pressure is 9 bar.

300 mbar in 125 m is equivalent to $\frac{300}{125}$ = 2.4 mbar/m

Join 9 bar on the air pressure line to 2.4 mbar/m on the pressure drop line and produce to cut reference line at X. Join X to 300 litres/sec and produce to cut pipe size line at approximately 61 mm.

Select pipe having a minimum bore of at least 61 mm (a 65 mm nominal bore pipe to BS 1387 has a bore of 69 mm and would therefore be suitable with some margin.

Table E20 Receivers for compressed air systems

Compressor capacity		Receiver dimensions					
		Diameter		Length		Volume	
(ft^3/min)	(m^3/min)	(in)	(m)	(ft)	(m)	(ft^3)	(m^3)
45	1.27	14	0.355	4	1.22	4.5	0.127
110	3.12	13	0.33	6	1.83	11	0.312
190	5.38	24	0.61	6	1.83	19	0.538
340	9.63	30	0.76	7	2.13	34	0.963
570	16.14	36	0.91	8	2.44	57	1.614
960	27.19	42	1.07	10	3.05	96	2.719
2115	59.90	48	1.22	12	3.66	151	4.276
3120	88.36	54	1.37	14	4.27	223	6.315
4400	124.61	60	1.52	16	4.88	314	8.892
6000	169.92	66	1.68	18	5.48	428	12.121

All capacities and volumes given in ft^3/m^3 of free air at atmospheric pressure

Table E21 Vapour chart

Temperature of air		Compressed air pressure in bars											
°C	°F	Litres of water											
		2	2.7	3.4	4	4.8	5.4	6	6.8	7.5	8.2	8.8	10
0	32	2.176	1.635	1.362	1.060	0.946	0.871	0.795	0.681	0.643	0.530	0.492	0.454
4.4	40	2.839	2.271	1.892	1.628	1.438	1.249	1.136	1.098	0.984	0.908	0.871	0.795
10	50	4.353	3.369	2.877	2.460	2.082	1.817	1.628	1.514	1.400	1.287	1.211	1.098
15.6	60	5.980	4.883	4.050	3.520	3.104	2.687	2.385	2.082	1.779	1.703	1.628	1.514
21.1	70	8.213	6.548	5.678	4.921	4.353	3.899	3.558	3.255	2.914	2.725	2.612	2.157
26.7	80	10.977	9.046	7.570	6.737	5.980	5.413	4.807	4.353	3.899	3.709	3.482	3.066
32.2	90	15.254	12.415	10.447	9.160	8.176	7.419	6.548	5.980	5.450	5.110	4.769	4.164
38.2	100	20.174	16.768	14.156	12.188	10.901	9.765	8.706	8.176	7.570	6.964	6.548	5.905
43.3	110	26.495	21.764	18.547	16.086	14.194	12.642	11.431	10.447	9.841	8.993	8.251	7.381
48.2	120	35.958	28.198	24.792	20.893	18.433	15.540	15.216	13.929	12.642	11.961	11.128	9.803

The above chart shows the amount of water which will accumulate every 8 hours in a compressed air system using 47 litres/sec. of compressed air at different air temperatures and air pressure. For example, at 5.4 bar, with a temperature of 32.2°C (90°F), a compressed air system would contain 7.419 litres of water in vapour form every 8 hours.

COMPRESSED AIR

Table E22 Equivalent volume of compressed air at common pressure

Volume of free litres	Equivalent volume (litres) when compared to gauge pressures of 4 bar	5 bar	7 bar
5	1.01	0.84	0.63
10	2.02	1.68	1.26
15	3.03	2.52	1.90
20	4.04	3.37	2.53
25	5.05	4.21	3.16
30	6.06	5.05	3.79
35	7.07	5.89	4.42
40	8.08	6.73	5.06
50	10.1	8.42	6.32
60	12.1	10.1	7.58
70	14.1	11.8	8.85
80	16.2	13.5	10.1
90	18.2	15.1	11.4
100	20.2	16.8	12.6
125	25.2	21.0	15.8
150	30.3	25.2	19.0
175	35.3	29.5	22.1
200	40.4	33.7	25.3
225	45.4	37.9	28.4
250	50.5	42.1	31.6
275	55.5	46.3	34.8
300	60.6	50.5	37.9
350	70.7	58.9	44.2
400	80.8	67.3	50.6
500	101.0	84.2	63.2
750	151.0	126.0	95.0
1000	202.0	168.0	126.0
1250	252.0	210.0	158.0

Table E24 Typical equipment consumption of compressed air

Appliance	Consumption free air (litres/sec)	Gauge pressure (bar)
Air motors per kW	16–22	–
Contractor's tools, breakers, diggers etc.	2–35	5.5
Controls, typical	0.005–0.01	1.0
Laboratories, bench outlets	5–15	1.5–5.5
Workshop tools, blast clearers, small	10–13	6.5
blast clearers, large	100–110	6.5
Percussive, light	2–8	5.5
Percussive, heavy	10–15	5.5
Rotary (eg drills etc.)	2–15	3.5–5.0
Spray guns	0.5–10	2–5

Graph E3 Compressed air – laboratories

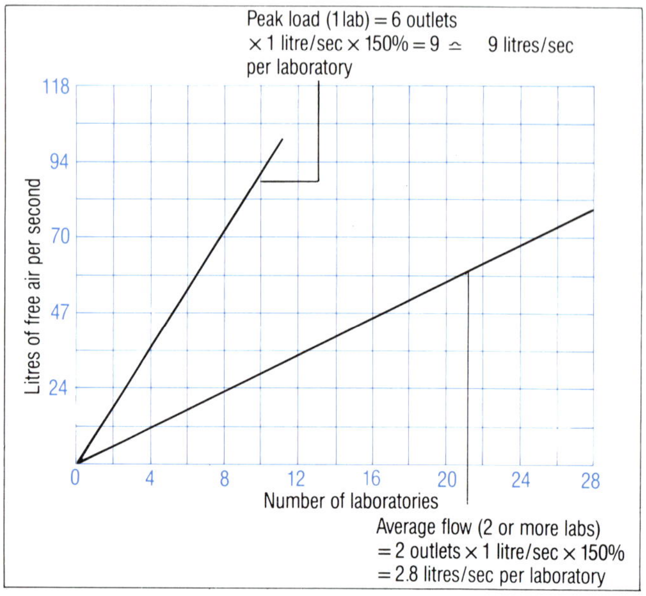

Table E23 Volume of compressed air carried by medium grade steel pipes, of minimum bore, to BS 1387 at given velocities

Velocity m/s	Flow of air (litres/sec.) through medium grade steel pipe to BS1387, minimum bore 15mm	20mm	25mm	32mm	40mm	50mm	65mm	75mm	100mm	125mm	150mm	200mm
3.0	0.6	1.1	1.7	3.0	4.1	6.5	10.9	15.1	25.7	39.2	56.2	98.5
3.5	0.7	1.3	2.0	3.5	4.7	7.6	12.7	17.6	30.0	45.7	65.5	115.0
4.0	0.8	1.4	2.3	4.0	5.4	8.7	14.6	20.1	34.2	52.3	74.9	131.0
4.5	0.9	1.6	2.6	4.5	6.1	9.8	16.4	22.6	38.5	58.8	84.2	147.0
5.0	1.0	1.8	2.8	5.0	6.8	10.8	18.2	25.1	42.8	65.4	93.6	164.0
5.5	1.1	2.0	3.1	5.5	7.4	11.9	20.0	27.6	47.1	71.9	103.0	181.0
6.0	1.2	2.1	3.4	6.0	8.1	13.0	21.8	30.1	51.3	78.5	112.0	197.0
6.5	1.3	2.3	3.7	6.5	8.8	14.1	23.7	32.6	55.6	85.0	122.0	213.0
7.0	1.4	2.5	4.0	7.0	9.5	15.1	25.5	35.1	59.9	91.5	131.0	230.0
7.5	1.5	2.7	4.3	7.5	10.1	16.2	27.3	37.6	64.2	98.0	140.0	246.0
8.0	1.6	2.8	4.5	8.0	10.8	17.3	29.1	40.1	68.5	105.0	150.0	263.0
8.5	1.7	3.0	4.8	8.5	11.5	18.4	31.0	42.6	72.8	111.0	159.0	278.0
9.0	1.8	3.2	5.1	9.0	12.2	19.5	32.8	45.1	77.1	118.0	169.0	296.0

Ascertainment of peak capacities for compressed air systems is largely a matter of judgement. It is necessary to consider the probable use factors, determined by the devices requiring compressed air, the estimated number of people who will be drawing air from the system and the pattern of air use, with special attention paid to periods of high load. For example, compressed air demand in a two-man laboratory may reach 9 litres/sec of free air per laboratory, depending on the number of air outlets (1/8 inch orifices) in use.

Table E19 (Discharge of air through orifices) gives free air discharge through orifices at various pressures. Where a number of small pipe branches are involved, application of Table E18 (Relative discharging capacities of steel tubes to BS 1387) is helpful in arriving at relative capacities of pipe headers.

Graph E3 (Compressed air – laboratories), a chart of air quantities in relation to the number of laboratories, was drawn in connection with the design of a specific installation. Similar charts are easily drawn for any installation as an aid to pipe sizing. The chart reflects specific conditions, and the designer's judgement in each case. A 150% contingency factor is used to cover any peak load, but this may be varied to meet specific requirements.

Figure E5 Typical layout of a compressed air plant

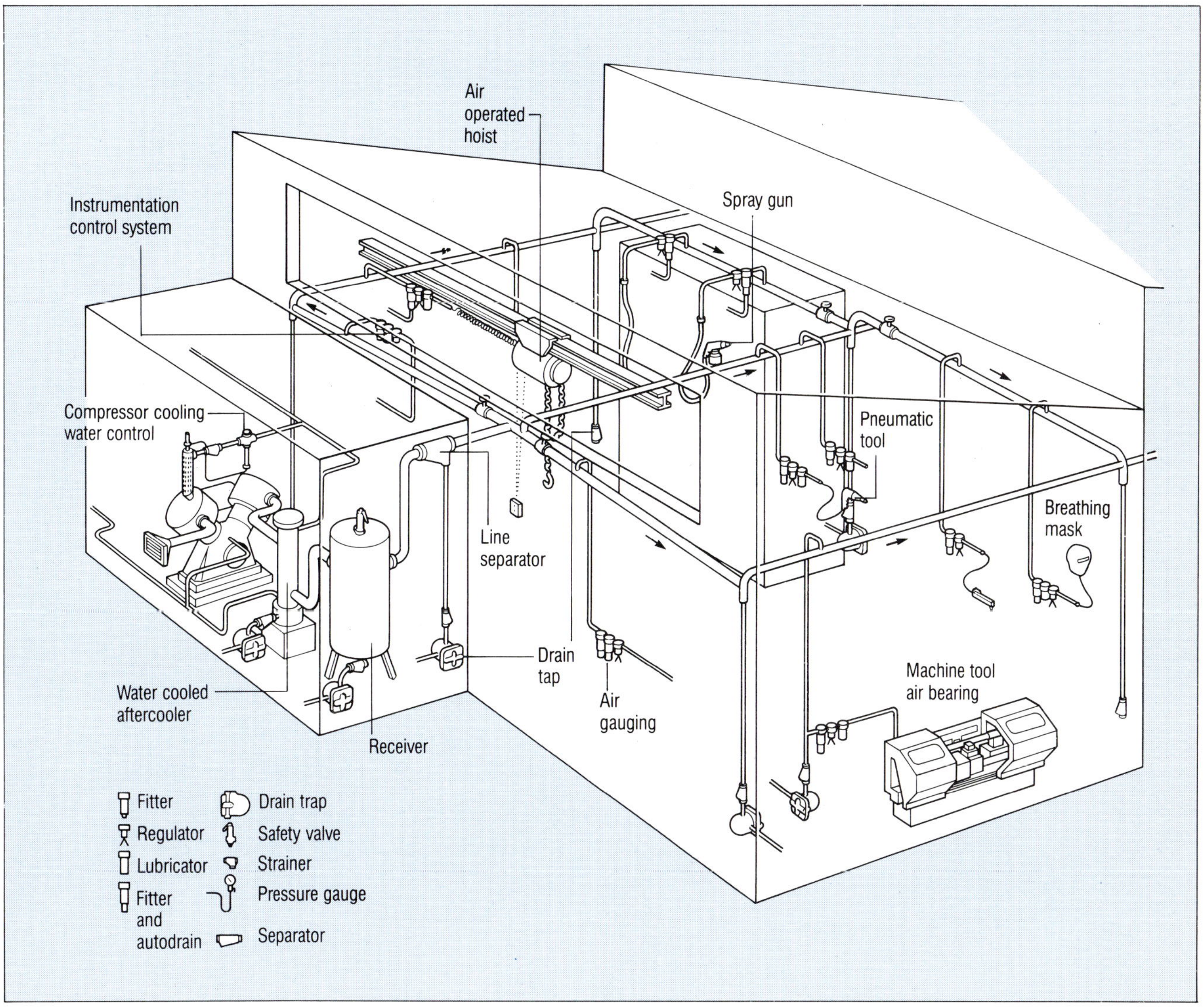

VACUUM

There are three elementary questions that are always asked in connection with this subject; what is it, what is it used for and by whom?

In layman's terms, it can be described as a piped distribution system from a prime mover (this generally takes the form of an electrically operated vacuum pump), terminating in a number of outlet points which give facilities for obtaining a vacuum or negative pressure service. Hospital personnel also refer to this type of service as the medical suction system.

The uses for such a service are vast and varied. There is the simple school laboratory system where students carry out elementary filtrations under reduced pressure to speed up the filtration time cycle. On the other hand, we have the complex research and industrial systems whereby a vacuum of better than 10 torr is required. In between these two extremes, there is the complete medical field for ward suction, operating theatre, dentistry and x-ray bolus. Most hospitals also have a separate laboratory system for pathology and pharmacy work.

This section does not cover in detail, the rather specialized high vacuum systems (below 1 torr) which for reasons such as outgassing of line material and vapour pressure problems, form the subject of more specialized publications. It is written with the object of familiarising the plumbing engineer with the basic facts

VACUUM

relating to the design and installation of a piped vacuum system. It is also important that the consultant who is asked to plan initially what is loosely termed a vacuum system, should know exactly what the client is going to use the system for, and it must firstly be determined what degree of vacuum is expected at the outlet points. This in itself is sometimes difficult to establish (very often because the user himself does not know), and is complicated further because vacuum can be identified in inches of mercury, millimetres of mercury or water gauge, and we also have the international term, torr (which for the purposes of this Guide, is equivalent to the millimetre of mercury).

Vacuum specialist firms always measure the degree of vacuum with atmosphere at 30in or 760 torr, and ultimate vacuum as 0, whereas the engineer thinks of vacuum in reverse, whereby atmospheric pressure is 0 and the ultimate vacuum is 760 torr or 30 inches of mercury. Also, manufacturers of vacuum gauges produce scales calibrated in both ways.

This guide therefore uses 0 as ultimate and 30in or 760 torr as atmosphere for the remainder of this summary.

Table E25 Comparative vacuum scale

Absolute vacuum

Hg mm	Hg ins	Torr
760	30	0
737	29	10
711	28	50
686	27	
660	26	100
635	25	
560	22	200
508	20	252
460	18.12	300
381	15	379
360	14.12	400
260	10.25	500
254	10	516
160	6.25	600
127	5	633
60	2.37	700
0	0	760

Atmosphere

The Hospital System

Apart from the scientific and industrial fields, the main usage of vacuum is in the medical and medical research field which covers five main areas: i) ward suction, ii) theatre suction, iii) x-ray bolus, iv) dental suction and v) pathology and pharmacy laboratories.

As any hospital engineer will confirm, the provision of an efficient central vacuum system will cut his maintenance work considerably.

Ward Suction

The free air displacement required at the bedside outlet varies of course, depending on the type of ward being served. For instance, in the respiratory ward, suction is required for quick removal of sputum etc, from the throat, and is not generally in use for long periods, whereas in the post-thoracotomy wards, the suction may be required for long periods of continuous duty. Therefore the outlet displacements can be anything from 1 litre per minute to an estimated maximum of 40 litres per minute (free air) at approximately 250 torr.

Theatre Suction

The free air displaced in the theatre is naturally much higher than for general ward usage, and it is quite possible for the surgeon and anaesthetist to require suction to be available at up to 80 litres per minute in the theatre at any one time, and also at a pressure of 250 torr.

X-ray Bolus

This is a technique using the suction system to evacuate bags filled with granuals that are placed in position around the limb to be X-rayed, and on applying the suction, it is found that the limb will be held firmly in the pre-set position. The free air displaced here is quite small once the initial evacuation has taken place, and the degree of vacuum need not exceed 150 torr.

Dental Suction

The usage here is mainly one of removal of saliva and water, the latter being sprayed on to the dentist's drill and the site of drilling.

There are at present two techniques used, the first involves a high degree of vacuum (30 torr) at low flows and known as the aspiration technique. This has one main disadvantage which is that the suction tube must be held very close to the tissues of the mouth in order to 'pick up' the fluid accumulating within the mouth, and sometimes these tisues can block the suction orifice and can both damage the mouth tissues and stop the water from being drawn away.

The second method, known as the high flow technique, employs suction at a great velocity, of the order of 120 miles/h, but the degree of vacuum required is only about 700 torr maximum. The 'pick up' bridge between fluid and sucker orifice in this case is approximately 12 mm, and no damage to mouth tissues occurs. It is therefore essential to establish which technique is being employed.

Pathology and Pharmacy Laboratories

Generally speaking, the pathology laboratory user and the pharmacist are the only case where a fairly high degree of vacuum would be required ie. 10-15 torr, and it should always be borne in mind when planning hospital systems, that the medical suction system should never be interconnected to the laboratory system.

The former requires a fairly high flow at relatively moderate suction pressure, ie. rarely better than 125 torr, whereas the latter requires a low flow rate at a relatively high degree of vacuum (10-15 torr). There is also the fear that bacteria from laboratory experiments could find their way to patients if the systems were linked.

When considering a pipeline installation to serve these types of area, especially ward and theatre suction and X-ray bolus, it must be borne in mind as a rough guide that to remove 1 litre of liquid rapidly will require a free air displacement at atmosphere of approximately 4 litres per minute.

The modern trend of providing an efficient suction system for ward and theatre usage is to have what is known as a twin standby pumping set. This generally consists of two prime movers interconnected to a common reservoir with all necessary cycling and safety gear, which enables the second pump to operate if either the first pump fails or cannot maintain the pipeline system within the predetermined pressure, in which case, both pumps will operate until the vacuum in the system builds-up, and the two pumps will cut-out in sequence.

Facilities are also provided by most manufacturers for alternating the duty pump. There is no reason why a series of pumps cannot be interconnected in this way and linked in pairs to pressure switches to enable further capacity to be provided as and when the new extensions are added to the system.

Siting of Pumps

Preferably, the central vacuum pipeline pumping equipment should be grouped in one area to save the hospital engineer and his staff precious time in maintaining hospital equipment at various points in the building, and the actual positioning of the vacuum units is important on three counts.

The first is the noise factor. Not all vacuum pumps are silent, so it is imperative that any noise or vibration is not transferred to areas where such inconvenience could not be tolerated. Noise carried through the walls of the pipeline can be eliminated by interposing a simple rubber sleeve connection close to the prime mover. Alternatively, vacuum bellows with demountable vacuum union connections are sometimes used but are naturally more expensive.

The second is the case of maintenance. The pumps will need to be positioned for accessibility of servicing and running of water cooling lines if necessary.

Thirdly, the exhaust of gases and perhaps equally important, the pump exhaust system. This should be run conveniently direct to atmosphere but away from windows, fresh air intake or any area where ignition would be possible should flammable vapours be discharged. It must be remembered that any rising exhaust line must be trapped to avoid condensates from running down into the pumping gear and more important, to prevent a back pressure. The length of the exhaust line must be related to the bore of the exhaust pipe; for example, it may be quite permissible to run a 15 mm bore exhaust line direct to atmosphere at low level through an exterior wall, but if the exhaust could only be piped to atmosphere via the roof at say 40 ft high, then the bore of pipe could easily be increased to 42 mm bore or greater. One answer to this problem is to site the problem gear on the top floor, but this is only possible if weight of pumps and availability of running services permit.

Pipelines

The actual pipelines that are provided must be designed to allow displacement of the required flow of air, and at the same time, be leak-tight at better that the ultimate vacuum required by the user.

For very low pressure or rough suction systems, ie. in the order of 450 torr, it is quite possible to obtain a satisfactory result by using gas-barrel tube with threaded fittings suitably sealed on the threads, but this material will be quite unsuitable for systems where better than 200 torr is required. One of the most reliable and well tried materials is copper tube with capillary fittings.

Some of the modern materials such as uPVC pipe and stainless steel tube are also quite suitable for most systems, but one has to be careful that the vapours and condensates pumped over are not solvents of the line material, and they will only be accepted where the risk of destruction by fire or heat does not apply. In general, copper tube is used by most vacuum specialist firms.

If the client is likely to pump over actual liquids, it is necessary to incorporate traps that would preclude such liquids from entering the pumping gear. On a multi-storey building, such traps can be conveniently placed at the base of each riser; also they should preferably be of a design whereby they can be emptied without interference of the main vacuum line. If the building is so designed that there are no main risers, ie. single storey buildings, then it is preferable that each branch should be individually trapped.

Laboratory Systems

The design of vacuum installations to serve university and college laboratory systems is far more critical; the demand of the user calls for a much more accurate control of the line pressure, because the student often has to carry out long term experiments at a controlled pressure and therefore cannot tolerate the wide pressure differentials that can be allowed with absolute safety on the medical suction system.

It is no secret that there exists a resistance by many heads of departments in some of our premier academic establishments, to the central vacuum system. They are quite rightly suspicious of replacing the well-tried water jet pump with which they could control their vacuum by adjustment to the water tap pressure. (This pressure should ideally be constant at between 2 and 2.8 bar). The fact is that even the latest water jet pumps will consume between 0.15 and 0.19 litres of water per second and the very suggestion of running a new university chemistry block with up to 400 laboratory places fitted with water jet pumps is just 'not on' with the Water Boards of most areas.

Even if the case is put that only a quarter of these outlets would be in use for say four hours per day, the possible water consumption is going to be about 270 m^3 per day. With the water problem getting worse each year, it is obvious that the central vacuum pipeline system is here to stay for sometime.

To overcome the water shortage, the obvious solution is to first look at a water recirculation system, but this has two main disadvantages. The ultimate vacuum obtainable is dependent upon the vapour pressure of water, and provided the water is maintained at an ambient temperature of 15°C, then a vacuum of 15 torr can be expected. However, the distribution pipes and tanks need to be internally treated and unless the recirculating water is filtered and cooled, the temperature will rise, thus affecting the ultimate vacuum obtained, ie. if the water temperature rose to 50°C, then the ultimate vacuum would be in the order of 100 torr as against the 15 torr at 15°C.

It may be permissible to allow a minimum number of water jet pump positions which could be reserved for pumping really corrosive vapours that would otherwise be detrimental to the working mechanism of the central vacuum pump, and this compromise may be accepted if it can be demonstrated that adequate protection is provided to prevent backflow in the water system.

Most laboratories will require a pressure at the bench outlets at least as good as a water jet pump, namely between 12 and 15 torr absolute, and the speed of evacuation found to be most suitable is about 6 litres per minute, which is a little faster than most water jet pumps.

The consultant or planning authority can generally base laboratory requirements on these figures, but a close check should be made during discussion with the client and establish quite clearly what pressure is to be provided, and if a pressure of better than 12 torr is wanted, say for instance 1 torr, then the problems increase considerably, and these cases will be discussed later.

We are now at the stage where the pipelines and accessories should be discussed, having established that a standard system is wanted for normal filtration through buncher flasks and possibly, some distillations etc, are to be carried out. If the system is to maintain a vacuum of 12 to 15 torr, then the lines, joints, isolation valves, traps and most important, the bench valves, must be proved free from vacuum leakage at better than 10 torr. Nowadays, many contracts are split into mains services and then bench furniture and services separately. This is not a convenient arrangement for most services, but in case of vacuum, it is not at all satisfactory.

If, upon completion of the laboratory, only 50 torr vacuum can be created at the bench outlet, where is the fault? The vacuum pump, the main distribution system, the bench pipe run or the bench

VACUUM

outlet? Such a situation could involve up to four different contractors, and to find a small leak in an already completed vacuum system can be a very tedious and expensive business.

One should check that the specification states quite clearly, the degree of vacuum at which the whole installation is to undergo a pressure rise test, and also the operating pressure at which the line will be maintained.

Bench Outlets

There are several good vacuum bench valves on the market in this country and the types that give a good flow control and easy replacement of working parts are preferred. For instance, never install metal-to-metal cone-type vacuum cocks with a grease seal where hydrocarbon vapours are present, because the vapours will quickly dissolve the grease and a pressure rise will take place, causing the pumping unit to cycle at frequent intervals.

Valves having a stuffing-box shaft seal are not considered suitable.

The push-in or bayonet type of outlet is very practical provided that flow control is not essential and is favoured on industrial projects. Its use in the teaching or research field is limited and its leak tightness in some cases is suspect.

The sizing of the bench runs is not difficult in this pressure range, and as a guide, a 15 mm copper line could serve up to four outlets on a 4 metre bench, a 22 mm line would serve up to eight outlets and a 28 mm line up to 16 outlets. A 15 mm line to each bench valve is preferred, and if elementary work is being carried out or the laboratory is to be used for a high density of students, then under-bench traps should be incorporated at a convenient position in each bench run. These traps should be designed to incorporate an isolation valve and an air admittance facility to enable the trap to be emptied without interference to the vacuum system. The laboratory steward would no doubt be responsible for emptying these traps, and it is most important that they are not installed in inaccessible positions where major work has to be carried out before the trap is exposed.

In senior and research laboratories, traps are not considered necessary because this calibre of student using the vacuum system would interpose his own trap on the bench top between the process being carried out and the bench valve, and this trap would be charged with a suitable desiccant to neutralize the particular vapour being pumped.

The sizes of the main runs do, of course, depend upon the total number of outlets to be served, and on the general layout of the building. If, for example, there is a tower block of laboratories four storeys high, having 50 outlets on the first two floors, 10 on the third and 40 on the top floor, a suitable arrangement would be a 54 mm diameter riser to the second floor and 42 mm extending to the top floor. This riser must be suitably trapped at the base.

The first two floors would have 42 mm subsidiary mains reducing at the bench runs, the third floor would only require a 28 mm diameter floor run but the top floor would probably be best served by a 42 mm main floor run.

As with the hospital system, copper tube with capillary fittings has been found trouble-free in this pressure range, but again, a rigid PVC line can be maintained at better than 10 torr, but it is essential to study the vapour content likely to be pumped to make sure that the material chosen for the line will not be adversely affected by vapours pumped over.

Pumping Units

Vacuum pump manufacturers in this country whose pumps can achieve the desired pressures are limited, and all have merits of their own. The three most essential points to look for in a vacuum pump to serve laboratories are (1) that it will achieve and maintain the pressure required, (2) that it is designed to discharge, or has facilities for dealing with corrosive vapours pumped over and (3) that its noise will not give cause for complaint.

It is not essential that a twin standby unit is provided, although many modern colleges are favouring this arrangement. However, it must be decided whether the prime mover is to be automatically controlled or manually operated.

The automatic unit is designed to provide a 'vacuum on tap' service at all times of the night and day, and is generally controlled by switches which operate from a vacuum bellows. Thus, the pump only runs when a pressure rise takes place in the line. The differential between the cutting-in and cutting-out pressure can be limited to about 5 torr if necessary. This type of controlled system is ideal for laboratories or research areas that have intermittent use.

The manually operated plant has to be turned on when vacuum is required, and this does mean that there is a time delay (which varies depending on pump speed and line capacity) in pumping out the system. It also means that a lower pressure can eventually be achieved provided that the line is leak tight. The main advantage with this type of system, as opposed to the automatic one is cost; also the manually operated unit is most suitable for large teaching laboratories where the vacuum service is only required for a short period each day but because of the high percentage of bench valves in use simultaneously, there is no real point in providing an automatic pumping unit. In fact, if automatic gear is provided in such areas, there is sometimes a risk that because of frequent cycling of the pumps, the starters and switch gear could become overloaded.

High Vacuum Systems

In certain laboratories, there may be a call for vacuum to be provided at better than 1 torr. An example would be for a rotary evaporator in pharmaceutical research work. This is a special requirement and to obtain such conditions, it will be necessary to take great care in the planning and installation of the system.

Although this pressure is still within the viscous flow range, it will be necessary to install the vacuum pump as close to the outlets as possible and keeping the bore of the pipelines as wide as possible. It is not practicable to try and obtain a 1 torr working system from a central pumping set as previously described, and each bench run would therefore have to be provided with its own pump and pipeline system. This presents two problems, the first is to make suitable arrangements to house the pump within the bench area without the noise level interfering with the user, and secondly, to run the pump exhaust to atmosphere at a convenient place.

As this type of system would generally consist of a small rotary oil-sealed high vacuum pump and motor and would be manually controlled, ie. no safety or cycling switches, it must be firmly established that the personnel using the system will interpose a desiccant flask between their process and the bench outlet which will preclude any liquids being accidentally pumped over and also that the vapours being pumped will be partially dried by the desiccant. If this simple precaution is taken, it will ensure that the small pump will not become contaminated by carelessness; this will add considerable life to the rotary pump.

When providing these small systems with individual pumps, the biggest hazard is to the working parts of the pump; for example, if soluble organic acids are to be pumped, the oil film in the pump gives no protection at all to the working parts and formic and acetic acids will cause quite severe corrosion depending on the number of outlets in use. The only way to prevent such acids from entering the pump is by the use of conical flasks on the bench tops, these containing sodium hydroxide pellets mixed with indicating soda lime.

The question of explosion risk in small rotary pumps is sometimes raised, especially when ether-air mixture is present, but the risk is very small because the temperature at which automatic ignition takes place would be about 180°C which is at least twice the maximum running temperature of most conventional vacuum pumps.

If requests are made for a central vacuum system calling for bench outlet pressures of better than 0.5 torr, then the problems are increased even further because at these pressures, we are moving from the viscous flow to molecular flow and such conditions are not generally accepted as economically practical on a pipeline system.

Certainly, the use of 90° elbows and sharp tees would not be permissible and the lines would possibly have to be in some other material such as glass, and these problems would cover outgassing of the line material, vapour content to be pumped, and speed of displacement required at any specific pressure.

Pipe Sizing

The difficulty experienced in trying to calculate pipeline sizes is bound up with the fact that is very rarely that the designer really knows what materials are going to be pumped in a particular pipeline system. It may be that the system will be required to deal with high vapour pressure materials which even at the pipeline pressure, produces very large quantities of vapour which obviously have to be handled by the mechanical pump and pipeline systems, and this makes the whole question of accurately sizing pipelines a very difficult one to resolve.

As a crude example, a gramme of water at, say 10 torr, will evolve about 100 litres of water vapour in the system, and other vapours will have a similar or even greater volumetric effect on the system. It has been found that the best approach is one based on discussion with the user, in order that the possible vapour content can be considered, and this, coupled with previous experience gained, should enable a sensible sizing of the lines to be possible.

The suggestions already given for pipe sizing may form a useful guide for those persons who have to design or install a piped vacuum system. It must be remembered however that there is no straightforward scale on pipe sizing for vacuum systems unless of course, the system is only going to handle clean, dry air, which for 99% of the time, will not be the case.

Summary

Figure E5 is indicative of air which is fairly dry, and consideration must be taken of excess moisture that may be generated. First establish usage and determine what pressures and displacements are required.

Clarify these usages, ie. medical or laboratory.

Discuss the various types of outlets, the siting of the pumps, the provision of an efficient exhaust system, and choose a line material which is both leak-tight at the desired pressure and one that is functional in use.

Pump Sizing for Individual Chamber Work

T = pump-down time (excavation time)
F = pump-down factor from Graph E4 for given pressure
S = pump speed (free air displacement)
V = system volume

Note

V and S must be consistent units (ie. litres and litres/min).

To calculate the time taken to reach certain pressure in a given system with a given pump:

$$T = \frac{F \times V}{S}$$

To calculate pump speed required to reach a given pressure in a given time:

$$S = \frac{F \times V}{T}$$

Example

Pump size required to excavate a chamber of 60 litres to 1 torr in 3 minutes:

$$\text{Pump speed } S = \frac{F \times V}{T}$$

From Graph E4, F = 7

$$\text{hence, } S = \frac{7 \times 60}{3} = 140 \text{ litres/min}$$

Therefore, pump with displacement of 140 litres/min and ultimate vacuum better than 1 torr is required.

Graph E4 Pump sizing

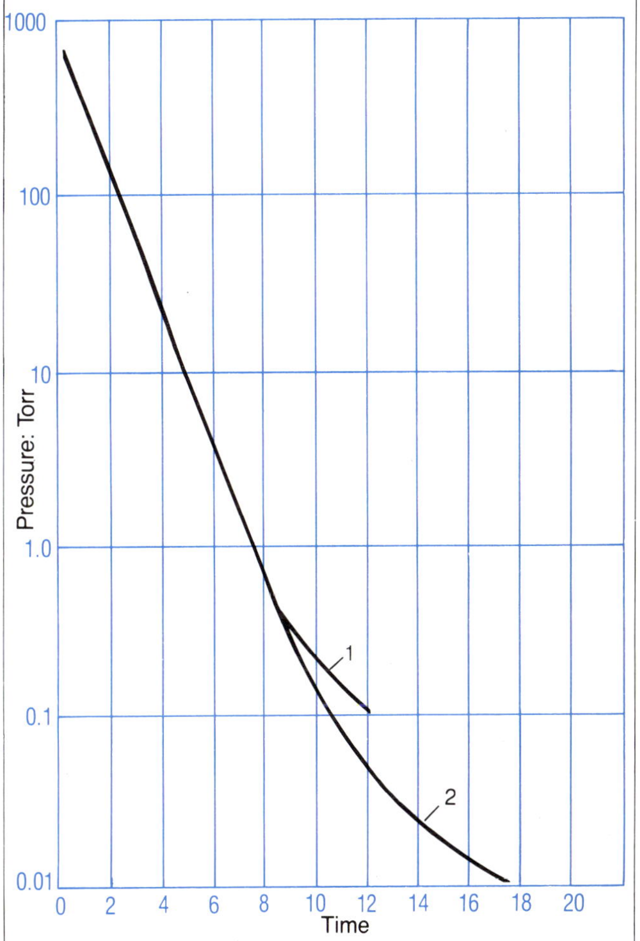

Graph E4 enables pump-down time or pump size to be evaluated for any clean, leak-tight rotary pumped system down to 0.1 torr.

Due allowance made for change of pump volumetric efficiency with reduction in pressure but impedance of connecting pipelines neglected.

For recommended plant layouts, consult specialized manufacturers and for medical vacuum, consult Hospital Technical Memorandum 22.

SECTION E
Piped Gas Services

MEDICAL GASES

The term medical gases covers the following:

(a) Medical compressed air (oil and moisture free)
(b) Vacuum suction
(c) Oxygen
(d) Nitrous oxide
(e) Nitrous oxide/oxygen mixture N_2O/O_2

Oxygen and vacuum are used extensively throughout most hospitals and are usually required in the following areas:

(1) In-patient departments
(2) Operating suites
(3) Maternity departments
(4) Accidents and emergency areas including out-patients

Vacuum and compressed air are required in dental suites, and operating suites for patient ventilators and operating surgical tools.

Nitrous oxide is required in operating suites, maternity units and accidents and emergency areas. In addition, N_2O/O_2 is usually required in maternity delivery rooms.

Note

For further information, refer to Hospital Technical Memorandum 22.

Graph E5 Vacuum pipe sizing, copper tube to BS 2871 Table X

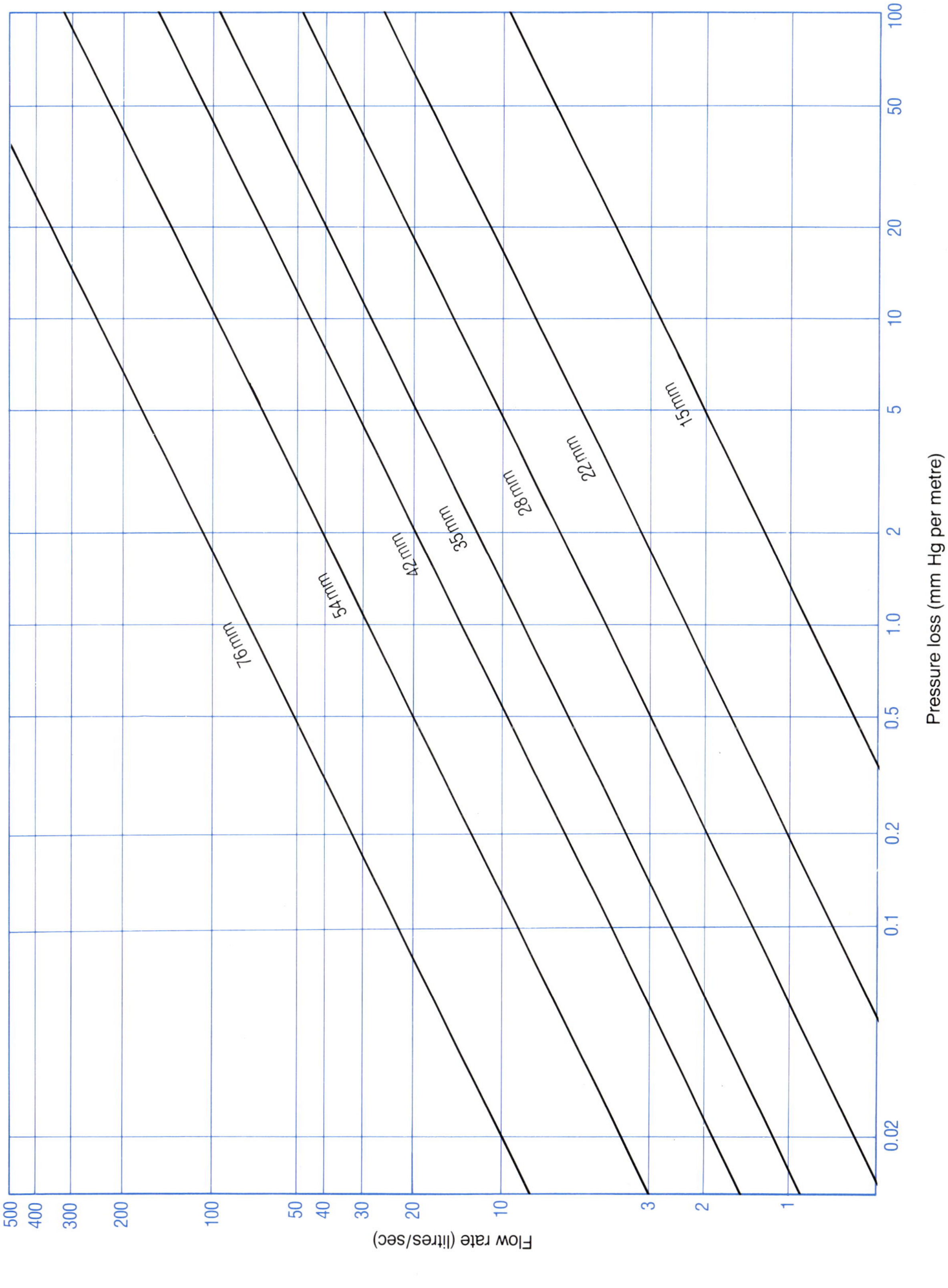

REFERENCES

Health and Safety Executive Publications

(1) HSE Series Booklet HS(G)34
The Storage of LPG at Fixed Installations.
This publication provides a general guide to safe practice in storing and handling LPG at fixed storage installations.
This guidance deals solely with installations where LPG is stored under pressure at ambient temperatures in fixed vessels larger than 150 litres.

(2) HSE Guidance Note CS4
The Keeping of LPG in Cylinders and Similar Containers.
This publication provides a guide to safe practice for storage and handling LPG containers. The recommendations are intended to minimise the risk of fire or explosion from LPG escaping from a leaking container and from the consequences of a fire, either at or near a cylinder store.

(3) HSE Guidance Note CS6
The Storage and Use of LPG on Construction Sites

(4) HSE Guidance Note CS8
Small Scale Storage and Display of LPG at Retail Premises.

Publications from the Liquefied Petroleum Gas Industry

Technical Association (UK) (LPGITA).

(1) Code of Practice No 1 Installation and Maintenance of Bulk LPG Storage at Consumer's Premises.
(2) Code of Practice No 3 Prevention and Control of Fire Involving LPG.
(3) An Introduction to LPG, obtainable from: Wm Culross & Son Ltd, Coupar Angus, Perthshire, PH13 9DF.

The Gas Safety (Installation & Use) Regulations 1984
Obtainable from: HM Stationery Office.

IEE Regulations for Electrical Installations
Obtainable from: The Institution of Electrical Engineers, PO Box 26, Hitchin, Herts, SC5 1RS.

BRITISH STANDARDS AND CODES OF PRACTICE

BS 341	Specification for valve fittings for compressed gas cylinders.
BS 341: Part 1: 1962 (1963)	Valves with taper stems (excluding valves used for breathing and medical purposes).
BS 669	Flexible hoses, end fittings and sockets for gas burning appliances.
BS 669: Part 1: 1984 (1986)	Specification for strip wound flexible hoses, covers, end fittings and sockets for domestic appliances burning 1st and 2nd family gases.
BS 746: 1987	Specification for gas meter unions and adaptors.
BS 864	Capillary and compression tube fittings of copper and copper alloy.
BS 864: Part 2: 1983 (1987)	Specification for capillary and compression fittings for copper tubes.
BS 864: Part 3: 1975	Compression fittings for polyethylene pipes.
BS 1123	Safety valves, gauges and fusible plugs for compressed air or inert gas installations.
BS 1123: Part 1: 1987	Code of practice for installation.
BS 1123: Part 2: 1976 (1987)	Specification for safety valves.
BS 1387: 1985	Specification for screwed and socket steel tubes and tubulars and for plain end steel tubes suitable for welding or for screwing to BS 21 pipe threads.
BS 1552: 1982	Specification for control plug cocks for low pressure gases.
BS 2871	Specification for copper and copper alloy tubes.
BS 2871: Part 1: 1971 (1976)	Copper tubes for water, gas and sanitation.
BS 2977: 1958 (1983)	Domestic lighting appliances for use with liquefied petroleum gases.
BS 3016: 1983 (1986)	Specification for pressure regulators and automatic change-over devices for liquefied petroleum gases.
BS 3212: 1975 (1983)	Specification. Flexible rubber tubing & hose (including connections where fitted & safety recommendations) for use in LPG vapour phase and LPG/air installations.
BS 3554	Specification for gas governors.
BS 3554: Part 1: 1971 (1972)	Independent governors for inlet pressures up to 25 mbar.
BS 3879: 1981 (1987)	Specification for portable appliances operating at vapour pressure from liquefied petroleum gas containers.
BS 4089: 1966	Specification for rubber hose and hose assemblies for liquefied petroleum gas lines.
BS 4096: 1967 (1968)	Specification for non-domestic space heaters burning liquefied petroleum gases.
BS 4161	Specification for gas meters.
BS 4161: Part 3: 1977 (1983)	Unit construction meters of 6 cubic meters (or 212 cu ft) per hour rating.
BS 4161: Part 5: 1977 (1978)	Positive displacement diaphragm meters for pressures up to 7 bar.
BS 4161: Part 6: 1979 (1980)	Rotary displacement and turbine meters for gas pressures up to 100 bar.
BS 4250	Liquefied petroleum gas.
BS 4250: Part 1: 1987	Specification for commercial butane and propane.
BS 5258	Safety of domestic gas appliances.
BS 5258: Part 1: 1986	Specification for central heating boilers and circulators.

BS 5258: Part 2: 1975 (1985)	Cooking appliances.
BS 5258: Part 4: 1987	Specification for fanned circulation ducted air heaters.
BS 5258: Part 5: 1975 (1985)	Gas fires.
BS 5258: Part 6: 1975	Refrigerators and food freezers.
BS 5258: Part 7: 1977	Storage water heaters.
BS 5258: Part 8: 1980	Combined appliances: gas fire/back boiler.
BS 5258: Part 10: 1980 (1983)	Flueless space heaters (excluding catalytic combustion heaters) (3rd family gases).
BS 5258: Part 11: 1980 (1983)	Flueless catalytic combustion heaters (3rd family gases).
BS 5258: Part 12: 1980 (1987)	Decorative gas log and other fuel effect appliances (2nd and 3rd family gases).
BS 5258: Part 13: 1986	Specification for convector heaters.
BS 5258: Part 14: 1984	Specification for barbeques (3rd family gases).
BS 5329: 1976 (1986)	Specification. Non-refillable metallic containers up to 1.4 litres capacity for liquefied petroleum gases.
BS 5482	Code of practice for domestic butane and propane gas burning installations.
BS 5482: Part 1: 1979	Installations in permanent dwellings.
BS 5482: Part 2: 1977 (1987)	Installations in caravans and non-permanent dwellings.
BS 5482: Part 3: 1979 (1986)	Installations in boats, yachts and other vessels.

Note

(1) Year after the Standard's number indicates the latest date of issue.
(2) Year in brackets gives the date of the latest amendment at the time of publication of this Guide.

PRINCIPAL CAUSES OF FIRES IN BUILDINGS

The three basic essentials required before any fire can start are:

(a) Fuel, ie. something to burn
(b) Air, ie. oxygen to sustain combustion
(c) Heat

The process of combustion or burning can be likened to a triangle as shown in Figure F1.

Figure F1 Fire triangle

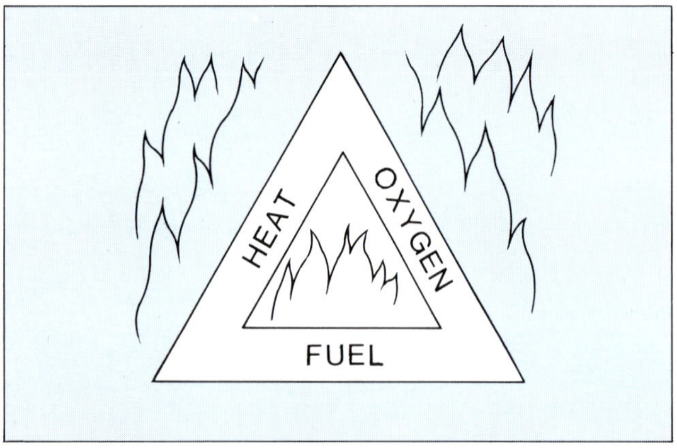

Removal of any one of the sides of the triangle will result in extinguishment of the fire and this is the principle on which all forms of fire extinguishment are based.

CLASSIFICATION OF FIRE RISKS

Fires are divided into three classes, A, B and C. Details of these together with the methods of extinguishment and extinguishing agents are shown in Table F1.

Table F1 Classification of fire risks

Class	Risk involved	Method of extinguishment	Extinguishing agent
A	Wood Textiles Paper All goods manufactured from above	Cooling	Water
B	Inflammable liquids	Smothering	Dry powder CO_2 gas Foam
C	Electrical equipment	Smothering with a non-conducive agent	Dry powder CO_2 gas

INSURANCE COMPANIES; ROLE AND INFLUENCE

Insurance companies play a major role in the provision of fire protection equipment especially automatic sprinkler installations by virtue of the reduction in premiums offered when approved equipment is installed.

To control the approval of equipment and to produce recognised standards of what is considered to be good fire protection engineering practice (or rules), the tariff Insurance Companies, ie. those Companies applying common premium rates, have joined together to organise and finance what is called the Fire Officers' Committee (FOC).

All equipment has to be submitted to the FOC for testing and approval and lists are published giving the names of approved suppliers and the equipment which they manufacture and install.

The majority of sprinkler installations in the UK are designed in accordance with the current edition of the FOC Rules and these rules are also applied throughout the world in countries where the insurance is carried by British companies. They also form the basis of the rules recently adopted by the European Insurance Organisation. The address of the FOC is:

Fire Officers' Committee
Aldermary House
Queen Street
London EC4

The FOC's Testing Station is at Borehamwood in Hertfordshire.

In addition to the FOC, the Insurance Companies have also formed the Fire Protection Association (FPA) to provide information, assistance and publicity material to industry on all aspects of Fire Prevention and Fire Protection. The FPA operates from the same offices as the FOC.

In the USA the principal organisations responsible for producing rules and standards are:

National Fire Protection Association (NFPA)
60 Batterymarch Street
Boston
Mass. 02110
USA

Factory Mutual Engineering Division (FM)
1151 Boston-Providence Turnpike
Norwood
Mass.
USA

National Board of Fire Underwriters (NBFU)
85 John Street
New York 38
USA

The NFPA publishes a series of standards in the form of a 10 Volume National Fire Code covering all types of Fire Protection but it does not approve equipment or installations. Equipment approval in the USA is carried out by the Testing Laboratories of FM and NBFU and they also publish rules and standards. Approval of individual installations is carried out by the Insurance Company concerned as in the UK.

In view of the large number of US owned firms in both this country and on the continent, FM have recently opened a branch office in London at the address below:

FM Insurance Co. Ltd
Palmerston House
Bishopsgate
London EC2

SPRINKLER INSTALLATIONS

General Description

A sprinkler installation consists basically of a network of piping connected to a suitable water supply and fixed at ceiling level throughout the building protected. Fusible sprinkler heads are screwed into the piping at intervals so that as a general rule each head covers approximately 12 m^2 of floor area (see clause on spacing requirements).

In the event of a fire, the heat generated causes the fusible element in the adjacent sprinkler heads to disintegrate and thus allows water to be discharged into the fire in the form of a fine spray.

At the same time as water is discharged onto the fire, water is also allowed to flow to a water turbine which in turn operates an alarm gong, thus giving an audible alarm that the system has operated. Provision can also be made for a connection to the local Fire Station to ensure prompt attendance by the local Fire Brigade or security service.

Thus, an automatic sprinkler installation combines the two functions of fighting the fire and sounding an alarm in the one installation.

Principal Types

Note

A copy of the FOC Rules should be obtained before embarking on the design of a sprinkler system.

Wet Installation

The most common type of installation used in heated buildings, the pipework remaining filled with water at all times. Preferred by insurance companies as water is discharged immediately a sprinkler head opens.

Alternate (Wet & Dry) Installation

This type is used in unheated buildings or buildings where heating is not maintained for 24 hours per day 7 days a week. The pipework is filled with water during the summer months (April to October) and with compressed air during the winter months (November to March). The compressed air is provided either by a small air compressor or by a connection from the factory air line. When the sprinkler head bursts, the air escapes and allows the water to flow in the systems.

Deluge Installation

This type of installation is used in high fire risk areas, eg. power stations, aircraft hangers and chemical plants to discharge water simultaneously over the whole of the risk.

Tail End System

This is a small 'alternate' (wet & dry) system attached to a 'wet' installation and protecting an unheated area or building, eg. loading bay. The water is held back during the winter months by an air valve in the same way as an 'alternate' installation and the compressed air supply is obtained in the same way.

Principal Components

The principal components of a typical sprinkler installation are shown in Figure F2 and are described in the notes which follow.

Figure F2 Typical sprinkler installation

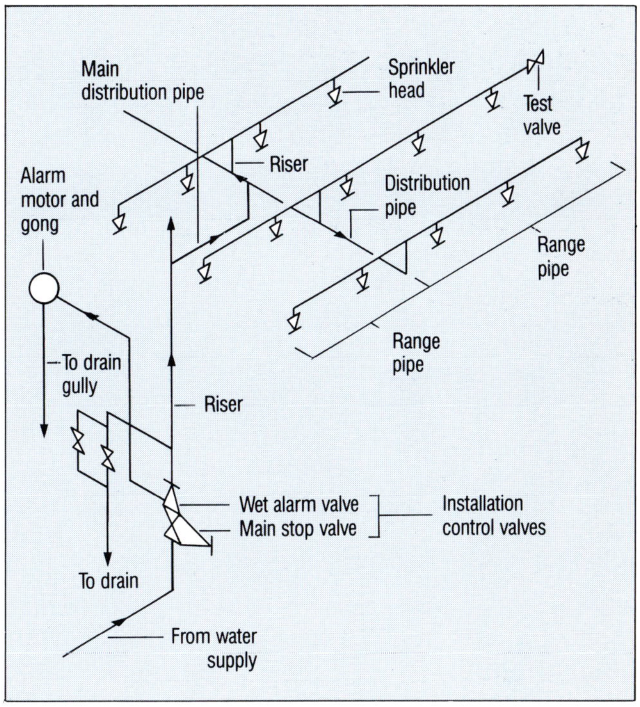

Wet Alarm Valve

Vertical pattern non-return valve with provision for water to flow to alarm gong when clack lifts.

Composite Alarm Valve

Identical to 'wet alarm valve' but with lever/trigger mechanism to hold clack down when installation is on air. Used only on 'alternate' installations.

Alarm Motor and Gong

Water turbine driven and supplied by a pipe connection from the alarm valve. Situated on an external wall of the premises to give an audible alarm.

Location Plate

Plate with 'Sprinkler Stop Valve inside' in raised letters fixed on external wall near to control valves to indicate position.

Main Stop Valve

Stop valve fitted under alarm valve and controlling the supply of water to the installation. Must be strapped and padlocked open.

Sprinkler Heads

There are two basic types, Conventional (Universal) and Spray Pattern. On 'wet' installations, sprinklers may be fitted in either upright or pendant position but on 'alternate' installations they must always be fitted in the upright position above the pipe to facilitate drainage.

Installation Test Valves

To enable the water supply proving tests required by the FOC Rules to be carried out, a test valve varying from 40 mm to 100 mm is fitted immediately above the alarm valve and piped away to a suitable drain point.

In addition, a 15 mm test valve is also fitted above the alarm valve to test the operation of the alarm gong with a water discharge equivalent to one sprinkler head operation and a 25 mm test valve is fitted at the position of the highest sprinkler head.

SPRINKLER INSTALLATIONS

Trunk Main

Main from water supplies to the underside of the installation control valves.

Riser Drop

Vertical pipes in mains or connecting ranges to distribution pipes. Determined by direction of water flow up or down.

Main Distribution Pipes

Pipes running from the main riser to the distribution pipes.

Distribution Pipes

Pipes directly feeding the range pipes.

Range Pipes

Pipes to which the sprinkler heads are attached either directly or through short arm pipes not exceeding 300 mm in length.

Automatic Sprinklers

Principles of Design

Sprinkler systems are invariably designed in accordance with:

(i) BS 5306
(ii) The Rules of the Fire Officers' Committee for Automatic Sprinkler Installations
(iii) Relevant Water Authority Byelaws

The Code of Practice makes general recommendations on design of component parts, materials used, inspection and maintenance. In all cases where insurance is involved, the systems must be installed and maintained according to the Fire Officers' Committee Rules. These rules were revised in 1973 following some years of research which was carried out in several different countries involving extensive tests at Fire Research Stations and practical tests in buildings. A study was also made of insurance statistics and the history of sprinkler systems. Based on this research, systems are now 'tailor made' to suit a particular occupancy and the system is designed to produce a certain density of water spray over a pre-determined area for a given period of time depending on the expected area of fire development in that particular occupancy.

Risk Categories

The Fire Officers' Committee arranges occupancies into risk categories, each having an accepted abbreviation. These are shown in Table F2.

Table F2 Fire Officers' Committee risk categories

Category	Abbreviation
Extra light hazard	XLH
Ordinary hazard: Group 1	OH1
Ordinary hazard: Group 2	OH2
Ordinary hazard: Group 3	OH3
Ordinary hazard: Group 3 Special	OH3(S)
Extra high hazard	XHH

(a) Extra Light Hazard (XLH)

These are premises containing hazards of the non-industrial type, ie. offices, libraries, etc. where the amount and combustibility of the contents is low.

(b) Ordinary Hazards (OH1, OH2, OH3, OH3(S))

Ordinary hazards are commercial and industrial premises involving the handling, processing and storage of a wide range of mainly combustible materials which are unlikely to burn intensely in the early stages of fire. A very wide range of fire risks fall into this category and it has been necessary to sub-divide them into four groups, as indicated above.

(c) Extra High Hazard (XHH)

This category covers commercial and industrial occupancies having abnormal fire loads:

(i) where the materials handled or processed are mainly of an extra hazardous nature likely to develop rapid and intensely-burning fires (process risks);
(ii) involving the high piling of goods (high-piled storage risks).

According to the hazardous nature of the stock and the height of the storage, 'high-piled storage risks' are sub-divided into four categories:

Category I
Category II
Category III
Category IV

(For specific detailed lists in each category, see Rules of Fire Officers' Committee for Automatic Sprinkler Installations, 29th Edition).

The term 'storage' includes the warehousing of the temporary depositing of goods or materials whilst undergoing processing.

Classes of System

Three classes of sprinkler system have been developed to suit the above risk categories and are called:

(a) extra light hazard system;
(b) ordinary hazard system and
(c) extra high hazard system.

Pipework for two or more different types of 'hazard system' may be connected to a common set of control valves provided the total number of sprinklers does not exceed the permitted maximum. Each of these systems has been designed to give an appropriate density of discharge over an assumed maximum area of operation in the highest and most hydraulicaliy remote parts of a protected building.

The density of discharge and the assumed maximum area of operation for the three classes are as follows:

(a) Extra Light Hazard System

Density of discharge + 2.25 mm/min.
Assumed maximum area of operation: 84 m^2.

For certain areas of 'extra light hazard' occupancies, such as attics, basements, boiler rooms, kitchens, laundries, storage areas and workrooms, the density of discharge may be increased to 5 mm/min.

(b) Ordinary Hazard System

Density of discharge + 5 mm/min.

Ordinary hazard systems have been graded into four groups on the basis of the assumed maximum area of operation:

Group 1 (light ordinary hazard OH1):	72 m^2
Group 2 (medium ordinary hazard OH2):	144 m^2
Group 3 (high ordinary hazard OH3):	216 m^2
Group 3 Special (OH3(S)):	360 m^2

(c) Extra High Hazard System

(1) Process risks

Density of discharge:	7.5 to 12.5 mm/min
Assumed maximum area of operation:	260 m^2

(2) High piled storage risks

Density of discharge:	7.5 to 30 mm/min
Assumed maximum area of operation:	260 to 300 m^2

Automatic Sprinklers

Water Supplies

Automatic sprinkler systems must be provided with a suitable and acceptable water supply. The Fire Officers' Committee Rules lay down the following as being accepted sources of water supply;

(i) town mains;
(ii) elevated private reservoirs;
(iii) gravity tanks;
(iv) automatic pump supply;
(v) pressure tanks.

Compliance with certain requirements is necessary in respect of each of these sources of supply. The town main is the most common, and when forming one of two or more supplies to a sprinkler system, is usually the primary or principal one.

General Requirements

The water supply must have a pressure and flow characteristic not less than that specified in the Fire Officers' Committee Rules. It must be automatic, thoroughly reliable and not subject to either frost or drought conditions that could seriously affect the supply. The supply should be under the control of the occupier of the building containing the installation or, where this is not practicable, the right of use of the supply must be suitably guaranteed.

The water must be free from any matter in suspension which would be liable to cause accumulation in the system pipework. The use of salt or brackish water is not normally allowed. In special circumstances where there is no suitable fresh water source available, consideration may be given to the use of salt or brackish water provided that the installation is normally charged with fresh water.

Grading of Systems According to Water Supplies Available

Sprinkler systems are divided into three grades according to the number of water supplies available.

Grade I

Grade I systems have duplicate water supplies, or one 'superior' water supply provided the total number of sprinklers (excluding those in concealed spaces) in the protected building or range of buildings does not exceed 2000 with not more than 200 sprinklers (excluding those in concealed spaces) in each separate risk.

Grade II

Grade II systems have one 'superior' water supply without the limitations on the number of sprinklers mentioned above.

Grade III

These have a single water supply from a town main or automatic pump. Grade III water supplies are not normally accepted for extra high hazard systems.

Superior Water Supplies

(a) Town Mains

The mains water supply must be fed from both ends by mains, each of which must be capable of sustaining the required pressure of flow. Duplicated connections from the main must be carried separately to the premises which contain the sprinkler installations, and there should be a stop valve on the main between the two branches.

In the event of fractures or partial breakdown of the main, operation of the stop valves ensures that the supply is maintained by that part of the main which is still functioning. The mains at each end must not be directly dependent on a common trunk main in the town main system, and this must be fed from more than one source. The main must be capable of furnishing at all times of the day and night the minimum pressure and flow requirements for the appropriate category of risk. If it is not possible to provide duplicate connections, special consideration may be given to the waiving of the requirements if there is a stop valve (secured open) on the town main immediately on each side of a single branch connection.

(b) Elevated Private Reservoir

This is simply defined as similar to a ground reservoir but situated at a higher level than the premises to be protected. Certain conditions regarding capacity must be complied with before this type of reservoir can be used as a source of supply to a sprinkler installation. The minimum capacity ranges from 9 m^3 to 875 m^3 depending on the class of system installed; this is on the understanding that the stored water is used entirely for the sprinkler system.

Where such reservoirs serve other than sprinkler installations, eg. water for trade and domestic purposes, there must be a constant capacity of at least:

500 m^3 in XLH categories:
1000 m^3 in OH categories:
1000 m^3 plus the stored capacity 225 to 875 m^3 in XHH categories (according to design density).

In certain cases smaller capacities may be accepted. The Fire Officers' Committee must approve all relevant details.

(c) Gravity Tank

A gravity tank is defined as a purpose-built container. It is erected on the site of the protected premises at such a height as to provide the requisite pressure and flow condition at the installation valves and needs to be adequately protected against freezing. Where the tank is not enclosed within a tower, the top must be covered in such a manner so as to exclude daylight and solid matter.

The main provisions are:

(i) The tank must have a minimum capacity of 9 m^3 for extra light hazard class rising to 875 m^3 for the extra high hazard class where the design density of discharge is 30 mm/min. Should the capacity of the tank exceed these requirements, it is permissible to draw upon the surplus for other purposes by means of an outlet pipe which must be taken out of the side of the tank above the level of the quantity to be reserved for the sprinkler installation.

(ii) The quantity of water required for the sprinkler installation must be automatically maintained. If the tank forms the sole supply to the system, the supply to the tank must be capable of refilling to the required capacity within six hours.

(iii) The tank must be fitted with an indicator showing the depth of water; the water must be kept clean and free from sediment.

(iv) The use of one tank to supply installations in two or more buildings under separate ownership is not allowed.

(d) Automatic Pump Supply

If a water supply is available with no head or only under limited pressure, a pump may be used to feed water into the installation at the required pressure. In such cases, it is stipulated that the supply shall be from a virtually inexhaustible source, such as a river, canal, lake or it may be from two limited capacity suction tanks. The primary suction tank must have a holding capacity equal to that required for the particular hazard class. The secondary suction tank may be of smaller capacity with automatic inflow provided it meets Fire Officers' Committee requirements, or the supply can be from a single limited capacity suction tank where the circumstances are considered favourable.

The most important provisions relating to pumps are as follows:

(i) With an automatic pump supply consisting of two automatic pumps, at least one must be driven by a compression ignition (diesel) engine with each pump capable of providing the required pressure and flow independently. With three automatic pumps, at

SPRINKLER INSTALLATIONS

least two must have a compression ignition drive. Any two pumps together must be capable of providing the necessary pressures and flows. In both arrangements they must be capable of operating in parallel, ie. with similar pressure and flow characteristics.

(ii) The pump must be housed in a readily accessible position in a sprinkler-protected building, or in the case of an electric motor driven pump, it must be housed in a separate building of non-combustible construction used for no purpose other than for the housing of fire protection water supplies. It must be adequately protected against mechanical damage. The temperature of the room should be maintained above 4°C (10°C where compression ignition engines are used).

(iii) Automatic priming equipment must be provided where necessary to ensure that the pump will be fully primed with water at all times.

(iv) The performance characteristics of the pumps should be such that the pressure falls progressively with the rate of demand. They must be capable of providing the rate of flow and pressure required at the highest and most remote parts of the protected premises. The output must be so controlled that there is not an excessive rate of discharge at the lowest level in areas close to the installation valves. To meet these conditions pumps must have performance characteristics complying with the requirements laid down.

(v) Where permitted by the water authority, a pump may draw directly from a town main, provided the latter is capable of supplying water at all times at the maximum rated output of the pump.

(vi) The pump should be fully operational within 30 seconds after starting.

(vii) The pump should have a direct drive and must start automatically. Means should be provided for manual starting and once started, the pump must run continuously until stopped manually.

(viii) Where an automatic pump forms the sole supply, a fall in water pressure in the sprinkler system which is intended to initiate the automatic starting of the pump, shall at the same time provide a visual and audible alarm at some suitable location; eg. in the gatehouse or by the installation control valves.

(ix) A test for the automatic starting of the pump must be carried out weekly.

(x) Pumps must be driven either by an electric motor or an approved compression ignition type engine. The electric supply must be obtained from a reliable source, preferably from a public supply. Where a compression ignition engine is used, provision must be made for two separate methods of engine starting. It is preferable that centrifugal pumps should be fixed at least 600mm below the minimum storage level of the water supply.

(xi) Any switches to the electric power feed to motors must be clearly labelled: "Sprinkler pump motor supply – not to be switched off in the event of fire".

(e) Pressure Tank

A pressure tank is an acceptable superior water supply for XLH and OH1 categories only, provided:

(i) The water capacity is not less than:
Sole supply: 7m^3 for XLH; 23m^3 for OH1
Duplicate supply: 7m^3 for XLH; 15m^3 for OH – all groups

(ii) There is an approved arrangement for maintaining automatically the required air pressure and water level into the tank under non-fire conditions.

Duplicate Water Supplies

A duplicate water supply, where required, must be capable of providing the same pressure, rate of flow and have the same capacity as that required for the primary supply. In case of extra light and ordinary hazard classes only, a pressure tank will be acceptable as one source of a duplicate water supply.

The following combination of water supplies are approved duplicate water supplies;

(i) Two town mains.

(ii) Town main and pressure tank (extra light and ordinary hazard classes only).

(iii) Town main and elevated reservoir or gravity tank.

(iv) Town main and automatic pump.

(v) Automatic pump and pressure tank (extra light and ordinary hazard classes only).

(vi) Automatic pump and elevated private reservoir or gravity tank, provided the latter does not form the source supply to the automatic pump.

(vii) Two elevated private reservoirs or gravity tanks.

(viii) Automatic pump supply drawing from:

(a) A virtually inexhaustible source, such as a river, canal, lake etc.

(b) Two limited capacity suction tanks; the primary suction tank must have a holding capacity of at least equal to that required for the relevant hazard class; the secondary suction tank may be of smaller capacity with automatic inflow provided it satisfies certain requirements. (Where limited capacity suction tanks are used the FOC rules state specific requirements as to the number, and certain requirements as to the type of pump to be used).

(c) A single suction tank, but consideration will only be given to this if the circumstances are thought to be favourable.

(ix) Elevated private reservoir and pressure tank (extra light and ordinary hazard).

Pressure Tank

A pressure tank (see Figure F3) is a cylindrical steel vessel with convex ends filled with air under pressure and water. It is acceptable as a sole water supply in the case of extra light hazard class and ordinary hazard class - Group 1, but only as one source of a duplicate water supply in the case of ordinary hazard class – Groups 2, 3 and 3(S).

(a) General

The general requirements for a pressure tank are:

(i) It must be housed in a readily accessible position in a sprinkler protected building or in a separate building of incombustible construction used for no purpose other than for the housing of fire protection water supplies. The tank must be adequately protected against mechanical damage. The temperature of the room should be maintained above 4°C.

(ii) When used as a single water supply, the tank must be provided with an approved arrangement for maintaining automatically the required air pressure and water level in the tank under non-fire conditions. The arrangement should include an approved warning system to indicate failure of the devices to restore the correct pressure and water level. This arrangement is also advocated in cases where the tank provides the duplicate supply.

(iii) The tank must be fitted with air pressure gauges and a gauge glass to show the level of the water. Stop valves and back pressure valves must be provided on both the water and air supply connections to the tank and they must be fixed as close to the tank as possible. (See Figure F3).

Figure F3 Pressure tank and town main with back pressure valves

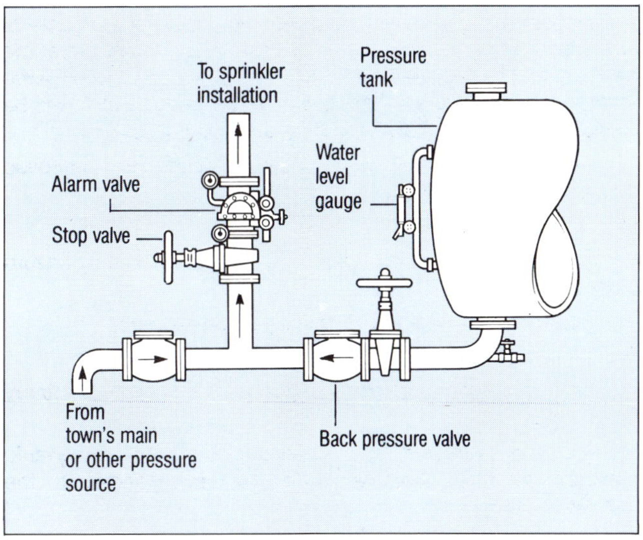

(iv) The minimum quantity of water to be maintained in the tank is:

When sole supply:	extra light hazard class:	7 m³
	ordinary hazard class-Group 1:	23 m³
When duplicate supply:	extra light hazard class:	7 m³
	ordinary hazard class-all Groups:	15 m³

In addition to these requirements, where a pressure tank forms the sole supply to an installation, connections are not allowed to be taken from the supply for any purpose other than sprinklers. If it forms the secondary supply, a pipe not exceeding 50 mm may be taken from the combined supply main to supply dystaulic hose reels for fire-fighting purposes only, subject to the pressure being replenished automatically as in (ii) above.

(b) Air pressure

The working air pressure for pressure tanks is shown in Table F3, and it will be seen that the pressure in the tanks stands at a comparatively high figure. Due to this high pressure, where it forms part of a duplicate water supply, it is often the first supply to be drawn upon and gives a strong initial discharge.

Pressure and Flow Requirements

The Fire Officers' Committee Rules lay down the minimum requirements for pressure and flow in a sprinkler system. These vary for each risk category.

(a) Extra Light and Ordinary Hazard Classes

For extra light and ordinary hazard classes, the figure is made up of;

(i) an actual pressure figure (at the installation control valve);

plus

(ii) a calculated pressure figure (based on the difference in height between the highest sprinklers and the valves at specified rates of water discharge).

Table F4 lists the relevant figures.

Table F3 Working air pressure for pressure tanks

Hazard class	Proportion of air in tank	Minimum air pressure to be maintained in tank when base is on a level with the highest sprinkler (bar)	Add for each metre or part thereof where the tank is below the highest sprinkler (bar)
Extra light	One-third	8.6	0.3
	One-half	5.4	0.2
	Two-thirds	3.8	0.15
Ordinary: Group 1	One-third	5.0	0.3
	One-half	3.0	0.2
	Two-thirds	2.0	0.15
Ordinary: Group 2	One-third	6.2	0.3
	One-half	3.8	0.2
	Two-thirds	2.6	0.15
Ordinary: Group 3	One-third	7.1	0.3
	One-half	4.4	0.2
	Two-thirds	3.0	0.15
Ordinary: Group 3(S)	One-third	8.0	0.3
	One-half	5.0	0.2
	Two-thirds	3.5	0.15

Table F4 Pressure and flow requirements

Risk category	(i) Actual pressure Minimum required running pressure (bar)	(ii) Plus calculated pressure Based on height between highest sprinkler and valves when rate of flow is: (dm³/min)
Extra light hazard	2.2	225
Ordinary: Group 1	1.0	375
	0.7	540
Ordinary: Group 2	1.4	725
	1.0	1000
Ordinary: Group 3	1.7	1100
	1.4	1350
Ordinary: Group 3(S)	2.0	1800
	1.5	2100

(b) Extra High Hazard Class

In the case of extra high hazard class (XHH), very full and specific figures are laid down in the Fire Officers' Committee Rules. The import of these figures is that the water supply must be capable of providing the required flow and corresponding running pressure, at the level of the highest sprinklers in the extra high hazard portion of the premises. The supply must also meet with the requisite density of discharge and area of operation specified for the particular class of occupancy.

Proving of Water Supplies

In the case of town mains, elevated private reservoirs and gravity tanks, facilities must be provided to enable proving tests to be carried out at the valves on each installation to verify that the water

SPRINKLER INSTALLATIONS

supply satisfies the requirements of pressure and flow specified for each hazard class. Water supplies from automatic pumps and pressure tanks are designed to meet the pressure and flow conditions appropriate to the hazard class and accordingly it is not necessary to require practical flow tests in these instances.

Fire Brigade Inlets

Sprinkler installations fed solely from water supplies of limited capacity, such as pressure tank or gravity supplies, must be fitted with a fire brigade inlet connection. This is to enable the fire brigade to pump water into the installation by the use of their own equipment. The fitting of fire brigade inlets to all other installations is also strongly recommended.

Siting of Sprinkler Heads

The following definitions are used to differentiate between the various pipework used on a sprinkler installation (see Figure F4).

Figure F4 Diagrammatic layout of the pipework of a sprinkler installation

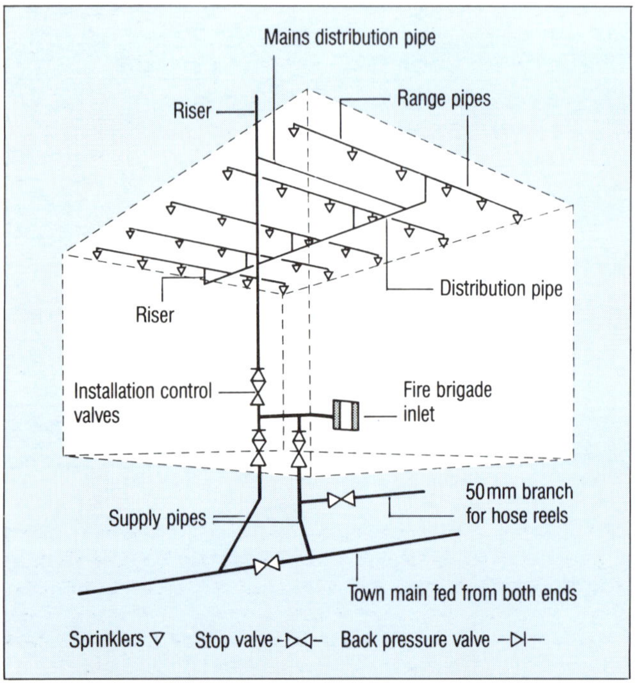

(i) Mains distribution pipes
Mains pipes feeding the distribution pipework.

(ii) Distribution pipes
Pipes directly feeding range pipes.

(iii) Range pipes
Pipes on which the sprinklers are attached either directly or through short arm pipes which do not exceed 300mm in length.

The number of sprinklers allowed on range pipes depends on the layout and size of pipe used, but does not exceed nine on any one pipe. The number of sprinklers fed by a distribution pipe is also determined by the size of the pipe, with a maximum of 48 heads fed by one distribution pipe. Pipe sizes are determined hydraulically, partly by pre-calculated pipe size tables and partly by hydraulic calculation. The area covered by a sprinkler and the distance between sprinklers on range pipes and adjacent rows of sprinklers is determined by the hazard class of the installation.

The maximum area covered by a sprinkler in the different classes is shown in Table F5.

Table F5 Maximum area covered by a sprinkler

Hazard class	General m^2	Special risk areas or storage tanks m^2
Extra light hazard	21	9
Ordinary hazard	12	9
Extra high hazard	9	7.5 to 10

Sprinkler design ensures that the water shall be deflected on to the ceiling and out in a wide circle which will overlap the distribution from the next head. The deflector should be between 75mm and 150mm below ceilings and roofs. Where this is not practicable, sprinklers may be installed at lower levels provided they are not more than 300mm below the underside of combustible ceilings or roofs.

To ensure that the efficiency of sprinkler protection is not lessened, a clear space of at least 500mm must always be maintained below the sprinkler deflectors throughout the room. For high piled combustible stock, an increased clearance of 1 metre or more must be provided. Roof trusses must at all times be accessible to water discharged from the sprinklers.

All parts of a building must be covered by sprinklers, otherwise fire can develop undetected for a period and become too large for the system to deal with effectively. Any roof space or floor space exceeding 800mm in depth must be sprinkler protected. Where holes are cut in floors to take machinery drives, conveyors, chutes and other vertical openings such as hoists, lifts and elevators, it is important that a sprinkler is sited above the opening on the upper floor in order that vertical spread of fire does not take place without early detection.

Multiple Controls

Heat sensitive sealed valve control outlets (see Figure F5) are used when it is desired to operate small groups of sprayers simultaneously – hence the term multiple control.

Figure F5 Diagram showing a multiple control system

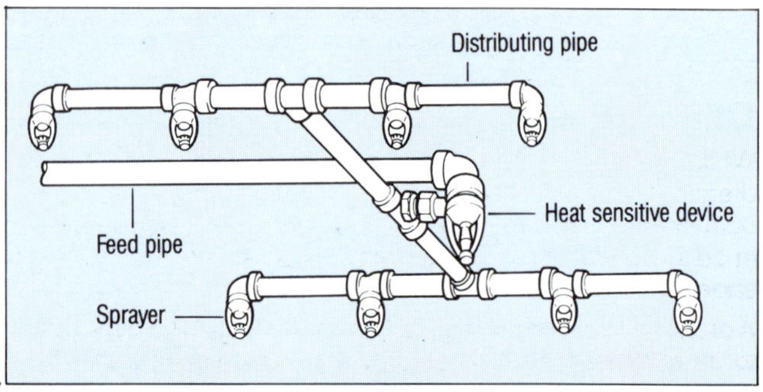

The heat sensitive device will be a glass bulb or soldered link or lever. When this fuses, water is delivered to open sprayers which cover the protected area. An example of a control is shown in Figure 6(a) and an open sprayer in Figure 6(b).

Coverage of a Sprinkler System

Where a sprinkler system is installed, it must cover the whole building, except where the omission of sprinklers is specifically allowed as an exception under the rules. Every building communicating directly or indirectly with, or adjoining the sprinklered building without separating walls must be sprinklered throughout, unless it is one of the permitted exceptions, and the separation is a separating wall with openings protected by fire-resisting doors or fire-resisting shutters.

Certain detached buildings within a specified distance of the sprinklered building which are considered to present an exposure hazard should also be protected by sprinklers. Alternatively the sprinkler protection in the protected building may be extended to provide external sprinkler protection over window and door openings and over any combustible sections of the wall opposite the exposure hazard.

Figure F6(a) An automatic control

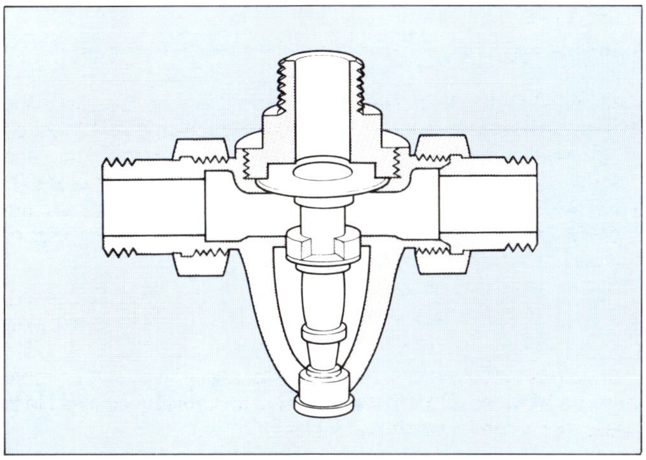

Figure F6(b) An open sprayer

Other Installations Using Water Drenchers

While a sprinkler system protects a building from internal fire, drenchers are placed on roofs and over windows and external openings to protect the building from damage by exposure to fire in adjacent premises. The layout of a typical drencher system is shown in Figure F7.

A drencher system is comprised of water-heads somewhat similar to those of sprinklers; these may be sealed or unsealed (open drenchers), but in the latter case, the water is turned on manually. In a few instances, drenchers may be controlled by quick-opening valves operated by loss of air pressure in a detector line system in a similar manner to high velocity water spray systems.

Drenchers are of three main types;

(a) Roof drenchers,
(b) Wall drenchers,
(c) Window drenchers.

(a) Roof Drenchers

Roof drenchers (Figure F8(a)) have a deflector rather similar to that of a sprinkler head. From the roof ridge, they throw a curtain of water upward which then runs down the roof. All parts of the roof and any skylights, windows or other openings must be protected.

(b) Wall or Curtain Drenchers

Wall or curtain drenchers (Figure F8(b)) throw water to one side only of the outlet, in the form of a flat curtain over those openings or portions of a building most likely to admit fire. In order to cover all combustible portions of a wall, it is the usual practice to put a line of drenchers just below the eaves if these contain flammable material, and to fit every window or opening on the top two storeys with a drencher. Those below this level, except the ground floor and basement, are fitted on every alternate storey.

The drenchers must be fitted so that the streams form a water curtain which must come in contact with the window 600mm from the top. A special use for this type of drencher is on the stage side of a theatre proscenium arch to protect the safety curtain.

Figure F7 Diagram showing a typical drencher system

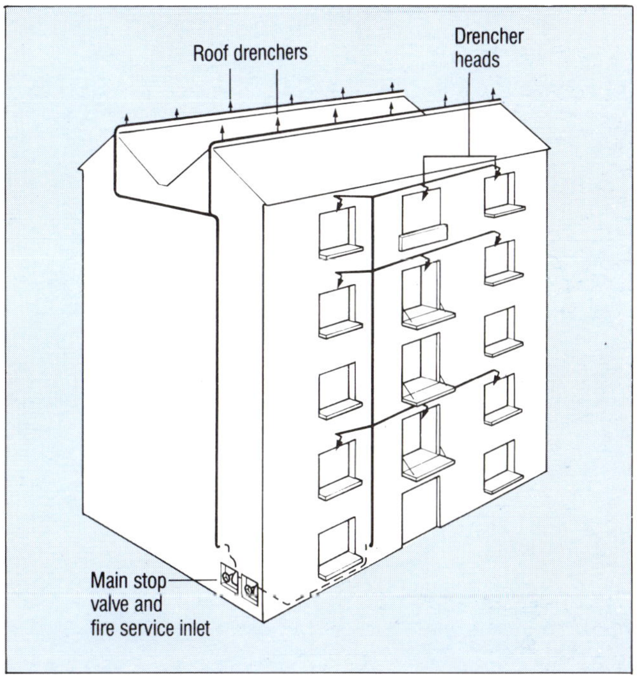

(c) Window Drenchers

As their name implies, window drenchers (Figure F8(c)) are used to protect window openings. They are placed horizontally, level with the top of the window, with a deflector 100mm from the surface of the wall providing a curtain of water to protect the glass. From the tail of the deflector, a jet is thrown inwards onto the glass near the top of the window, while two streams are directed at an angle of 45° to the lower corners.

(d) Water Supplies

The installation should be connected to a nominally unlimited water supply with a pressure sufficient to give at least 0.34 bar at the level of the highest drencher with the 50mm drain valve fully open.

A fire brigade inlet should be provided on the basis of one connection for installations of 55 heads and under, and two connections for installations of over 55 heads. These inlets should be fitted with a non-return valve as should the normal supply pipe.

(e) Valves

The controlling valves must be located in accessible positions on or near ground level and away from fire risk of the exposing hazard. Protection from frost for the supply pipe and valves is essential. A padlocked or riveted strap must be used to secure the valves in the appropriate position.

The position of each valve and the drenchers it controls must be clearly indicated by a wall plate. Not more than 72 drenchers may be controlled by a single controlling valve.

SPRINKLER INSTALLATIONS

***Figure F8** Types of drencher; (a) Roof drencher (b) Wall or curtain (c) Window drencher*

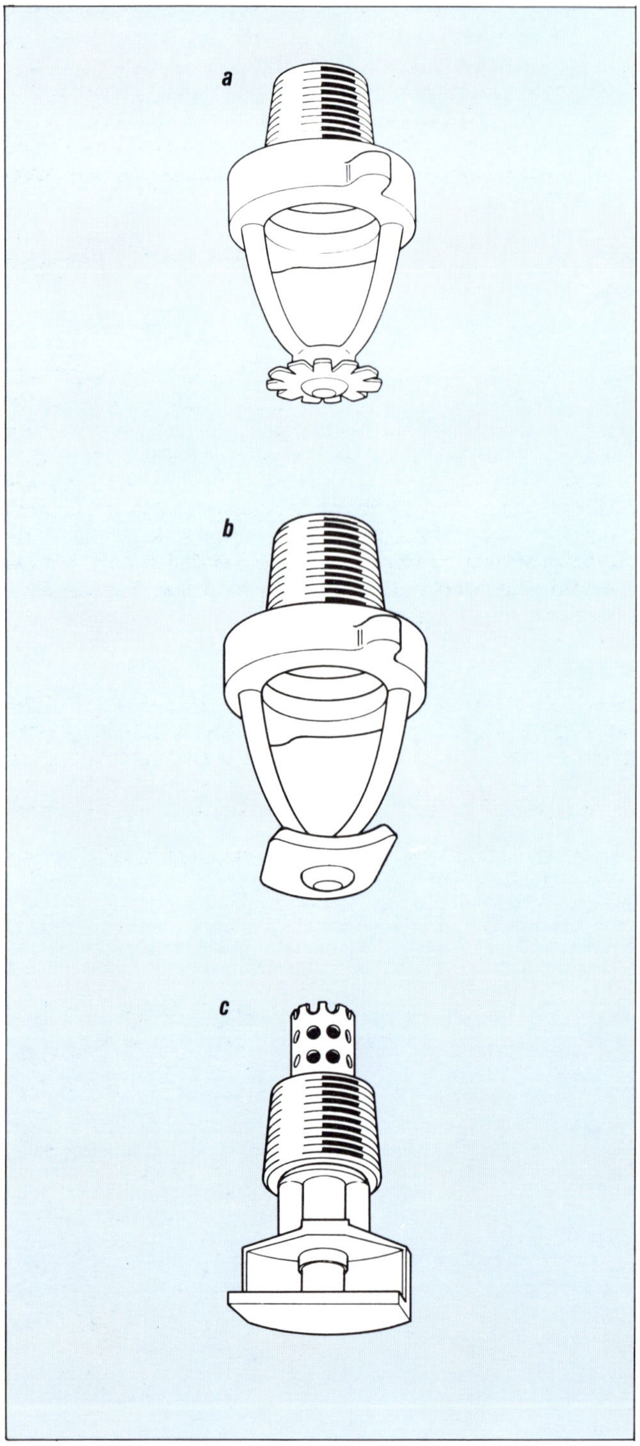

(f) Spacing of Drencher Heads

Drenchers fitted on the top row below the eaves and those on the apex of the roof, must have maximum horizontal spacing of 2.5 m. Windows or other openings or combustible materials in walls exceeding 2.5 m in width must be protected by two or more drenchers not more than 2.5 m apart, and not more than 1.25 m from the window jambs. Windows separated by not more than 600 m of incombustible material may be treated as one window. Not more than 12 drenchers may be fixed on any horizontal line of pipe, and not more than six on one side of the vertical feed pipe.

(g) Discharge

Drenchers may be either open or sealed. Open drenchers are operated simultaneously by the opening of the main valve, while the sealed type are individually actuated in the same way as a sprinkler head. Sealed drenchers differ little from sprinkler heads except in the shape of the deflector plate. They normally operate on the alternate system, and are more economical in the use of water than open drenchers, since only those heads operate which are required, and in consequence, the pressure is maintained more efficiently. Multiple control layouts may however be used.

(h) Drainage

All pipes and fittings above the controlling valves must be so arranged that the water can be drained away. A 20 mm drain tap and pipe must be fitted immediately above each controlling valve. In an open drencher system, the drain taps must always be kept open except when the drenchers are in operation. A full-way 50 mm waste valve and pipe must also be installed below the controlling valve or valves, so that the running pressure tests can be carried out at any time.

Water Spray Projector Systems
Extinction of Oil Fires by Water

When water is used as the sole means of extinction of oil fires, it is normally applied by means of specially installed fixed fire-fighting equipment closely resembling a sprinker system. Precise information as to the way in which burning oil is extinguished is still incomplete but three main factors are known to be involved (see the Manual, Book 1. 'Elements of combustion and extinction').

These are:

(i) cooling;

(ii) dilution of oxygen supplies and

(iii) dilution (or removal) of the liquid (fuel).

Cooling

Oil burns as a vapour distilling from the surface of the liquid involved in fire. Cooling of the liquid reduces the rate of vaporization and consequently the rate at which fuel can reach the fire. When water is applied to the burning surface, the oil is cooled by contact with it, the heat absorbed by the water raises the temperature and converts part at least into steam.

The latent heat of vaporization of water (2260 kj/kg) is such that it is of little importance whether the water projected on to the oil is hot or cold – the cooling effect is caused primarily by its conversion to steam. When water strikes the hot surface of the oil, considerable disturbance is caused and the underlying cool oil is mixed with the hot oil which is thereby cooled, thus again reducing the rate of vapour formation.

Dilution of Oxygen Supplies

The steam formed by the vaporization of the applied water displaces air from the zone of combustion and this tends to smother the fire. Furthermore, in water spray projector systems, the whole of the probable fire area is surrounded by projectors, all of which come into operation on an outbreak of fire. The steam generated, together with the products of combustion seeking to escape from the fire, tend to be driven back into the zone of combustion, and thus still further reduce the oxygen available for combustion.

Dilution of the Liquid

If a flammable liquid which will mix with water is progressively diluted, a stage will be reached where so much water is present that the liquid will no longer burn. Petroleum oils, however, cannot mix with water but under certain conditions are capable of forming an emulsion which may consist either of globules of oil suspended in water, or globules of water suspended in oil. The nature of emulsion varies widely, as does its stability; for instance, an emulsion will only last in spirit, such as petrol, whilst it is being formed, whereas in heavy oils it may take several hours to break down. The formation of an emulsion demands energy which must be provided by the water striking the surface of the oil, thus high velocity is required to give high energy to the water drops. Some of the energy may also come from the heat content of the hot oil. This, in evaporating some of the water supplied, causes turbulence which emulsifies the rest. The flammability of an oil-in-water emulsion (that is one of which the drops of oil are surrounded by water) is low because the heat from the fire must pass through the envelope of water surrounding each drop of oil before the latter can be vaporized and add its fuel to the fire.

The relative importance of each of these three factors in the extinction of an oil fire varies considerably, but all three undoubtedly play a part in the operation of a water-spray projector system.

Types of System

There are two basic types of water-spray systems installed as fixed equipment. One of these is used to extinguish fires and usually referred to as a 'water-spray projector system'; the other is used to provide protection to plant, equipment and to prevent explosions, and is usually referred to as a 'water-spray protector system'.

High Velocity System

This system applies water in the form of a conical spray consisting of droplets of water travelling at high velocity; the three principles

Figure F9 High velocity water-spray projectors
(a) Wormald Fire Systems type (formerly Mather & Platt)
(b) Atlas Fire Engineering type

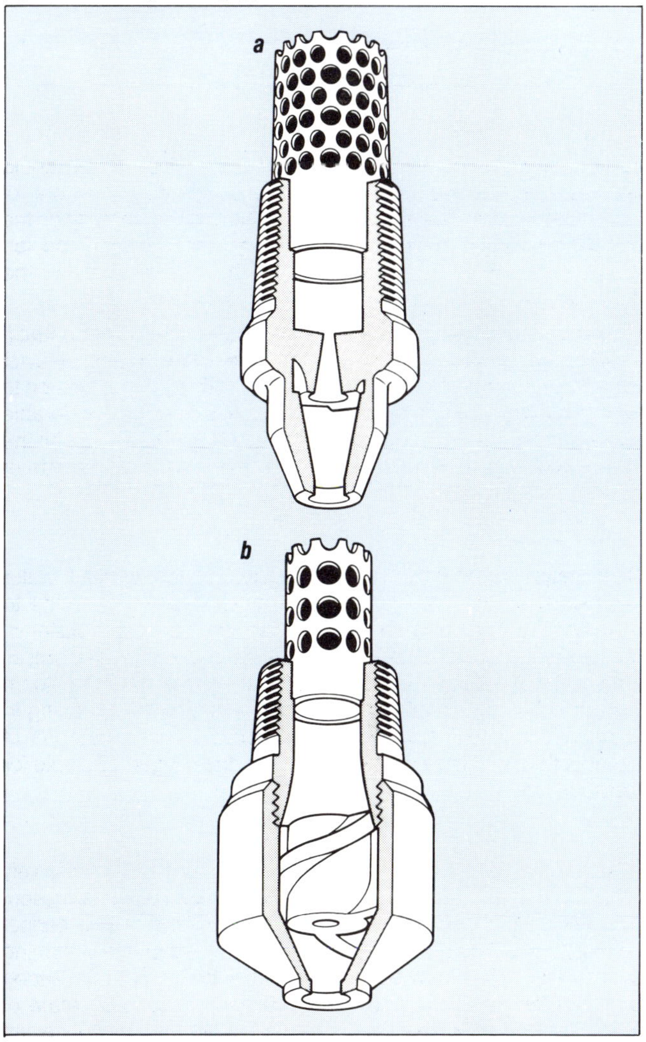

of extinguishment are employed, namely, emulsification, cooling and smothering.

The high velocity system is used on fires involving medium and heavy oils or similar flammable liquids. Droplets of water travelling at high velocity bombard the surface of the oil to form an emulsion of oil and water that will not support combustion. The effect of this emulsion is to convert a flammable liquid into one that will not burn. The emulsion then formed is unstable and after the water has been shut off, the oil will begin to separate from the water. The length of time that this takes depends primarily on the viscosity of the oil, the higher the viscosity, the longer it will take. The addition of the water to the burning oil also cools it and reduces the rate of vaporization.

Thirdly, whilst the water droplets are passing through the flame zone, some of the water is turned into steam, diluting the oxygen feeding the fire thus in effect smothering it.

Projectors

The conical spray of water, consisting of droplets travelling at high velocity is discharged through specially designed high velocity projectors (see Figure F9). These are made in various sizes and are designed to give an even distribution of water over the area covered. The different sizes give a combination of differing flow rates and angles. The equipment is equally suitable for indoor or outdoor use because the pipework can be so designed that no water enters the installation until a fire is detected. The risk of water freezing in the pipework is therefore, eliminated.

Methods of Operation

High velocity water spray projector systems can be either automatic or manual in operation or, in some cases, manual only. Manual control is usually by hand-operated valves placed outside the probable fire zone, or by valves with remote control from one or more points. There are two main forms of automatic operation.

Figure F10 Illustration of a water-spray projector system (inset): detail of the automatic control from below

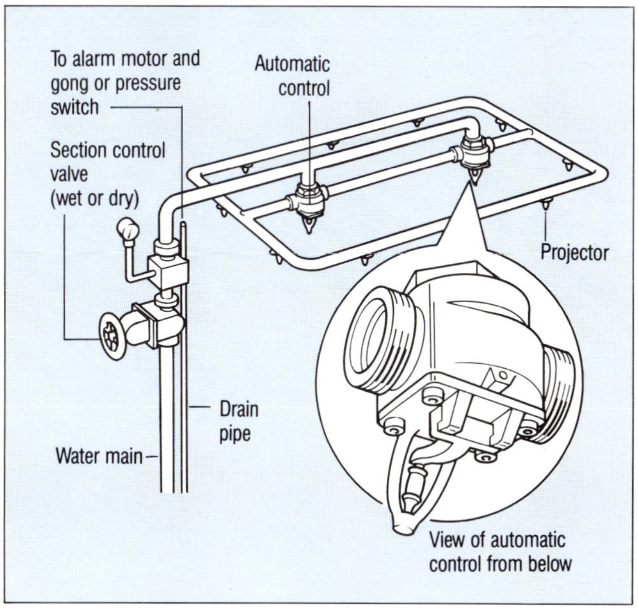

The glass bulb automatic control (Figure F10) coupled to a small group of projectors, is used on situations where fires in their incipient stages are likely to be small. The projectors are arranged in small groups to cover the risk, each group being connected by pipework to a control (shown inset in Figure F10).

When fire causes a control to operate, water is discharged simultaneously through the projectors in the group. When this type of equipment is installed in situations where the temperature

SPRINKLER INSTALLATIONS

may fall below freezing point, the pipework between the controllling valves and the automatic control is charged with compressed air and the system functions in a similar way to a dry sprinkler installation. Where fires are likely to be larger or to spread rapidly over an extended area, a larger number of projectors are designed to operate simultaneously.

A typical application of this method of control is illustrated in Figure F11.

Figure F11 Diagram showing the arrangement of the projectors to protect a transformer

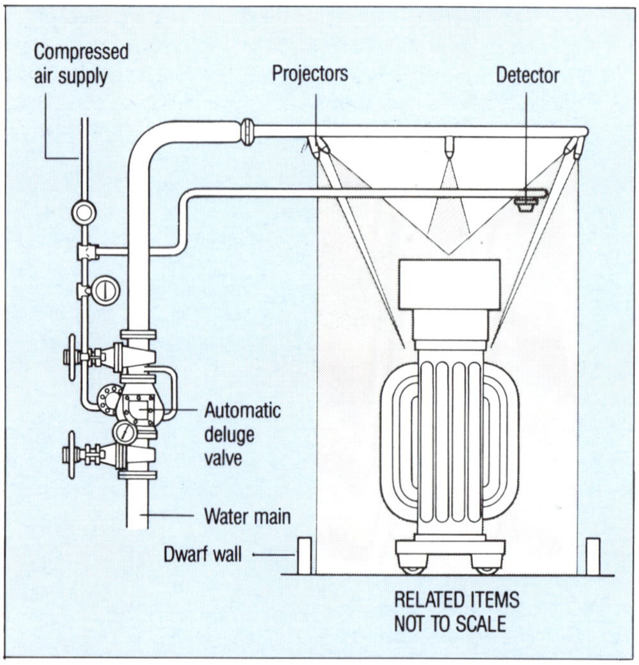

Projectors mounted on empty pipework command the whole of the exterior of the transformer and its conservation tank, and also the floor area around the transformer. Glass bulb detectors mounted on independent pipework containing compressed air are so positioned that wherever a fire may originate, one at least will operate and allow the compressed air in the pipework to escape. The escaping compressed air causes a rapid fall in pressure on the diaphragm in the automatic deluge valve, to which both systems of pipework are connected. The movement of the diaphragm causes the deluge valve to open and water to discharge through the projectors.

Alarms

An alarm is a normal part of a water-spray projector system. It is usually of a type very similar to that used in a sprinkler system; ie. a loud sounding gong operated by a water motor which is driven by a small flow of water diverted at the installation controlling valves when open. In addition, an electric alarm may be provided to give warning at some control point of the outbreak of fire and its location.

Combined High Velocity Systems

Combined high velocity systems and sprinkler protection may be used in some industrial processes which involve the use of flammable liquids. A glass bulb control forms the fire detecting element and automatic valve for a group of projectors and open sprinklers of a special type which are designed to distribute the water discharge over a wide ceiling area. When the control operates, water is discharged from the projectors onto the burning liquid and from the sprinklers on to the ceiling and adjacent walls.

Medium Velocity Systems

When a fire occurs, this system applies water in finely divided droplets travelling at medium velocity; it gives protection to tanks, structures and factory equipment by cooling, by controlled burning of flammable liquids and by dilution of explosive gases. Cooling is achieved by the water flowing over the external surface of the exposed vessel and supporting structure, thus reducing their temperature and preventing absorption of heat.

Protection by controlled burning is obtained in three ways; firstly by directly cooling the surface of the liquid, thus reducing the rate of vaporization; secondly by producing steam which dilutes the air and vapour feeding the fire, and thirdly by introducing droplets into the flame thus cooling it and reducing the transfer of heat back to the liquid surface.

When highly flammable gases leak from their containers to form a mixture with air, a critical mixture will form so that any source of ignition may cause an explosion of fire. If the spray system is operated when a leakage has been detected and before fire has started, it produces air turbulence which promotes quicker dilution of the gases by air and water vapour. Fire can be prevented by employing this method to produce quickly a mixture which is too weak to burn. Even if a fire occurs, then the spray system may prevent a dangerous increase of pressure inside the containing vessels and protect the supporting structures against failure caused by high temperature.

Medium velocity systems are very similar in operation and layout to the high velocity systems previously described. The sprayers discharge a cone of water spray consisting of medium-size droplets of water with a range of different sizes and discharge angles. In most installations the system can be discharged automatically and manually, although some systems may be found which only operate manually. There are three main forms of automatic operation; the first of these is the glass bulb sealed sprayer, very similar to the sprinkler head. The second method, used where a group of sprayers are required to operate simultaneously is an automatic control as used in the high velocity system, and the third is by the use of two systems of pipework using the automatic deluge valve.

Systems Using Steam

Steam, the oldest of the smothering agents, is not now widely used except in ships' holds,though it is found occasionally in industry - particularly for the protection of flammable liquids. To be immediately effective steam must be directed onto the fire in such a way that it blankets the fire. When used for space flooding it will take considerably much longer to be effective because in the early stages of discharge the steam condenses until the whole of the protected space has heated up.

Except when steam from a hand-held type of branch is being used, it is necessary for the space protected to be reasonably airtight. Fixed piping is generally used to connect the boiler with the nozzle arranged to direct the steam onto the risk, it being usual to provide a main valve, either itself designed to open slowly or else fitted with a by-pass which must first be opened. The purpose of this precaution is to give employees working in the area warning that steam is being used and to leave the area, whilst also protecting the pipe against severe water hammer. It is not uncommon to locate the valves outside the protected compartment, so that the steam can be turned on without danger to the operator. In some installations a sprinkler-type layout with fusible heads is used, water being kept in the pipes between the steam and the heads in order to avoid condensation and to prevent the steam from actuating the sprinkler.

Manual steam jet systems, consisting of flexible tubing connected to pipework and fitted with a branch, are frequently found, the control valve for the branch usually being fitted where the tubing leaves the main pipe.

Steam may be used to protect the following risks:

(i) ships' holds (depending on cargo);
(ii) flammable liquid users (benzol plants, refineries, dry cleaners etc).
(iii) oil quenching tanks, driers, waterproofing processes.

HOSE REEL INSTALLATIONS

A hose reel consists basically of a steel drum or reel on which is rolled a length of suitable rubber hose with a shut-off nozzle at the end, the whole unit being connected to a suitable water supply. They are installed to provide the occupants of a building with a first aid means of fighting a fire whilst awaiting the arrival of the local Fire Brigade but they are not normally used by the Fire Service personnel unless the fire happens to be a small localised one, as the amount of water discharged by each reel, approximately 23 litres/min is not sufficient to extinguish a large fire.

Statutory Regulations

Hose reel installations are normally provided in accordance with the requirements of the statutory regulations including:

1. The Factories Act 1961
2. The Shops, Offices and Railway Premises Act 1963
3. The Building Regulations 1985
4. The London Building Act 1939 (Section 20)
5. The Fire Precautions Act 1971

In addition to the above, Para 1600 of the 29th Edition of the FOC Rules requires that hand extinguishing appliances be provided in all sprinklered buildings and this includes hose reels. The FOC's detailed requirements for hose reel installations are contained in Appendix D to the Rules for Automatic Fire Alarms.

Positioning of Hose Reels

As hose reels are intended for use by the building occupants, it is of vital importance that they are placed in readily accessible positions so that they can be used without exposing the personnel to danger. This means that they must as a general rule be fixed along escape routes or adjacent to fire exits so that personnel escaping from an outbreak of fire will pass them on their way to safety and can thus use them without having their means of escape cut off.

In office blocks especially the multi-storey type, the hose reels must be fitted inside the actual office accommodation which as a general rule means that they are fitted adjacent to the fire exit doors into the lift or stair lobbies. This enables the hose reel to be used without opening the smoke stop doors of the lobby and thus, prevents the lobby from becoming filled with smoke. In industrial premises, it is not always possible to site hose reels adjacent to the fire exits owing to the fact that the width of the building would prevent the hose from reaching a fire in the centre. In these circumstances it is necessary to position the hose reels in the centre of the building usually on the columns or stanchions but care must be taken to ensure that they can be used safely in the event of fire.

Types of Hose Reel

There are two basic types of hose reel, fixed or swinging. The fixed type of reel should normally have its centre line at least 1.5 m above the floor. A swinging type hose reel can be swung through 180° and be mounted at any convenient height the average being 900 mm to 1.05 m above the floor.

Hose reels are normally fitted with 20 mm dia. reinforced non-kink rubber hose but hose reels fitted with 25 mm dia. are available if required. The usual lengths of hose are 23 m, 30 m, 36 m and 45 m.

Each reel is also fitted with a 5 mm lever operated shut-off nozzle at the end of the hose.

In determining the length of hose to be used the critical factor is the requirement that all areas of the building must be covered and no part must be more than 6 m from the hose reel nozzle when the hose is uncoiled.

Manually operated hose reels are fitted with a wheelhead type isolating valve on the inlet to the reel which is closed when the reel is not in use to prevent the hose from being continuously subject to water pressure. Before running out the hose it is necessary to turn on the isolating valve and a warning notice to this effect must be positioned adjacent to the reel.

Automatic hose reels are fitted with a valve which is opened or closed by the revolving action of the reel when the hose is pulled out or rewound and with this type it is only necessary to pull out the hose and open the shut-off nozzle at the end to allow water to be discharged. It is however, normal practice to fit a lockshield type isolating valve on the feed pipe so that individual reels can be shut down for maintenance purposes without isolating the whole system.

Fire Protection Discharge from Jets

Table F5 was derived from the following formula:

Quantity = Velocity × area

$$q = C_d \times V \times A \times 1000$$

putting $C_d = 0.96$ – coefficient of discharge

H in m
d in mm
V in m/s = $\sqrt{2gH}$
q in litres/sec

$$q = 0.96 \times \sqrt{2gH} \times \frac{\pi}{4} \times \frac{d^2}{10^6} \times 10^3$$

This reduces the formula:

$$q = \frac{3.35d^2}{1000} \times \sqrt{H}$$

Example when d = 8 mm
H = 9 m

$$q = \frac{3.35 \times 8^2 \times \sqrt{9}}{1000} = \frac{3.35 \times 64 \times 3}{1000} = \frac{10.05 \times 6.4}{1000}$$

q = 0.64 l/sec

Table F6 Height of jet (max) in metres

Head on jet in metres	Diameter of orifice in millimetres 3.2	6.3	9.5	12.7	16.8	19.0	25.4
3.0	2.67	2.86	2.93	2.96	2.97	2.99	3.00
6.0	4.57	5.33	5.59	5.72	5.79	5.85	5.91
9.0	5.79	7.44	8.00	8.29	8.46	8.63	8.72
12.0	6.10	9.14	10.15	10.67	10.97	11.28	11.43
15.2	—	10.49	12.07	12.86	13.41	13.72	14.05
18.3	—	11.43	13.72	14.84	15.54	15.83	16.50
21.3	—	11.89	15.24	16.76	17.68	18.29	19.02
24.4	—	12.19	16.15	18.29	19.51	20.42	21.34
27.4	—	—	17.07	19.81	21.34	22.25	23.47
30.4	—	—	17.68	21.03	22.86	24.08	25.60

HOSE REEL INSTALLATIONS

Table F7 Discharge from jets in litres per second

Head on jet in metres	Diameter of orifice in millimetres 3.2	4.8	6.3	8.0	9.5	11.5	12.7	16.8	19.0	22.0	25.4
1.5	0.04	0.09	0.16	0.25	0.36	0.49	0.65	1.00	1.46	1.99	2.60
3.0	0.06	0.12	0.23	0.36	0.52	0.70	0.92	1.43	2.07	2.81	3.67
4.5	0.07	0.16	0.28	0.44	0.63	0.86	1.12	1.76	2.53	3.45	4.50
6.0	0.08	0.18	0.32	0.51	0.73	0.98	1.30	2.03	2.70	3.99	5.20
7.6	0.09	0.20	0.36	0.57	0.82	1.11	1.45	2.27	3.27	4.46	5.82
9.0	0.10	0.22	0.40	0.62	0.89	1.22	1.59	2.44	3.58	4.88	6.37
10.6	0.11	0.24	0.43	0.67	0.97	1.32	1.72	2.69	3.87	5.27	6.89
12.0	0.12	0.26	0.46	0.72	1.03	1.41	1.87	2.87	4.14	5.63	7.36
13.7	0.125	0.27	0.49	0.77	1.10	1.53	1.95	3.05	4.39	6.11	7.80
15.2	0.13	0.29	0.51	0.80	1.15	1.61	2.05	3.21	4.63	6.43	8.26
18.3	0.14	0.31	0.56	0.88	1.27	1.73	2.25	3.52	5.07	6.90	9.02
21.3	0.15	0.34	0.60	0.95	1.37	1.86	2.43	3.80	5.48	7.46	9.77
24.4	0.16	0.36	0.65	1.02	1.46	1.99	2.60	4.06	5.85	7.96	10.38
27.4	0.17	0.39	0.69	1.08	1.55	2.05	2.76	4.24	6.21	8.41	10.99
30.4	0.18	0.40	0.73	1.14	1.64	2.23	2.91	4.55	6.55	8.86	11.59

Pipework

Hose reel systems are classed by the Water Authorities as domestic cold water services owing to the fact that water can be freely drawn off through the reels.

All hose reels must be fitted with a union between the isolating stop valve on the feed pipe and the inlet to the reel. This enables individual reels to be shut-off and dismantled for maintenance and overhaul without shutting down the entire system.

The drop pipe to the hose reel from the distribution main is normally 25 mm dia. with a 25 mm × 20 mm reducing elbow at the bottom, the feed pipe to the reel being 20 mm diameter.

Some Water Authorities require that all valves are tested and stamped before installation.

Care should be exercised when positioning hose reels on drinking water systems that a sufficient flow of water passes along the pipework (oversized to meet the hose reel requirements) to ensure fresh potable water at all drinking and culinary terminal fittings.

Water Supplies

The following water supply systems are acceptable for hose reel installations and are the ones most commonly used:

1. Direct connection from town's main (or mains).
2. Connection from sprinkler installation trunk main.
3. Pump supply (usually with suction break tank).
4. Connection from boosted domestic water service.
5. Connection from hydrant main or wet fire riser.

The normal requirement for a hose reel system water supply is that it should be capable of providing a flow rate of least 2.27 l/s at a running pressure of 2 bar at the level of the top hose reel in the system. With a pressure drop of approximately 0.5 bar through the reel this is sufficient to produce a 6 m jet of water to comply with the requirements. The flow rate of 2.27 l/s will permit up to six reels to operate efficiently (0.38 l/s per reel), although if a fire has reached a size requiring this number of hose reels to be brought into action it is unlikely to be extinguished solely by their use.

Special Provision in Respect of Pump Supplies

A typical pump supply is shown in Figure F12.

Figure F12 Typical arrangement of hose reel system

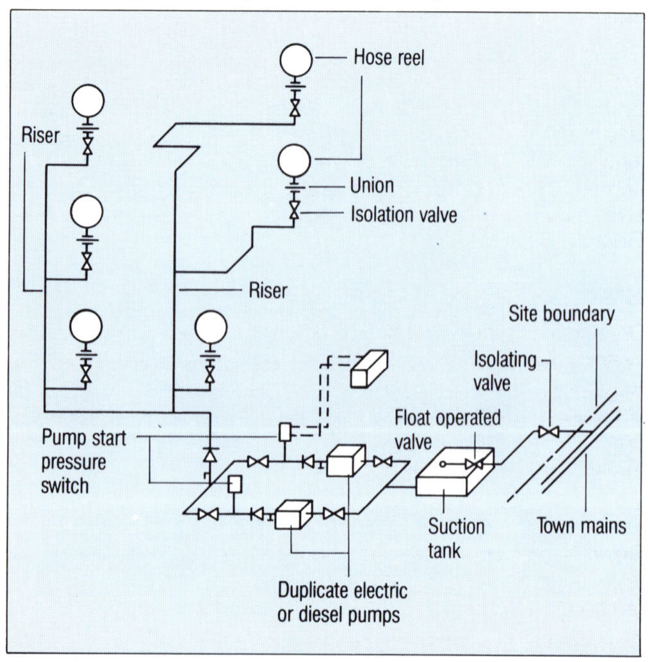

As most local Water Authorities will not now permit direct connections from their mains to be boosted, it is often necessary to provide a suction break tank supplied from the town's main via a suitable float operated valve. The required capacity of the break tank varies. As a general guide a capacity of 1250 litres or 1.2 m^3 is a common requirement, but the Regional Water Authority should be contacted to establish their requirements.

Duplicate pumps are normally provided and these may be controlled as follows:

Automatic 'On' & Manual 'Off'

This system of control is identical to that used for sprinkler system pumps in that the pump is started automatically by a pressure switch and stopped manually by a 'Stop/Reset' push button on the starter. In addition, an electrical alarm bell is provided to ring all the time the pump is running and thus provide a remote warning of system operation.

Automatic 'On' & 'Off'

With this system the pumps are started by a pressure switch but a flow switch is also provided in the pump delivery line to ensure that the pump continues to run all the time there is a flow of water through the system.

Testing and Approval

Completed installations are normally required to be tested in the presence, and to the satisfaction, of the local Fire Authority and be approved by both the local Fire and Water Authorities.

DRY RISERS, WET RISERS AND FOAM SYSTEMS

DRY RISERS

Basically, a dry riser consists of an empty or dry pipe rising vertically up a building with hydrant valve outlets on each floor and at roof level. An inlet (Breeching Piece) is provided at ground level in an external wall to enable the Fire Brigade to pump water into the riser from the nearest suitable hydrant.

Dry risers are provided solely for use by the Fire Service personnel and they are not intended as first aid fire fighting equipment for use by the building occupants in the same way as a hose reel system. Use of a dry riser in the event of fire avoids the necessity of running long lengths of canvas hose up the staircase of a building and thus enables the Fire Brigade to tackle the blaze much more effectively and in the case of very tall buildings much more quickly (see Figure F13).

Statutory Regulations

Dry risers are normally installed only when required by the Local Water Authority and therefore there are no insurance requirements relating to them. Most local Fire Brigades publish a standard for dry risers installed in their area. They are normally only installed in buildings up to 61 m high; wet risers being required in buildings exceeding this height.

Size and Positioning of Risers

Risers should be 100 mm dia. where only one hydrant valve outlet is provided on each floor. When two outlets are provided on a floor fed from the same riser then the diameter should be 150 mm.

Dry risers should normally be positioned in the ventilated lobby approach staircase or within a staircase enclosure. This enables Fire Service personnel to couple up their hose to the riser outlet in a smoke-free area and it is not necessary to open the smoke stop doors until the last moment when the water is turned on.

Figure F13 Typical arrangement of dry riser

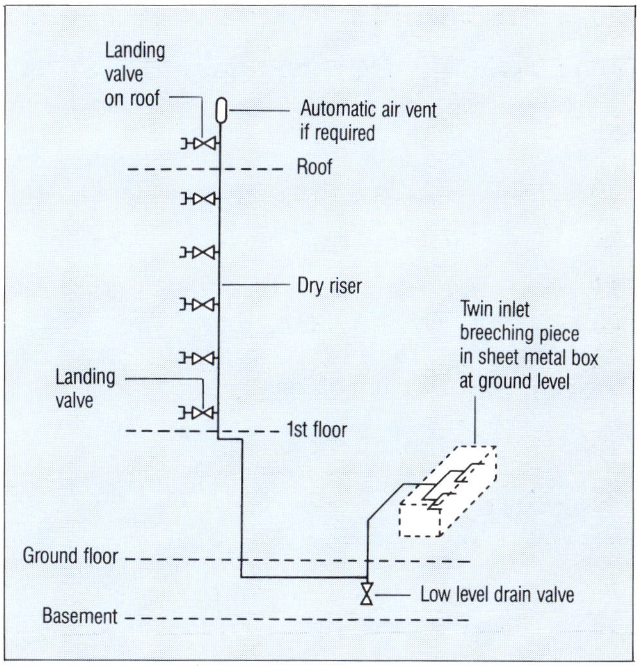

Inlets

A 100 mm riser should be fitted with twin inlets and a 150 mm riser should have four inlets. Each inlet consists of a 65 mm dia. male instantaneous coupling to BS 336 with a non-return valve and a blank cap secured with a short length of chain. The inlets are normally grouped together in a single casting called a "breeching pipe" or "inlet breeching" which has a single flanged or screwed outlet. The "inlet breeching" must be mounted in an external wall with its centre line not more than 762 mm above pavement level. When positioning inlets due regard must be paid to accessibility, the positions of adjacent street hydrants and danger from falling glass, etc, in the event of a fire.

Inlet breechings are normally contained in sheet metal inlet boxes with wired glass doors in accordance with BS 3980. The doors are secured with spring locks so that they can be opened from inside by smashing the glass and releasing the catch on the lock.

Outlets

Dry riser outlets are 65 mm dia. gunmetal gate pattern landing valves with flanged inlets and female instantaneous outlets fitted with plugs secured by short chains. They should be mounted with their centre lines between 910 mm and 1.06 m above finished floor level.

Outlets should be provided for every 929 m^3 of floor area on every floor from first floor level to the roof. An exception to this requirement is made in the case of blocks of flats where outlets are normally only required on every other floor from first floor to roof.

Some Water Authorities require that all inlet breechings and landing valves are tested and stamped before installation. Automatic Air Relief Vents are sometimes required by the Local Fire Authority to be fitted at the top of the riser.

Owing to the fact that dry risers project above the roof level of buildings it is necessary for them to be efficiently earthed to prevent damage from lightning.

Installation

Dry risers should be installed progressively as a building is constructed in order to provide protection during building operations. In buildings over 30.48 m in height the riser must be installed when the building exceeds 18.28 m in height.

DRY RISERS

Testing and Approval

Completed installations are normally required to be tested in the presence, and to the satisfaction, of the Local Fire Authority and be approved by both the Local Fire and Water Authorities.

WET RISERS

A wet riser is a dry riser with a permanent water supply system (usually pumped) added on (see Figure F14).

Figure F14 Typical arrangement of wet riser

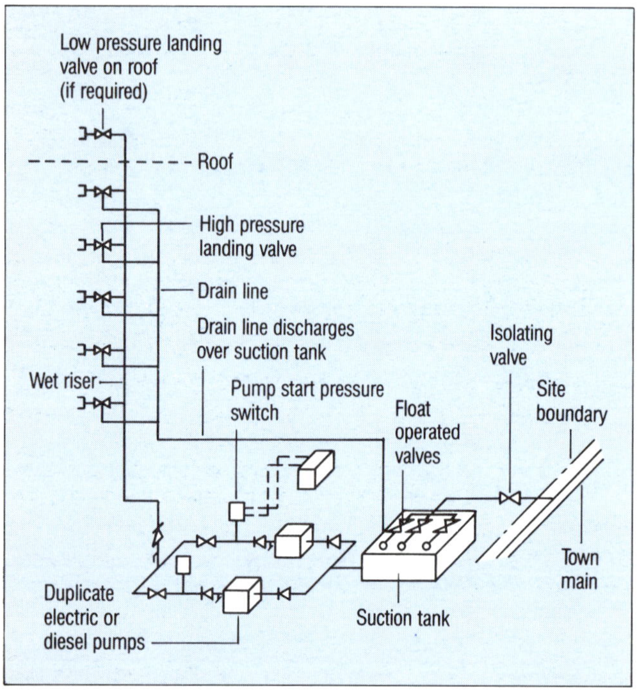

Statutory Regulations

As is the case with dry risers there are no insurance requirements relating to wet risers. Also, as they are only installed in buildings exceeding 61 m in height and as there are very few standard requirements, it is advisable to obtain the Local Authority requirements in each instance.

Size and Positioning

Again as with a dry riser, a wet rising main should be positioned either in the ventilated lobby approach staircase or within the staircase enclosure itself.

The riser should be 100mm dia. normally with outlets on all floors. Normally only one outlet per floor will be required but where two are necessary a second riser should be provided at the opposite end of the building.

Water Supplies

Generally, a wet riser supply system should be capable of maintaining a minimum running pressure at the top outlet at roof level of 4 bar at a flow of 22.7 l/s. The maximum running pressure permitted with only one outlet in operation is 5 bar. To maintain the above pressure and flow rates it is necessary to employ pumping equipment, usually duplicate electric and diesel fire pumps. In view of the very high pressures involved (up to 16.5 bar at the pump delivery) direct boosting of the town's mains is not permitted and therefore it is necessary for the pumps to be supplied from a suction break tank which must have a minimum actual capacity of 45.45 m^3. An automatic inflow from a town's main having a flow rate of at least 7.6 l/s must be provided to refill the tank.

Outlets

Wet riser outlets are 65mm dia. high pressure landing valves with flanged inlets and female instantaneous outlets fitted with plugs secured by short chains. They should be mounted with their centre lines between 910mm and 1.06m above finished floor level and one outlet should be provided for every 929 m^2 of floor area.

In order to reduce the running pressure in the canvas hose line, high pressure landing valves are fitted with an adjustable butterfly valve in the inlet which acts in a similar manner to an orifice plate and reduces the pressure under flow conditions when the valve is open between 4 bar and 5 bar. The butterfly can be adjusted by means of a nut on the outside of the landing valve body and by use of a special test pipe and gauge, can be altered to give the correct running pressure when the landing valve is fitted in position.

It is also necessary to limit the static pressure in the canvas hose to a maximum of 6.6 bar if water is shut off by closing the branch pipe nozzle at the end of the hose. This is achieved by means of a spring loaded pressure relief valve incorporated in the outlet from the landing valve. The discharge from the relief valve is piped via a 50mm dia. connection into a 100mm dia. drain pipe running vertically down the building alongside the main riser and discharging over the suction tank. In early wet riser installations a lot of trouble was experienced through inadequate drainage facilities and it is important to ensure that the drain pipework is short and direct with the minimum number of bends.

Pumping Equipment

The arrangement of the duplicate electric and diesel fire pumps, starters and suction tank is identical to that used for sprinkler installations with the exception of the method of control. In a similar way to hose reel systems, the electric pump can be either "Auto On" and "Manual Off" or "Auto On" and "Auto Off" but in both cases the diesel pump (which is the standby) is "Auto On" and "Manual Off". The "Auto Off" facility is provided by incorporating a flow switch in the electric pump delivery line.

Installation

The requirements for installation during construction of a building are the same as from a dry riser. If when the riser is initially installed it is impossible to commission the pumping equipment, then it must be fitted out as a temporary dry riser with a temporary breeching inlet in a suitable position at ground level.

As with a dry riser it is necessary to efficiently earth wet risers to prevent damage from lightning.

Testing and Approval

Completed installations are normally required to be tested in the presence, and to the satisfaction, of the Local Fire Authority and be approved by both the Local Fire and Water Authorities.

FOAM SYSTEMS

Foam is a frothy product, similar in appearance to soap suds and it is so light that it will float on the surface of liquid. The properties desired in foam are that it should be tough, tenacious, long lasting and flow freely. It should also resist heat, wind and rain, and retain its water content for a long time.

Foam has three constituents; foam compound, water and air, and its production takes place in two stages. First the liquid foam compound is induced into the water stream and the aeration takes place. The two stages (induction and aeration) are sometimes carried out in one apparatus but often separate units are employed, in which case the mixture of foam compound and water is pumped through a pipe or hose to the foam maker (sometimes known as an aerator box or aspirator box).

Application

Foam may be used for extinguishing fires of oils, spirits, paints, molten fats and similar liquids. Foam may also be used for items of plant or complete buildings.

To appreciate the value of foam when used to fight fires involving liquids, it must be remembered that the liquid itself does not burn; it is the mixture of air and the vapour given off from the liquid which burns. Combustion does not occur actually at the surface of the liquid because the proportion of vapour to air is too great to form a combustible mixture. The fire occurs slightly above the surface of the liquid.

Liquid stored in open or closed tanks is dealt with by applying foam to the surface of the liquid, in such quantities as to cover the entire surface of the burning liquid. This blanket of foam excludes air and prevents the formation and possible reignition of further vapour.

Foam should not be used on live electrical equipment because it is a conductor of electricity.

Types of System

Portable Extinguishers

Extinguishers generally have a nominal capacity of up to approximately 17 kg and should be capable of expelling a continuous discharge of foam in the form of a jet until the whole of the contents have been discharged.

Mobile Foam Units

May be used in situations where the quantity of foam likely to be required is greater than could be provided by portable extinguishers, foam trailers, tenders or engines. These units have the disadvantage in that they are always ready for immediate use, but as they cannot be refilled quickly, their use is limited to the time taken to discharge the original contents.

Fixed Foam Installations

Where large quantities of foam are required for areas such as oil storage tanks, boiler houses, etc, it is normal practice to install a system of pipework to deliver the foam to the particular risk.

The oil storage tank is a risk commonly protected by the application of foam and the installation would be designed (generally in accordance with the National Fire Protection Association Standard No: 11) to provide a blanket of foam to the burning oil surface within the storage tank. Assuming an adequate supply of water at the required pressure is available from an underground main, a typical installation would comprise a valved connection from the water main via an induction unit or venturi to the oil storage tank.

Foam Branch Pipes

These are generally carried on Fire Brigade appliances and comprise a specially designed branch pipe for use with a 65 mm hose. A small tank of foam compound (sometimes in the form of a knapsack) is connected by a flexible pipe to the branch pipe. The flow of water through the branch pipe induces the foam compound into the water stream, aeration takes place within the branch pipe and a jet of foam capable of a range of 9 m from water pressure of 3.3 bar up to 30 m from water pressure 8.3 bar.

Foam Inlets

Rooms which contain oil-fired boilers, oil storage tanks, oil filled electrical equipment or other materials or apparatus for which foam is a suitable extinguishing medium may be fitted with pourers so placed that a foam blanket can be formed over the equipment and floor of the room. Pipework from the pourers would be connected back to a convenient point in the open (generally in an outside wall), the fire brigade can then connect a foam making branch pipe by means of an adaptor and pump foam into the room or basement area to deal with the fire.

Design Data

Minimum recommendations for foam systems have been established by the National Fire Protection Association (NFPA) and are contained in their standard No. 11. Application rates higher than these will be beneficial and reduce extinguishing time but as a general rule the NFPA standard is accepted as a good basis for design purposes.

High Expansion Foam

Foam has an expansion ratio of between 6 and 8 to 1. High expansion foam has an expansion rate of between 600 and 1,000 to 1. The method of generating high expansion foam differs from the method employed with ordinary foam systems in that a fan is incorporated to provide the large quantity of air necessary for making high expansion foam.

High expansion foam units may be installed as a permanent system with manual operation, automatic operation or both. They may also be carried as part of a Fire Brigade's equipment.

CARBON DIOXIDE SYSTEMS

Carbon dioxide (CO_2) is a readily available commercial product and it is most familiar, perhaps as the gas which aerates soda water.

At normal pressures and temperatures, carbon dioxide is an odourless and colourless inert gas with a density of approximately 50% greater than air.

As a medium for extinguishing fires carbon dioxide has certain features which should be considered.

Application

CO_2 extinguishes fires by reducing the oxygen content and/or disrupts the fuel to air ratio to a point where further combustion is prevented. The normal proportion of oxygen in air is 20% and ordinarily, a reduction to 15% would extinguish inflammable liquids.

Clean and Penetrating

It is a non-corrosive inert gas with virtually no adverse effects on most materials and it does not contaminate foodstuffs. It will penetrate into otherwise inaccessible areas and will disperse leaving no trace whatsoever. The gas is not poisonous, but if breathed in quantity it can cause suffocation simply by excluding oxygen from the lungs.

Non-conductivity

It has good dialectric strength and can be applied safely to live electrical equipment.

Uses and Limitations

It is effective on small isolated fires of inflammable liquids, greases and fats, paint dips, enamelling ovens, paint spray booths, varnish kettles etc.

It is well suited for use in electrical switchrooms, sub-stations and electrical apparatus generally, it is also used in situations where the use of water would cause severe damage, eg. museums, libraries and computer rooms etc.

CARBON DIOXIDE SYSTEMS

Carbon dioxide should not be used to extinguish fires involving:

(a) Reactive metals such as sodium, potassium, magnesium, titanium and zirconium.

(b) Chemicals containing their own oxygen supply such as cellulose nitrate.

(c) Metal hydrides.

Storage

There are three standard sizes of CO_2 cylinders, these having capacities of approximately 23kg, 36kg and 46kg of CO_2 content. Smaller capacity cylinders are available for use as portable extinguishers.

When CO_2 is highly compressed it liquefies and it is generally stored in liquid form in cylinders under a pressure of 46-56 bar. Each cylinder incoporates a safety disc designed to rupture at a pressure of 173-186 bar. The pressure exerted by the CO_2 within the cylinder varies with changes in temperature. For example the CO_2 content of the cylinder at 0°C would be 33 bar whereas at 60°C the pressure would be 165 bar. Due to the possible explosion hazard when CO_2 cylinders are heated, it is desirable to cool by means of water spray protection to CO_2 storage areas, although CO_2 is in itself an extinguishing agent.

Another way which CO_2 can be stored is to employ a single low pressure container for the required quantity and since the pressure exerted by CO_2 varies over a wide range with changes in normal atmospheric temperature, it is desirable to control the temperature of the contained CO_2 within relatively narrow limits so that the storage vessel can be a reasonable thickness and a more economic proposition.

A fully insulated low pressure vessel would be equipped with a small refrigeration plant to control the temperature and therefore the pressure of the liquid CO_2.

This method of storage for CO_2 has been in operation for many years in industry for purposes other than fire extinguishing but the system is more widely used now for fire protection purposes.

Types of System

Portable Extinguishers

Extinguishers are generally made in sizes ranging from those containing 1.14kg of liquid CO_2, whose gross weight is about 4.8kg, to those containing 6.8kg of CO_2, whose gross weight is about 23kg. The extinguisher consists of a steel cylinder containing the liquefied gas, fitted with a sealing disc and piercing mechanism or valve.

Mobile Supply

Wheeled units contain one of two large cylinders which are connected to a length of hose and a discharge "horn" or discharge similar to a portable extinguisher.

Hand Hose Systems

These systems comprise one or more large cylinders mounted side by side in a rack and connected together by a pipe manifold. Each cylinder is provided with a lever operated piercing mechanism. The pipe manifold is connected direct to one end of the hose reel and a distributor with control valve is attached to the other end of the reel.

Local Application (or direct impingement)

A local application generally consists of a fixed supply of CO_2 permanently connected to a system of fixed pipework and nozzles designed to discharge CO_2 directly to a fire which may occur in an area which has no enclosure round it and where portable extinguishers could not be used effectively, either because of difficulty in approaching the hazard or because sufficient gas could not be brought into action quickly. Examples of hazards that may be protected by local application include: dip tanks, quench tanks, spray booths, oil baths, printing machinery where flammable liquids are involved, kitchen range hoods, deep fat fryers etc. The gas would be stored in a battery of cylinders that are interconnected with steel piping. This piping is mounted around the hazard, and is fitted with discharging nozzles located at strategic points. When CO_2 is released from the cylinder it travels through the fixed pipework to discharge via the nozzles onto the hazard.

Total Flooding

A total flooding system consists of a fixed installation of permanently connected CO_2 cylinders with piping and discharge nozzles similar to that described above. This type of system may be used where there is a permanent enclosure. The enclosure around the hazard must be adequate to enable the required concentration of CO_2 to be built up and maintained for a period of time required to ensure complete and permanent extinguishment.

A reasonably well enclosed space is required to minimise the loss of CO_2. Openings should be closed or be capable of being closed to prevent the continuing loss of the extinguishing medium. Where openings must be closed at the time of a fire, this can be arranged by providing suitable steel shutters or fire resisting curtains, which are automatically tripped shut by CO_2 or electrically. In some cases it is possible to provide a CO_2 gas screen across an opening and a quantity of CO_2 must be reckoned in the calculation for this purpose.

Safety Precautions

Carbon dioxide can kill by smothering. It is not a poison and has no damaging effect on body tissue. It can be likened to drowning. In most cases the actual danger to personnel is where the protected area is such that CO_2 may enter unsuspected, such as basement, pits, electrical apparatus rooms etc.

Warning notices, emergency lighting signs etc may be displayed and the provision of adequate passageways, exit routes and keeping them clear are all desirable features. But apart from these basic precautions, certain safeguards must be incorporated in the CO_2 system.

Where CO_2 can be triggered off automatically by detectors, some means must be incorporated so that the detectors are isolated from the CO_2 release mechanism when personnel wish to enter the protected area.

The means of "locking-off" the automatic part of the system can be mechanical or electrical isolation of the trigger circuit. The exact design of the "lock-off" depends on the route taken by the personnel but the object is to provide a safety arrangement which will ensure that the system is rendered safe before personnel enter the protected space. It is also important to provide visual indication at entrances to protected areas so that it may be seen to be safe to enter. Should a fire occur in a protected space while the automatic system is "locked-off" it is often essential that a warning alarm is sounded. Good house-keeping is essential if these systems are provided with a means for "locking-off" automatic parts of the system. The issue and return of keys or locking pins etc, must be adequately controlled.

Testing and Maintenance

In keeping with all fire protection installations the CO_2 system must be tested and maintained regularly. Because CO_2 does not harm the protected areas it is very easy to carry out simulated fire conditions to test the operational side of the installation. Inspection and maintenance should be carried out regularly by a competent qualified inspector.

HALOGENATED EXTINGUISHING AGENT SYSTEMS

These systems use a gas and act as extinguishing agents in a similar way to carbon dioxide. The systems incorporate the use of Bromochlorodifluoromethane (BCF) or Bromotrifluoromethane (BTM). Extinguishing systems using these gases are becoming popular and further information can be obtained from the National Fire Protection Association, 470 Atlantic Avenue, Boston MA 02210. USA.

PRIVATE HYDRANTS

Private hydrants are often installed in premises with extensive yards, sidings, storage areas etc. where the nearest statutory hydrant is a considerable distance from the risk, or where the nature of the risk requires large quantities of water to be immediately to hand.

(a) Connections

These hydrants may be connected to the service main to the premises, if this is of large enough capacity, to a separate branch from the town main, to a ring main which is connected to the town main at two points or, occasionally, by a single connection.

Ring mains are also installed without any connections to town mains, being supplied from private water supplies such as overhead tanks, reservoirs, lakes, canals etc. Some premises with a supply from lakes, canals etc may also use the town main as a supplementary or primary supply in a fire situation. In this event, the arrangement of valves in the system must ensure that there can be no possibility of contamination of the town main.

A ring main installation has many obvious advantages, the most important of which is that any hydrant is fed by both arms of the ring and that, since a division valve is fitted in both connections with the town main or other water supply and, sometimes, at intermediate points it may be possible to isolate a damaged section and this allows a portion of the ring main to remain in action. Where premises are equipped with a sprinkler system as well as private hydrants, separate branches should be provided for each.

(b) Hydrant Markings

Increasing use is being made of the standard hydrant indicator plate (see the Manual, Book 7, 'Hydraulics and water supplies') to mark the position of private hydrants, although various individual markings may still be found. The hydrants themselves are of various patterns, the most common being the standard underground hydrant, less common are pillar hydrants and wall hydrants.

(c) Outlets

The outlets of private hydrants usually conform to British Standard 750, although other types may be found. Where private hydrants are non-standard, adaptors should be provided at the premises to enable fire brigade equipment to be utilized.

(d) By-pass Valves

As private mains usually supply the domestic needs of the premises they are almost always fitted with a water meter so that the water undertaking can record the consumption. Where water is fed into industrial premises for business purposes through a meter, it is common practice for a by-pass to be fitted. If the water on the factory side of the meter is required for fire fighting the meter can be by-passed by opening the by-pass valve, thus eliminating frictional resistance through the meter. In addition, the water used for fire fighting does not register on the meter. The location of the valve controlling a meter by-pass should be indicated by a standard by-pass indicator plate (see Book 7, 'Hydraulics and water supplies'). The valve is usually wire-locked in the closed position and when a hydrant is used for fire fighting the valve should be opened fully to enable the maximum flow to be obtained.

REFERENCES

Manual of Firemanship
Book 1 – Elements of combustion and extinction.
Book 7 – Hydraulics and water supplies.
Book 9 – Fire protection of buildings.
Obtainable from:
HM Stationery Office.

Fire Officers' Committee Rules
29th Edition, 1973
and
Fire Protection Association publications.
Obtainable from:
Aldermary House, Queen Street, London EC4.

BRITISH STANDARDS AND CODES OF PRACTICE

BS 336: 1980 (1982)	Specification for fire hose couplings and ancilliary equipment.
BS 750: 1984	Specification for underground fire hydrants and surface box frames and covers.
BS 1635: 1970	Graphical symbols and abbreviations for fire protection drawings.
BS 3165: 1986	Specification for rubber and plastics suction hoses and hose assemblies for fire fighting purposes.
BS 3169: 1986	Specification for first aid reel hoses for fire brigade purposes.
BS 3251: 1976	Specification. Indicator plates for fire hydrants and emergency water supplies.
BS 5041	Fire hydrant systems equipment.
BS 5041: Part 1: 1987	Specification for landing valves for wet risers.
BS 5041: Part 2: 1987	Specification for landing valves for dry risers.
BS 5041: Part 3: 1975 (1987)	Specification for inlet breechings for dry riser inlets.
BS 5041: Part 4: 1975 (1987)	Specification for boxes for landing valves for dry risers.
BS 5041: Part 5: 1974 (1987)	Specification for boxes for foam inlets and dry riser inlets.
BS 5274: 1985	Specification for fire hose reels (water) for fixed installations.

BRITISH STANDARDS AND CODES OF PRACTICE

BS 5306	Fire extinguishing installations and equipment on premises.
BS 5306: Part 0: 1986 (1987)	Guide for the selection of installed systems and other fire equipment.
BS 5306: Part 1: 1976 (1984)	Hydrant systems, hose reels and foam inlets.
BS 5306: Part 2: 1979 (1987)	Sprinkler systems.
BS 5306: Part 3: 1985 (1987)	Code of practice for selection, installation and maintenance of portable fire extinguishers.
BS 5306: Part 4: 1986	Specification for carbon dioxide systems.
BS 5306: Part 5	Halon systems.
BS 5306: Section 5.1: 1982 (1984)	Halon 1301 total flooding systems.
BS 5306: Section 5.2: 1984	Halon 1211 total flooding systems.
BS 6391: 1983 (1986)	Specification for non-percolating lay flat delivery hoses and hose assemblies for fire fighting purposes.

STEAM

Steam is water which has taken in heat and passed through to the complete vaporization stage.

At atmospheric pressure, water and steam have the same temperature ie. 100°C.

As in practice, a differential pressure is required to cause steam to flow along a pipe steam is normally supplied at a pressure, some way above atmospheric pressure. Above atmospheric pressure the steam temperature rises; this temperature gradient makes the steam more useful as a heating agent.

There are three quantities of heat in steam:

(a) Specific Enthalphy of water – this is the heat which is added to the water up to the point at which it is converted into steam.

(b) Specific Enthalphy of evaporation – this is sometimes referred to as the 'hidden heat' which must be added to the water at the vapour change stage to convert all the water into steam at the same temperature.

(c) Specific Enthalphy of steam – this is the sum of the specific enthalphy of water and specific enthalphy of evaporation.

Graph G1 shows the amount of heat available in steam and condensate at various steam pressures.

Graph G1 Sensible, latent and total heat in steam, 0 – 14 bar

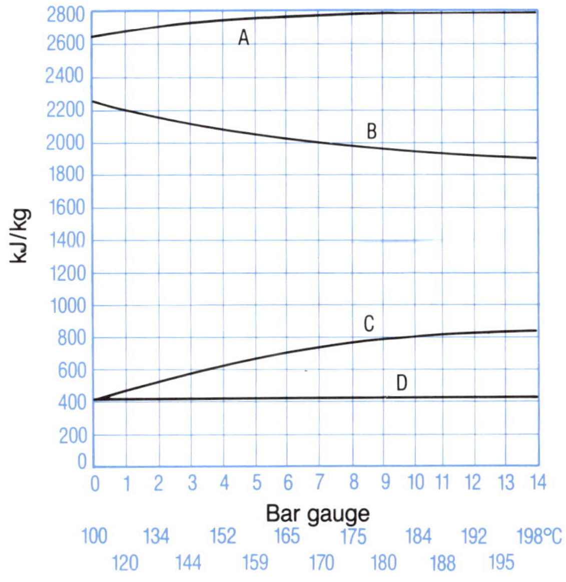

Notes adjacent to A, B, C & D

A. Specific enthalphy of steam.
B. Specific enthalphy of evaporation available at different pressures.
C. Heat in condensate at steam temperature.
D. Heat in condensate at atmopsheric pressure.

Steam Tables

Owing to somewhat complex effects which occur at various steam pressures, the actual quantity of heat to be supplied or available in steam can only be determined from steam tables.

The steam table, Table G1, indicates the specific enthalphy of water, evaporation and steam available in steam at pressures up to 27 bar (1 bar = 14.5 lb/in^2).

Design Considerations

Steam will only flow along a pipeline when there is a sufficient pressure differential between the point of supply and the outlet. The first consideration therefore is distribution pressure. Per unit of weight, low pressure steam is a better heat carrier than high pressure and is easy to control. However, low pressure systems require larger diameter pipes than high pressure systems utilising smaller diameter pipework with reductions at the equipment outlet and these are generally preferred.

When undertaking the sizing of steam mains, a differential principle is expressed in the formula:

$$F = \frac{(P_1 - P_2)}{L}$$

and equates the necessary final pressure required.

where P_1 = a factor based on the initial pressure
P_2 = a factor based on a final pressure
L = an effective length of service pipe which is inclusive of frictional allowance for fittings
F = pressure drop factor determined by above formula.

Table G2, indicates the relationship between pressure factors and the working gauge pressure expressed in bar.

Required flow rates of steam over a wide range of pressures can be evaluated from Table G3. The pressure factors applicable to design can be determined from the following worked example:

Example

A steam main, after allowing for all resistances has an effective length of 165 metres. The initial gauge pressure (P_1) is 7 bar and the final outlet pressure (P_2) after due allowance for radiation loss, is 6.6 bar.

Figure G1 Diagrammatic steam main

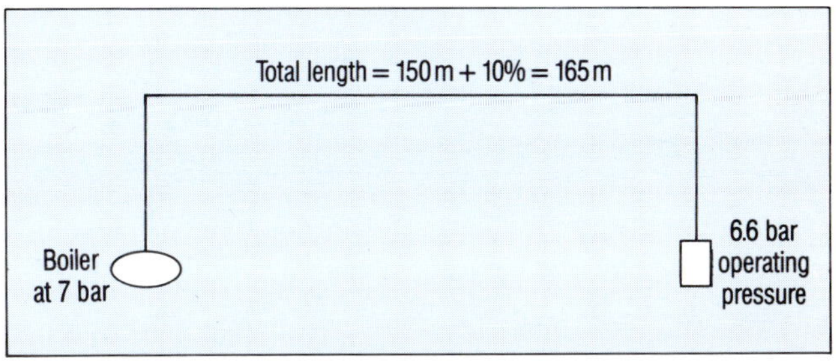

Using Table G2, we can calculate a suitable pipe diameter for a steam demand of 270 kg per hour required by the unit above.

P_1 at 7.0 bar has a pressure factor of 56.38

P_2 at 6.6 bar has a pressure factor of 51.05

From $\frac{P_1 - P_2}{L}$ F equals $\frac{56.38 - 51.05}{165} = 0.0323$

At 0.0323 the steam flow rate from a 2 inch (50 mm) diameter pipe read from Table G3 is 501.00 kg/h.

Checking that the supply available in 1 1/2 inch (40 mm) diameter pipe is below our needs, a 2 inch (50 mm) diameter is the correct size and allows for a working margin.

STEAM

Table G1 Steam tables

Pressure			Temperature	Specific Enthalpy			Specific Volume Steam
				Water (h_f)	Evaporation (h_{fg})	Steam (h_g)	
bar		kPa	°C	kJ/kg	kJ/kg	kJ/kg	m^3/kg
0.30	absolute	30.0	69.10	289.23	2336.1	2625.3	5.229
0.50	absolute	50.0	81.33	340.49	2305.4	2645.9	3.240
0.75	absolute	75.0	91.78	384.39	2278.6	2663.0	2.217
0.95	absolute	95.0	98.20	411.43	2261.8	2673.2	1.777
0	gauge	0	100.00	419.04	2257.0	2676.0	1.673
0.10	gauge	10.0	102.66	430.2	2250.2	2680.4	1.533
0.20		20.0	105.10	440.8	2243.4	2684.2	1.414
0.30		30.0	107.39	450.4	2237.2	2687.6	1.312
0.40		40.0	109.55	459.7	2231.3	2691.0	1.225
0.50		50.0	111.61	468.3	2225.6	2693.9	1.149
0.60		60.0	113.56	476.4	2220.4	2696.8	1.083
0.70		70.0	115.40	484.1	2215.4	2699.5	1.024
0.80		80.0	117.14	491.6	2210.5	2702.1	0.971
0.90		90.0	118.80	498.9	2205.6	2704.5	0.923
1.00		100.0	120.42	505.6	2201.1	2706.7	0.881
1.10		110.0	121.96	512.2	2197.0	2709.2	0.841
1.20		120.0	123.46	518.7	2192.8	2711.5	0.806
1.30		130.0	124.90	524.6	2188.7	2713.3	0.773
1.40		140.0	126.28	530.5	2184.8	2715.3	0.743
1.50		150.0	127.62	536.1	2181.0	2717.1	0.714
1.60		160.0	128.89	541.6	2177.3	2718.9	0.689
1.70		170.0	130.13	547.1	2173.7	2720.8	0.665
1.80		180.0	131.37	552.3	2170.1	2722.4	0.643
1.90		190.0	132.54	557.3	2166.7	2724.0	0.622
2.00		200.0	133.69	562.2	2163.3	2725.5	0.603
2.20		220.0	135.88	571.7	2156.9	2728.6	0.568
2.40		240.0	138.01	580.7	2150.7	2731.4	0.536
2.60		260.0	140.00	589.2	2144.7	2733.9	0.509
2.80		280.0	141.92	597.4	2139.0	2736.4	0.483
3.00		300.0	143.75	605.3	2133.4	2738.7	0.461
3.20		320.0	145.46	612.9	2128.1	2741.0	0.440
3.40		340.0	147.20	620.0	2122.9	2742.9	0.422
3.60		360.0	148.84	627.1	2117.8	2744.9	0.405
3.80		380.0	150.44	634.0	2112.9	2746.9	0.389
4.00		400.0	151.96	640.7	2108.1	2748.8	0.374
4.50		450.0	155.55	656.3	2096.7	2753.0	0.342
5.00		500.0	158.92	670.9	2086.0	2756.9	0.315
5.50		550.0	162.08	684.6	2075.7	2760.3	0.292
6.00		600.0	165.04	697.5	2066.0	2763.5	0.272
6.50		650.0	167.83	709.7	2056.8	2766.5	0.255
7.00		700.0	170.50	721.4	2047.7	2769.1	0.240
7.50		750.0	173.02	732.5	2039.2	2771.7	0.227
8.00		800.0	175.43	743.1	2030.9	2774.0	0.215
8.50		850.0	177.75	753.3	2022.9	2776.2	0.204
9.00		900.0	179.97	763.0	2015.1	2778.1	0.194
9.50		950.0	182.10	772.5	2007.5	2780.0	0.185
10.00		1000.0	184.13	781.6	2000.1	2781.7	0.177
10.50		1050.0	186.05	790.1	1993.0	2783.3	0.171
11.00		1100.0	188.02	798.8	1986.0	2784.8	0.163
11.50		1150.0	189.82	807.1	1979.1	2786.3	0.157
12.00		1200.0	191.68	815.1	1972.5	2787.6	0.151
12.50		1250.0	193.43	822.9	1965.4	2788.8	0.148
13.00		1300.0	195.10	830.4	1959.6	2790.0	0.141
13.50		1350.0	196.62	837.9	1953.2	2791.1	0.136
14.00		1400.0	198.35	845.1	1947.1	2792.2	0.132
14.50		1450.0	199.92	852.1	1941.0	2793.1	0.128
15.00		1500.0	201.45	859.0	1935.0	2794.0	0.124
15.50		1550.0	202.92	865.7	1928.8	2794.9	0.119
16.00		1600.0	204.38	872.3	1923.4	2795.7	0.117
17.00		1700.0	207.17	885.0	1912.1	2797.1	0.110
18.00		1800.0	209.90	897.2	1901.3	2798.5	0.105
19.00		1900.0	212.47	909.0	1890.5	2799.5	0.100
20.00		2000.0	214.96	920.3	1880.2	2800.5	0.0949
21.00		2100.0	217.35	931.3	1870.1	2801.4	0.0906
22.00		2200.0	219.65	941.9	1860.1	2802.0	0.0868
23.00		2300.0	221.85	952.2	1850.4	2802.6	0.0832
24.00		2400.0	224.02	962.2	1840.9	2803.1	0.0797
25.00		2500.0	226.12	972.1	1831.4	2803.5	0.0768
26.00		2600.0	228.15	981.6	1822.2	2803.8	0.0740
27.00		2700.0	230.14	990.7	1813.3	2804.0	0.0714

Table G2 Pressure factors for pipe sizing

Pressure bar abs	Volume m^3/kg	Pressure Factor
0.05	28.192	0.0301
0.10	14.674	0.0115
0.15	10.022	0.0253
0.20	7.649	0.0442
0.25	6.204	0.0681
0.30	5.229	0.0970
0.35	4.530	0.1308
0.40	3.993	0.1694
0.45	3.580	0.2128
0.50	3.240	0.2610
0.55	2.964	0.3140
0.60	2.732	0.3716
0.65	2.535	0.4340
0.70	2.365	0.5010
0.75	2.217	0.5727
0.80	2.087	0.6489
0.85	1.972	0.7298
0.90	1.869	0.8153
0.95	1.777	0.9053
1.013	1.673	1.025
bar gauge		
0	1.673	1.025
0.05	1.601	1.126
0.10	1.533	1.230
0.15	1.471	1.339
0.20	1.414	1.453
0.25	1.361	1.572
0.30	1.312	1.694
0.35	1.268	1.822
0.40	1.225	1.953
0.45	1.186	2.090
0.50	1.149	2.230
0.55	1.115	2.375
0.60	1.083	2.525
0.65	1.051	2.679
0.70	1.024	2.837
0.75	0.997	2.999
0.80	0.971	3.166
0.85	0.946	3.338
0.90	0.923	3.514
0.95	0.901	3.694
1.00	0.881	3.878
1.05	0.860	4.067
1.10	0.841	4.260
1.15	0.823	4.458
1.20	0.806	4.660
1.25	0.788	4.866
1.30	0.733	5.076
1.35	0.757	5.291
1.40	0.743	5.510
1.45	0.728	5.734
1.50	0.714	5.961
1.55	0.701	6.193
1.60	0.689	6.429
1.65	0.677	6.670
1.70	0.665	6.915
1.75	0.654	7.164
1.80	0.643	7.417
1.85	0.632	7.675
1.90	0.622	7.937
1.95	0.612	8.203
2.00	0.603	8.473
2.05	0.594	8.748
2.10	0.585	9.026

Pressure bar gauge	Volume m^3/kg	Pressure Factor
2.15	0.576	9.309
2.20	0.568	9.597
2.25	0.560	9.888
2.30	0.552	10.18
2.35	0.544	10.48
2.40	0.536	10.79
2.45	0.529	11.10
2.50	0.522	11.41
2.55	0.515	11.72
2.60	0.509	12.05
2.65	0.502	12.37
2.70	0.496	12.70
2.75	0.489	13.03
2.80	0.483	13.37
2.85	0.477	13.71
2.90	0.471	14.06
2.95	0.466	14.41
3.00	0.461	14.76
3.10	0.451	15.48
3.20	0.440	16.22
3.30	0.431	16.98
3.40	0.422	17.75
3.50	0.413	18.54
3.60	0.405	19.34
3.70	0.396	20.16
3.80	0.389	21.00
3.90	0.381	21.85
4.00	0.374	22.72
4.10	0.367	23.61
4.20	0.361	24.51
4.30	0.355	25.43
4.40	0.348	26.36
4.50	0.342	27.32
4.60	0.336	28.28
4.70	0.330	29.27
4.80	0.325	30.27
4.90	0.320	31.29
5.00	0.315	32.32
5.10	0.310	33.37
5.20	0.305	34.44
5.30	0.301	35.52
5.40	0.296	36.62
5.50	0.292	37.73
5.60	0.288	38.86
5.70	0.284	40.01
5.80	0.280	41.17
5.90	0.276	42.35
6.00	0.272	43.54
6.10	0.269	44.76
6.20	0.265	45.98
6.30	0.261	47.23
6.40	0.258	48.48
6.50	0.255	49.76
6.60	0.252	51.05
6.70	0.249	52.36
6.80	0.246	53.68
6.90	0.243	55.02
7.00	0.240	56.38
7.10	0.237	57.75
7.20	0.235	59.13
7.30	0.232	60.54
7.40	0.229	61.96
7.50	0.227	63.39
7.60	0.224	64.84

Pressure bar gauge	Volume m^3/kg	Pressure Factor
7.70	0.222	66.31
7.80	0.219	67.79
7.90	0.217	69.29
8.00	0.215	70.80
8.10	0.212	72.33
8.20	0.210	73.88
8.30	0.208	75.44
8.40	0.206	77.02
8.50	0.204	78.61
8.60	0.202	80.22
8.70	0.200	81.84
8.80	0.198	83.49
8.90	0.196	85.14
9.00	0.194	86.81
9.10	0.192	88.50
9.20	0.191	90.20
9.30	0.189	91.92
9.40	0.187	93.66
9.50	0.185	95.41
9.60	0.184	97.18
9.70	0.182	98.96
9.80	0.181	100.75
9.90	0.179	102.57
10.00	0.177	104.40
10.20	0.174	108.10
10.40	0.172	111.87
10.60	0.169	115.70
10.80	0.166	119.59
11.00	0.163	123.54
11.20	0.161	127.56
11.40	0.158	131.64
11.60	0.156	135.78
11.80	0.153	139.98
12.00	0.151	144.25
12.20	0.149	148.57
12.40	0.147	152.96
12.60	0.145	157.41
12.80	0.143	161.92
13.00	0.141	166.50
13.20	0.139	171.13
13.40	0.135	175.83
13.60	0.133	180.58
13.80	0.132	185.40
14.00	0.130	190.29
14.20	0.128	195.23
14.40	0.127	200.23
14.60	0.125	205.30
14.80	0.124	210.42
15.00	0.122	215.61
15.20	0.121	220.86
15.40	0.119	226.17
15.60	0.118	231.54
15.80	0.117	236.97
16.00	0.115	242.46
16.20	0.114	248.01
16.40	0.113	253.62
16.60	0.111	259.30
16.80	0.110	265.03
17.00	0.109	270.83
17.20	0.108	276.69
17.40	0.107	282.60
17.60	0.106	288.58
17.80	0.105	294.62
18.00	0.104	300.72

STEAM

Table G3 Pipeline capacity and pressure drop factors

F		Pipe Sizes 15mm ½″	20mm ¾″	25mm 1″	32mm 1¼″	40mm 1½″	50mm 2″	65mm 2½″	80mm 3″	100mm 4″	125mm 5″	150mm 6″	175mm 7″	200mm 8″	225mm 9″	250mm 10″	300mm 12″
0.00016	a						30.40	55.41	90.72	199.1	360.4	598.2	890.0	1275	1755	2329	3800
	v						4.30	4.86	5.55	6.82	7.90	9.16	10.05	10.94	11.94	12.77	14.54
0.00020	a					16.18	34.32	62.77	103.0	225.6	407.0	662.0	1005	1437	1966	2623	4276
	v					3.96	4.85	5.51	6.31	7.72	8.92	10.13	11.34	12.33	13.37	14.38	16.36
0.00025	a				10.84	17.92	38.19	69.31	113.2	249.9	450.3	735.5	1008	1678	2183	2904	4715
	v				3.74	4.39	5.40	6.08	6.92	8.56	9.87	11.26	12.5	14.40	14.85	15.92	18.04
0.00030	a				11.95	19.31	41.83	75.85	124.1	271.2	491.9	804.5	1209	1733	2390	3172	5149
	v				4.13	4.73	5.92	6.65	7.60	9.29	10.79	12.31	13.65	14.87	16.26	17.39	19.7
0.00035	a			6.86	12.44	20.59	43.76	80.24	130.0	285.3	519.2	845.3	1279	1823	2497	3346	5406
	v			3.88	4.30	5.04	6.21	7.04	7.96	9.77	11.38	12.94	14.44	15.64	17.00	18.34	20.69
0.00045	a		3.62	7.94	14.56	23.39	50.75	92.68	150.9	333.2	604.6	979.7	1478	2118	2913	3884	6267
	v		3.54	4.49	5.03	5.73	7.18	8.13	9.24	11.42	13.26	15.00	16.68	18.18	19.82	21.29	23.99
0.00055	a		4.04	8.99	16.18	26.52	57.09	103.8	170.8	337.1	674.2	1101	1663	2382	3281	4338	7057
	v		3.96	5.09	5.59	6.49	8.08	9.10	10.46	12.78	14.78	16.85	18.22	20.44	22.32	23.78	27.01
0.00065	a		4.46	9.56	17.76	29.14	62.38	113.8	186.7	409.8	739.9	1207	1823	2595	3597	4781	7741
	v		4.37	5.41	6.13	7.14	8.82	9.98	11.43	14.04	16.22	18.48	20.58	22.27	24.47	26.21	29.62
0.00075	a		4.87	10.57	19.31	31.72	68.04	124.1	203.2	445.9	804.5	1315	1977	2836	3908	5172	8367
	v		4.77	5.98	6.67	7.77	9.62	10.88	12.44	15.28	17.64	20.13	22.32	24.34	26.59	28.35	32.02
0.00085	a		5.52	11.98	21.88	35.95	77.11	104.7	230.2	505.4	911.8	1490	2240	3215	4429	5861	9482
	v		5.41	6.78	7.56	8.80	10.91	12.34	14.09	17.32	19.99	22.81	25.29	27.59	30.13	32.13	36.29
0.00100	a	1.96	5.84	12.75	23.50	38.25	81.89	148.6	245.2	539.4	968.5	1579	2403	3383	4707	6228	10052
	v	4.10	5.72	7.21	8.12	9.37	11.58	13.03	15.01	18.48	21.24	24.17	27.13	29.03	32.02	34.14	38.47
0.00125	a	2.10	6.26	13.57	24.96	40.72	87.57	159.8	261.8	577.9	1038	1699	2544	3634	5035	6655	10639
	v	4.39	6.13	7.68	8.62	9.97	12.39	14.02	16.03	19.80	22.76	26.01	28.72	31.19	34.26	36.48	40.71
0.00150	a	2.39	7.35	15.17	28.04	45.97	98.84	179.3	295.1	652.8	1172	1908	2896	4091	5631	7493	11999
	v	5.00	7.20	8.58	9.68	11.26	13.98	15.72	18.07	22.37	25.70	29.21	32.69	35.11	38.31	41.08	45.92
0.00175	a	2.48	7.51	16.30	29.61	49.34	103.4	188.8	311.1	686.5	1270	2017	3046	4291	5921	7852	13087
	v	5.19	7.36	9.22	10.23	12.08	14.63	16.56	19.05	23.52	27.85	30.88	34.39	36.83	40.28	43.04	50.08
0.0020	a	2.84	8.58	18.63	33.83	56.39	118.2	215.8	355.5	784.6	1451.	2305	3482	4904	6767	8974	14956
	v	5.94	8.40	10.54	11.68	13.81	16.72	18.93	21.77	26.88	31.82	35.28	39.31	42.09	46.04	49.19	57.24
0.0025	a	3.16	9.48	20.75	37.25	61.30	132.0	240.5	391.3	881.7	1556	2546	3819	5422	7544	10090	16503
	v	6.61	9.29	11.74	12.86	15.01	18.67	21.09	23.96	30.21	34.12	38.97	43.11	46.53	51.33	55.31	63.16
0.0030	a	3.44	10.34	22.5	40.45	66.66	143.4	262.0	429.8	942.4	1701	2767	4183	6068	8275	11033	18021
	v	7.20	10.13	12.73	13.97	16.33	20.29	22.98	26.32	32.29	37.30	42.36	47.22	52.08	56.30	60.48	68.97
0.0040	a	4.17	12.50	26.97	48.55	80.91	173.1	313.8	514.9	1128	2040	3330	5051	7208	9905	13240	21625
	v	8.73	12.25	15.26	16.77	19.82	24.49	27.52	31.53	38.65	44.73	50.97	57.02	61.86	67.39	72.58	82.76
0.0050	a	4.71	14.12	30.40	54.92	90.23	196.1	354	578.6	1275	2305	3727	5737	8189	11278	14858	24469
	v	9.86	13.83	17.20	18.97	20.10	27.74	31.05	35.43	43.68	50.54	57.05	64.76	70.28	76.73	81.45	93.64
0.0060	a	5.25	15.69	35.80	60.31	99.05	215.8	392.3	647.3	1412	2550	4148	6277	9072	12406	16476	26970
	v	10.99	15.37	20.26	20.83	24.26	30.53	34.41	39.63	48.38	55.92	63.50	70.86	77.86	84.40	90.32	103.21
0.0080	a	6.08	18.34	39.23	70.12	116.2	251.5	456.0	750.3	1648	2976	4879	7355	10543	14417	19173	31384
	v	12.72	17.97	22.20	24.22	28.46	35.58	40.00	45.95	56.46	65.26	74.69	83.03	90.48	98.09	105.1	120.1
0.0100	a	6.86	20.64	44.13	79.44	130.4	283.9	514.9	845.9	1863	3334	5492	8336	11867	16280	21576	35307
	v	14.36	20.22	24.97	27.44	31.94	40.16	45.16	51.80	63.83	73.11	84.07	94.11	101.8	110.8	118.28	135.1
0.0125	a	7.35	22.20	47.28	81.00	140.1	302.1	547.3	901.9	1983	3589	5867	8844	2697	17426	23074	37785
	v	15.38	21.75	26.75	27.98	34.31	42.74	48.00	55.22	67.94	78.70	89.81	99.84	109.0	118.5	126.5	144.6
0.0150	a	8.27	25.00	53.33	95.62	157.2	342.0	620.6	1020	2230	4045	6620 .	10022	4251	19584	25974	42616
	v	17.31	24.49	30.18	33.03	38.50	48.38	54.43	62.46	76.40	88.70	101.3	113.1	22.3	133.2	142.4	163.09
0.0175	a	8.58	26.39	55.78	100.4	165.6	360.4	665.1	1073	2360	4291	6994	10512	5017	20595	27461	44194
	v	17.95	25.85	31.56	34.68	40.65	50.99	58.34	65.70	80.52	94.09	107.1	118.7	128.9	140.1	150.5	169.1
0.020	a	9.80	30.16	63.75	114.7	189.3	411.9	760.1	1226	2697	4904	7993	12014	7163	23538	31384	50508
	v	20.51	29.55	36.07	39.62	46.36	58.27	66.67	75.01	92.41	107.5	122.3	135.6	147.3	160.1	172.0	193.3
0.025	a	10.99	33.48	70.73	127.3	209.8	459.7	834.6	1367	2970	5422	8817	13296	9332	26357	34750	56581
	v	23.00	32.80	40.02	43.97	51.39	65.03	73.20	83.70	101.7	118.9	135.0	150.1	65.9	179.3	190.5	216.5
0.030	a	12.00	36.78	77.23	137.9	229.9	501.1	919.4	1480	3264	5884	9792	14481	20917	28595	37697	62522
	v	25.11	36.03	43.70	47.63	56.31	70.89	80.64	90.62	111.8	129.0	149.9	163.5	179.5	194.5	206.6	239.3
0.040	a	14.46	44.13	93.17	169.2	279.5	600.7	1093	1790	3923	7110	11622	17457	25254	34571	45604	75026
	v	30.26	43.23	52.72	58.44	68.46	84.98	95.87	109.6	134.4	155.9	177.9	197.2	216.7	235.2	250.0	287.1
0.050	a	16.43	49.53	104.4	191.2	313.8	676.7	1231	2020	4413	8042	13044	19370	23441	39229	51489	85324
	v	34.38	48.52	59.08	66.04	76.86	95.73	108.0	123.7	151.2	176.3	199.7	218.7	244.1	266.9	282.3	326.5
0.06	a	18.14	52.96	115.7	210.8	343.2	750.3	1373	2231	4855	8827	14368	21282	31384	43152	57373	
	v	37.96	51.88	65.47	72.81	84.06	106.1	120.4	136.6	166.3	193.5	219.9	240.2	269.3	293.6	314.5	
0.08	a	21.08	62.28	134.8	245.2	402.1	872.8	1594	2599	5688	10249	16672	24518	36532			
	v	44.11	61.02	76.28	86.69	98.49	123.5	139.8	159.1	194.9	224.7	255.2	276.8	313.5			
0.10	a	24.03	70.12	152.0	277.0	456.0	980.7	1804	2942	6424	11524	18879	27461				
	v	50.29	68.70	86.01	95.67	111.7	138.7	158.2	180.1	220.1	252.7	289.0	310.0				
0.12	a	25.99	77.48	167.7	306.5	500.2	1079	1986	3236	7110	12700	20841					
	v	54.39	75.91	94.90	105.9	122.5	152.6	174.2	198.1	243.6	278.5	319.0					
0.15	a	28.50	84.13	183.9	334.2	551.7	1195	2161	3494	7769							
	v	59.64	82.42	104.1	115.4	135.1	169.0	189.5	213.9	266.2							
0.20	a	34.32	102.0	220.7	402.1	662.0	1427	2599	4217	9317							
	v	71.82	99.93	124.9	138.9	162.1	201.9	228.0	258.2	319.2							
0.25	a	37.72	112.7	245.2	447.9	735.5	1565	2876	4668								
	v	78.94	110.4	138.7	154.7	180.1	221.4	252.3	285.8								
0.30	a	41.37	122.7	266.6	487.3	804.5	1710	3126	5057								
	v	86.58	120.2	150.9	168.3	197.0	241.9	274.2	309.6								
0.35	a	43.34	128.7	283.2	514.9	841.0	1802	3261									
	v	90.70	126.1	160.2	177.8	206.0	254.9	286.0									
0.40	a	49.93	147.1	323.6	588.4	961.1	2059	3727									
	v	104.5	144.1	183.1	203.2	235.4	291.3	326.9									
0.45	a	50.31	150.0	326.6	600.2	979.7	2083										
	v	105.3	146.9	184.8	207.3	239.9	294.7										
0.50	a	55.90	166.7	362.9	666.9	1089	2314										
	v	117.0	163.3	205.3	230.3	266.7	327.4										

Table G3 (Cont'd)

F		15mm ½"	20mm ¾"	25mm 1"	32mm 1¼"	40mm 1½"	50mm 2"	65mm 2½"	80mm 3"	100mm 4"	125mm 5"	150mm 6"	175mm 7"	200mm 8"	225mm 9"	250mm 10"	300mm 12"
0.6	a	62.28	185.3	402.1	735.5	1201											
	v	130.3	181.5	227.5	254.0	294.1											
0.7	a	63.07	188.8	407.6	750.9												
	v	132.0	185.0	230.6	259.3												
0.8	a	72.08	215.8	465.8	858.1												
	v	150.8	211.4	263.6	296.4												
0.9	a	73.28	218.4	476.6													
	v	153.3	214.0	269.7													

(Pipe Sizes)

a = kg/h capacity v = m/s velocity with a volume of 1 m^3/kg

Pipeline Velocity in Steam Mains

The velocity of flow in a steam main for a given flow rate in kg/h depends upon the inlet pressure and the specific volume of steam in m^3/kg.

The specific volume of steam will be found in steam tables at the required inlet pressure.

The following simplified formula will give results of sufficient accuracy for most practical purposes.

$$V = \frac{354 \times \text{kg/h} \times \text{specific volume m}^3\text{/kg}}{d^2}$$

Example

A steam main of 2in (50mm) diameter will provide a flow rate of 500kg/h with an inlet pressure of 6 bar.

The velocity of flow in m/s in the pipeline can be calculated thus:

$$V = \frac{354 \times 500 \times 0.272}{50^2}$$

V = 19.25m/sec

Condensate

Having established the requirements of the flow pipework, the next consideration is condensate and how to deal with it.

When steam releases its latent heat, the result is hot condensate which may contain up to 25% of the original heat required to generate steam. Most systems simply return it direct to the boiler house for re-use and help reduce the cost of fuel. It can also be utilized to provide other hot water sources such as to pre-heat hot water storage cylinders.

The location of equipment and constructional design of a building rarely allow the condensate to flow back to its natural source under gravity conditions and assistance must be provided.

Condensate is a constant product of steam as it flows through the distribution pipework as well as the unit being supplied. To reduce condensate in the branch lines, the connections should be taken from the top of the distribution main and will ensure a dry steam supply.

As waterlogging is one of the main causes of reduced output from a steam line the condensate must be removed as efficiently and quickly as possible. To avoid problems during the systems life, steam lines must be arranged to ensure that condensate is transferred along the pipe to a convenient discharge drain point. A fall of 1 in 250 has proved an adequate gradient to assist the condensate flow.

Drain point locations will depend on line size, frequency of start up and low points where condensate can collect. Generally, intervals of 30–50m are acceptable for drain pocket location; changes in pipe direction are an effective position. Figure G2 highlights pockets on a typical pipeline of long horizontal length and the inclusion of steam traps which accompany each drain before entering the condensate return line.

Figure G2 Condensate removal from distribution main

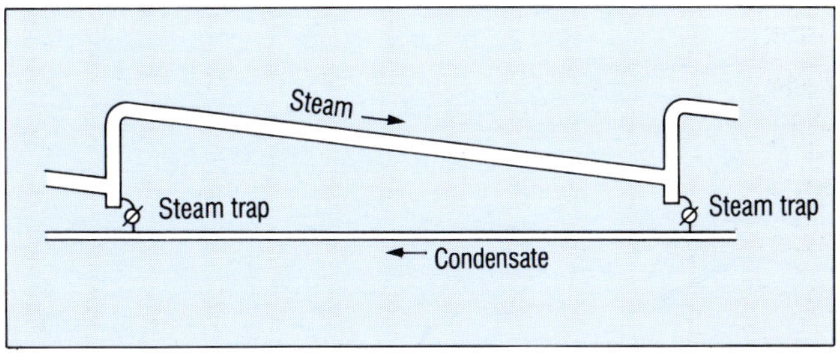

Large diameter lines or lines subject to frequent start-up may be subject to air entrapment. An air vent complete with pocket and steam trap should be provided at the end of any steam line. Figures G3 and G4 are examples of air venting on distribution lines and condensate removal from branch lines.

Figure G3 Air venting on distribution lines

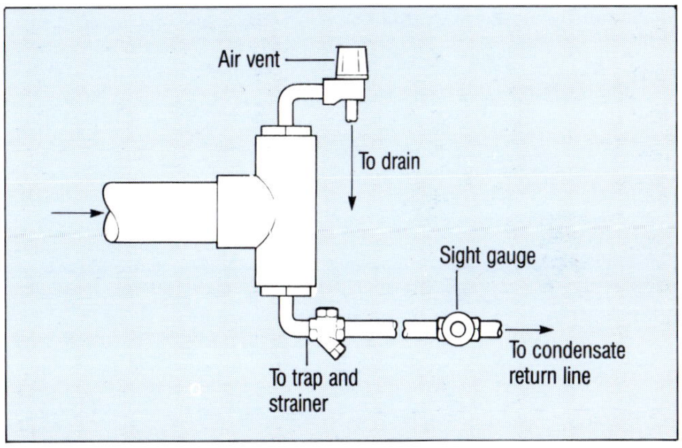

Figure G4 Condensate removal from branch lines

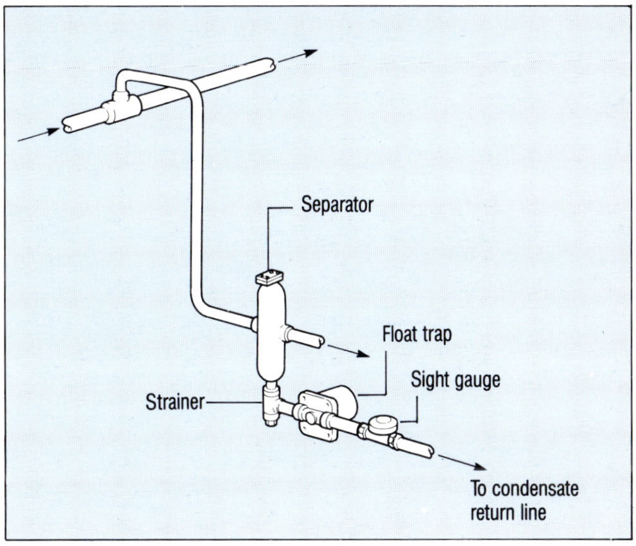

STEAM

At start up the plant is cold. Steam will condensate rapidly and the steam consumption may be two or three times the normal running rate. Under these conditions, the condensate pipework will be required to pass two or three times the normal condensate rate. As the plant warms up, so the amount of condensate reduces to the running load. Experience has shown that pipes sized for the above conditions will have an adequate capacity to handle the normal running conditions. For this kind of exercise the start-up load should be taken as not less than twice the running load. This provision will be sufficient for steam pressures in the range 7–10bar. Above this pressure range, some additional capacity in the condense lines may however be desirable.

The flow rates and pressure drops given in Tables G4 and G5 may be found convenient for pipe sizing condensate lines.

Example

A plant having a normal running load of 1000kg/h and a starting load of 2000kg/h is to be adopted for condensate line sizing. Assuming a steam trap is fitted not more than 3m from the condensate receiver tank, determine a suitable diameter of pipe for the condensate pipeline.

Referring to Table G4, it will be seen that a 32mm pipe can handle 2000kg/h at a pressure drop of 1.40mb per metre of travel.

The overall pressure drop is therefore 1.40 x 3 = 4.20mb and this small amount of back pressure would be acceptable. If on the other hand, the condensate has to travel over a distance of say 500m, then the back pressure will be 500 x 1.4 = 700mb and allowing for additional resistance, would impose a total back pressure in the region of 800mb, to which must be added the back pressure caused by any lift in the line.

Table G4 Flow of water in steel pipes (kg/hr)

Starting load ie. Running load × 2
Approx Frictional Resistance (mbar per m of travel)
Heavy steel tube

Tube diameter mm	0.3 (30 Pa)	0.5 (50 Pa)	0.6 (60 Pa)	0.8 (80 Pa)	1.0 (100 Pa)	1.4 (140 Pa)
15	95	130	140	160	180	220
20	220	290	320	370	420	500
25	410	540	600	690	790	940
32	890	1180	1300	1500	1700	2040
40	1360	1790	2000	2290	2590	3100
50	2630	3450	3810	4390	4990	6000
65	5350	6950	7730	8900	10150	12100
80	8320	10900	12000	13800	15650	18700
100	17000	22200	24500	28200	31900	38000

Steam Traps

Condensate should run away from the trap by gravity but there are occasions when it is necessary to 'lift' it. This is achieved by the pressure available at the trap outlet and therefore, no steam trap will actually 'lift' the condensate.

It is handy to remember in initial design that for each 0.11bar (11 KPà) pressure at the trap, the condensate can be lifted 1 metre. It should be remembered when selecting steam traps that water hammer and attendant noise with the risk of mechanical damage are inevitable when lifting condensate.

A steam trap is an essential piece of equipment which must be fitted to the condense outlet of the steam heated equipment before discharging into the condensate return. The basic function of the steam trap is to release condensate and to hold back steam in the equipment.

There are three basic designs of steam trap which are most widely used:

(1) Thermostatic Traps

These traps are bi-metallic devices or liquid-filled bellows which, because of their construction, are able to 'sense' the difference in temperature between condensate and steam. When condensate is flowing, the trap remains open but when the higher temperature steam enters the trap, it closes and will not allow the steam to pass through. When the steam condenses and loses temperature, the sensing element allows the trap valve to open and the cycle is repeated.

(2) Mechanical Traps

This type of trap relies for its principle of action, on the difference in density between steam and water acting on a ballfloat which rises in the presence of condensate and thus opens the valve which then permits condensate to pass through the trap. The inverted bucket in this type of trap, floats when steam is admitted to the trap and thus this action closes the trap valve.

(3) Thermodynamic Traps

Some traps rely on the fact that when condensate at near steam temperature is reduced in pressure at an orifice, flash steam is produced. The most common trap of this type is the thermodynamic trap which uses the increased velocity of the flash steam to close a disc against a seat. The trap opens on a cyclic basis when steam pressure in the control chamber is dissipated due to radiation. Figures G5, G6 and G7 indicate the working principles of the traps described.

Figure G5 Float trap with thermostatic air vent

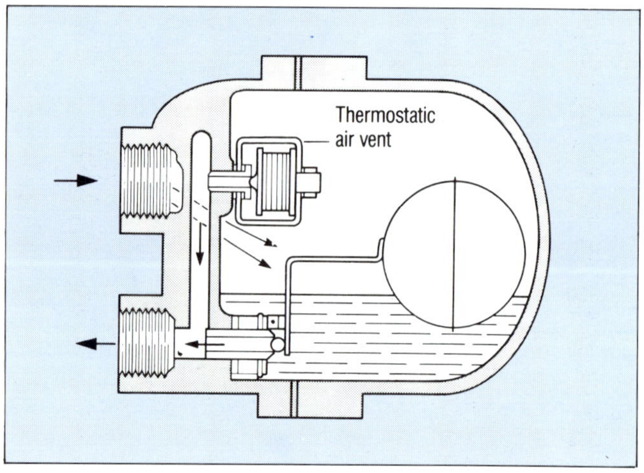

Table G5 Flow of water in heavy steel pipes

Pa per metre	mbar per m	15mm	20mm	25mm	32mm	40mm kg/h	50mm	65mm	80mm	100mm
28	0.28	90	209	380	865	1320	2554	5194	8079	16511
29	0.29	92	214	400	878	1340	2590	5271	8196	16756
30	0.30	93	218	403	890	1361	2631	5348	8314	17000
33	0.33	97	226	414	930	1420	2744	5579	8677	17736
39	0.39	107	249	469	1028	1565	3025	6142	9526	19514
40	0.40	108	253	476	1040	1583	3062	6214	9639	19736
43	0.43	113	263	496	1079	1646	3180	6454	10024	20457
45	0.45	116	270	508	1107	1687	3261	6618	10297	21002
47	0.47	119	277	521	1134	1728	3338	6777	10523	21500
50	0.50	123	286	538	1172	1787	3447	6949	10859	22154
53	0.53	127	296	557	1211	1846	3565	7235	11249	22907
55	0.55	130	302	569	1238	1887	3638	7380	11476	23360
57	0.57	133	308	580	1261	1923	3710	7525	11703	23814
59	0.59	135	314	591	1288	1959	3783	7666	11884	24268
61	0.61	138	320	602	1311	1996	3851	7806	12111	24721
64	0.64	141	327	615	1338	2041	3933	7970	12383	25220
67	0.67	146	337	634	1379	2100	4051	8210	12746	25991
69	0.69	148	343	645	1402	2136	4119	8342	12973	26400
70	0.70	149	345	649	1411	2150	4146	8432	13041	26563
71	0.71	150	348	655	1424	2168	4182	8473	13154	26808
73	0.73	152	354	665	1447	2200	4246	8600	13336	27216
75	0.75	154	358	673	1458	2227	4291	8695	13517	27506
76	0.76	155	359	675	1465	2236	4305	8723	13563	27579
77	0.77	157	365	685	1488	2268	4368	8850	13744	27987
78	0.78	158	366	689	1497	2282	4390	8900	13812	28132
80	0.80	160	370	695	1510	2300	4427	8972	13925	28350
82	0.82	162	375	704	1529	2331	4491	9072	14407	28758
88	0.88	168	391	733	1590	2427	4536	9453	14651	29865
90	0.90	170	395	740	1606	2449	4717	9548	14787	30142
98	0.98	179	414	777	1696	2567	4944	10025	15513	31616
100	1.00	180	418	785	1701	2590	4990	10115	15649	31879
114	1.14	194	450	845	1832	2790	5366	10841	16828	34247
118	1.18	198	457	857	1860	2830	5443	11022	17055	34746
120	1.20	199	462	867	1880	2860	5502	11113	17282	35120
131	1.31	209	484	907	1996	2994	5761	11657	18053	36742
137	1.37	215	497	931	2018	3071	5906	11948	18507	37667
140	1.40	216	502	939	2037	3103	5965	12066	18688	38012
147	1.47	224	516	966	2096	3189	6128	12383	19187	39055
157	1.57	231	534	1002	2168	3298	6337	12814	19822	40361
160	1.60	234	541	1011	2195	3334	6409	12973	20049	40797
163	1.63	237	546	1025	2218	3370	6477	13109	20278	41232
176	1.76	246	570	1066	2309	3511	6740	13608	21092	42938
180	1.80	249	576	1075	2331	3547	6808	13744	21319	43364
196	1.96	261	603	1129	2440	3710	7130	14379	22317	45360
200	2.00	265	611	1143	2472	3760	7221	14560	22589	45931
212	2.12	273	629	1179	2549	3874	7434	15014	23270	47265
216	2.16	275	634	1188	2567	3905	7493	15132	23451	47637
220	2.20	278	641	1200	2595	3942	7570	15277	23678	48014
229	2.29	284	655	1225	2649	4028	7729	15604	24177	49125
235	2.35	288	664	1243	2689	4086	7843	15840	24522	49832
240	2.40	292	672	1256	2719	4129	7927	16017	24780	50363
245	2.45	295	679	1270	2749	4173	8010	16193	25039	50894
255	2.55	301	694	1297	2806	4260	8176	16520	25556	51928
260	2.60	304	701	1311	2834	4304	8260	16683	25814	52445
261	2.61	305	703	1315	2844	4318	8287	16738	25900	52618
274	2.74	313	721	1348	2917	4331	8501	17173	26554	54069
277	2.77	315	727	1356	2935	4459	8555	17282	26717	54432
280	2.80	317	730	1363	2948	4479	8593	17357	26839	54636
294	2.94	325	749	1402	3025	4595	8813	17781	27533	55793
300	3.00	328	757	1414	3055	4641	8900	17956	27803	56428
310	3.10	335	771	1438	3112	4726	9063	18280	28308	57507
314	3.14	336	775	1446	3129	4752	9074	18380	28459	57879
320	3.20	340	784	1462	3163	4825	9204	18579	28767	58424
327	3.27	344	792	1479	3198	4853	9299	18776	29076	58968
333	3.33	348	801	1495	3230	4904	9408	18979	29366	59612
340	3.40	351	809	1511	3263	4955	9516	19178	29656	60057
343	3.43	353	813	1520	3279	4980	9571	19278	29801	60329
353	3.53	358	825	1541	3328	5054	9707	19550	30237	61236
359	3.59	362	834	1556	3361	5103	9798	19732	30522	62143
360	3.60	363	835	1558	3365	5109	9809	19754	30564	62211

STEAM

Figure G6 Inverted bucket trap

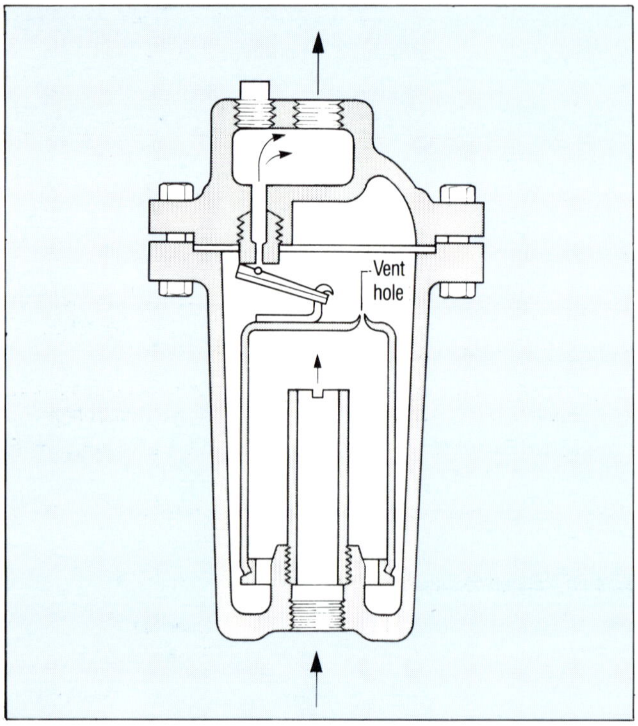

Figure G7 Operation of bimetal steam trap

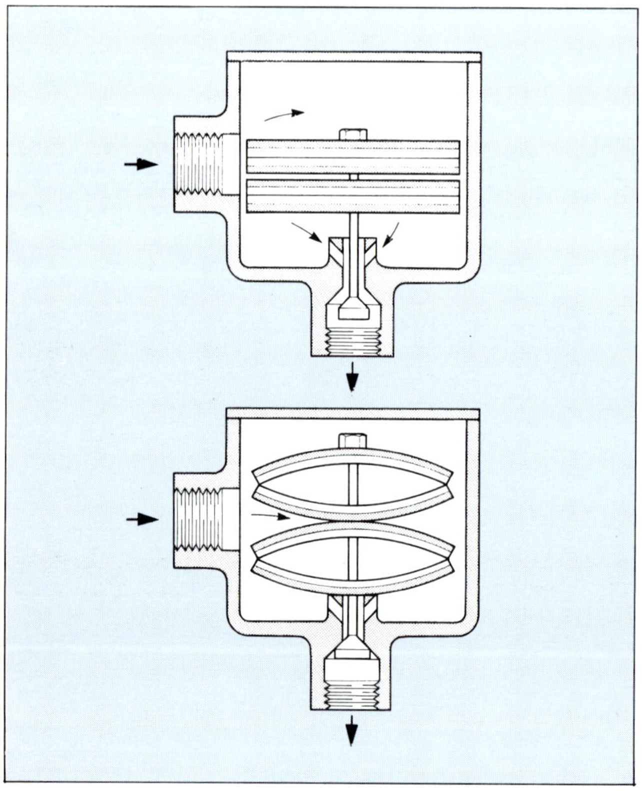

There are a number of other more complex designs of steam trap which can be used in special circumstances. These traps and their special functions lie within the realms of the specialist steam applications engineer.

Steam Separators

Although drain points are adequate to deal with condensate on pipeline walls, they cannot remove moisture from the steam itself. This is achieved by a series of simple baffle plates and are arranged to force the steam to change direction and are formed in an item of equipment called a separator. Dry steam passes through with little difficulty but allows the water droplets to drop to the bottom of the unit to a drain outlet and run to a drain point and a steam trap, thus providing better steam for power and process.

Strainers

The internal passageway of the pipework is continually confronted with deposits that cause blockages. These may take the form of rust, carbonate deposits in hard water districts and any general debris entering during installation. These deposits can easily block pipework and control equipment, jam open valves as well as scoring the face of valves and equipment due to the high velocity at which steam operates.

The most common practice is to provide and install a pipeline strainer upstream to any equipment, steam trap and valve set. A typical strainer is shown in Figure G8.

Figure G8 Strainer

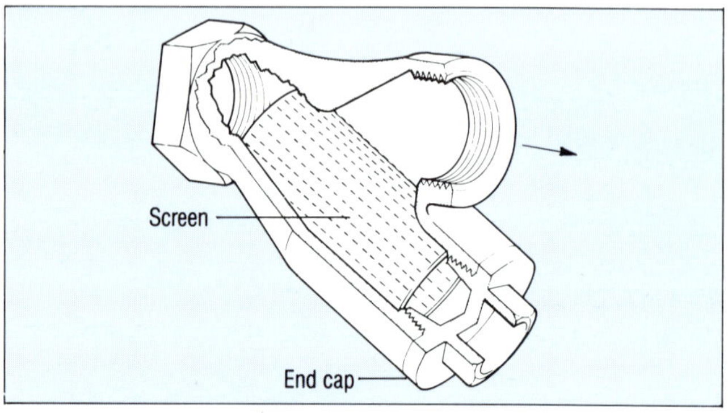

Steam Trap Checking

Even when adequately protected by a strainer, a steam trap may fail to operate for a number of reasons. Failure in the closed position would mean rapid reduction of plant output and remedial action must be taken. Conversely, if the trap fails in the open position or partially open position, output may not be affected but live steam could be passed to waste.

The traditional method of checking for failure is the installation of a sight glass downstream of each trap to enable viewing of the flow discharge. Sensoring equipment is now available as an alternative to the visual sight glass.

Figure G9 illustrates a simple installation of the items of equipment described.

Figure G9 Steam trap checking equipment

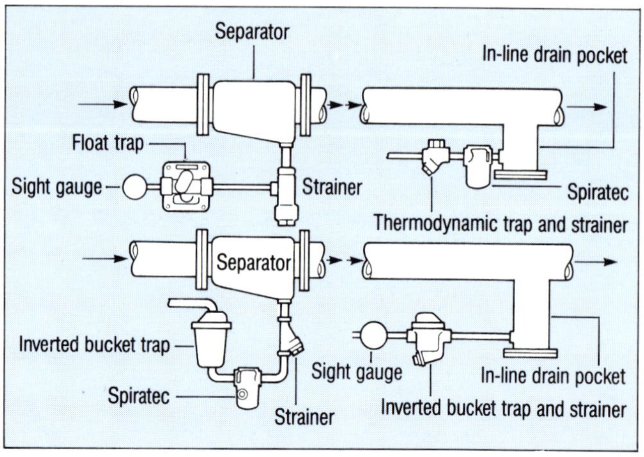

Expansion of a Steam Pipeline

As with any matter that is heated, expansion occurs and as steam services utilize high temperatures, distortion due to pipeline expansion is a serious problem. Sufficient allowance must therefore be provided with adequate arrangements for expansion along the pipelines.

The accommodation of expansion will vary depending on the pipework location and size. Expansion may be taken up at changes of direction, by pipe loops, stressing of the pipework or 'cold draw' and purpose manufactured mechanical expansion bellows. Mechanical bellows are most commonly used mainly because they require less space than loops and are more conveniently installed.

Figures G10 to G13 illustrate different methods of allowing for expansion, (also see Section H, Pipework Expansion).

Figure G10 Loop

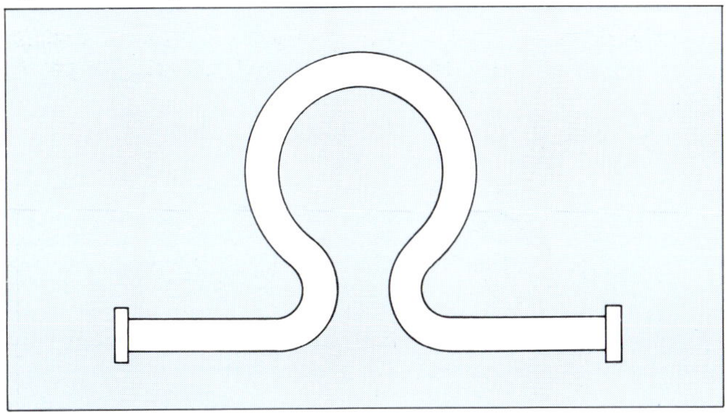

Figure G11 Alternative loop

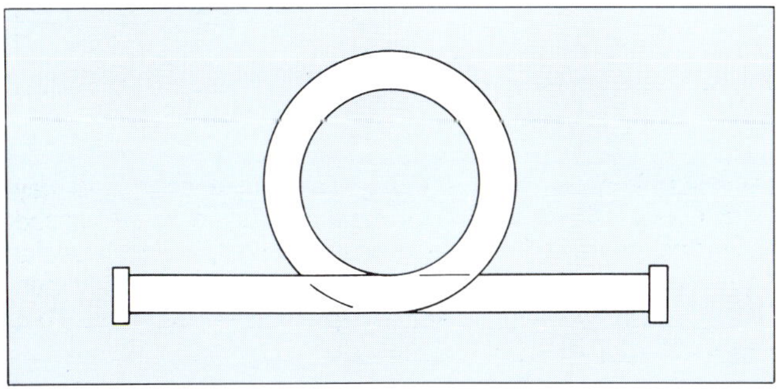

Figure G12 Bellows

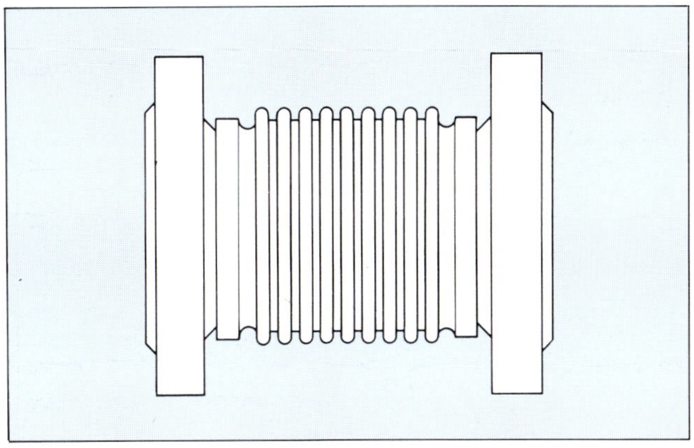

Figure G13 Sliding joint

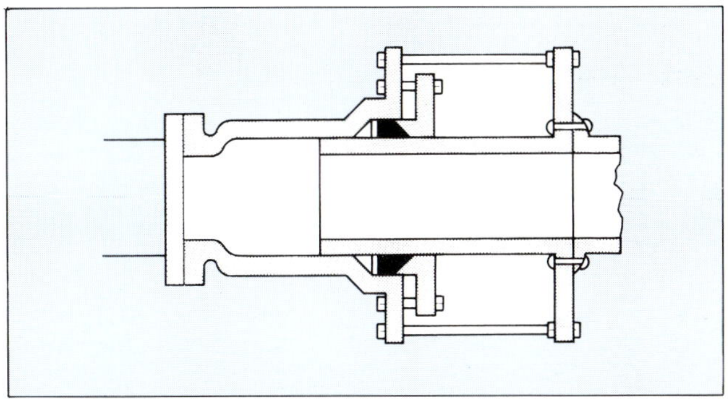

The loops shown in Figure G10 and G11 should preferably be fitted in the horizontal plane in order to avoid the accumulation of condensate on the upstream side.

It is essential that guides are used with expansion joints and anchor points are used in conjunction with all expansion fittings.

Heavy quality steel pipes have a linear coefficient of expansion of 1.25×10^{-5} °C per unit length.

Table G6 is an example of the linear coefficient factors of expansion in steel pipe

Table G6 Expansion of pipes

Final temperature °C	Expansion per 30 m mm
66	19
93	29
121	41
149	50
177	61
204	74
232	84
260	97

Expansion (mm) = $1.25 \times 10^{-5} \times$ °C (diff) × 1000/m

Example

Assume a steam main 30 metres long has a temperature rise from 16°C to 260°C. Calculate the linear expansion to be taken up by an expansion unit.

$$\text{Expansion (mm)} = 1.25 \times 10^{-5} \times 30 \times (260 - 16) \times 1000$$
$$= 1.25 \times 30 \times 2.44$$
$$= 91.5\,\text{mm}$$

The heat emission from bare horizontal pipes with ambient temperatures between 10°C and 21°C and still air conditions are shown in Table G7.

STEAM

Table G7 Heat emission from pipes

Temp. Diff. Steam to Air	15mm	20mm	25mm	32mm	40mm	Pipe Size 50mm	65mm	80mm	100mm	150mm
°C						W/m				
56	54	65	79	103	108	132	155	188	233	324
67	68	82	100	122	136	168	198	236	296	410
78	83	100	122	149	166	203	241	298	360	500
89	99	120	146	179	205	246	289	346	434	601
100	116	140	169	208	234	285	337	400	501	696
111	134	164	198	241	271	334	392	469	598	816
125	159	191	233	285	321	394	464	555	698	969
139	184	224	272	333	373	458	540	622	815	1133
153	210	255	312	382	429	528	623	747	939	1305
167	241	292	357	437	489	602	713	838	1093	1492
180	274	329	408	494	556	676	808	959	1190	1660
194	309	372	461	566	634	758	909	1080	1303	1852

Table G8 Pipeline capacities at specific velocities

Pressure bar	Velocity m/s	15mm	20mm	25mm	32mm	40mm	kg/h 50mm	65mm	80mm	100mm	125mm	150mm
0.4	15	7	14	24	37	52	99	145	213	394	648	917
	25	10	25	40	62	92	162	265	384	675	972	1457
	40	17	35	64	102	142	265	403	576	1037	1670	2303
0.7	15	7	16	25	40	59	109	166	250	431	680	1006
	25	12	25	45	72	100	182	287	430	716	1145	1575
	40	18	37	68	106	167	298	428	630	1108	1712	2417
1.0	15	8	17	29	43	65	112	182	260	470	694	1020
	25	12	26	48	72	100	193	300	445	730	1160	1660
	40	19	39	71	112	172	311	465	640	1150	1800	2500
2.0	15	12	25	45	70	100	182	280	410	715	1125	1580
	25	19	43	70	112	162	295	428	656	1215	1755	2520
	40	30	64	115	178	275	475	745	1010	1895	2925	4175
3.0	15	16	37	60	93	127	245	385	535	925	1505	2040
	25	26	56	100	152	225	425	632	910	1580	2480	3440
	40	41	87	157	250	357	595	1025	1460	2540	4050	5940
4.0	15	19	42	70	108	156	281	432	635	1166	1685	2460
	25	30	63	115	180	270	450	742	1080	1980	2925	4225
	40	49	116	197	295	456	796	1247	1825	3120	4940	7050
5.0	15	22	49	87	128	187	352	526	770	1295	2105	2835
	25	36	81	135	211	308	548	885	1265	2110	3540	5150
	40	59	131	225	338	495	855	1350	1890	3510	5400	7870
6.0	15	26	59	105	153	225	425	632	925	1555	2525	3400
	25	43	97	162	253	370	658	1065	1520	2530	4250	6175
	40	71	157	270	405	595	1025	1620	2270	4210	6475	9445
7.0	15	29	63	110	165	260	445	705	952	1815	2765	3990
	25	49	114	190	288	450	785	1205	1750	3025	4815	6900
	40	76	117	303	455	690	1210	1865	2520	4585	7560	10880
8.0	15	32	70	126	190	285	475	800	1125	1990	3025	4540
	25	54	122	205	320	465	810	1260	1870	3240	5220	7120
	40	84	192	327	510	730	1370	2065	3120	5135	8395	12470
10.0	15	41	95	155	250	372	626	1012	1465	2495	3995	5860
	25	66	145	257	405	562	990	1530	2205	3825	6295	8995
	40	104	216	408	615	910	1635	2545	3600	6230	9880	14390
14.0	15	50	121	205	310	465	810	1270	1870	3220	5215	7390
	25	85	195	331	520	740	1375	2080	3120	5200	8500	12560
	40	126	305	555	825	1210	2195	3425	4735	8510	13050	18630

BRITISH STANDARDS AND CODES OF PRACTICE

BS 806: 1986 (1987)	Specification for design and construction of ferrous piping installations for and in connection with land boilers.
BS 6022: 1983	Classification of automatic steam traps.
BS 6023: 1981	Glossary of technical terms for automatic steam traps.
BS 6024: 1981	Specification for marking automatic steam traps.
BS 6025: 1981	Specification for face-to-face dimensions of flanged automatic steam traps.
BS 6129:	Code of Practice for the selection and application of bellows expansion joints for use in pressure systems.
BS 6129: Part 1: 1981	Metallic bellows expansion joints.

Note

(1) Year after Standard's number indicates the latest date of issue.

(2) Year in brackets gives date of latest amendment at the time of publication of the Guide.

CONSIDERATIONS FOR DESIGN AND INSTALLATION

All plastics and metals used in the manufacture of pipework expand and contract along their lengths due to temperature variations. The extent of the change in dimension can be calculated using the following formula:

Formula (a)

$$\Delta L = l \times \alpha \times \Delta t$$

where:

ΔL = the change in length of the pipe (m)
α = the coefficient of linear expansion (m/m°C)
Δt = the change in temperature of the pipe.

Values of the coefficient of linear expansion for various materials are detailed in Table H1. Due to variations in plastic materials, it is recommended that the manufacturer of plastic pipes be consulted for the precise coefficient.

Table H1 Coefficients of linear expansion

Materials	m/m°C $\times 10^{-6}$
Cast iron	10.2
Copper	16.9
Mild steel	11.3
PVC, normal impact	55.1
PVC, high impact	75.1
Polythene, low density	225.0
Polythene, high density	140.2
ABS (acrylonitrile butadiene styrene)	110.2
Aluminium	25.6
Lead	29.0

Example 1

What is the increase in length and the final length of a straight copper pipe 15 metres long when subjected to a temperature increase from 20°C to 75°C.

From Table H1, the coefficient of linear expansion for copper is 16.9 $\times 10^{-6}$ m/m°C.

From formula (a) $\Delta L = l \times \alpha \times \Delta t$
$= 15 \times 16.9 \times 10^{-6} \times (75-20)$
$= 0.014$ m (14 mm)

Therefore the pipe will expand 0.014 m and the final length of the pipe will be 15 m + 0.014 m = 15.014 m

Example 2

What is the increase in length and the final length of a straight normal impact PVC pipe 15 metres long when subjected to a temperature increase from 20°C to 75°C.

From Table H1, the coefficient of linear expansion for normal impact PVC is 55.1 x 10^{-6} m/m°C.

From formula (a) $\Delta L = l \times \alpha \times \Delta t$
$= 15 \times 55.1 \times 10^{-6} \times (75-20)$
$= 0.046$ m (46 mm)

Therefore the pipe will expand 0.046m and the final length of the pipe will be 15 m + 0.046 m = 15.046 m

It can be seen from Examples 1 and 2 that normal impact PVC expand approximately 3.3 times that of copper. Upon cooling down the pipe will return to its original length.

The expansion and contraction of pipework create pressures which can cause damage to the pipework, supports, fittings, other components in the pipework system and even the building structure.

The pipework expansion may create forces and pressures which must be controlled by careful design and installation to ensure that damage does not result. Important points to ensure the pipework system is allowed to move laterally where required include:

(a) Ensuring that the pipework is either provided with sleeves or neatly drilled holes through walls, floors and ceilings to enable unrestricted movement.

(b) Providing felt pads or similar between pipework and notched joists.

(c) Ensuring that the design of pipe clips and supports do not restrict the movement of the pipe particularly on long pipe runs.

(d) The pipe clips and supports are carefully positioned to enable the pipe to move.

(e) Where it is necessary to bury the pipework in the screed, provision must be incorporated to prevent damage. This could take the form of fibreglass wrap around the pipe and the whole surrounded with expanded metal or similar.

(f) Ensure that any branches connected to expanding pipework, ie. connections to radiators, are of adequate length to provide natural flexibility to prevent the connection shearing.

(g) Ensure that expanding pipework will not clash with structure.

The above requirements are also essential to prevent possible expansion and construction noises.

Note

Pipework should not be buried in the screed if alternative routing can be provided.

Where expansion is excessive and pipework cannot be allowed to flex naturally, consideration must be given to ensure that the pipe movement is controlled by providing anchors. These are usually fitted to the middle of the pipe run to enable the pipe to move in two directions.

Figure H1 Typical anchor positions

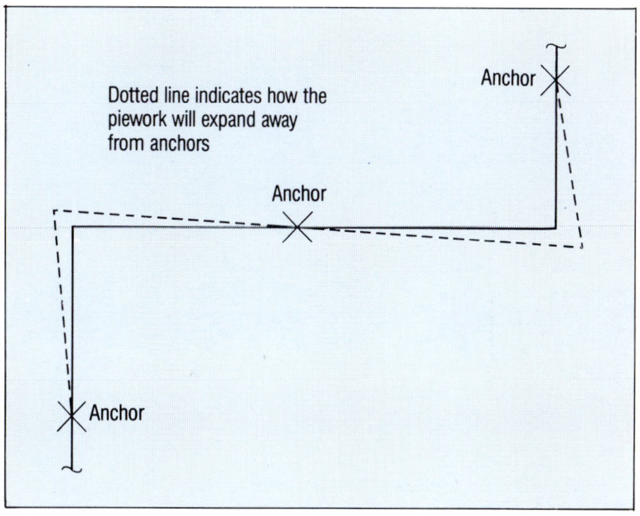

If expansion and contraction of the pipework cannot be controlled with the use of anchors alone then consideration must be given to providing an artificial method to enable movement to take place. These include:

(a) Routing pipework with additional bends.

(b) Providing purpose designed loops or horseshoe sections in the pipework.

CONSIDERATIONS FOR DESIGN AND INSTALLATION

Figure H2 Expansion loop

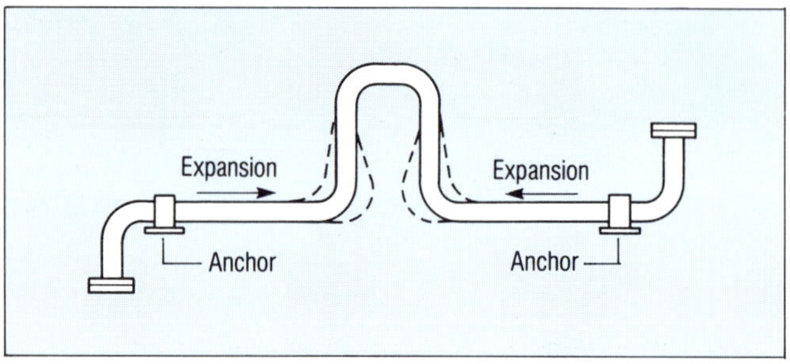

(c) Installing expansion bellows or compensators.

Note

The manufacturer of the bellows and compensators should be consulted for their particular recommendations and requirements of anchors and guides.

Cold Draw

When installed into the pipe, it is normal for the bellows to be 'cold drawn' (ie. stretched) by half the total calculated expansion as Figure H3. The bellows will then absorb the total expansion, half in tension and half in compression from its neutral position. This keeps the stress in the bellows material to a minimum for a given expansion and gives maximum bellows life.

Figure H3 Expansion bellows

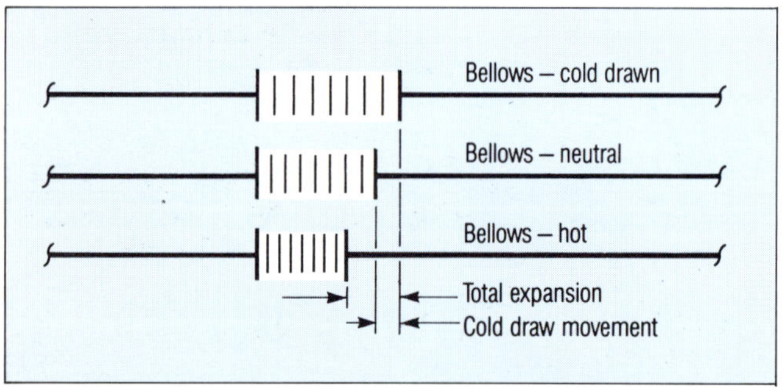

It may be found convenient, particularly on large installations where there are many bellows, to cold-draw by half the total rated expansion of the type of bellows. This is also satisfactory but will obviously not give quite as good a bellows in life. In steam and high temperature heating applications, it is not necessary to allow for ambient temperature variation (day-night, summer-winter) when installing the bellows, since this variation is very small compared with the working temperature range. However, on low temperature applications, allowances must be made when calculating cold draw. Where very small movements, vibration or subsidence are to be absorbed, the bellows manufacturers often recommend no cold draw.

Pre-cold drawn units are available in which the cold draw is already applied by the manufacturer.

Methods of applying cold draw are shown below.

Flanged Units

A wooden spacer is fitted (see Figure H4), equal to the cold draw required. This is removed when the pipe is fully installed and anchored and the flanges drawn together to provide the cold draw on the bellows.

Figure H4 Flanged unit cold draw

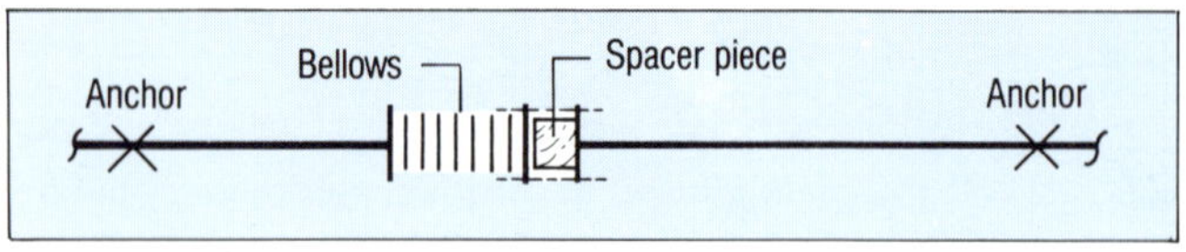

Welded End Units

The cold draw is taken at a point away from the bellows, either at a flange or by pulling up the pipe at a welded joint.

Figure H5 Welded end unit cold draw

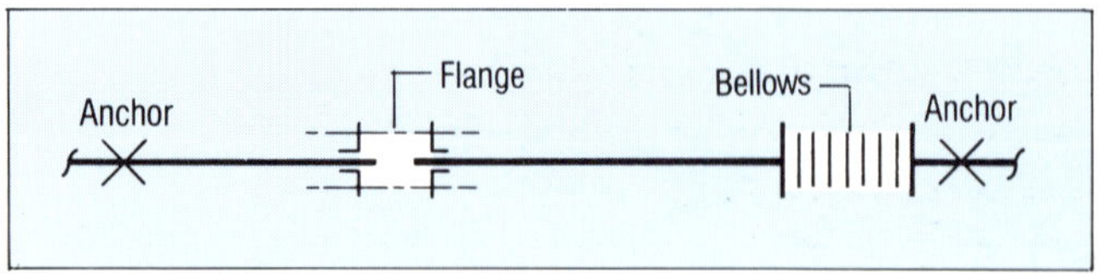

Screwed End Units

Wherever a screwed end unit is used, it is essential to fit a union connection near the bellows so that the bellows is not subject to torsion during installation. A guide should be fitted on each side of the bellows and close to this first guide, the union connection.

Always consult the bellows manufacturer regarding cold draw requirements.

The recommended fitting procedure is as follows:

(a) Fit a short length of pipe, screwed at each end, into one end of the bellows; the length to be sufficient to pass through the first guide and project about 100 mm beyond it.

(b) Screw the other end of the bellows into the pipe, taking every precaution to prevent twisting of the bellows.

(c) Fit a union to the other end of the short length of pipe and adjust the free pipe end so that the correct gap will be left when the female end of the union is fitted.

(d) Stretch the bellows by hand and fit the first thread of the union connection.

(e) Holding bellows firmly to prevent twisting, tighten up union.

Anchor Design

When anchors are installed to control expansion, certain forces will be exerted on the anchor. It is therefore essential to ensure that the anchor is adequately secured to the pipe, that the anchor fixings to the structure are adequate and that the structure itself is adequate to accept the forces. Where necessary, the advice of a structural engineer should be obtained.

Figure H6 Anchor for copper pipework (light load)

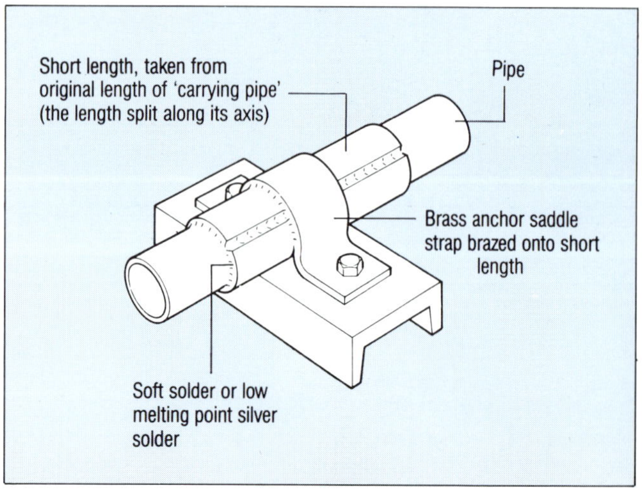

Figure H7 Anchor for steel pipework (light load)

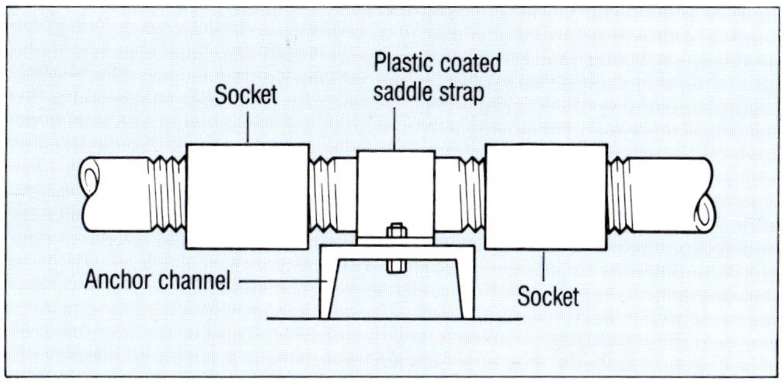

Figure H8 Light to medium load anchor

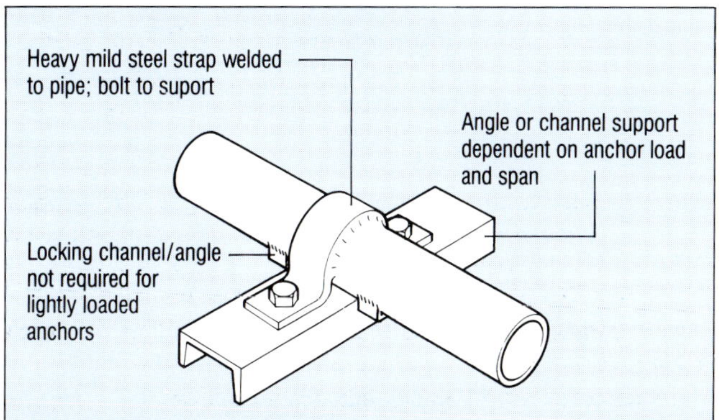

This anchor may be used for copper pipes using brass strap. Alternatively, a mild steel strap with asbestos tape, wrapped around pipe and clamped tight.

Figure H9 Typical duct application

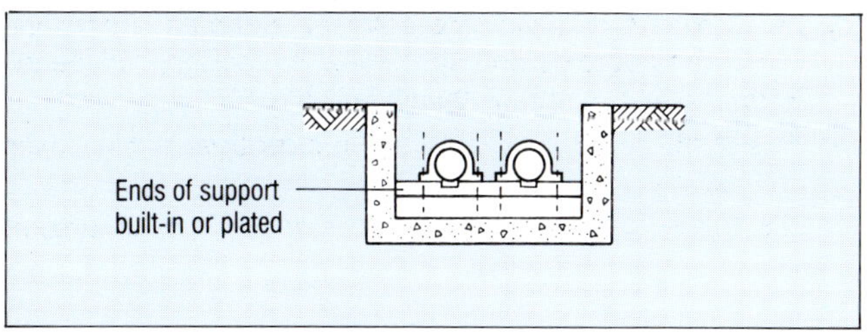

Figure H10 Typical riser application

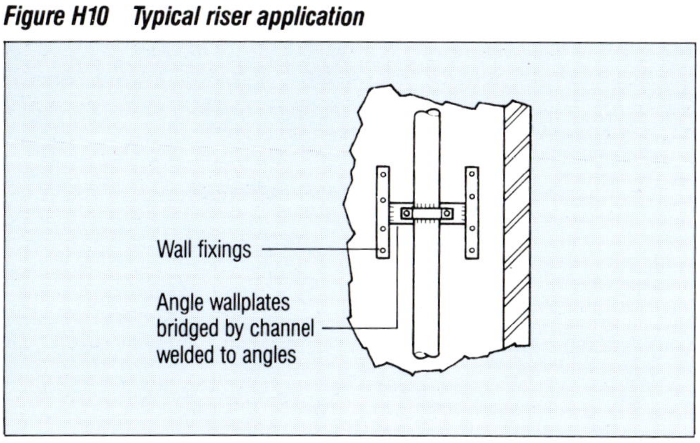

Figure H11 Alternative light to medium load anchor

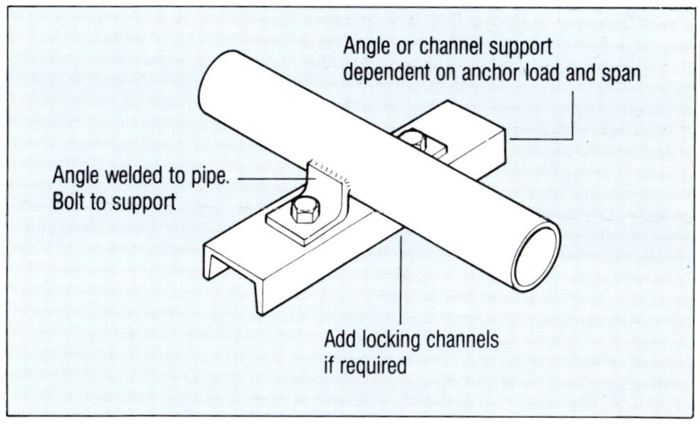

Figure H12 Application using angle welded to pipe

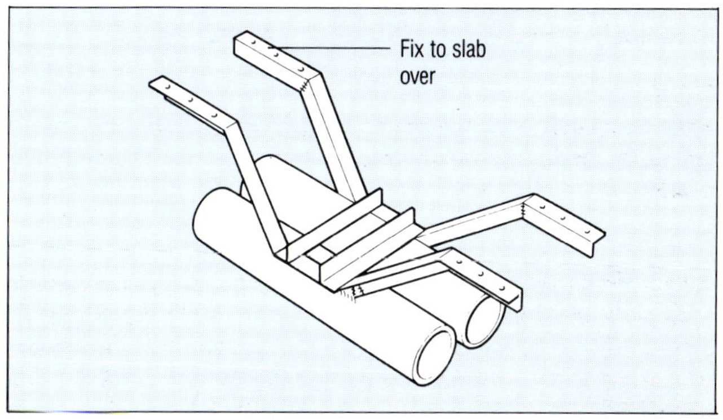

Figure H13 Typical heavy load anchor

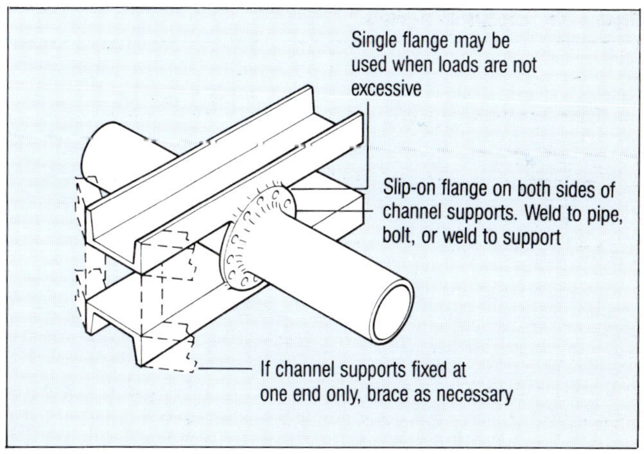

Figure H14 Typical service walkway application

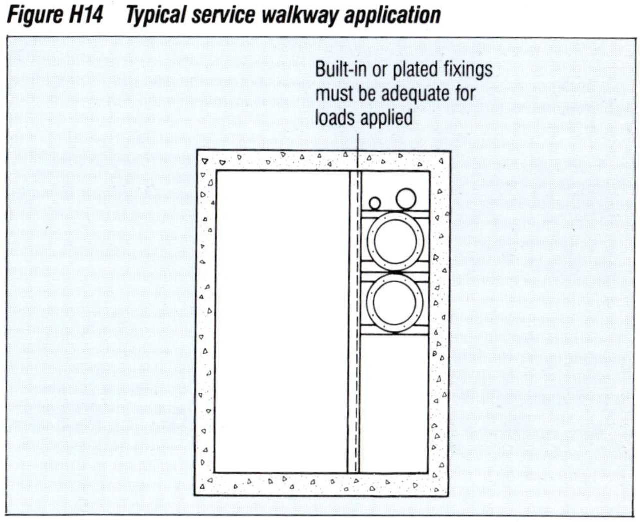

CONSIDERATIONS FOR DESIGN AND INSTALLATION

It is important to ensure that adequate provision is incorporated to enable the pipework to physically expand.

The following main factors have to be taken into consideration when determining forces on anchors:

(i) The effect of test pressure on the cross-sectional area of the bellows.

(ii) The elasticity force to stretch and compress the bellows through the working movement.

(iii) Frictional forces between the pipe and its supports, acting against the direction of movement.

(iv) The slope of the pipe, affecting the lower anchor (vitally important if the pipe is vertical).

(v) Pulsation and flow effects in the pipe.

(vi) Differential forces due to changes in pipe diameter.

Friction coefficients between pipework and supports vary considerably, but the methods shown in Figures H15 (a), (b), (c) and (d) may be used as a general guide for steel on steel.

Figure H15(a) Point contact

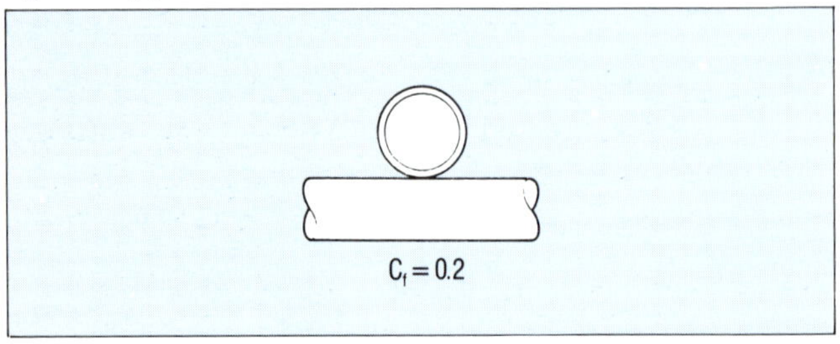

Figure H15(b) Edge contact

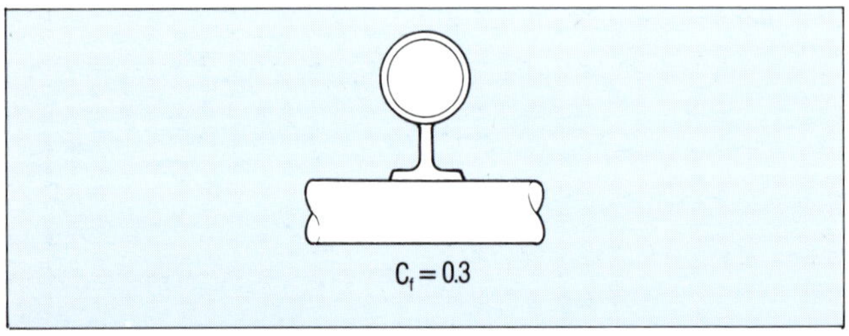

Figure H15(c) Line contact

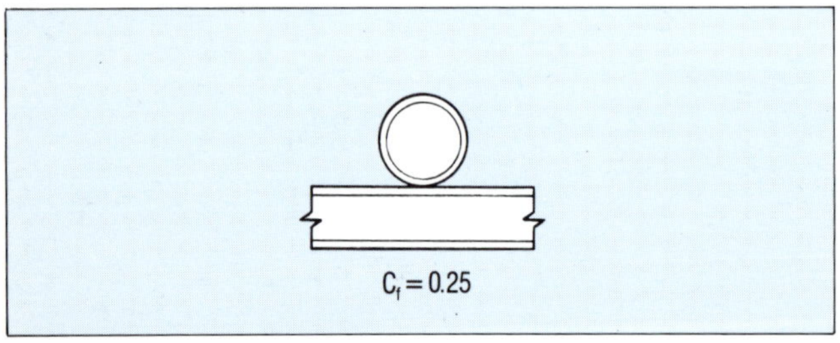

Figure H15(d) Face-to-face contact

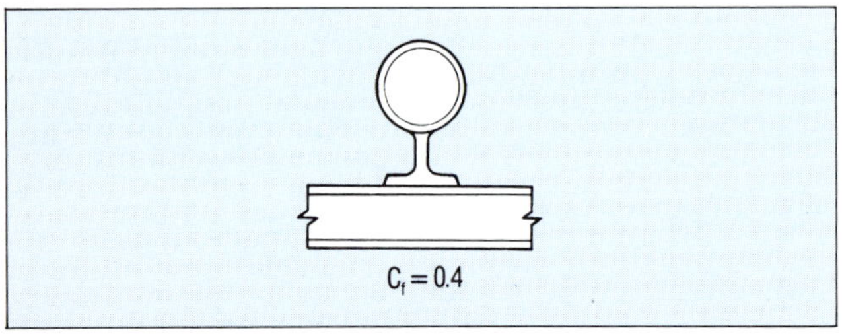

The frictional resistance (kg) = dead load at point of support (kg) × coefficient of friction. This resistance acts against the direction of movement, ie. it can act either way, depending on whether the pipe is warming up or cooling down.

Should the above figures give unacceptably high forces, consider the use of special PTFE support pads which gives figures as low as 0.03.

The following example illustrates the calculation to determine the force imposed on an anchor on the pipework installation shown in the diagram, which incorporates bellows under normal working and test conditions.

The designer must obtain the relevant data from the manufacturers.

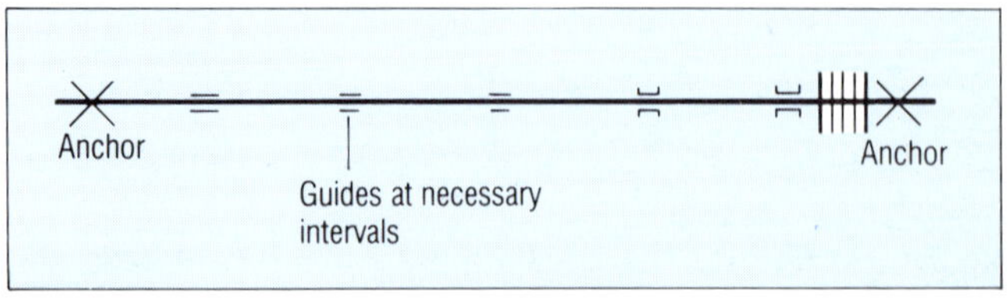

Example

Pipe weight	= 22 kg/m
Water weight	= 19.5 kg/m
Lagging weight	= 6 kg/m
Effective area of bellows	= 230 cm^2

Thrust to compress or extend bellows through full movement
= 6700 N (manufacturers data)

Coefficient of friction between pipe and supports	= 0.3
System working pressure	= 550 KPa
System test pressure	= 830 KPa
Length of pipes	= 42.5 m

Pipework Systems Under Working Conditions

Thrust due to internal pressure on bellows
= working pressure x effective area
= 550 × 230 × 10^{-1} N
= 12650 N pushing outwards on anchors

Force due to bellows stiffness (bellows compressed in working position)
= 6700 N pushing outwards on anchors

Frictional resistance to pipe movement over its supports
= coefficient of friction × (total wt. pipe + water + lagging)
= 0.3 × [(22 × 42.5) + (19.5 × 42.5) + (6 × 42.5)] kg
(To convert to Newtons, 1 kg = 9.807 N)
= 0.3 × [(22 × 42.5) + (19.5 × 42.5) + (6 × 42.5)] × 9.807 N
= approximately 6000 N against the direction of movement, ie. pushing outwards on anchors.

Total thrust on anchor = 12650 + 6700 + 6000 = 25350 N pushing outwards on anchors.

Pipework Systems Under Test Conditions

Thrust due to internal pressure on bellows

= test pressure × effective area

= $830 \times 230 \times 10^{-1}$ N

= 19,090 N pushing outwards on anchors

Force due to bellows stiffness (bellows stretched in cold-draw position)

= 6700 N pulling inwards on anchors

Frictional resistance = NIL as pipe only moves under temperature effect

Total thrust on anchor = 19090 − 6700

= 12,390 N pushing outwards on anchors

It can be seen in this instance that the greatest force acting on the anchor is under normal working conditions, this however is not always the case.

Pipe Guides and Supports

Wherever pipework will expand longitudinally or move due to expansion, it is essential to ensure that all supports facilitate movement except for purpose-designed anchors. If the pipework is incorrectly restrained, damage to the pipe supports or structure may result due to excessive friction or forces.

When a fitting such as a bend, valve or distance piece is installed into a pipe, it acts just like the pipe itself, holding the internal pressure but not pushing or pulling on the pipe.

When an axial bellows is fitted into a pipe, it reacts to the internal pressure in just the same way as a piston acting in a cylinder. By introducing internal pressure into the pipe, the bellows tries to open out lengthwise.

When the pipe gets hot, it expands towards the bellows and tries to compress it. Meanwhile, the pressurised bellows is trying to open out lengthwise. The expanding pipe therefore has to overcome this pressure force as well as the stiffness of the bellows and the friction of the pipe supports. Hence the need for firm anchors at each end of the pipe run, and careful guiding not only on each side of the bellows but along the pipe run.

The pipe between bellows and anchors is often in compression, and unless the pipe is guided carefully, and runs accurately in a straight line from anchor to anchor, the pipe may bow out sideways at some point. This will pull the bellows with it and may cause failure as indicated in Figure H16.

Figure H16 Pipeline failure may occur without pipe guides

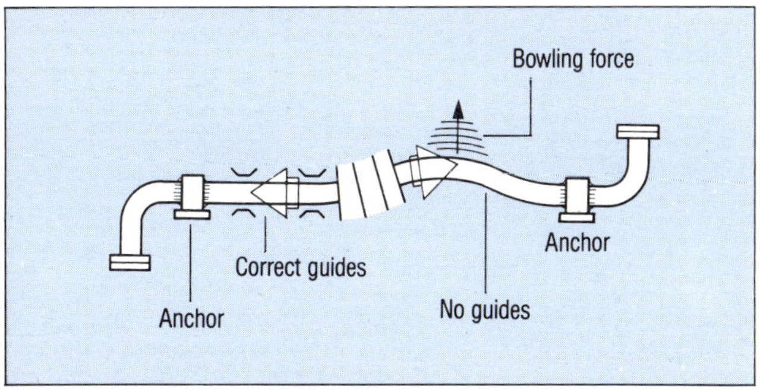

An axial bellows installation should therefore look like the illustration in Figure H17, showing alternative arrangements with the bellows in the middle or at the end of the line.

Confirmation should be obtained from the bellows manufacturer regarding any special requirements.

Figure H17(a) Bellows positioned centrally in an axial installation

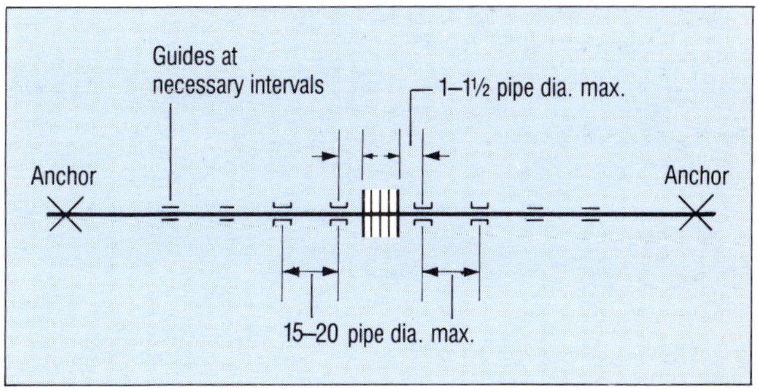

Figure H17(b) Bellows positioned at end of line in an axial installation

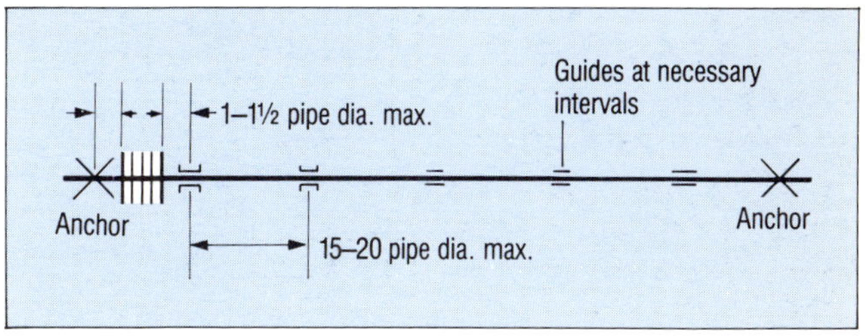

A pipe guide is any form of constraint which allows the pipe true axial movement along its length, but prevents offset movement either horizontal or vertical.

It will be appreciated that when the pipe is suspended on hangers, it is free to swing and cannot be considered as guided. Therefore, never use axial compensators in a hung pipe.

Some typical pipe guides are indicated in Figures H18 to H28.

Figure H18 Strap-type guide

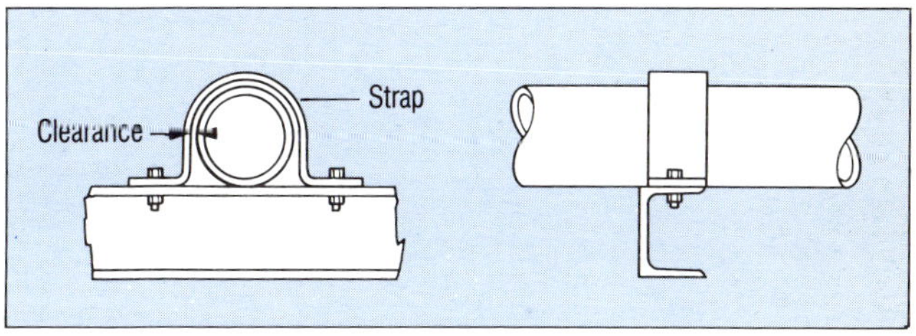

Figure H19 Strap-type guide with tube roller

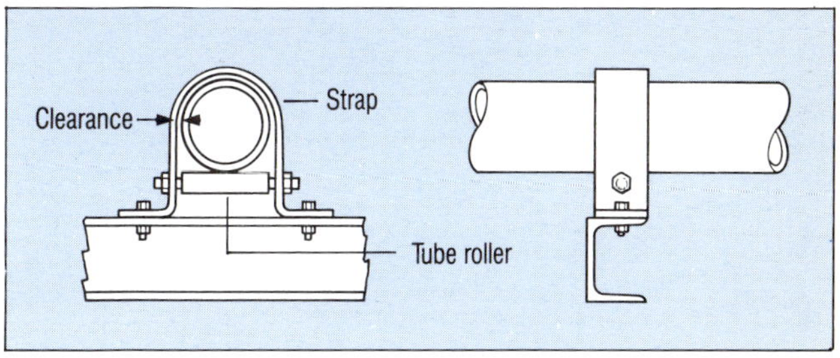

Figure H20 Strap-type guide with roller and chair

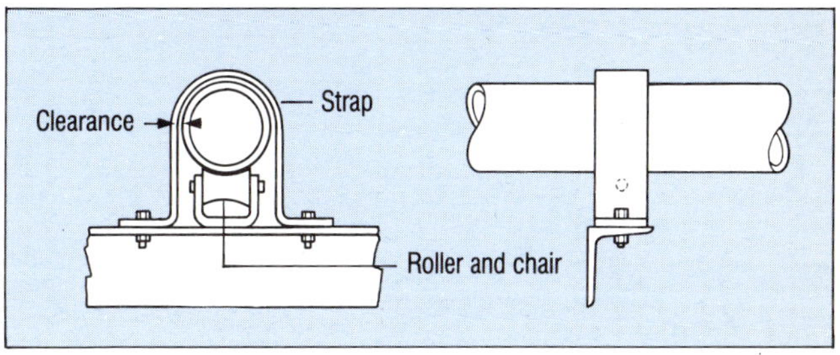

CONSIDERATIONS FOR DESIGN AND INSTALLATION

Figure H21 Guide using angle rollers on box section

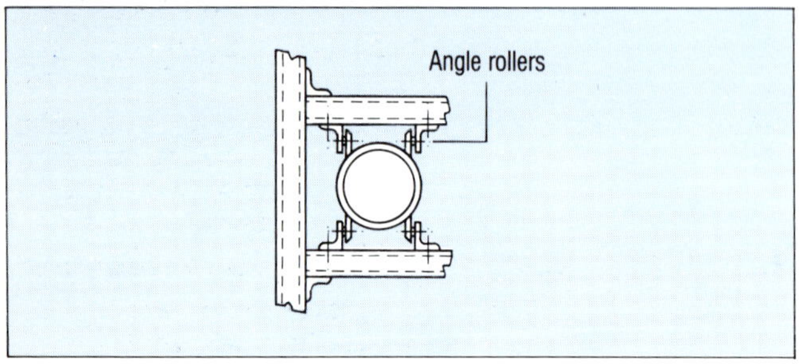

Figure H22 Strap-type guide using angle rollers on box section

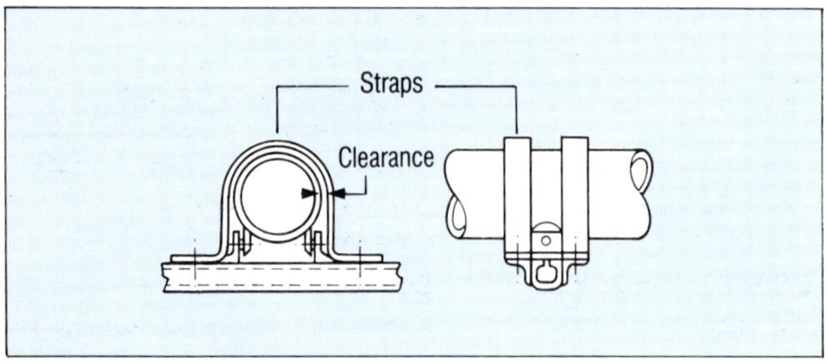

Figure H23 Tee support with angle guide

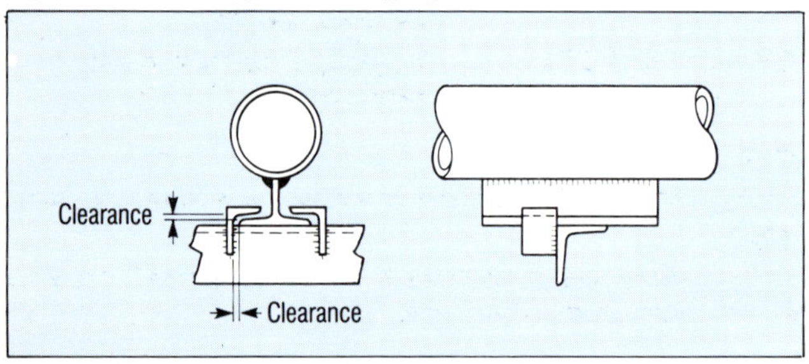

Figure H24 Tee support with roller and angle guide

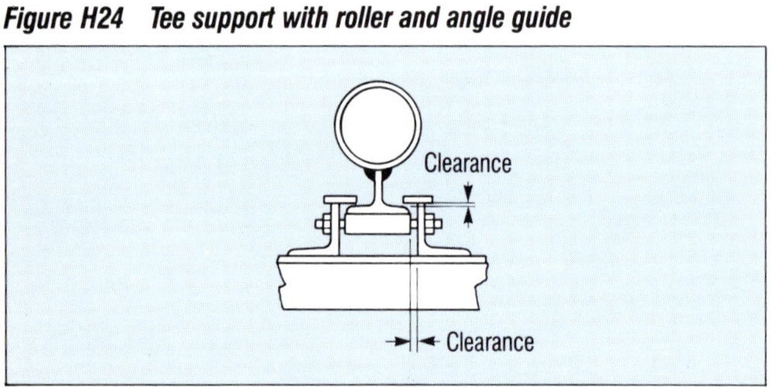

Figure H25 Guides for large bore and/or high pressure pipes

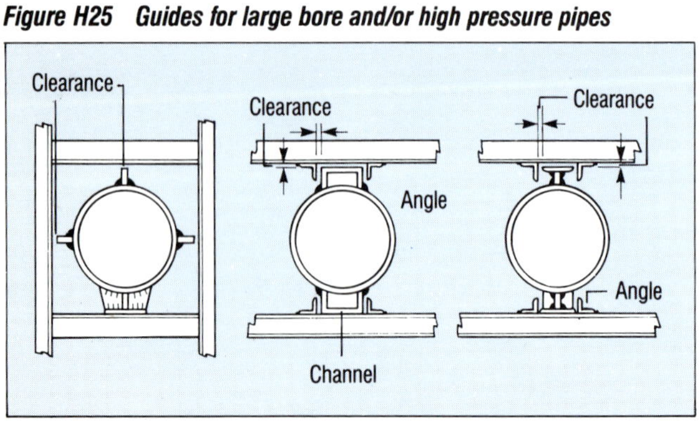

Figure H26 Tube-type guide

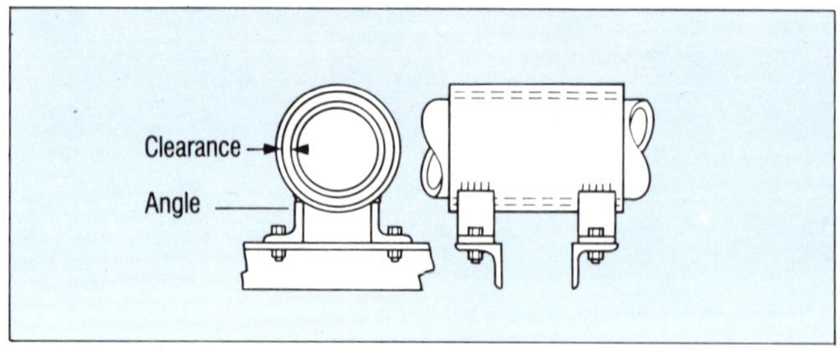

Figure H27 Tube-type guide with rods
(will reduce friction and excessive clearance)

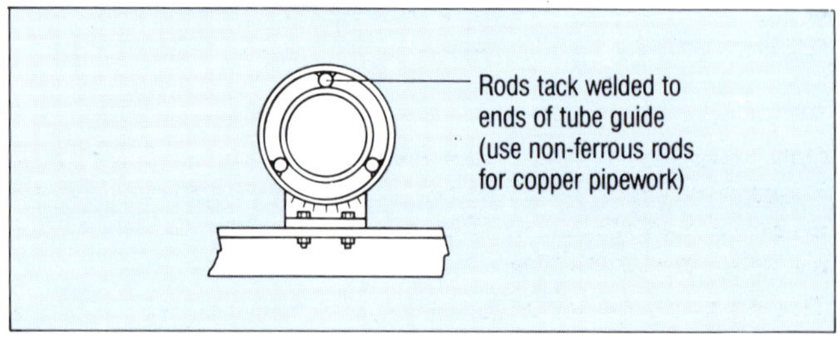

Figure H28 Slip-on flange type guide

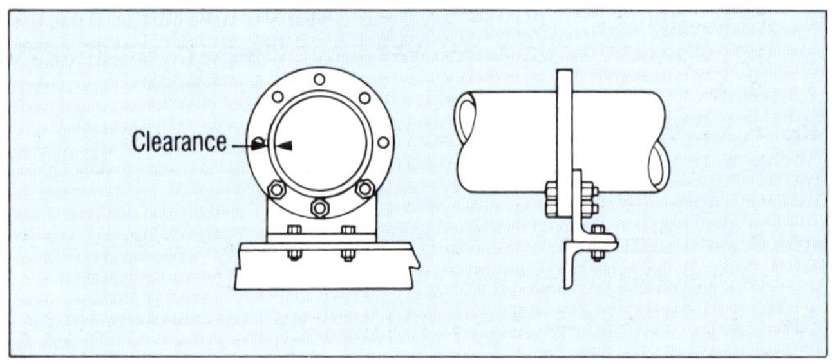

For guides nearest to the bellows, the clearances shown in Figures H18 to H28 should not be greater than 1.6 mm for pipe diameter up to 100 mm and not greater than 3.2 mm for larger pipes. For the guides along the pipe run, clearance of 3.2 mm and 6.4 mm respectively should be used.

Guides Nearest the Bellows

The function of guides nearest the bellows is to ensure true axial movement on to the bellows. This could be achieved by using a tubular type guide as Figure H29 of such length, in relation to its diameter, that the necessary clearances to permit axial movement do not allow appreciable angular offset movement. This generally works out as a length of five to six pipe diameters.

Figure H29 Guides near bellows

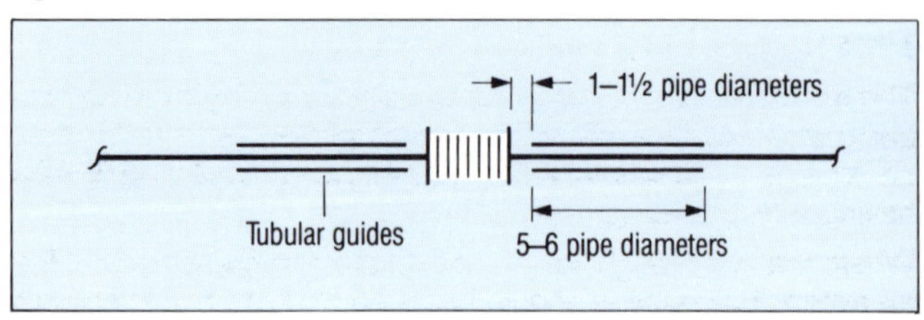

A tubular guide is normally only fitted to small pipes and, in the great majority of cases, straps and rollers are employed. These are short and individually cannot control angular movement of the pipe through them. To ensure alignment of the pipe onto the bellows, an arrangement using additional guides is considered suitable (see Figure H30).

Figure H30 Installation of additional guides

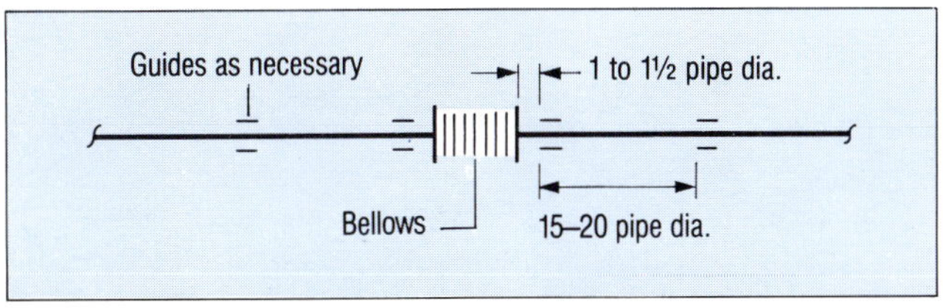

Guides Along the Remaining Pipe Run

It is a widespread misconception that, as long as the bellows is guided as above, the rest of the pipe can be supported on hangers, and left unguided. As the remaining pipe run is subjected to compressive forces, it may blow out unless suitably guided at regular intervals.

Expansion Bellows

Only axial type bellows have been considered in this section. These are bellows which take up expansion by compression and extension along the axis. Other types of bellows are available and manufacturers should be consulted.

The following points should be noted when using expansion bellows:

(a) When axial bellows are used, the pipe should never be hung or suspended freely. The pipe supports must guide the pipe and allow only axial movement along the entire length of the pipe.

(b) It is vital to ensure that the pipe is guided carefully on each side of the bellows. The guide must allow only axial movement onto the bellows.

(c) It is vital to ensure that the pipe is guided carefully at the necessary intervals along the whole pipe run.

(d) Only one bellows unit should be installed between two pipe anchors.

(e) The line between the two anchors should be straight, in plan and in elevation, with no sets or bends.

(f) Pipe anchors should be sufficiently strong.

(g) Bellows must be stretched by half the total expansion movement (cold draw). This does not apply to special applications or 'pre-cold drawn' units.

An example would be a gas pipe over a bridge, with a maximum summer temperature of 30°C (in the sun) and a minimum temperature of −20°C giving a total of about 50 mm expansion over a 100 metre long bridge. If the unit was installed in a moderate winter temperature of 5°C then correct cold pull would be zero.

(h) Screwed end units must be held firm when installing to prevent twisting of bellows.

Plastic Soil and Waste Pipes

The most common method of providing the facility for expansion and contraction in plastic soil and waste pipe systems is the use of an expansion coupling incorporating socket and spigot joints having rubber sealing rings.

Where the total straight length of a discharge pipe exceeds 1.8 m, an expansion coupling should be introduced at 1.8 m intervals to connect pipes together.

The pipework should be anchored and supported in accordance with the manufacturers recommendations which will depend on the system employed.

BRITISH STANDARDS AND CODES OF PRACTICE

BS 5449	Code of Practice for central heating for domestic premises.
BS 5449: Part 1: 1977 (1981)	Forced circulation hot water systems.
BS 6129	Code of Practice for the selection and application of bellows expansion joints for use in pressure systems.
BS 6129: Part 1: 1981	Metallic bellows expansion joints.
BS 6700	Specification for design, installation, testing and maintenance of services supplying water for domestic use within buildings and their curtilages.
BS 6880	Code of Practice for low temperature hot water systems of output greater than 45 kilowatts.
BS 6880: Part 1: 1988	Fundamental design considerations.
BS 6880: Part 2: 1988	Selection of equipment.
BS 6880: Part 3: 1988	Installation, commissioning and maintenance.
CP 342	Code of Practice for centralized hot water supply.
CP 342: Part 2: 1974	Buildings other than individual dwellings.

Note

(1) Year after the Standard's number indicates the latest date of issue.

(2) Year in brackets gives the date of the latest amendment at the time of publication of this Guide.

INTRODUCTION

The following information provides designers with guidelines for designing plumbing services for the disabled community. It includes information on access to buildings which apply whether the building is large or merely a small toilet. More specifically it gives details of measurements required for WC compartments.

In the case of the compartment, passage ways and door sizes, the figures given are the minimum size required. Additional space is recommended wherever possible.

APPROACHES TO BUILDINGS

Facilities for the disabled should be marked with the internationally agreed disabled sign.

Figure I1 Disabled logo

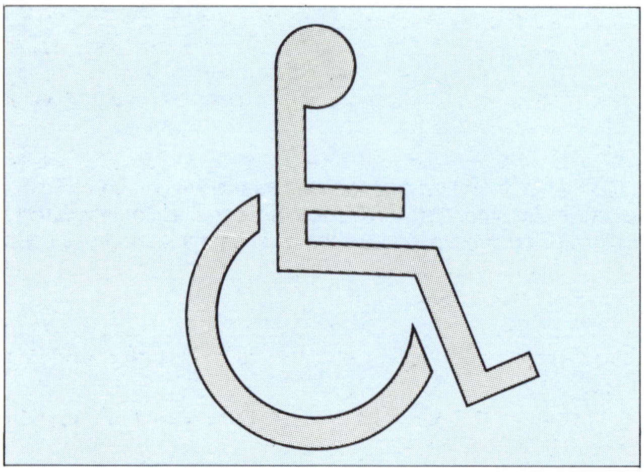

Parking Spaces

Parking spaces should be close to an accessible entrance and preferably be on the same level (ramps are permissible but may be tiring for wheelchair or zimmer users). If possible, they should be undercover with covered access to the building and there should be adequate extra space for the transfer from car to wheelchair.

Dropped Kerbs

These should be 1400mm wide with a gradient of at least 1 in 12 and should be placed where there is a merging of roadway and pavement.

Stepped Access

This is not advised but if unavoidable, goings should not be less than 280mm and uniform with risers not higher than 150mm. The vertical rise of any flight of stepped access should not be greater than 1200mm.

Level/Ramped Access

Should be provided at one entrance from the adjacent street and car parking spaces with access to the toilet facilities. Ramps should be a minimum width of 1200mm with a maximum gradient of 1 in 12 and having an unobstructed platform of 1200 square millimetres. Resting places should be provided at intervals if the approach ramps are longer than 3 metres and near the maximum gradient of 1 in 12.

Handrails

Handrails should be provided on both sides of a ramp, or in the centre if its width exceeds 2000mm, and/or where the gradient exceeds 1 in 15. They should be easy to grip firmly with a circular diameter 45-50mm.

Figure I2 Handrail dimensions

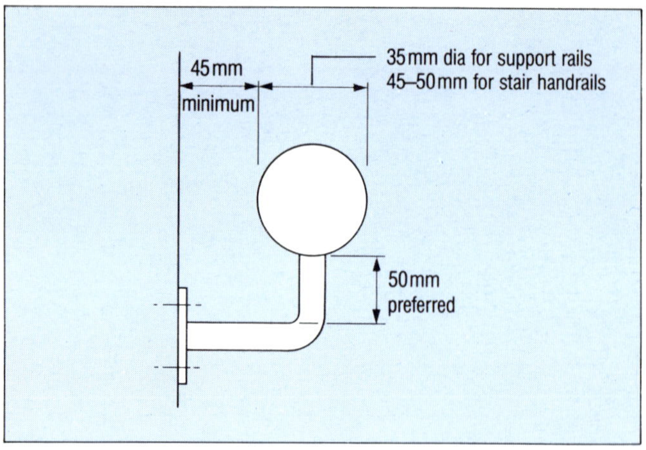

Entrance Doors

These should have a clear opening not less than 980mm wide and an auxiliary side hung door the same width should be provided if they are revolving. The doors should be easily swung and any spring or hydraulic damper provided must allow easy access for the wheelchair user.

Automatic Entrance Doors

These are preferable in certain circumstances but they should have a minimum clear distance of 2100mm between sets for sliding doors and 2500mm for swing types.

INSIDE BUILDINGS

Double Swing Doors

These are preferred in circulation areas.

Glass Doors and Other Glass Areas

Should be made easily identifiable to those with defective vision.

They should never be used in areas/institutions where wheelchair users are prominent.

Sliding Doors

May be necessary in bathrooms and WC compartments. They should not be less than 800mm wide. They can however have disadvantages as adjacent wall space is rendered useless, there is poor acoustic insulation and the typical sliding gear used with low cost doors does not easily withstand disabled usage and may be pulled across the line of the track. People in wheelchairs prefer swing doors to sliding doors.

Internal Doors

These should have a clear opening of not less than 900mm. On the leading face of side hung doors there should be a clear wall space or panel equivalent of 300mm.

Kick Plates

These are not essential but are recommended to be fixed to the trailing face of side hung doors as they can help to minimize damage to the building finishes.

Thresholds

Should be flush without steps or upstands.

Spring Closers

These should be avoided but delayed action types are preferred if they are essential.

Figure 13 Dimensions – internal doors and adjacent areas

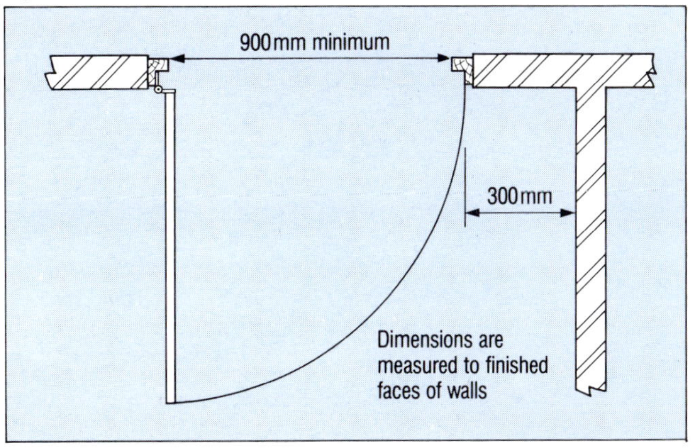

Door Handles

Door handles should be approximately 1040mm above floor level and of the lever type.

Switches

Switches for lights should align with the door handles and be no higher than 1200mm above the floor level. For lifts, touch light controls are preferable, with embossed digits to enable identification by blind people.

Internal Circulation Areas

There should be a wheelchair access at the same level as an accessible entrance door, or at the same level as a lift. Changes in level by ramps are permissible.

Passage Ways

Should not be less than 1200mm wide.

Floors

Floors should have a slip resistant surface (wet or dry). There should be no simulation of steps in the patterning.

Hazards

Hazards such as level variations and projections or obstructions such as floor drainage gullies and upstanding access and manhole covers must be avoided.

Level Changes

Should be easily identifiable by lighting or colour texture contrasts.

WC COMPARTMENTS IN PUBLIC BUILDINGS

At least one unisex public lavatory accommodation should be provided in:

Entertainment Facilities

Concert halls, theatres, cinemas, recreational buildings, leisure and sport facilities.

Travel Facilities

Transport buildings, principal railway stations, road service stations, air terminals.

Commercial Buildings

Shopping centres and large department stores. At least one unisex staff lavatory accommodation should be provided in large office buildings and buildings related to other types of employment.

In addition to the above unisex facilities where one assistant can aid the disabled person, supplementary facilities should be provided in large buildings.

Points to Consider for WC's

(a) If the upstairs WC is inaccessible, a stairlift or through-floor lift should be considered if it enables other rooms to be used. Also, if there is sufficient space in the WC the door could be re-hung; a sliding door should not be fitted. If the WC is in an adjacent room, an additional WC could be incorporated into the bathroom.

(b) If the only WC is external to the building, consideration should be given to either covered access to it or a one sided protected walkway installation. Grant help may be available through the Environmental Health Department (for private housing) or the Local Authority (for council housing).

(c) Strict building regulations apply if a new WC is installed and these must be checked with the Environmental Health Department. Utilizing space downstairs may necessitate the fitting of extra doors, additional ventilation and special consideration for understairs installation.

(d) Where there is no suitable downstairs space available, consider installing a ground floor bathroom or shower room extension incorporating a WC suite. All work must comply with the Building Regulations. Allowances must be made for the specific needs or future needs of the disabled person, such as level access, manoeuvering space etc.

(e) When installing a new WC installation, check that the seat height is suitable for the user, securing it on a raised plinth if necessary.

(f) A wall-hung WC pan must be fitted with a spacer box. This type of pan allows for complete floor cleaning.

(g) A low-level cistern affords some back support and may be useful for those with poor balance but can impede others who have stiff hips. It is very difficult and often impossible for disabled people to flush a low level cistern using a handle behind their seated position.

(h) A slim-line cistern provides more space for those with stiff hips or for a helper assisting with body cleansing.

(j) The WC appliance with a warm water douche and warm air drying facility can provide independence in personal cleansing for some severely disabled users.

Access to WC Compartments

All recommendations for passage width, floor types etc are detailed in the preceding paragraphs. It is most important that a route is available to the toilet area having suitably sized doors and openings etc. of minimum clearance of 900mm wide along which the disabled person travels to the compartments. Remember no steps, ridges or steep changing floor levels.

The WC compartment should be provided with access to allow wheelchair approach to all the facilities which should include the WC and flushing unit, hand basins with taps, towel dispensers (or hand dryers), sanitary towel disposal units and mirrors.

Access Through Lobbies

This is not recommended but where it is essential, room should be allowed for a wheelchair to manoeuvre. Standard wheelchairs need 1800 by 1400mm to turn 180°. Note should be made that wheelchair lengths vary between 945mm and 1150mm with the width between 550mm and 675mm. It must also be remembered that the disabled persons feet extend in front of the wheelchair and their arms and especially their hands and fingers to operate the wheels, extend to the side of the wheelchair.

WC FACILITIES FOR WHEELCHAIR DISABLED

Dimensions

The WC compartment should be not less than 2000mm by 1500mm.

Door

This should be 1000mm wide, open outwards (so that the doorway is not blocked if a person inside falls), have a pull rail and handle, or it should slide. It should only open inwards if there is 1100mm by 700mm of unobstructed space inside to allow a free door swing with the wheelchair out of its line. Doors swinging inwards cause problems to the disabled user as the wheelchair has to be manoeuvred backwards at the same time as pulling the door open.

The door must be unlockable from the outside in the event of emergency and assistance needed by an attendant.

Support Rails

50mm diameter support rails should be provided on sidewalls adjacent to the WC with a hinged horizontal rail of 300mm minimum length on the exposed side. If the design is for an individual disabled person the rails may be placed according to their specific requirements.

The mirror can equally be positioned above the basin to save movement within the cubicle. The basin can also be used to support make-up and toiletry items.

Toilets

These should be placed with the seat of the WC at 450mm above floor level to allow easy transfer from wheelchairs. Preferably the WC should be placed to allow a lateral transfer from the wheelchair, which is placed to one side, or possibly a frontal transfer, in which case a pedestal unit should recede towards floor level to ease transfer. Side transfer should be designed in preference to front.

The flushing device must be easy to operate and must not be higher than 1200mm above floor level. It may be possible to place the flushing device on a side wall within easy reach.

It is poor practice to design the flushing operating device/handle behind the disabled person.

If users are likely to be too severely handicapped to be able to clean themselves the installation of a proprietary WC/bidet (closomat or medic-loo type of unit) should be considered. These have a warm douche and warm air for drying. These units need a relaxation from the water authority before installation.

Toilet Paper Holder

Place on the side closest to the WC seat within easy reach. It is advisable to design the holder to accomodate two rolls of paper.

Hand Rinse Basin

Where practical, this should be placed so that it is usable whilst seated on the WC pan. If it is not within reach the individual has to transfer back into the chair before washing their hands and so fouls the chair's arm rests and wheel runs used for propulsion. The basin should be at a height of 750mm.

Do not install shallow basins with pedestals. Vanity units are suitable for wheelchair users but built into the wall types are most useful and practicable. It is possible to get an adjustable height basin, such as the "zoom" basin, or items which tip at the front, such as the "Ifo tilting" wash basin.

Hand Dryers

These are preferable to other less hygienic methods of drying the hands, such as paper towels or roller towels. They should be set at a temperature of not more than 35°C and have a cut-out switch when this is exceeded. There are a number of automatic dryers on the market that work by remote sensor; there is also a selection of less recommended models operated by "soft touch".

Figure 14 Dimensions of WC compartments and facilities

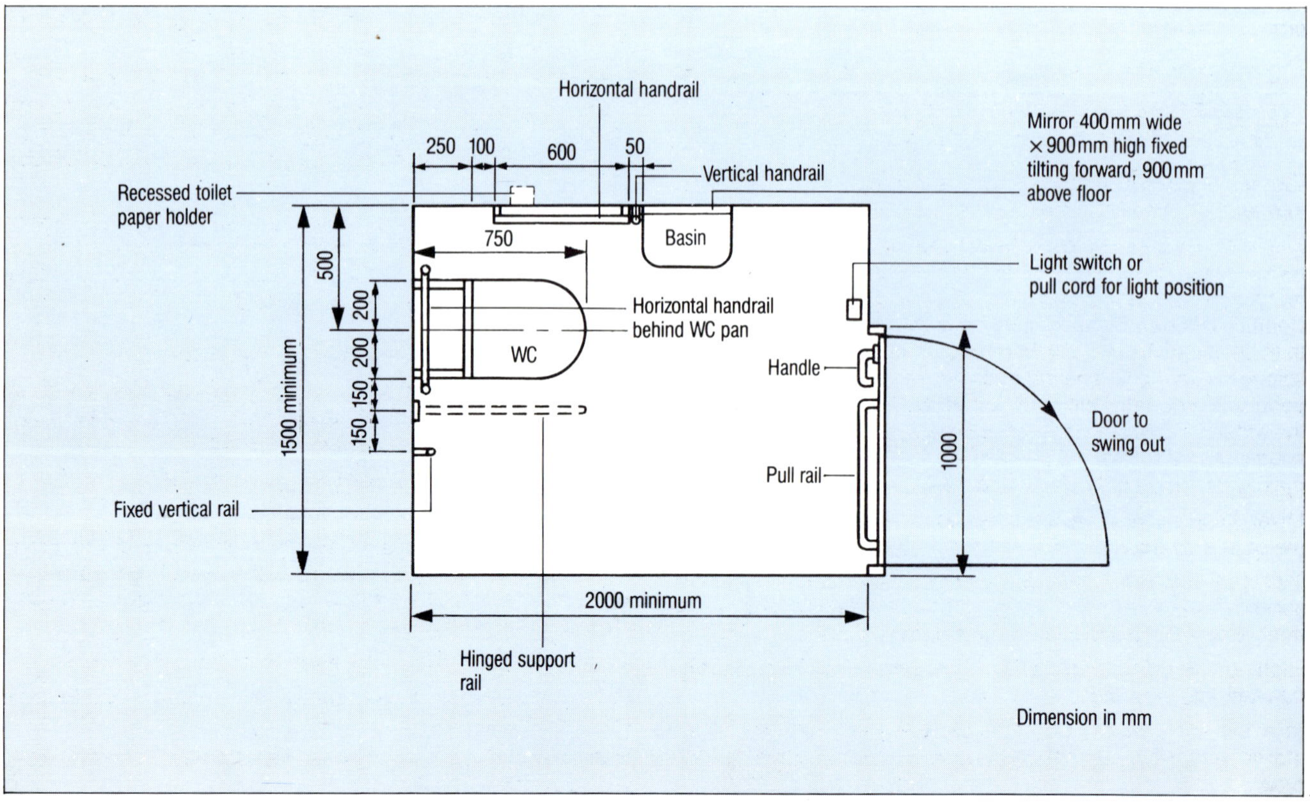

Mirrors

The lower edge should not be higher than 900mm above floor level. Angled mirrors are available and more suitable than flat mounted ones.

WC Seats

These should be of a robust material and must not deform when sat on. The seats should be manufactured to British Standards Specification and kitemarked. The "flimsy" type of WC seat which deforms and moves when sat on must not be used as it could cause accidents during transfer operation.

A raised seat should be adjusted carefully to fit the WC securely and the user or a helper should check frequently that the fittings are secure.

Restricted Space

In a building whose size makes it impractical for a unisex compartment to be installed, the recommended minimum dimensions should be not less than 2000 by 1500mm. This can result in a limited manoeuvring space but better lateral transfer from wheelchair to WC pan set.

Additional Facilities

Where there is already a unisex toilet compartment but wheelchair users require additional toilet accommodation it is suggested that a WC compartment 2000 by 1500mm minimum dimensions is installed.

WC Macerating Boxes and WC Pans

These units cannot dispose of sanitary towels or incontinence towels and must not be fitted to WC pans used by the disabled community.

WC FACILITIES FOR AMBULANT DISABLED

Supplementary facilities for ambulant disabled should be provided in large buildings as for the wheelchair disabled.

WASHBASIN FITTINGS

Taps may be paired, with cross-tops, levers, short (or long), quarter turn levers, single lever mixer types or electronic taps.

Electronically operated taps are operated by the user placing their hands under the spout which activates a remote sensor. The remote sensor should be adjusted to suit the application. Combined mixed water temperatures should be set at not more than 38°C and the use of a thermostatic valve to maintain an even safe temperature is required to prevent scalding. Some disabled people are not sufficiently sensitive to temperature to realise when the water is too hot and for such applications, the advice should be sought from the disabled persons medical professional as to the maximum temperature the mixed water should be set.

Mixer fittings with swivel arms can be moved to the side to leave the bowl area free.

Taps can be wall-mounted which necessitates reaching forward, set into the basin or into the surround. Controls can be fitted to the front of a vanity unit to allow remote control of the tap outlets.

High-necked fittings provide more space when hand-washing and can facilitate cleaning.

In a domestic facility, choose taps which are suitable for the individual, carefully assessing the action required to operate each type.

Long-handled lever taps require less power to operate but need a wider arc of movement than short-handled levers.

Pressure handles vary in the pressure needed and must be held in position to operate. The springs must be reduced in tension for the disabled.

Mixer taps may be dangerous if heat sensation is impaired or if the person is confused.

Foot-operated pressure pads may be useful for some with upper limb disabilities, but they require good balance to use. The are not suitable for wheelchair users.

Knee-action valves are available for those who cannot manage taps and these can be operated from wheelchairs.

Tap turners are a cheaper alternative to fitting new taps for general home use and can be useful when away from home.

Basin plugs with short chains tangle least and may be easier for the disabled user to reach than the longer type.

Central column fitting plugs can be operated with a push and pull movement. Control knobs are easier to handle than a chain but are smaller than most tap handles which cause arthritic disabled people severe problems.

WC LEVELS IN PRIVATE FACILITIES

Considerations

The required overall height should be checked carefully for the individual user.

The comfort of the seat for the individual user should be checked.

Wooden seats may be preferred for comfort by some users and may be more stable than plastic seats for some types of transfer.

Most raised seats can be removed for normal use of the WC by able bodied and for cleaning.

Check that the seat fittings are easy to clean.

Some users may require carefully positioned grabrails or a WC support frame to help them transfer on and off the seat.

A padded WC seat may be of help to those with sensitive skins and those prone to pressure sores. Regular dusting with talcum powder helps to prevent the seat from becoming hot and sticky. A raised seat with one cut-out side may be suitable for a person with a stiff hip.

A person with two stiff lower limbs or extremely weak upper limbs may be helped by a self-lifting seat or angled seat fixed to a frame support. Both should be adjusted carefully to the individual and are unsuitable for those with poor balance.

Selection of Raised Toilet Seats

The seat should preferably be fixed to the WC pan by brackets; in certain situations removable seats are useful but seats that are not fixed are less secure and can be dangerous.

The method of fixing should be easy to carry out.

The fixing bracket should not damage the WC pan.

The seat height must be assessed for each individual and a seat should be supplied which, when placed on the WC pan, provides this height. Most disabled people require a seat height of between 475-550mm.

A convex seat rim is more comfortable than a concave one.

The surface finish of the seat should not adhere to the skin even after prolonged sitting.

To allow easy access for personal cleansing the aperture length should preferably be 300mm.

FACILITIES FOR PRIVATE USE AND NURSING HOMES

Bathroom Layout

The layout will depend on various factors including the severity of the disability and the amount of assistance necessary, the size and type of house and the potential space, the financial assistance possible, the needs of the family, the method of transfer and the type of hoist and other appliances.

The height of the fittings may not be ideal for the disabled user but a compromise may be necessary to reconcile their needs with those of other members of the family.

Because of the longer time taken by the disabled member to carry out their toilet needs, separate WC and washing facilities are recommended.

Grants may be available for major adaptions and applications should be made to the Local Authority's Environmental Health Department for private housing and to the local council's Housing Department for council housing.

Space

Doors may need repositioning to provide quick, direct access from the bedroom or living room to the WC compartment without negotiating corners and landings.

An ambulant person usually prefers the security of a relatively confined space with carefully positioned and easily reached grabrails.

A wheelchair user will require space for manoeuvring and it may be necessary to remove a non-loading bearing wall between the bathroom and WC.

Space will be needed to manoeuvre a portable hoist and the bath panel may be removed to accomodate the hoist legs. Where space is inadequate, a floor-fixed base plate into which the hoist mast can be fitted, could be suitable.

The ceiling joists will need strengthening before an overhead hoist track is fitted.

A straight track from the bedroom to the bathroom spanning the bed, WC and bath may be ideal for a severely disabled person.

The Electricity Board should be consulted before the installation of an electric hoist.

An electric hoist must operate from 24 volts voltage supplied through an approved double-wound isolating transformer. The transformer must be installed outside the bathroom, properly enclosed with a permanently wired connection to the electricity supply.

Space will be necessary to hang towels and clothes.

Doors

Depending on the space available doors may open inwards or outwards. All doors should have emergency release fittings.

Sliding doors are not recommended but may be suitable alternatives where space is limited.

All locking doors must be fitted with a lever-operated indicator bolt with an external emergency release.

Door openings must be a minimum of 900mm wide to accommodate the wheelchair user.

Flooring

Anti-slip flooring such as vinyl sheeting incorporating abrasive grains of aluminium oxide or anti-slip surface tiles are suitable. Carpets designed for bathroom use may be appropriate in some cases.

A damp-proof membrane should be laid beneath the flooring and extended 80mm up the wall to prevent seepage.

In shower areas, either coved tiles or PVC covering at walls and upstands should be fitted.

Mats should be removed as they are a hazard for an ambulant person and obstruct a wheelchair.

Fittings

Sanitaryware must be appropriate for the user and correctly positioned for their use.

An emergency call system may be necessary.

Users with disabled hands should be provided with pull-ring fittings to pull cords for switches.

Shaver points and mirrors should be 1200mm above the floor level and accessible for the user.

Suitable soap holders and twin toilet roll holders must be accessible and at the correct height for the user.

A storage area for toilet accessories must be provided.

Grabrails

Grabrails will be needed alongside the WC, bath, shower and the basin. A front rail at the basin may be needed to prevent the basin rim being used as a support.

Rails must be positioned for the individual and placed so that accidental misuse of the heated towel rail is avoided as this could be dangerous.

Heating

The bathroom temperature should be maintained at a higher level than the rest of the home for the disabled person's comfort (see reference to Heating for the Disabled in this Section).

Only heaters approved for bathroom use should be used. They should be fixed above shoulder height to avoid the risk of burns, especially to those with impaired temperature sensation.

If there is condensation an extractor fan is recommended.

Bathing
Points to Consider Before Alteration

Investigate all possible alternatives before altering bath or shower facilities. An adaption of a preferred arrangement will often be more acceptable than a totally new one, especially to the elderly.

If access to an upstairs bathroom is a problem, consider a stairlift or through-floor lift especially if the person is able to use other rooms upstairs. This is often more economical than building an extension and a lift can be re-sited if the person moves premises.

The bathroom door can be re-hung to open outwards if equipment is bulky or if wheelchair manoeuvring space is needed.

Alternatives should be considered carefully. A bath with a low side may make getting in easier but provides insufficient support when getting out.

The bathroom floor should be checked carefully for suitability if a special bath is being considered.

Fitting a non-slip mat in the bath and teaching the person to turn on to their hands and knees after draining the bath water may enable them to get out without further aids.

A combination of bathrails, bathboard, bath seat and non-slip mat may assist many people who have difficulty getting into and out of the bath.

If a portable hoist is used there must be adequate space to manoeuvre in the bathroom. Space is needed beneath the bath to accommodate the hoist legs.

Baths and Accessories
General Considerations

Baths must comply to the British Standards specifications:

Baths for domestic purposes made from cast acrylic sheets BS 4305

Cast iron baths for domestic purposes: BS 1189.

Sheet steel baths for domestic purposes: BS1390.

Enamelled steel and cast iron baths will take boards and seats but acrylic baths are insufficiently robust to do so.

A bath seat should not be used inside an acrylic bath.

Baths are usually 1700mm or 1500mm in length and 700, 750 or 800mm wide. Longer and wider baths are available.

It is recommended that all baths have anti-slip bases. Handgrips are recommended as either built-in or fitted as optional extras.

Grabrails are fitted on some baths but these should be checked to ensure that they do not impede the disabled user getting into and out of the bath.

It is recommended that quarter turn lever taps may be the most suitable.

Corner taps may be practical for some but if fitted on the outer side, may impede getting into and out of the bath. They are unsuitable if a plastic insert is used.

The tap closest to the bath edge must be the cold tap to help the blind and for the safety of children.

A walk-in bath may suit those who are liable to fall or are afraid of falling. It is particularly practical for institutional use and may prevent the attendants hurting their backs.

The bath can be used as a sit-in bath or as a shower cubicle.

The user must be able to sit in the normal sitting position.

It is recommended that a water thermostat is fitted.

The sensation of rising water may be frightening to some elderly or confused people.

An adjustable height bath is intended for use in hospital or in residential accommodation. It saves the attendant from unnecessary bending and may help prevent back strain.

An ambulant person can step into the bath at its lowest level and be raised to a convenient height for the helper bathing him.

A person who is unable to get into or out of the bath unaided should be hoisted using the particular model the bath was designed to accommodate.

A tilting bath may reduce the need for assistance and increase the independence of some people confined to a wheelchair.

A wheelchair user should check that he can transfer easily into and out of the tilting bath, using a transfer board if necessary.

If the user cannot close the door and operate the controls of the tilting bath a helper must be available to do so.

Check that the person is not likely to suffer dizziness when changing from horizontal to vertical.

Baths with Built-in Seats

The small square bath is designed to fit into a bathroom with limited space. It can be used as a shower area by a seated person. This bath is made from cast acrylic sheet and has a slip-resistant finish. A seat is built-in and handgrips are fitted on each side.

Bath Inserts

A rigid insert may be more suitable than a bathboard combined with bath seat if the person has great difficulty in getting out of an ordinary bath or lacks balance or co-ordination.

A bath board is useful for those with difficulty getting over the side of the bath, including those with lower limb disabilities, balance or co-ordination problems or stiff joints. The person sits on the board, feet outside the bath, slides backwards on the board, turns to swing the legs over into the bath and slides to the middle of the board.

A non-slip bath should always be used where available.

Those who lack limb strength or have poor balance may prefer to sit on the bath and use an over-bath shower. The curtain will need to hang outside the bath so suitable flooring is required with a drain gully in the floor.

A seat is useful for a person who is unable to sit in the bottom of a bath because of balance, mobility or lower limb problems; strong arms are needed to lower to the seat and onto the bottom of the bath. The seat can be used alone or in conjunction with a bath board. A bath rail may be necessary. The user can sit on the seat to bath or shower if he is unable to lower himself into the bottom of the bath. Some makes have mesh or slatted seats for water drainage. Only a seat which hangs from the bath rims or stands on the bottom of the bath is suitable in an acrylic bath.

Bath Lifts

Some bath lifts can be operated independently by the bather, others are assistant-operated.

Installing a lifting device may be more cost-effective than fitting a special bath or shower. Some are suitable for installing in small bathrooms and most can be re-sited if the person leaves the house.

The disabled person or assisting helper must have sufficient arm movement and strength to operate a mechanical device effectively.

Some lifting devices require the user to sit with bent hips and straight knees. If this is a problem a bathboard and bath seat or a bath insert may be more appropriate.

Cushion-type lifting devices are simple to use. They are portable and can be used by community nurses to bath patients at home. They are unsuitable for those with poor balance; devices with rigid seats are safer.

If the device is operated electrically, a suitable power point will be required immediately outside the bathroom. Those involved in using the device must be made fully aware of the safety factors recommended by the manufacturer.

Before a floor-fixed device is installed the floor should be checked for suitability. Periodically, the user or helper should check the device for stability.

Portable patient lifters can be used in the bathroom provided adequate manoeuvring space is available and there is space under the bath for the legs of the hoist. Alternatively, a base plate can be floor-fixed at the side or end of the bath and the hoist mast fitted into it.

An electrically-operated hoist can be used in the bathroom; the ceiling joists must be inspected and their structural strength confirmed that they are adequate to carry the tracking. Slings used must allow the water to drain and it is recommended that a spare set be supplied to allow for drying.

Showering

Discussion should be given to the choice between an over-bath shower and shower cubicle. Showering does not always overcome all problems of disabled bathers. Some, particularly the elderly disabled dislike showers and prefer a strip-wash.

Grants may be available for part of the cost of installing a shower from Environmental Health Departments (for private housing) or the Local Authority (for council housing).

If new downstairs facilities are essential, consider the site of the shower with the WC and washbasin, allowing sufficient space for a wheelchair and a helper as necessary.

Heating, lighting and ventilation of the shower areas should be as

FACILITIES FOR PRIVATE USE AND NURSING HOMES

detailed previously and the installation of an emergency call system is recommended.

Shower trays designed for the disabled do not have a step-over ledge. Check the model and install a grabrail for safety by the entrance. A non-slip sloping floor area is much safer than a shower tray and is essential for wheelchair use, though it is more expensive to install.

Sloping floors must be non-slip. Non-slip mats are dangerous in such areas. Adhesive safety treads can be used.

It may be difficult to install a sloping floor area in an upstairs bathroom.

The shower curtain or door should be easy for the disabled user to manage. Some magnetic door catches on cubicles require considerable dexterity. A shower curtain must hang inside the shower unit which reduces the area that can be used.

The shower mixing valve should be temperature pre-set at 38°C and thermostatically controlled. This is especially important for those with sensory problems when the temperature must be pre-set lower to suit their medical needs.

An instantaneous electric heater must contain an anti-scald device to cope with variations in water pressure. There should be visual method of checking the water temperature for those with sensory problems. The unit must be temperature pre-set to suit the users medical needs.

The controls must be easy to operate and within safe reach of the user or helper. It is recommended that the shower be used for hairwashing and conveniently sited for this application.

A shower head fitted to a flexible tube with wall-mountings at alternative heights or a shower head with a sliding bracket on a wall-mounted tube are recommended in preference to a fixed spray position.

A tray for soap and sponges should be accessible to the user.

Floor drain gratings should be flush with the floor. The grating material should be nickel bronze or stainless steel.

Bidets

A bidet is recommended for independent personal cleansing. The installation must conform to the Water Byelaws.

The seat height of the bidet must be suitable for the disabled ambulant person; special models with a higher seat are available. Grabrails will be necessary.

For a wheelchair user, the seat height must be the same as the wheelchair seat. The user can sit facing forwards or backwards on the bidet as convenient.

Urinals

Specially adapted wheelchairs, cushions or seat cushions may help those using urinals. Generally the wheelchair user will not be able to use wall hung or slab urinal.

Grabrails

Grabrails are fitted in the bathroom to provide support to the disabled user when washing at the basin, getting into and out of the bath and on and off the WC. They should be positioned so that the washbasin, toilet roll holder or towel rail are never used for support through habit or emergency.

Rails should be positioned to suit the individual. Standard layouts and combination may not be satisfactory.

Walls and partition walls must be of sufficient structural strength to support the load exerted on the rail. Reinforcement may be necessary.

Installation should be carried out by a qualified person.

Grabrails fitted in a bathroom or WC must have an anti-slip finish.

Hinged grabrails should be checked regularly to ensure that the hinge is in good working order.

Disposal of Incontinence and Sanitary Towels

For reasons of hygiene and to prohibit the spread of disease by contamination, macerators or incinerators should be installed instead of bins. Maintenance personnel must be protected against contracting diseases; incinerators are therefore preferable. The use of macerators are only acceptable providing maintenance personnel are given training and protection against AIDS and other related diseases.

Drainage

If no disposal unit is made available, the incontinence and sanitary towels will be placed into the WC pan for disposal as no other facility is provided. This is not good practice. The designer should however design the drainage system to accommodate these items when the disposal units are not functioning.

All changes of direction should be designed with long radius easy bends and sufficient access points, readily available in practical positions for use by maintenance personnel.

Ensure that drainage gradients are not minimum but generous for self cleansing velocity of the solids from the drain. Before using reduced water capacity flushing units, calculate the volume and discharge rates required to keep the drainage system from the toilet areas used by the disabled community clean of soiled items.

ADVANCED WC CUBICLE DESIGN FOR WHEELCHAIR DISABLED

This uses the recommended guidelines for the size of the WC compartment and facilities. The WC compartment is illuminated by an automatic light incorporating an infra red device which switches "on" when a disabled person enters the room and only switches "off" after the person has left the room.

The 'PSD' which flushes automatically, is located to one side of the room, for maximum use of the area, opposite the door to allow a lateral transfer from the wheelchair to the WC. The toilet flushes by a programmed remote sensor after the disabled person has completed their use.

The tap and hand dryer are operated by automatic sensors. The sanitary fittings are placed within easy reach of the disabled person whilst they are seated on the WC pan.

Support rails of the standard size are placed in the positions shown in Figure I4. The whole cubicle has a fail safe mechanism.

CENTRAL HEATING FOR THE DISABLED

Design Criteria

Care must be exercised when designing and installing heating systems for the disabled to ensure that all components requiring adjustment or attention by the occupant are readily accessible and can be easily operated or adjusted by the disabled user. All components must be selected and designed to ensure that no injury or harm can result to the disabled user during normal use or due to possible foreseeable accidents. The following represent the main criteria to consider.

Special consideration may be required with regard to the design internal temperature. This is particularly important for those users who do not move frequently and would therefore require a warmer environment. Each installation must be considered individually to determine the required temperature for the user. In some instances, higher than standard design ventilation rates will be required, these could be provided either by natural ventilation ie. openable windows or mechanical ventilation. In those instances, the additional heat load must be included within the design of the heating system.

The standard design heating flow and return temperatures of around 82°C and 71°C may result in heat emitter surface temperatures which could cause harm to users who do not have complete sensitivity to heat. In these instances it will be necessary to either decrease the water temperatures or select heat emitters such as convectors or low surface temperature radiators. If water temperatures are decreased it is imperative that heat output from the emitters are corrected. It may also be necessary to conceal or guard pipework to protect the user from the possibility of burns.

The system should be designed to ensure that all air and hydrogen dispels naturally through the open vent to avoid the necessity to vent radiators. Correct pipework gradients are therefore essential.

Boilers

Boilers should be fully automatic in operation, requiring the minimum of adjustment by the user. They should be situated to ensure that all controls or components requiring adjustment or attention by the user are at such a height to facilitate easy access and that no obstructions, such as return edges of walls or kitchen cabinets, are placed close to the boiler to restrict or prevent the user gaining access to the controls. Doors or flaps concealing the controls must be provided with an adequate method to enable the user to open and close them easily. Control knobs and levers must be smooth in operation and of a design to enable the user to operate them easily. The controls must be so situated to ensure that the user could not be injured from the hot surfaces in the boiler. Boilers with sharp edges must be avoided.

Solid Fuel

It is recommended that where possible, solid fuel boilers should not be provided for disabled users. They must only be considered when the user would be capable of transferring fuel, removing ashes, lighting and stoking the fire. The associated dangers of this type of boiler must not be overlooked particularly the necessity of the user coming into close contact with flames, heat and hot fuel and ashes. Where solid fuel boilers are used it is recommended that the fuel feed should be hopper fed requiring the minimum of attention. The bunker itself may require special consideration to ensure that the fuel can be readily racked to the outlet; it may be necessary to raise the bunker to provide adequate access. Consideration should be given to providing a method of ash removal direct to outside so that it is not necessary for the user to carry hot ashes through the premises.

Gas/Oil Fired

These are generally more reliable, require less attention and are easier to operate than solid fuel boilers. The ignition system should be of the fully automatic self-igniting type to avoid the necessity of re-lighting the pilot in the event of failure. If this cannot be provided then a pilot having piezo or electronic ignition should be used. Pilots requiring manual ignition by flame should be avoided. The fuel can be oil, natural or liquid petroleum gas. If banks of liquid petroleum gas bottles are provided having standby or changeover provision, the valving arrangement must be readily accessible and easily operated by the user. The main isolating gas or oil cock must also be readily accessible for use in an emergency.

Off-Peak Electric Storage Heaters

Off-peak electric heating is probably the safest and most easily operated method of heating for the severely disabled user. The cost of off-peak electricity may be more expensive than other fuels but the installation and maintenance costs are generally much lower for storage heaters than other forms of heating. All these factors must be taken into consideration to justify the installation if safety is not the prime consideration. The control of heat output from storage heaters is not as precise as other methods of heating and may cause overheating or underheating at times particularly during autumn and spring when external temperatures vary considerably from day to day.

Automatic Controls

The automatic control system should be simple to operate and reliable in service. Room thermostats and thermostatic radiator valves should be selected with a method of adjustment that can be readily altered by the user. Special consideration may be necessary with regard to the mounting height and position of the thermostat. The temperature range of the thermostat must be adequate to ensure that a comfortable environment can be maintained. This is particularly important for those users who do not move frequently and therefore require a higher room temperature. The user would not normally adjust the domestic hot water cylinder thermostat and therefore no special provisions are required. The initial setting of the cylinder thermostat must receive special consideration to ensure that the stored water is maintained at a temperature that will not cause harm to the user, particularly those who do not have complete sensitivity to heat. Consideration should be given to using digital time switches which many users would find easier to adjust than the electro-mechanical type.

Heat Emitters

Consideration must be given to selecting radiators and other heat emitters without sharp edges or protrusions on which the user could be injured due to an accidental knock or fall. The protrusions could include radiator valves, air vent and plugged connections. Radiators and convectors are available with the facility to conceal angle valves or with bottom or side connections which could be used in conjunction with straight valves. Alternatively, some users would find top and bottom opposite end connections more desirable with the isolating valve or thermostatic valve at the top connection to prevent the necessity to bend down for adjustment. The radiator valves used should have an easy and smooth action to enable the user to isolate the radiator if required. As previously mentioned, it may be necessary to ensure that the heat emitters have a low surface temperature to prevent the possibility of the user receiving burns.

Where it is known that the occupant will require the aid of a wheelchair, it is essential to ensure that the heat emitters and pipework are installed in such a manner that damage is unlikely to result to the services in the event of an accidental collision. It may be necessary to provide a protective guard around the services.

Where possible, heat emitters should be positioned under windows to reduce cold down draughts and the cold radiant effect. Where this is not possible they should be installed adjacent to the windows. Radiators should also be positioned adjacent to external doors to reduce the effect of cold draughts.

REFERENCES

The Building Regulations 1985
HMSO

Spaces in the Home, Bathrooms and WC's
HMSO

Disabled Living Foundation
380-384 Harrow Road
London, W9 2HO
Tel: 01-289 6111

Equipment for the Disabled
Mary Marlborough Lodge
Nuffield Orthopaedic Centre
Headington
Oxford, OX3 7LD
Tel: 0865 750103

Plumbing Services for the Disabled Community
Cherry Tree Offices
Home Farm
Chislehurst
Kent
Tel: 01-468 7767

The Royal Association for Disability and Rehabilitation (RADAR)
25 Mortimer Street
London, W1N 8AB
Tel: 01-637 5400

Equipment Manufacturers:
Armitage Shanks Ltd
Armitage
Rugeley
Staffs, WS15 4BT
Tel: 0543 490253

Ifo Products (UK) Ltd
c/o The Geoff Marsh Partnership
12 Appletons
Hadlow
Tonbridge
Kent, TN11 0DT
Tel: 0732 850744

Meynell Valves Ltd
Bushbury
Wolverhampton, WV10 9LB
Tel: 0902 28621

BRITISH STANDARDS AND CODES OF PRACTICE

BS 5619:1978 (1981)	Code of Practice for design of housing for the convenience of disabled people.
BS 5810:1979	Code of Practice for access for the disabled to buildings.

Note

(1) Also refer to the list of British Standards and Codes of Practice at the end of Section B (Domestic Central Heating)

(2) Year after the Standard's number indicates the latest date of issue.

(3) Year in brackets is the date of the latest amendment at the time of publication of this Guide.

ENGINEERING IMPLICATIONS

It has long been known that hepatitis A, salmonellosis, dysentery and other gastrointestinal infections may result from drinking water from contaminated potable supplies. Fortunately, it is a rare event for potable supplies to be contaminated in this country but it is a potential hazard, particularly in all those hospitals still supplying drinking water from their own reservoirs, wells or holding tanks.

In recent years it has been recognised that cooling waters and domestic water systems in hospitals may also serve as sources of certain infections.[1,2] Legionnaires' disease, caused by the bacterium Legionella pneumophila, is one such infection which occurs both in outbreaks and endemically in hospitals. Hot water systems have been shown to be the most important source, with the infection being acquired not by ingestion but through the inhalation of water droplets. Other organisms such as Acinetobacter lwoffii, Micrococcus varians and Pseudomonas stutzeri may also colonise hospital domestic hot water systems and produce disease in patients whose normal defences are weakened by illness or treatment.[3]

In general, the prevention of such infections from hospital water systems depends on the maintenance of good engineering housekeeping practices. The aim of this Section is to set out in some detail those steps which might be taken when a hospital outbreak occurs and a water system is identified as the source by epidemiological and microbiological studies. These proposed control measures were drawn up in consultation with microbioligists and epidemiologists and are based largely on experience gathered in the control of the outbreak of Legionnaires' disease in Kingston District General Hospital in 1980[4,5] and several outbreaks in hotels both in the United Kingdom and overseas.

Hot water systems in hospitals and hotels have been shown to be the most important source of epidemic Legionnaires' disease in the United Kingdom and probably elsewhere in Europe.[2,6] The infection is acquired from the inhalation of the causative bacterium Legionella pneumophila carried in small water droplets produced by running taps and showers. Fortunately this appears to be a rare infection and most people appear not to be susceptible.

Surveys have shown that Legionella pneumophila and related bacteria are found very commonly in hot and cold water systems as well as in cooling waters.[7,8] It is not clear yet which factors are associated with the development of a high concentration of the bacterium in water, although temperature does seem to be important. Colonisation of Legionella pneumophila is most likely to occur in water systems which are stagnant and have a temperature between 20°C and 50°C. The most critical temperature is 36°C. Every effort should be made to avoid stagnant water conditions and to store and supply water outside these critical temperatures. The bacterium grow in systems supplied with mains water which is chlorinated to 0.5ppm but is susceptible to higher concentrations of free residual chlorine.

As contaminated aerosols produced from cooling towers, shower heads, taps and similar apparatus are thought to be the vehicles of infection, special care should be taken in the maintenance of this equipment, in its cleaning, sterilization, temperature control, and testing.

OUTBREAK CONTROL

Duties of the Maintenance Department

Once the Control of Infection Team have decided to control the growth of Legionella pneumophila in the hospital water system, because of epidemic Legionnaires' disease, the steps below should be taken by the engineers in the Maintenance Department.

Their permanent duties are needed to eliminate the recurrence of Legionnaires' disease in buildings and are summarized as follows:

(a) All cold water supplies (other than drinking water) and water storage should be chlorinated to ensure that water delivery to every cold water outlet in the building contains a constant chlorine density of between 1–2ppm free residual chlorine.

(b) All cold water storage tanks should be regularly inspected, maintained and sterilized in the manner specified in this Section.

(c) All hot water supplies and hot water storage cylinders (calorifiers) should have all the water controlled within the temperature limits of between 55–60°C at the tap outlets. Because there is a small risk of scalding at these temperatures, a warning notice (see Figure J1) should be displayed at each hot water outlet.

Figure J1 Hot water warning notice

**BEWARE
Very hot water
coming
from the tap**

(d) All apparatus likely to produce contaminated aerosols such as humidifiers, cooling towers etc., should be scheduled and a detailed system of cleaning and disinfecting by chlorination carried out at regular intervals.

(e) Engineers should carry out regular testing and inspection of hot and cold water outlets to standards set out above. All such testing should be recorded in an agreed manner.

(f) In conjunction with other staff, such as administrators, production personnel, senior officers, cleaners etc., a system should be initiated which ensures that any department, rooms and areas in the building, if left unoccupied for a week or more, should be thoroughly tested before bringing back into service.

(g) Special attention to steps e) and f) should be observed in respect of showers and spray taps. For drinking water, all drinking outlets in the building should have been checked that they are connected to the drinking water mains and have been clearly labelled. All drinking water mains in the building must be connected directly to the mains water to ensure that they are not fed through any storage or tank system. It has been determined in the United Kingdom that no chlorination or form of sterilization is required for existing drinking water mains or outlets except for the initial sterilization in the commissioning of a new building or pipework. It is important that when requests for additional drinking water outlets are received, the maintenance staff should check that connections are in fact made to drinking water mains and that the outlets are labelled as such. All new pipework should be sterilized before being put into use. (In the National Health Service this is recommended in HTM.27).

Cold Water Storage Systems and Tanks

When necessary, modifications to pipe and storage systems should take place so that only one unit for continuous chlorination need be installed.

OUTBREAK CONTROL

Wherever possible, the cold water storage system and storage tanks should store water at below 20°C. Engineers should eliminate the possibility of any conditions which produce abnormally high temperature rises. This can be done by painting the top of the tanks with reflective paint, spraying the top of the tank with cold water from a mains source, lagging around the tanks themselves or lagging the roof over the top of the tank. A daily record should be kept of the temperature of the water nearest to the top of the tank.

It is essential that cold water storage systems and all storage tanks should be thoroughly cleaned out at least once per year. Cleaning should take place in the first week of April to precede the birds' nesting season and during this period, all covers and fittings should be checked and renewed as necessary, to include making arrangements to prevent birds etc. entering the overflow pipe. This means installing wire netting or a wire balloon over the outlet to the tank overflow.

The cold water storage system is sometimes associated with pressurized vessels and storage tanks, which should be heavily chlorinated initially to 20ppm free residual chlorine (FRC) and held in this state for a minimum period of two hours. All equipment should then be drained to waste and the system refilled. A continuous system of chlorine injection can be constantly maintained and should provide a concentration of between 1 and 2ppm free residual chlorine at every cold water outlet. An official log book shall be maintained by plant engineers and the reading of chlorine and temperature present in the cold water storage system tested and recorded daily. At least once a week the engineer should examine and sign the log book and in the event of the chlorine level falling below 1ppm, the system should be inspected and the fault rectified.

Testing should be carried out using a comparator tester and the engineer should ensure that all members of his staff are trained to carry out this test efficiently.

In addition to testing for chlorine presence, a test should be made for free ammonia in the mains supply water prior to site chlorination. This test should be carried out daily using a Hessler test instrument on the mains tap water and if a colour change to brown is recorded, the engineer should then contact the local water authority and request a lower density of free ammonia be inserted into the water system.

Where applicable, tanks and lids are to be WRc approved and are to comply with Byelaw 30. Lids made from wood, clear plastic or polystyrene must not be used.

Continuous Chlorination Plant

The continuous chlorination plant should be located in a suitable position and this should be permanent. The use of chlorine gas is potentially hazardous so it is recommended that sodium hypochlorite solution be used for chlorination. The engineer should arrange for the inclusion in his Planned Preventive Maintenance scheme of daily inspections of the chlorination plant. The engineer should also order and obtain all spare parts, chemicals etc. necessary to ensure that the equipment can be rapidly returned to service in the event of a breakdown.

Note

Sodium hypochlorite solution should be stored in a cool place and could lose its strength if stored in excess of one month.

Thorough Testing of all Outlets

It is recommended that during the first week of April in each year, a thorough test of every hot and cold water outlet, including all thermostatic mixing valves, etc. takes place and a suitable log retained. In the case of cold water outlets, these should indicate between 1-2ppm free residual chlorine strength within two minutes of running to waste. In the case of hot water outlets, these should indicate temperatures of between 55-60°C within two minutes of running to waste. All thermostatic mixing valves, shower heads and blender spray taps etc. should be tested by first running hot water to waste without recording temperatures for a minimum of one minute, then running cold to waste, when a chlorine strength of between 1-2ppm free residual chlorine should be obtained.

Recording of test results must be carried out by the supervising engineer to the approved schedule. Any outlet which fails these test conditions should be recorded and details submitted to the engineer who will rectify the fault.

In addition to the testing proposed above, more frequent testing of selected outlets, both hot and cold, should be carried out in a similar manner. Chlorine levels and hot water temperatures should be tested once each week to maintain observation on the current situation.

As before, the recording of test results should be carried out by the supervising engineer to an approved schedule of outlets, determined by a responsible body of people. Any outlets which fail the test conditions must be recorded and brought to the attention of the engineer.

Water Softening and Hot Water Storage Systems

If the site has a main central storage cistern, this should be modified so that if possible, one compartment is allocated to contain soft water only for the hot water supply system.

If necessary, existing pipework should be altered so that mains water is connected directly to the water softening plant and then to the continuous chlorine injection unit to provide the softened water with chlorine to approximately 1-2ppm at the outlets. The softened and chlorinated water should be delivered to the appropriate compartment of the central storage cistern prior to distribution to calorifiers. The softened water storage compartment should be cleaned and chlorinated to 20ppm every year in the first week of April in a similar manner to that specified for the raw water compartments. In addition, all associated pipework, pumps and equipment which can be segregated from hot water calorifiers etc., should be similarly chlorinated each year.

Note

The water softener unit itself should NOT be subjected to chlorination of any kind as it damages the resin bed.

At each of the calorifiers, the water should be stored at about 60°C and suitable thermometers installed in flow and return pipework to aim for a maximum temperature deviation of 5°C.

Where these conditions cannot be obtained, a schedule of calorifiers should be prepared, clearly indicating discrepancies and their effect on hot water outlet temperatures and the faults then rectified.

Stagnant Water Positions – Occupied

An inspection should be carried out to determine which shower positions controlled by thermostatic mixing valves, hot and cold taps etc., are infrequently used. A decision should then be taken as to which of these can be dismantled and removed, including all associated pipework.

All shower positions in the building should be regularly run to waste weekly, whether used frequently or not.

Each shower fitting should be run for five minutes, during which in the first two minutes hot water should be passed, and for the remaining time, cold chlorinated water. This would be under the supervision of a nominated person in each department, and a list of the staff nominated kept by the Administrator.

All shower mixer units should have 'drain-out' valves fitted.

Stagnant Water Positions – Unoccupied

From time to time, cases arise where departments or individual rooms in the building are left unoccupied for various reasons. This may be the result of staff shortages, changes in management policy or for general upgrading purposes. It is essential that a policy be advised so that such accommodation cannot be returned to general use until a full test of hot and cold water outlets has been carried out by the engineering staff.

It is suggested that the administrator organises a 'Permit for Re-occupation' to the accommodation, based on a clearance certificate signed by the various officers concerned. The engineer will be required to certify that all hot and cold water outlets, including thermostatic outlets have been tested in the manner previously specified.

Domestic Hot Water Calorifiers

Where Legionella pneumophila has been identified, it is recommended that the calorifiers are thoroughly cleansed and the following guide-lines be followed:

(1) Isolate the calorifier from the system. If more than one is involved, select the one that is most affected.

(2) Do not disturb the other calorifiers, ie. putting them on or off line.

(3) To make the calorifiers safe, one of the following may be done:
(a) superchlorination with sodium hypochlorite, or
(b) pasteurization.

(4) (a) To achieve 3(a), attach a clean PVC hosepipe to the draincock and make sure the end of the hose is right in the drain, (this is to prevent the inhalation of infected areosol), and drain off as much water as is required to make room for the hypochlorite solution to be added. This should be measured according to the amount of water in the calorifier to achieve 20-30ppm free residual chlorine and should stand for a minimum of 24 hours or longer if possible, then let the water drain. Proceed as No. 5.

(b) To achieve 3(b) make sure the domestic hot water flow valve is closed (DHWF) and leave the remaining valves open, ie. the domestic hot water return valve (DHWR) and the cold water supply make-up valve (CWS). The primary heating source, ie. steam flow and return must be closed. Drill small holes approx 15mm dia. into the lagging at points over the top and bottom of the calorifier so that a surface temperature probe can be inserted to the metal of the calorifier casing.

The recommended precautions must be taken if the lagging is made from asbestos or an asbestos based material.

Remove or disconnect any 'over temperature control stats'. Raise the temperature of the water in the calorifier by turning on the steam so that a temperature of 70°C is reached all over the casing of the calorifiers, then let the calorifier stand for a minimum of 24 hours or longer if possible. This is to enable the heat to penetrate through the scale and sludge and kill off any legionellae. It will be noticed that the temperature will fall very slowly owing to the lagging. Drain to waste, as before, with the hosepipe inserted right into the drain.

(5) Remove primary heating coil(s) for insurance requirements and thoroughly and mechanically clean out.

(6) Reassemble all heating coils etc. in the calorifier and fill up with water. Make sure that the domestic hot water flow valve REMAINS CLOSED.

(7) Reconnect the high limit stats and turn the heat back on, the temperature control to be set at approximately 60°C.

(8) Let the calorifier stand for a minimum of 24 hours or longer with the heat on and the header valve still closed. This is to let the sediment in the calorifier settle to the bottom.

(9) When the calorifier is to go 'on line', the heater valve should be opened very, very slowly. This is to reduce the amount of turbulence in the calorifier to stop any residual sludge being stirred up and pumped through the system.

(10) The hot water coming out of the tap outlets should be at a temperature between 55 and 60°C at the furthest point from the calorifier which could mean that the hot water storage temperature could be higher than 60°C.

(11) If bacterial growth can still be found at the tap and shower outlets, the washers should be changed on these terminal fittings to a type approved by the WRc[9].

Cooling Towers

If it is necessary to disconnect or remove air conditioning cooling towers, either temporarily or permanently, the following procedure should be followed:

(a) If required by the microbiology departments, a sample of water and sludge should be taken from the cooling tower under the supervision of the engineer, with staff taking full safety precautions.

(b) Add chlorine to cooling tower water to provide a density of at least 15ppm, depending on the size of the tower and amount of sludge.

(c) Circulate cooling tower water for a minimum of two hours with the fan off.

(d) Extract a further sample at the end of this period.

(e) Drain-off water. Refill, recirculate and drain-off.

(f) Clean inner surface of cooling tower by hosing down and remove all internal scale, sludge and fitments for cleaning.

(g) Refill system and add chlorine to at least 4-10ppm.

(h) Circulate for two hours.

(i) Drain-off.

(j) Refill tank and put back into operation.

(k) Ensure chemical and injection pumps are working correctly to control scale and algae, etc.

Humidifiers

There are a number of different types of humidifiers, ie. steam, battery-spray, spinning disc, and simple 'pouring water' humidifiers. These are normally found in the heating and ventilation systems of big office blocks, computer rooms, and sometimes in hospitals, particularly in operating theatre areas.

Up until now, none of these humidification systems has been implicated as the source of legionella infection.

Approved Materials

All terminal fittings and other materials in contact with potable water installations must be of an approved type and as listed in the Water Research Centre's Fittings Directory.

REFERENCES

1. Broome CV. State of the Art Lecture: Current Issues in Epidemiology of Legionellosis, 1983. 1984, Legionella: Proceedings of the 2nd International Symposium., American Society of Microbiology, Washington, D.C.
2. Bartlett CLR. State of Art Lecture: Potable water as Reservoir and Means of Transmission. 1984, Legionella: Proceedings of the 2nd International Symposium., American Society of Microbiology, Washington, D.C.
3. Meers PD. Bacterial Multiplication in Hospital Water. Proceedings of Symposium on Hospital Water Supplies., University of Technology, Loughborough 1981.
4. Fisher-Hoch SP, Bartlett CLR, Tobin JOH, Gillett MB, Nelson AM, Pritchard JE, Smith MG, Swann RA, Talbot JM, Thomas JA. Investigation and Control of an Outbreak of Legionnaires' Disease in a District General Hospital. Lancet 1981; 932-936.
5. Fisher-Hoch SP, Smith MG, Colbourne JS. Legionella pneumophila in Hospital Water Cylinders.
6. Bartlett CLR, Bibby LF. Epidemic Legionellosis in England and Wales 1979-1982. Zbl Bakt I Abt Orig A 1983; 255:64-70.
7. Tobin JOH, Swann RA, Bartlett CLR. Isolation of Legionella Pneumophila from Water Systems: Methods and Preliminary Results. BMJ 1981; 282:515-517.
8. Dennis PJ, Taylor JA, Fitzgeorge RB, Bartlett CLR, Barrow GI. Legionella Pneumophila in Water Plumbing Systems. Lancet 1982; i:949-951.
9. Colbourne JS, Pratt DJ, Smith MG, Fisher-Hoch SP, Harper D. Sources of Legionella Pneumophila in a Hospital Plumbing. Lancet 1984; i:210-13.

DESIGN CONSIDERATIONS

To design the hydraulic system for a unit with moving, active water, a very firm idea must first be developed as to how the unit is to look in order to achieve the appearance the client has specified.

Variation or types of water display need not necessarily be separated from each other; there can be mixed forms and combinations of the different types. To utilize these combinations effectively, one needs to be aware of the differences.

Every water feature displays its optical and acoustical appeal against the background of its environment, be it an element of the garden, a park or town landscape. The designer must take this into account.

Taking note of ten basic rules will help the designer to avoid the bad mistakes which are repeatedly found in practice. If there is creative fantasy in a confident inter-relationship with the engineering, very attractive, enduring fountain units or water displays are a certainty.

The Ten Most Important Rules

(1) Not under trees.
Fallen leaves, insects and bird droppings contaminate the water and can clog the suction filters in the pump circuit.

(2) Not in positions exposed to high winds.
Unfavourable wind conditions disturb the desired water pattern and can take the water in the fountains over the edge of the basin. The points of water columns/jets should not be greater than the distance to the nearest edge of the basin.

(3) Water-saving basin depth.
To reduce water use during changes required for cleaning the basin and to prevent accidents to children, the basin and thus the water depth should be kept as shallow as possible.

(4) Protection by cover gratings
The cavities in the fountain basin which are necessary for accommodating the equipment should be covered by gratings.

(5) Overflow/outlet to the drain.
For easy emptying and cleaning the fountain basin, an overflow/outlet to the drainage should always be provided.

(6) Freshwater supply.
A permanently installed freshwater pipe with automatic topping up and pump dry-running device should always be provided, especially for public installations.

(7) Securing the power supply cables.
Power supply cables should be installed in cable ducts in the fountain basin; this being essential for public installations.

(8) Selecting the correct pump.
Fountain pumps are subject to heavy endurance requirements and must therefore be specially prepared for their intended application.

(9) Selecting the correct electrical switchgear.
The electrical switchgear must be matched to the pumps and lighting circuits and be provided with the special protective and safety equipment for fountain installations.

(10) Protection during winter.
Precautionary measures for the winter should be taken at the planning stage. Above all, it must be possible to protect the pumps from frost, by removal if necessary.

Construction Principles and Structural Components

The construction principles of fountain technology arise from the desired objectives which always involve putting water into motion in order to produce beautiful, eye-catching water patterns. To achieve this, adequate quantities of water must be available. In the ideal case, this will be a natural water reservoir in the form of a pond or lake. Usually however, a water reservoir must be constructed artificially and be filled from public mains water. Larger basins are mostly of concrete or brick but there are also ready made basins of considerable size made from GRP or similar synthetics.

Lakes and ponds are mostly modelled as more or less deep hollows in the ground and sealed with a membrane of bitumen or clay.

Once the water reservoir is settled, the movement of the water and the water display must be considered.

The water movement can consist of a single, hardly perceptible flow of water; flowing water in the form of a stream; a spring; a multi-jet fountain; a transparent water bell; a powerful high fountain or a rushing waterfall.

If the designer has decided on a certain water pattern and therefore on a specific nozzle or nozzle combination, the selection of the 'driving force', the fountain pump then follows.

The simplest solution is to use a submersible motor pump directly in the basin. The most important criteria for selecting the pump is its power and efficiency. How much water can it deliver, what is the water discharge period and at what height? What is the corrosion resistance and endurance characteristics?

With these three main elements, the water reservoir, fountain attachment and pump, a fully serviceable fountain installation can be provided. It is important with any fountain installation, that these three main elements have an appropriate and harmonious relationship to each other. The size and form of the water pattern should be adapted to the shape and size of the basin.

Types of Fountains and Water Displays

The design options for fountains and water displays are multitudinous with practically no limits placed on the artist or designer. However, three main areas can easily be defined. On one, there are installations in which the element water plays the dominating role and the practical, sculptural design follows behind. In another, there are those in which the artistic design and formation of the fountain is of central importance and the water has only a supporting function. The third and main area is formed by an installation in which the water and the known design play an almost equal role and mutually reinforces each other in its effect.

The more individually fountains and water displays are designed, the more important it is in the planning to consider the hydraulics. Undercalculated water volumes or incorrectly dimensioned components and leads make the installation defective for its whole life.

Previously, fountains and water displays were supplied and operated from springs or a drinking water system and once the water passed through the system, it was lost. Almost exclusively today, recirculating systems are employed. Once a basin is filled, the water is used over a longer period and recirculated again and again by means of the system pump. Only after it is contaminated is the water drained off, the fountain unit cleaned and refilled. The intervals between changes can be substantially extended by using a filtration system. Water loss due to evaporation or splash is usually replaced automatically from the drinking water make-up system.

The pumps used are either submersible motor pumps or dry mounted centrifugal pumps which are installed in a pump chamber or in a cellar.

DESIGN CONSIDERATIONS

Figure K1 A simple fountain unit using a submersible pump.

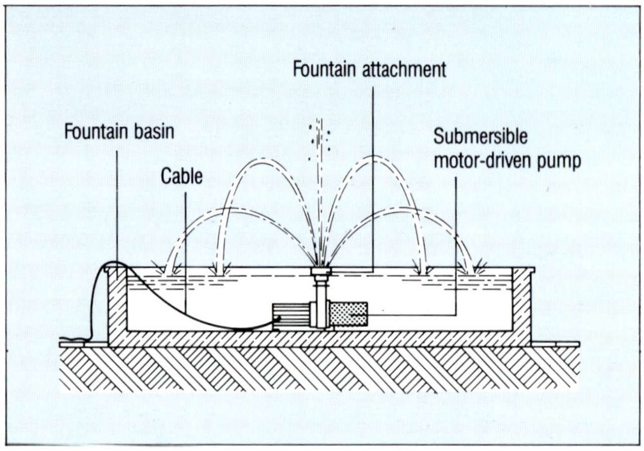

Figure K2 Fountain installation incorporating a dry-mounted centrifugal pump

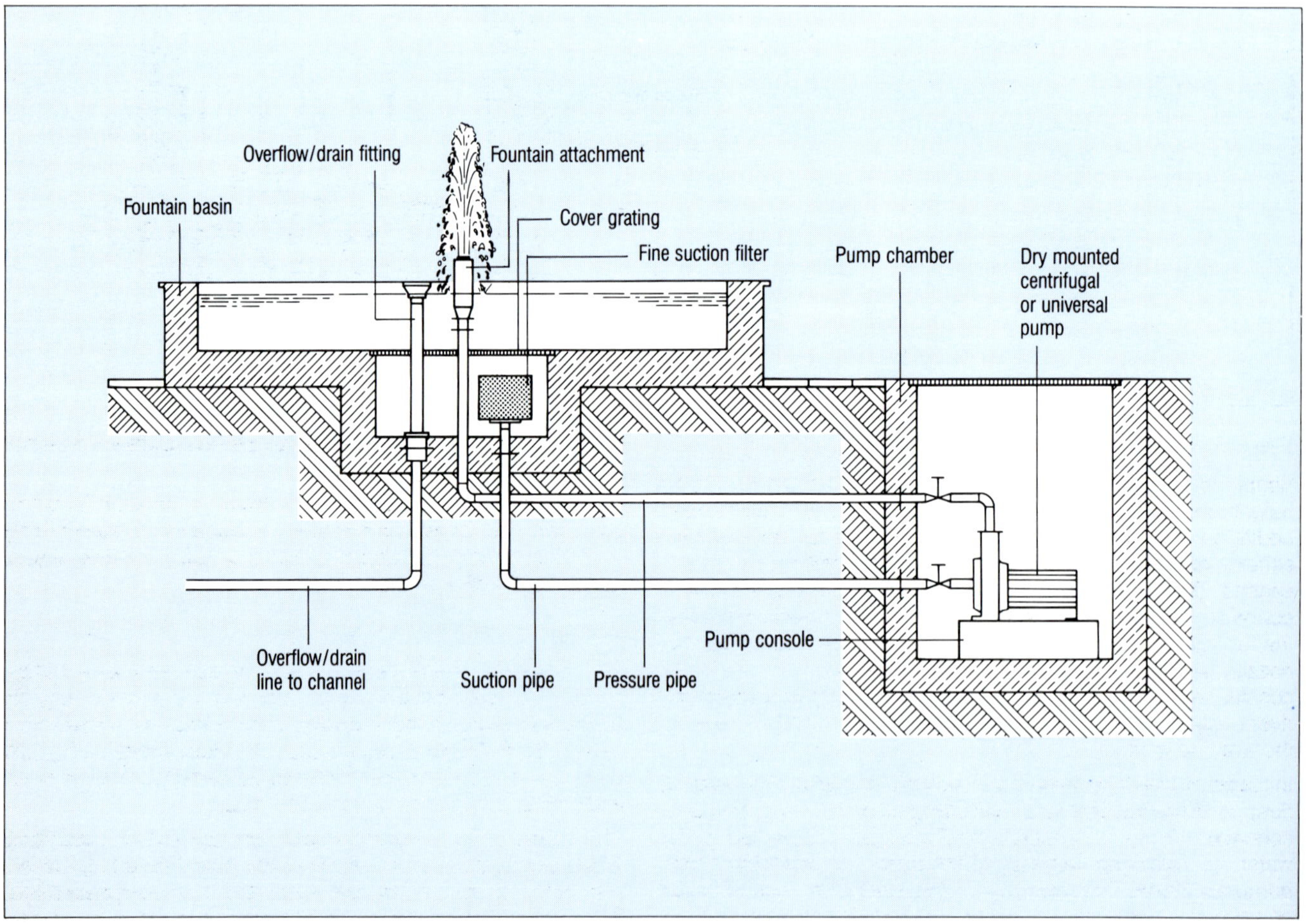

Figure K3 Fountain installation with an open underground reservoir

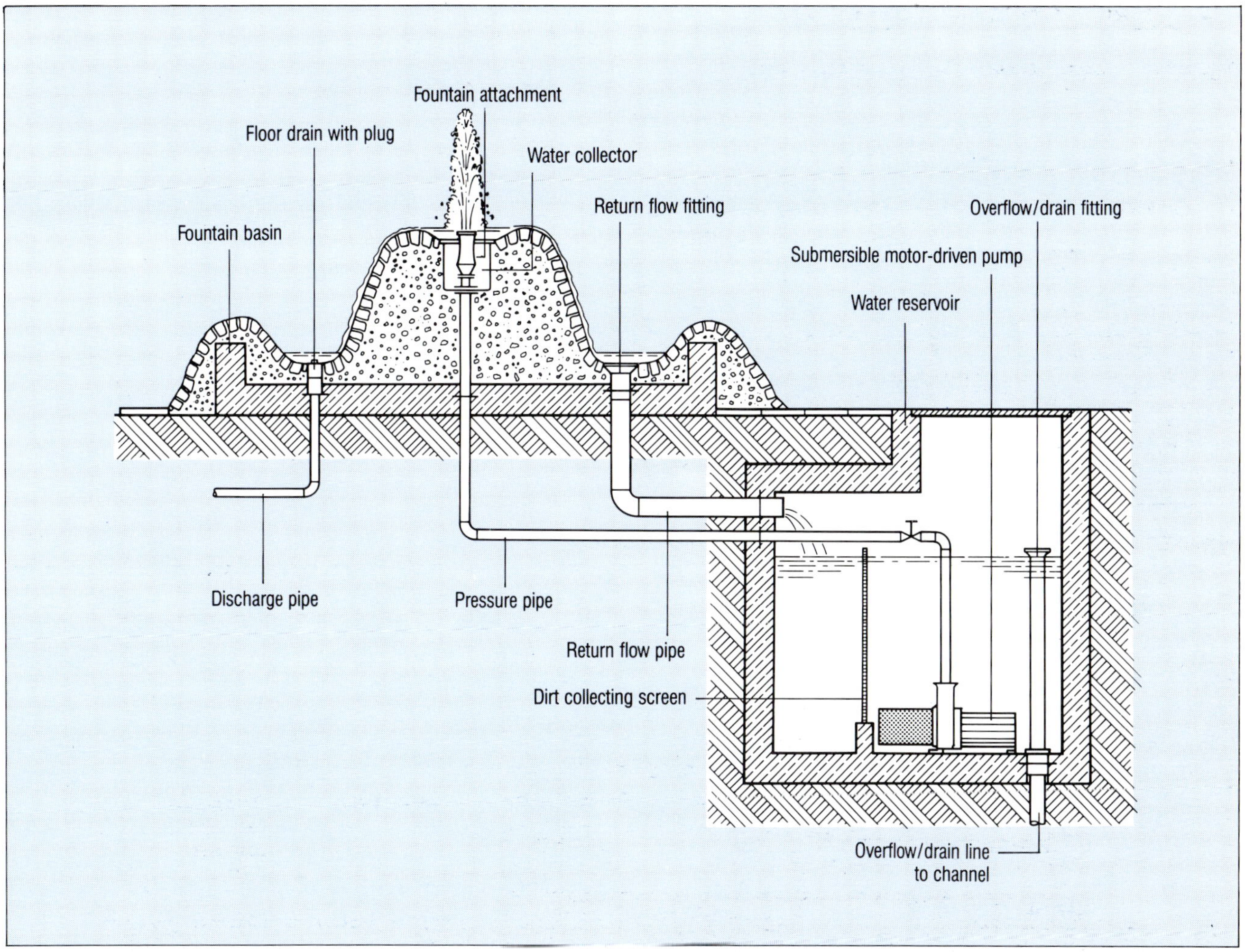

Fountain Installations

Simple fountains are generally single water columns to varying heights with a column diameter of the appropriate thickness. In addition, there are groups of nozzles which can be combined either in one multi-jet head or in several individual nozzles widely spaced. The height and 'throw' radius of the jets require a corresponding large water area and depth. The general rule is that the fountain should not be higher than the distance between the nozzle and the nearest edge of the basin. An artificially constructed basin, a pond or lake, a dammed-up and unmoving river course, a lagoon or even a quiet ocean bay can be used as the water reservoir.

In cases where water levels are not constant or where the water is deep, floating installations are usually constructed. In other cases, concrete columns or concrete platforms can be erected in the water for mounting fountain attachments, submersible motor pumps and lights. Depending on the requirements, 'special' full jet or a hollow jet nozzles are suitable as fountain attachments.

The supply to the nozzle attachment or group of nozzle attachments is made either with submersible motor pumps directly in the water or, with a pressured pipe and a pump, dry-mounted in a sump or pit close to the bank. In the case of very high fountain heights which require extra large motor outputs, special transformer stations are often necessary adjacent to the installation.

To enable the fountain height to be adapted to the prevailing winds, automatic controls are available.

Because of the special problems connected with high fountains, the advice, planning and services of experienced specialists is essential. Some manufacturers produce nozzles and attachments for jet heights of over 100 metres.

Watercourse Installations

Watercourse installations are usually copies from nature, but are sometimes quite conciously conceived as 'concrete stepped' installations. The principle is always that water has to flow from a source at a higher level over a more or less long and winding water course, to a lake, pool or water basin. An important aspect in this is the enrichment of the water with oxygen.

The widths of the watercourse can vary considerably and at its widest points, a uniformly thin film water flow should be provided. The result is that, at the narrowest points of the watercourse, thicker sheets of water form and the side edges must be shaded deeper. In addition to the watercourse width, the geodetic delivery head (difference in the height between the lower water level and the upper supply point of the source) is of decisive importance for the design of the pump. The pump forces the water from the lower water reservoir, through a pressure pipe to the upper source. The supply can take place almost imperceptibly as a still source by means of foam-effect nozzles.

THE ART OF WATER DISPLAY

Waterfall Arrangements

A curtain of water tumbling down to form a waterfall can have varying widths and heights. As the thickness of the water film depends on these two factors, so will the quantity of water that the pump is required to circulate.

Water is pumped by a concealed unit from a lower basin, through one or more pressure lines, into a trough with a shaped overflow, the width and height of which determine the pattern of the waterfall. Most arrangements require an adjustable lip at the beginning of the overflow and a 'weir' edge on the other side. With wide waterfalls, the lip should be adjusted perfectly level to prevent the water film from breaking up. It is also important that leaves and dirt do not collect on either the 'fall-away' or lip. This can be prevented by means of a cover.

Arrangements frequently combine fountains with waterfalls, with a succession of basins placed on intermediate pedestals one above the other, getting smaller in diameter as they get higher. Water, fed into the top basin, flows in steps down to the lower, larger basins, resulting on a neatly proportioned display of constant water curtains.

Important Aspects of Waterfall Design

(a) Trough should be adequately dimensioned.

(b) Lip and weir edge should be kept clear.

(c) Avoid siting under trees (leaves and twigs).

(d) Ensure pump is correctly sized.

Small Fountains

Today, almost anyone can afford or find space to enjoy a small fountain or waterfall display, and depending on the area available, fountains can be obtained for the garden, the patio or even inside the home.

The choice covers numerous variations from a complete range of fountains, and also includes items such as garden pools, pond liners, submerged pumps and underwater lighting, together with a host of fittings combinable at will to create a truly personalized water display.

Outdoors, in the garden or elsewhere, pools are usually set in the ground. For safety and children's sake, the water surface should be guarded by a galvanized screen or grating. For patios and living areas, decorative arrangements are available in wood or ceramic. Pebbles, volcanic rock and plants lend a distinctive touch to these fountains. Outdoor water displays are designed either as free-standing units or to be placed on a table or windowsill. Not only do they have a decorative and soothing effect, but they also make a dramatic improvement in room atmosphere.

None of these fountains require a continuous water supply. The basin or reservoir is simply filled with water and the pump, usually a submerged motorized unit, circulates the water to a fountain, a sculptured figure or an overflow; from there, water falls or pours back into the container, ready for a new cycle.

Water Sculptures

As an alternative to a fountain, or to complement and enliven an arrangement, figures and gargoyles are available as water inlets or outlets. The operation of a water sculpture is the same as for a fountain. A water basin or reservoir is required, from which a circulator (fountain pump) sends water along a hose or other suitable duct to the water sculpture inlet. From the spout, water must then be conveyed back to the reservoir to ensure that the closed-circuit system functions continuously. A distinction should be made between traditional gargoyles frequently adorning old village and town fountains, and waterspouts of more modern style which are often integrated into various sculptured designs to form a living part of a work of art.

Animal sculptures with integral waterspouts are becoming increasingly popular for home use. Besides the more common simple figures such as ducks, frogs etc., the choice now extends to artistic wildlife representation, with built-in spouts made of reconstituted stone and bronze.

When choosing sculptures and waterspouts, care should be taken to ensure that their style, colour and size integrates well with the overall setting. If children have access to the arrangement, sculptures should be quite sturdy and completely free from protruding and sharp edges.

Construction and Maintenance

The development of butyl/plastic sheeting simplified much of the construction of pools and lakes and also reservoirs, but it did tend to suggest that all that was required by way of preparatory work was to dig out the shape and lay the liner in welded sheets together with the adhesive available for that purpose.

Unfortunately, there were problems that had not been forseen, basically concerning the removal of hard debris such as stones, tree roots etc. and also the necessity to perforate the sheeting to run pipework into the pool or lake which resulted in seepage which progressively reduced the pool or lake depth and gradually weakened the surrounding earthworks causing in some cases, the earthworks to collapse.

The real answer to both these problems that relate to all liner pools whatever their size and depth, is to be sure by either raking and sieving to a depth of approximately 600mm for small pools and of course, considerably more the larger the pool or lake. Another method would be to build-up the excavated base with sand which would be the cheapest method relative to the pool or lake size.

Mention was made above of the necessity of perforating the sheeting for pipework. This can be achieved for most pools or lakes by cutting a hole through the liner in accordance with the overall pipe size and clamping the membrane between two gasketed flanges. In some cases, the liner could be sealed to the flanges with a suitable mastic.

Large ponds, lakes etc. can be made effectively more waterproof with 'Volclay', a natural fine volcanic dust, approximately 30 million years old that can be supplied in bags like cement. The Volclay can be mixed with soil or soft sand, when it becomes like vaseline and watertight so that if the earth/Volclay membrane is damaged, it will self heal. It is also very good for ornamental lakes and streams that are to be planted, the roots can grow through the membrane without causing leaks.

Cascades

This type of water feature needs careful planning relative to the water volume over the weir edge and storage content of the bottom pool. It is often forgotten that when pumping ceases, the top weir or weirs will continue to run until the static level is reached in the top or secondary pool, which means that the bottom pool from which the water was initially pumped will overflow if enough freeboard has not been allowed. There is also the question of rainfall to be considered, it follows therefore that a suitable overflow system should be installed above the static water level to allow for rainfall build-up.

Finally, the quantity of water to be pumped to produce a successful weir or cascade feature is often guesswork, which results in a raging torrent from an oversized pump which can be problematical, or a pitiful dribble. As a guide to a successful feature, the following calculation could be used:

13mm depth of water flowing over a 300mm wide linear weir edge will require 1.8 litres per second of water.

This 'rule of thumb' formula can be multiplied or divided as required to satisfy the needs of the client's final design.

DUCT SIZING

The sizing of low velocity air flow in ducts to internal WC's and bathrooms is dependent on the following relationships:

$$Q = 6.3 \times 10^{-7} \times \sqrt{\frac{d^5 \times h}{L}}$$

where

Q = flow rate in m^3/sec
d = diameter of duct in mm
h = pressure drop in mm Wg
L = length of duct in metres

By transposition and simplifying we get also:

$$d = 305 \times \sqrt[5]{\frac{Q^2 \times L}{h}}$$

Example

An air duct is required to handle 0.20 m^3/sec at a pressure drop not exceeding 0.10 mm per 10 m run.

Calculate a suitable diameter for the duct.

$$d = 305 \times \sqrt[5]{\frac{0.20^2 \times 10}{0.10}}$$

$$= 305 \times 1.3195 = 402.449$$

$$d = 400\,\text{mm (nominal)}$$

Note

This calculated result agrees with flow chart (Graph L3).

Extract Ventilation

Example

8 Bathrooms + WC's (100 mm branches)

Main Duct

$$\sqrt{D^5} = N \times \sqrt{d^5}$$

$$= d \times \sqrt[5]{N^2}$$

$$= 100 \times \sqrt[5]{8^2} = 229.73\,\text{mm}$$

$$D = 250\,\text{mm (nearest)}$$

Volume Flow

$8 \times 45\,m^3/h = 0.10\,m^3/sec$, $D = 250$ mm

From Graph L3 at 0.10 m^3/sec,

friction = 0.35 mm Wg/10

At 30 m run = 1.05 mm Wg

and V = 2.2 m/s

Branch Ducts (100 mm diameter, each 1.5 m run)

$8 \times 1.5\,m = 12.0\,m$ run

Air flow rate = 0.0125 m^3/sec at 100 mm dia. duct

Flow resistance = 0.60 mm Wg/10

$12.0 \times 0.60/10 = 0.72$ mm Wg

and V = 1.60 m/s

Branch Duct – Resistances
(see Graph L2)

Exhaust vent + bends,

(say) K = 6 (each branch)

= 0.9 mm Wg (at 1.60 m/s)

$= 8 \times 0.9$

= 7.2 mm Wg

Total Losses

$1.05 + 0.72 + 7.2 = 8.97 = 87.995\,N/m^2$

$= 88\,N/m^2$

Fan head allowed = 125 N/m^2

Provision for adjustment at exhaust outlets in bathrooms and in main trunking:

$$\frac{125}{88} = 1.42 \text{ or } 42\%$$

With this arrangement, an adequate air flow from bathrooms will be maintained.

Duct Fittings – Resistance

The use of bends, elbows, junctions and inlet grilles as part of ventilation duct work means that a fairly simple and rational method must be used to take these various resistances into account.

The 'velocity head' method is widely used in connection with air, and water flow problems:

$$H = K \times \frac{V^2}{2g} \text{ (water)}$$

$$H = K \times \frac{V^2}{2g} \times \frac{P}{W} \text{ (air)}$$

where, P is the density of air
W is the density of water } same temperature

For one velocity head, one m/s

$$H = \frac{1 \times 1^2}{2 \times 9.81} \times \frac{1.205}{998.23} \times 1000$$

$$H = 0.062V^2 \text{ mm Wg}$$

$$H = \frac{V^2}{16.25} \text{ mm Wg}$$

Example

$K = 1 \qquad V = 2\,m/s$

$$H = 1 \times \frac{V^2}{16.25} = \frac{2^2}{16.25} = 0.246\,\text{mm}$$

$$H = 0.25\,\text{mm from Graph L2}$$

Table L1 Ventilation head resistance factors (K) for bathroom and WC ventilation ducts

Fitting	K	Fitting	K
Bends 90°		*Double bends*	
Radius 4 × D	0.15	Radius 2D	0.20
1 × D	0.25		
Reducer		*Double elbows*	1.00
Square or Circular	0.25		
Branch piece		*Branch piece*	
inclined		radiused	
30°	0.15	2D	0.25
45°	0.25	1D	0.35
60°	0.45	1/2D	0.75
Extract gratings	3–5		
50% free area to room	1.50	Entrance to duct from room	0.50

Graph L1 Relationship of capacity of rectangular and circular ducts (equal capacity and pressure)

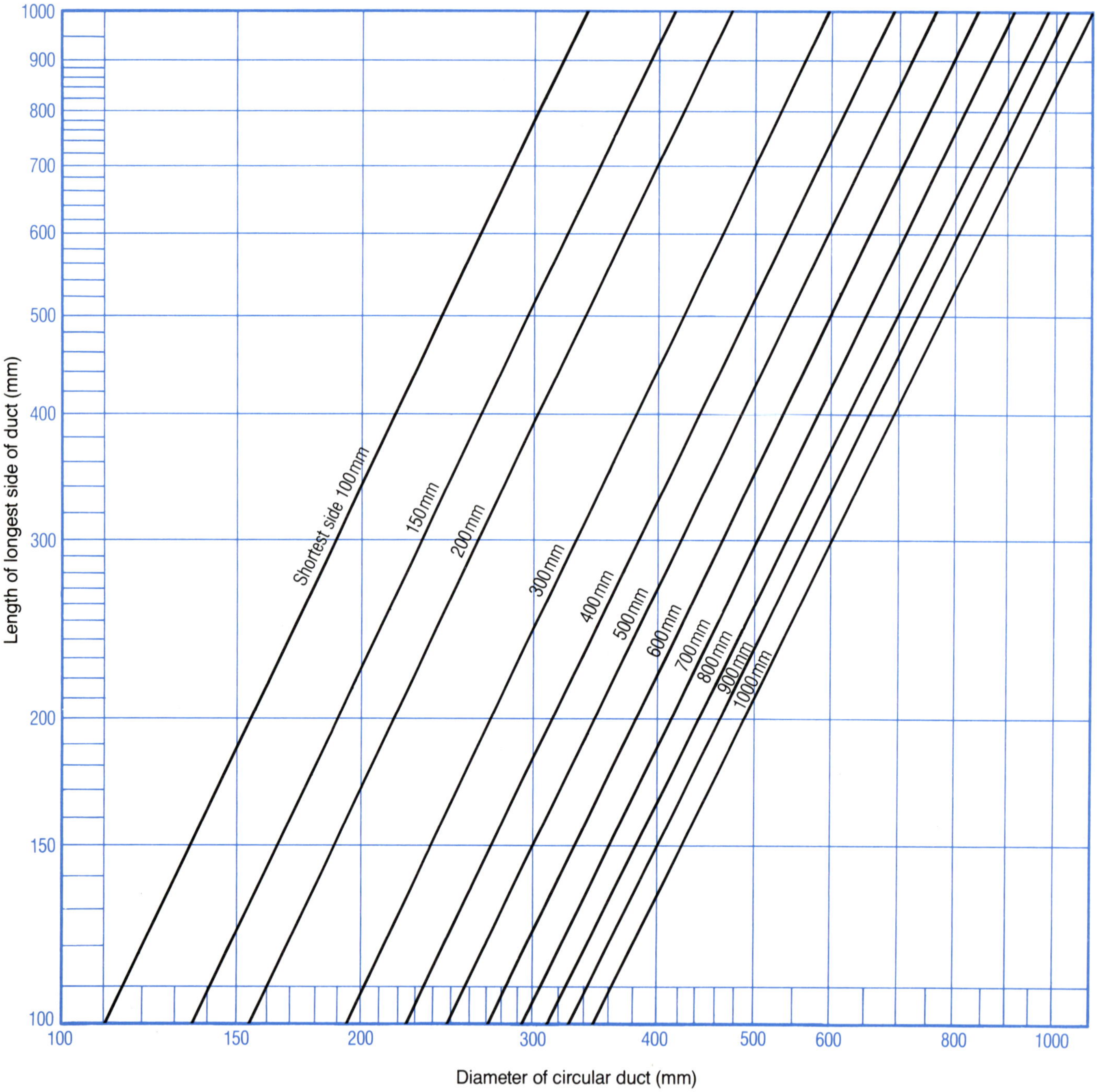

Graph L2 Relationship of K, V and head loss for duct fitting resistances

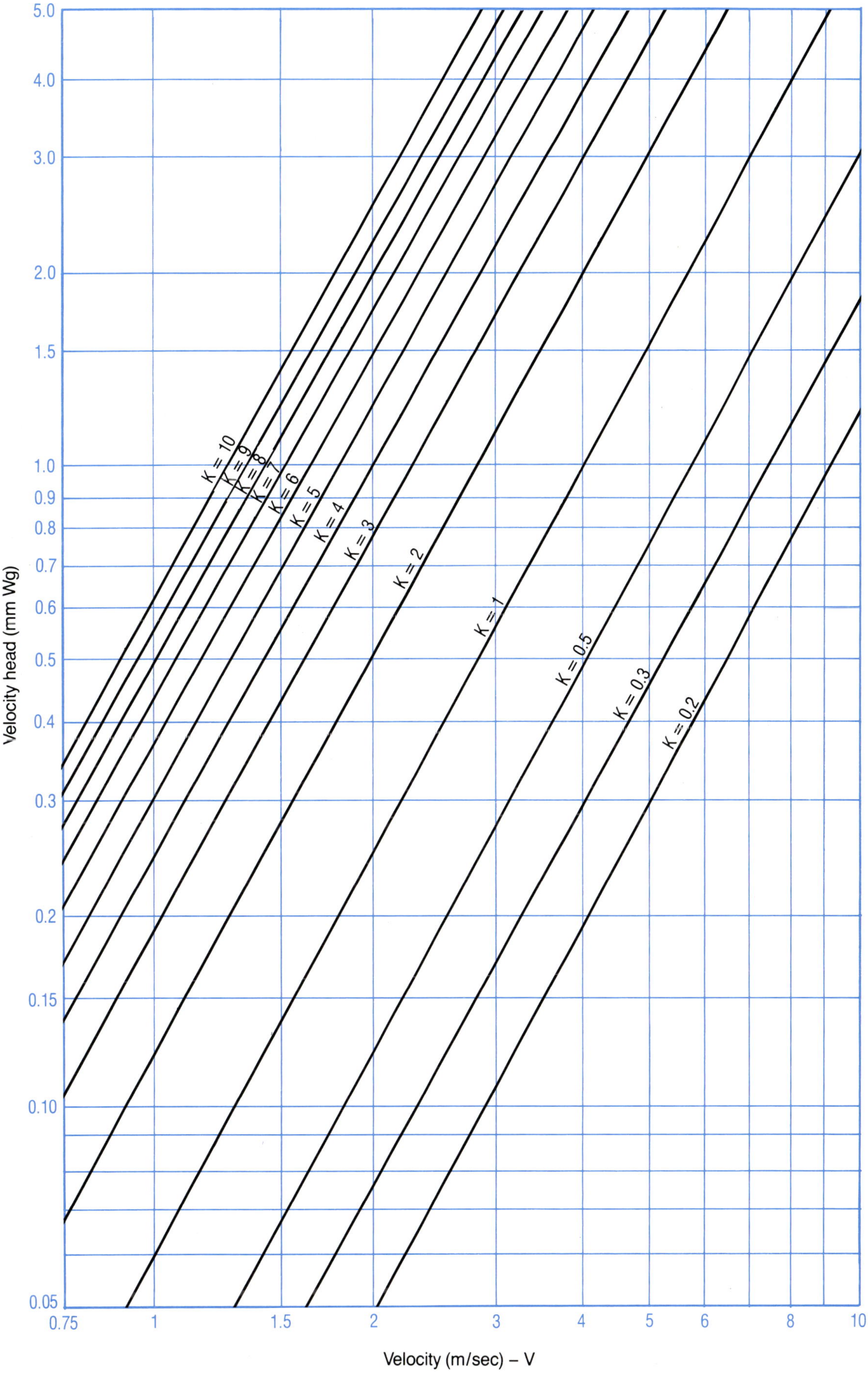

Graph L3 Sizing chart – ventilation ducts for bathrooms and WC's

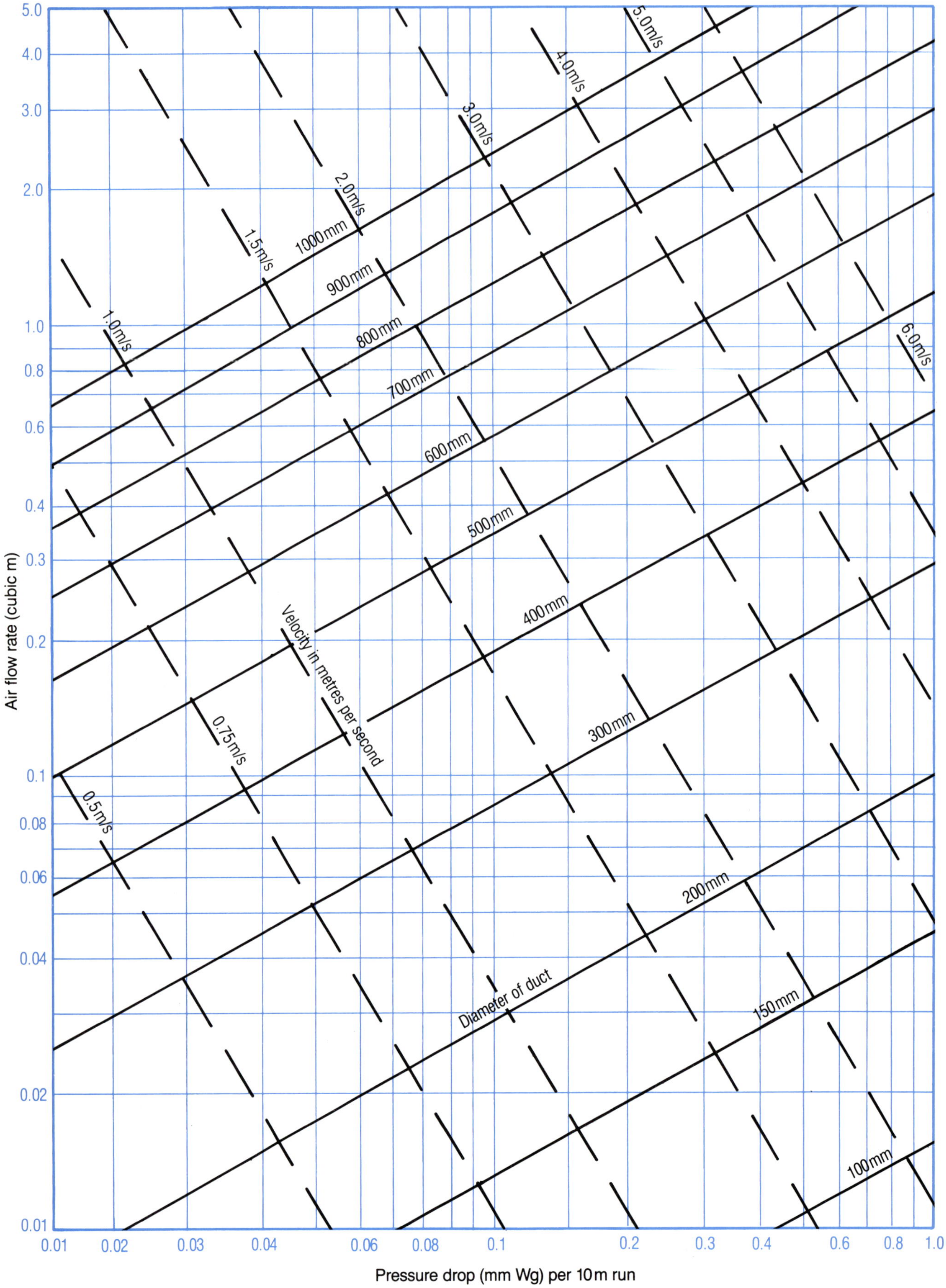

Ideal System

The ideal system of heating and ventilating required to maintain comfort and efficiency can be summarized as follows:

(1) To provide good ventilation without reliance on open windows.
(2) To provide quickly responsive heating equipment to climatic changes.
(3) To provide cooling in summer and at other times of the year when solar and other gains cause unpleasant overheating.
(4) To provide a pleasant circulation of filtered air within a building free from draughts.
(5) To maintain near uniform relative humidity.
(6) To provide some element of individual control.
(7) To require minimum maintenance.

FANS

The definition of a 'fan' is a machine which propels air continuously by aerodynamic action.

There are three basic types of fan:

(a) Centrifugal.
(b) Propeller.
(c) Axial flow.

Centrifugal Fans

Figure L1 Centrifugal fan

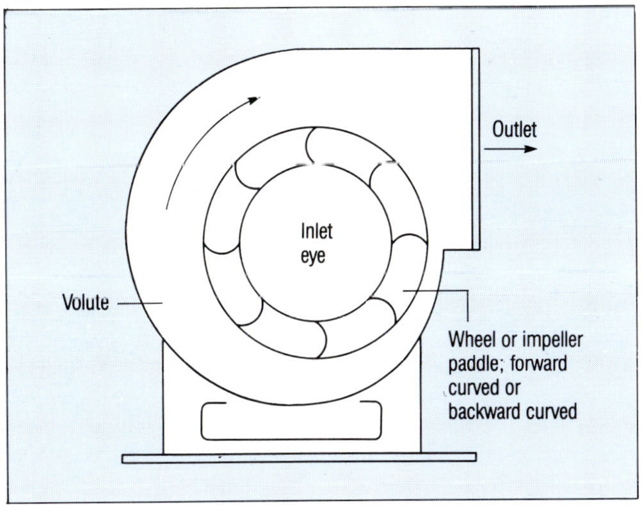

Centrifugal fans are further sub-divided into three types:

(i) Paddle or straight radial blade.
(ii) Forward curved blade.
(iii) Backward curved blade.

Paddle Blade

Figure L2 Paddle blade

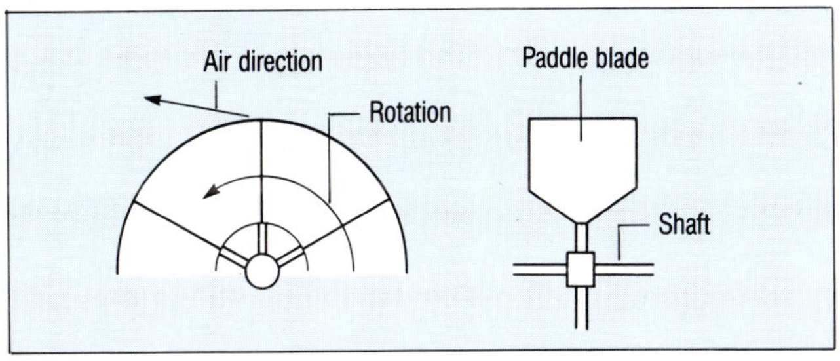

The paddle type fan is the simplest possible construction, robust in design and low in cost. These fans are suitable for dust removal work because matter carried into the airstream does not cling to the blades. They are also used for handling high temperature gases or those of an abrasive or corrosive nature. Six or twelve blades are normally used.

Forward Curved Blade

Figure L3 Forward curved blade

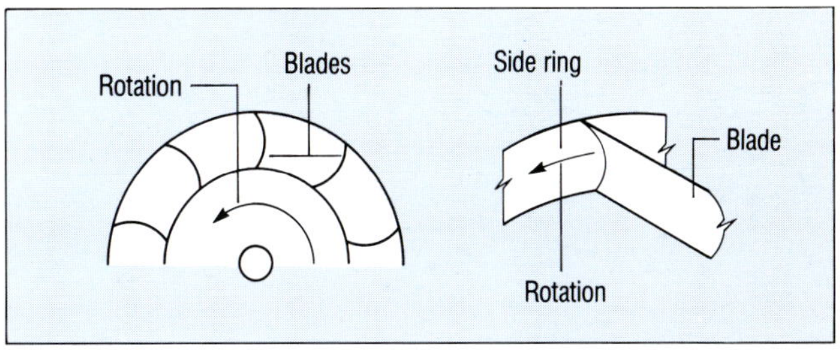

The concave side of the blade faces the direction of rotation and thereby increases the efficiency of the impeller. The blades are shallower radially, but a large number are used, giving rise to the name 'multi-vane fans'. The side rings carrying the blades are carried by a spider. The forward curve of the blade acts as a scoop, and as a result large volumes of air can be moved by relatively small impellers working at moderate speeds.

Backward Curved Blade

Figure L4 Backward curved blade

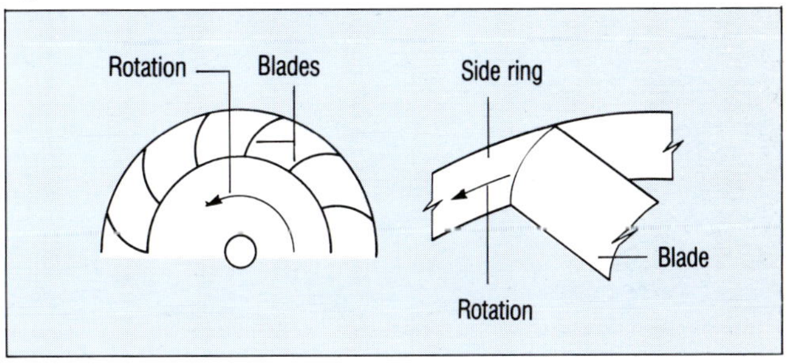

Backward curved blade fans have the highest efficiency in the centrifugal type range.

The convex side faces the direction of motion. This improves the air flow through the blades and reduces shock and eddy losses. These fans can operate at higher tip speeds than other types of centrifugal fan. The blades are longer and heavier than other types so that stiffening rings and large shafts are necessary. They have a lower air output for a given wheel diameter than the forward blade type, the efficiency is however high. Special types, developing very high pressure, are used as forced draught fans for boilers.

Backward curved blade fans are not suitable for handling gases of an unduly dusty nature, particularly where moisture is present. This tends to collect on the back of the blades and throw the impeller out of balance.

To reduce the cost and bulk, double width wheels with inlets on both sides of the casing are employed when large volumes of air are to be handled. Double inlet fans give twice the volume of similar single inlet fans of the same wheel diameter and speed.

FANS

Propeller Fan

Figure L5 Propeller fan

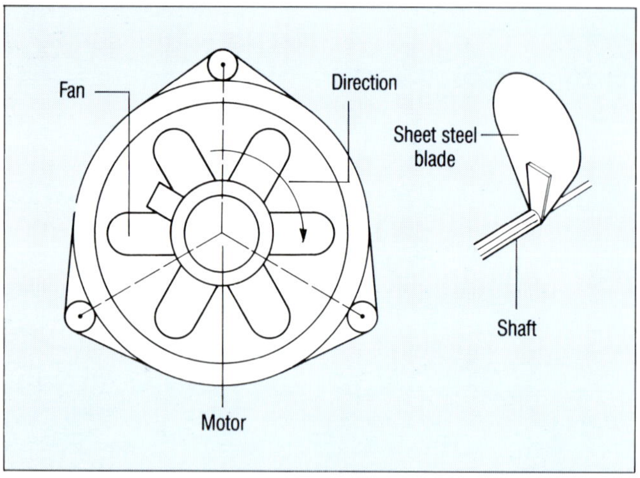

Propeller fans are used where the resistance to air flow is low, ie. where there is no duct system or where it is very short. In the majority of cases they move air through a hole in a wall. Their main advantage is that they can handle large volumes of air economically due to low capital cost.

The basic design is that of a ship's propeller or in some cases an aircraft propeller blade profile is used. The most efficient shape for the blade is broad curved, running at moderate speeds; it is also most quiet at this condition.

In modern ventilation practice it is usual to mount the fan directly on to the spindle of an electric motor. They are mainly used as inlet or exhaust fans, but not however combined with ducting as they cannot operate against a resistance.

Axial Flow Fan

Figure L6(a) Axial flow fan

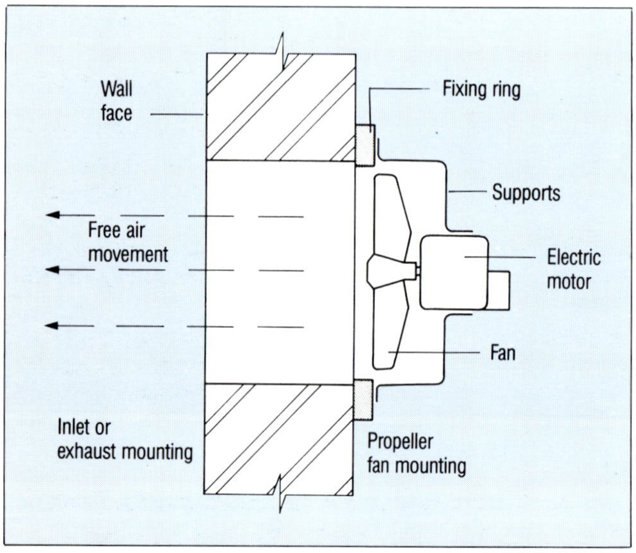

Figure L6(b) Axial flow fan mounting, non guide vane type

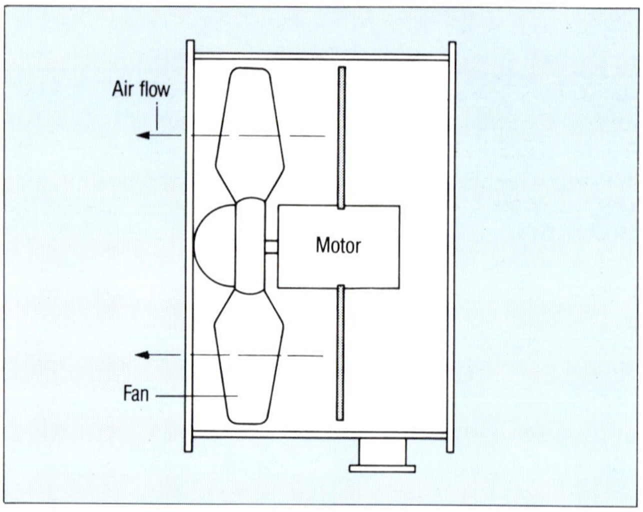

Figure L6(c) Axial flow fan with guide vanes

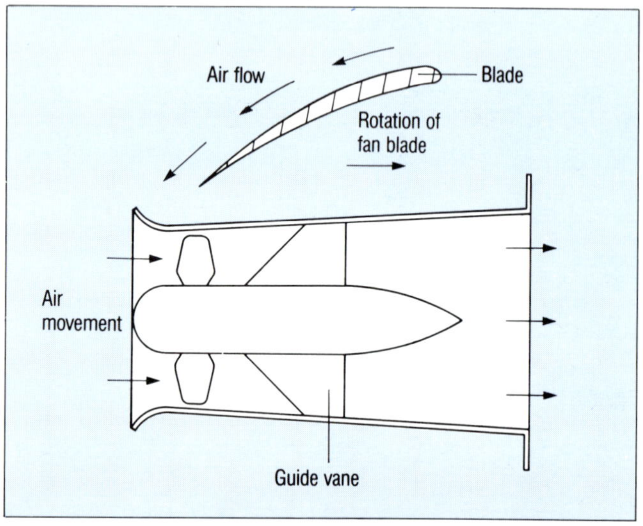

Axial flow fans have been widely used in recent years in applications where centrifugal fans would formerly have been used. This is mainly due to their high efficiency, compactness and simplicity of installation.

The fan blades are of aerofoil section rotating in a cylindrical casing. The air flow through the fan is virtually parallel to the impeller shaft, ie. axially. Efficiencies range from 70% to 90% in the case of special fans.

The straight-through-flow enables the fan to be inserted directly into a straight length of ducting, making the connections between fan and ducting simpler. The overall size of an axial flow fan is considerably less than a centrifugal fan of equivalent duty.

Guide vanes are used with this type of fan arrangement to remove rotation from the air and thus get maximum pressure development. Contra rotating axial flow fan units are also used. The pressure developed by these units is about three times that of a non-guide vane axial flow fan.

Regulations and Requirements

Building Regulations 1985 – Sanitary Conveniences

WC compartments shall be provided with either:

(a) window or other means of ventilation opening to the external air, the area capable of being opened shall be not less than 1/20 of the floor area;

(b) mechanical means of ventilation which provides for not less than three air changes/h and discharges directly to the open air.

Inner London Authorities

For bathrooms and WC's not provided with openings to the external air (at least 0.18 m² is required to comply), the District Surveyor shall insist on mechanical ventilation. This must be capable of supplying at least 22 m³/h of fresh air either per occupant of the building or per 4.7 m² of floor area whichever is the greater.

In blocks of flats, one extract fan is usually provided for a group of bathrooms, but over a certain height strict control is exercised by the Local District Surveyor.

Figure L7 Roof mounted supply duct with recirculating damper control

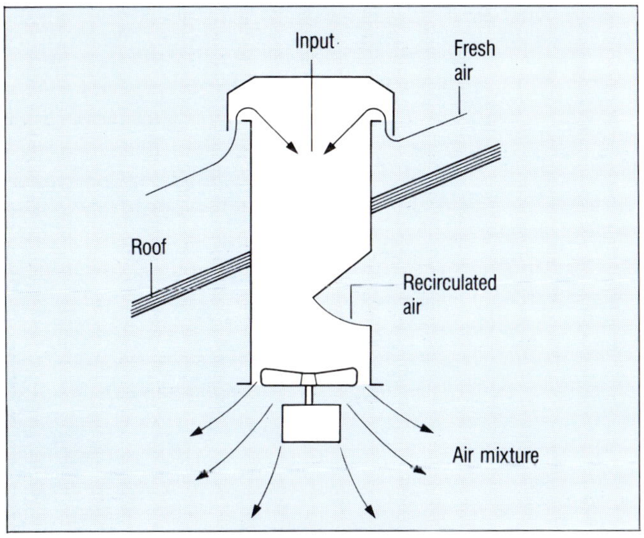

Figure L8 Roof mounted extract fan - natural inlet

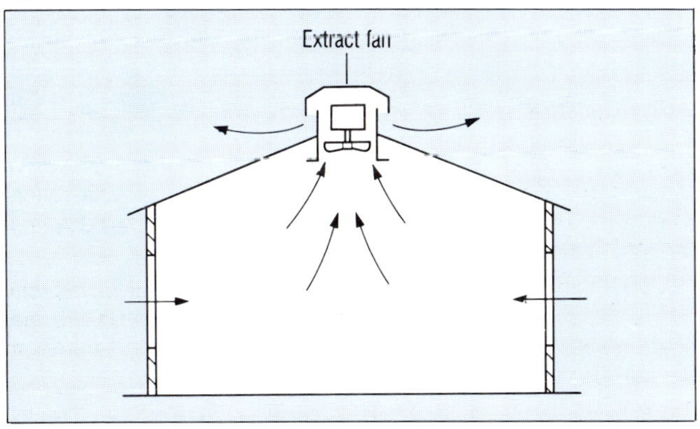

Figure L9 Recommended proportions for a plain canopy hood

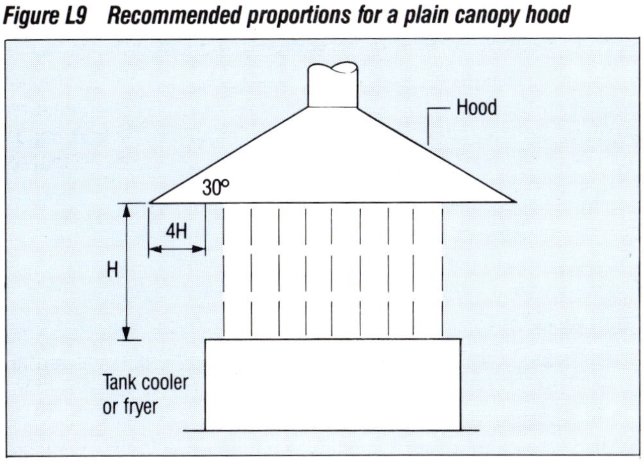

Figure L10 Combined supply and extract system with propeller fans

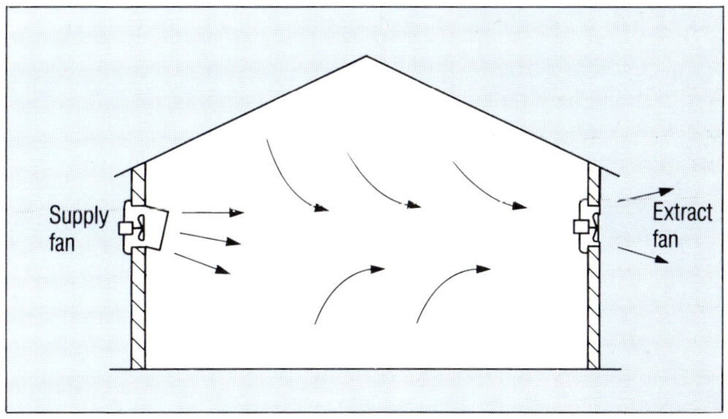

Figure L11 Cross ventilation for public rooms

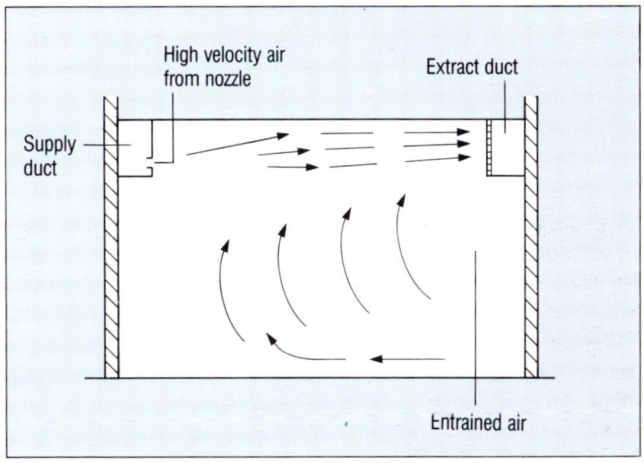

BRITISH STANDARDS AND CODES OF PRACTICE

CP 413: 1973 (1980)	Code of Practice for ducts for building services.
BS 5285: 1975 (1987)	Specification. Performance of AC electric ventilating fans and regulators for non industrial use.
BS 5643: 1984	Glossary of refrigeration heating, ventilating and air conditioning terms.
BS 5720: 1979	Code of Practice for mechanical ventilation and air conditioning in buildings.
BS 5925: 1980	Code of Practice for design of buildings: ventilation principles and designing for natural ventilation.

Note

(1) Year after Standard's number indicates the latest date of issue.

(2) Year in brackets gives the date of the latest amendment at the time of publication of this Guide.

Electrical Earthing and Bonding in Water Installations

INTRODUCTION

The practice of earthing electrical systems to main water services has, in the past, provided the electricity industry with a convenient and seemingly reliable protection against the risks of shock and fire in consumer's premises. There is no doubt however, that it caused occasional damage to water pipes and sometimes put waterworks personnel and plumbers at risk when, for example, earthing continuity was interrupted by the disconnection of pipework for repair or replacement. Despite being approved by only a limited number of water undertakings, it became a sufficiently common practice to warrant some agreement as to the conditions under which earthing connections should be made.

Protective Multiple Earthing

As previously used earthing systems have become increasingly unsatisfactory, Protective Multiple Earthing (PME) has come to the fore in recent years, enabling electricity boards to provide the consumer with a reliable earthing terminal connected to the neutral conductor at the supply intake position. For protection against earth-leakage currents, electrical connections are made between the earthing terminal and all metal structures, metal pipes and other metalwork which:

(a) are or may reasonably be expected to become in electrical contact with the general mass of earth, and

(b) are so situated that any person may reasonably be expected to make simultaneous contact with:

 (i) on the one hand, such structures, pipes or other metalwork (or any metalwork in electrical contact therewith), and

 (ii) on the other hand, any exposed non-current carrying metalwork of the consumer's installation (or any metalwork in electrical contact therewith).

Figure M1 Protective multiple earthing – method of bonding

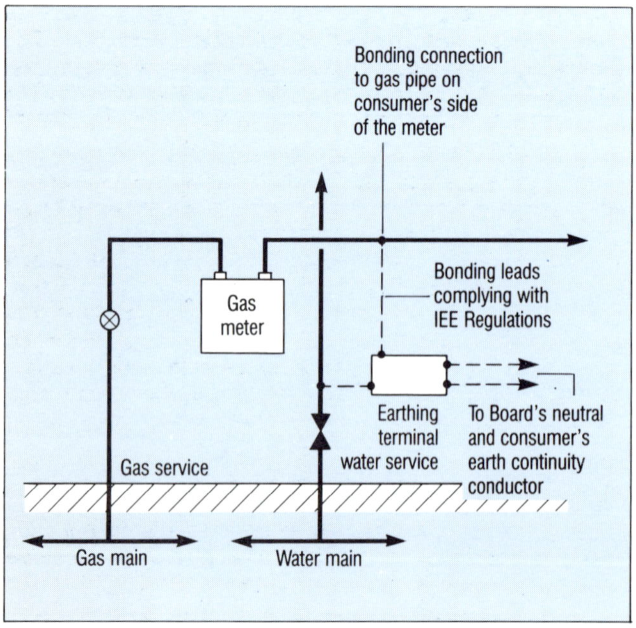

Bonding

These necessary safety connections between two or more conductors (whether or not normally used for conducting electricity) are described as "equipotential bonding". They are intended to ensure that all metal within reach is at the same potential, thus making it impossible for a person touching metal parts to suffer a shock because of a potential difference between the parts.

The IEE Regulations stipulate that the consumer's earthing terminal be bonded to the metalwork of incoming gas and water services, as close as practicable to the point at which those services enter the premises. But where there is a section or insert of insulating non-metallic pipework at the point of entry, the bonding connections have to be made to the metalwork on the consumer's side of that section or insert and, in the case of a gas service, on the consumer's side of the gas meter.

In particular, these main bonding connections have to be carried out after the consumer's earthing terminal has been connected to the effective means of earthing. Additional regulations require that all extraneous fixed metalwork which cannot be effectively segregated from the exposed earthed metalwork of all electical apparatus must be bonded thereto, at possible points of contact between both types of metalwork. Included in the fixed metalwork of a plumbing installation that may have to be bonded in these circumstances are exposed pipes, sinks, baths, radiators and tanks.

Bonding connections to metal pipes are usually made with British Standard approved earthing clamps and copper bonding leads, the latter being subject to minimum sizes and resistance values specified in the IEE Regulations. Figure M2 shows an Area Electricity Board's basic requirements for main bonding in a system of Protective Multiple Earthing.

Figure M2 Protective multiple earthing–area electricity board's requirements for main bonding

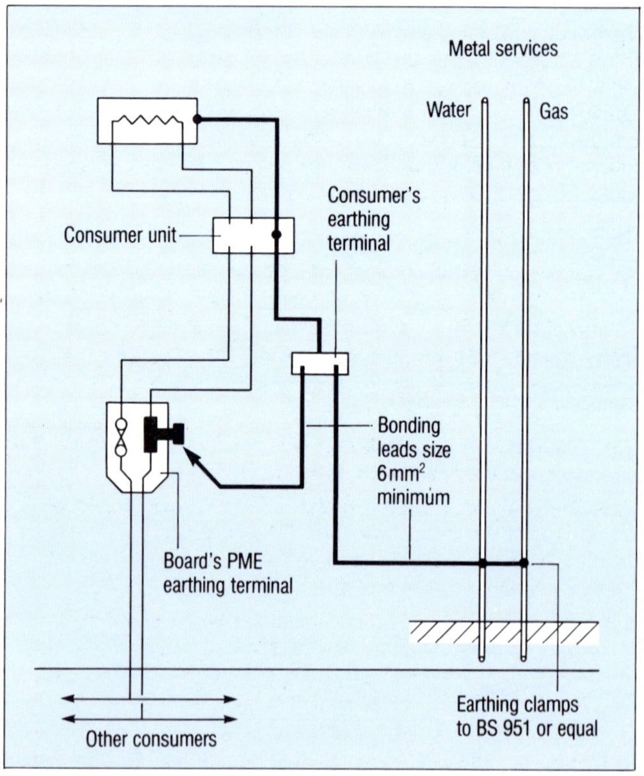

Problems with Plastics

The National Inspection Council for Electrical Installation Contracting (NICEIC) has frequently drawn attention to the increasing difficulties of maintaining earthing continuity in domestic water systems, due to the widespread use of non-metallic pipes and components.

Plastic "push-fit" couplings used for connecting copper tubing on hot and cold water systems are a particular problem. If installed in an existing copper pipeline they may result in a loss of effective bonding, where reliance has previously been placed on the metal-to-metal joints of compression fittings. Their use throughout a new system, however, does not necessarily mean that every individual length of pipe would need to be bonded. Some lengths would be concealed beneath floors, while others would be remote from any electrical apparatus. Nevertheless, any lengths of pipework that could not be effectively segregated from the exposed earthed metalwork of the electrical installation would require supplementary bonding.

Supplementary Bonding

Supplementary bonding connections may also be necessary when plastic taps are installed. Consider, for example, the case of a plastic tap controlling the water supply to an instantaneous electric water heater or shower unit which has elements directly immersed in the water.

The IEE Regulations require that the water supply enters such a heater through a metallic pipe, effectively earthed by some means independent of the earth-continuity conductor to the heater. This is usually achieved by bonding the continuous metal pipework from the heater intake to the main earthing terminal. But if the metal pipe is interrupted by a plastic tap it becomes necessary to bond across the tap or provide a bonding connection from the final section of metal pipe at the heater intake, back to the main earthing terminal. However, it is now possible to obtain plastic-bodied stopvalves incorporating a metallic bonding strip designed to maintain earthing continuity in the pipeline.

Supplementary bonding shall be provided to comply with Clause 413-7 of the IEE Regulations sized in accordance with Clauses 547-4 to 547-7.

The bonds shall be green PVC sheathed, and in concealed parts of the installation shall be concealed as far as is practicable in voids, partition walls or buried direct in plastered walls. Generally, bonding lugs will not be available on extraneous conductive parts and allowance shall be made for purpose-made clamps or clips, drilling or utilizing fixing or connecting bolts with the permission of the Mechanical Services Contractor.

The following shall be used as guidance to requirements:

(1) Between the hot and cold water pipes and metallic waste at each basin sink, bath or shower, and to the basin sink, bath or shower tray if metal.

(2) From the connection in (1) to any radiator, WC water pipe or other conductive part in a bathroom or shower and within 2m of the basin or sink in other spaces.

(3) From each radiator to its connecting pipework if the pipework does not provide a sound metal to metal contact.

(4) From the metal conduit, metal cable sheath or metal box (not the internal earth terminal) of the socket outlet system in each space to each radiator, sink, basin or other fixed conductive parts in that space.

(5) From each metal conduit, metal cable sheath or metal box (not the internal earth terminal) of a cooker circuit, heater or water heater circuit or any other power circuit in each space to each radiator, sink, basin or other conductive part within 2m of the power equipment.

(6) Between any fixed conductive parts within 2m of each other unless bonded indirectly by the requirements of (1) to (5) above.

(7) In every space with a suspended tiled (not plastered) ceiling from the metal conduit, metal cable sheath, metal box or metal luminaire of the lighting circuit in the ceiling void at one point to the metal ceiling frame, and at one point to each ventilation duct system entering the space. (Bonding across any ductwork flexible connections to grilles or diffusers will be by the Mechanical Services Contractor).

(8) Bond together all metal pipes, ducts, trunking, tray plate etc., at the top, and at the bottom, of each service riser.

(9) Bond across the flexible ductwork at each end of each air handling unit and axial fan.

(10) Bond across the flexible coupling from the motor to the impeller casing on each pump.

(11) Bond at one point across any system of anti-vibration mountings which separates metallically any rotating plant from its frame.

(12) Bond together and to the control panel, every metal pipe, duct or trunking in each plant room, near the point where they enter or leave the plant room.

(13) On metal tanks, bond from the tank to each metal pipe connecting to the tank unless the connections form a sound metal to metal contact and bond across metal pipework entering and leaving a plastic tank.

Note

Mechanical/Plumbing contractor shall provide copies of all services or other appropriate drawings to the Electrical Contractor to enable adequate and correct earth bonding of all services to be carried out.

Drainage Systems

Internal soil/vent pipes constructed of metal can pose similar problems when metal-to-metal contact between individual lengths of pipe is interrupted by special joints of non-conducting material. Although such pipes are often enclosed in ducts, plumbers are occasionally asked to provide bonding across the joints, presumably because the soil/vent pipe cannot be effectively segregated from possible contact with exposed electrical apparatus. There is also the possibility that current leaking from a damaged or defective underground supply cable could be conducted through the mass of earth to a nearby buried section of an all-metal drainage system. Here again some electrical risk could be involved if the internal pipework is fitted with non-metallic joints, and local bonding connections might be required.

New Regulations

If the present rules on bonding are correctly understood and applied by individual designers and installers there should be no significant difference in the number of essential bonding connections found in similar types of electrical installation. But it appears that many specifiers call for bonding far beyond the minimum requirements of the IEE Regulations, though it is often doubtful whether this produces a proportionately safer installation.

Briefly, bonding will be required between the exposed metalwork of the electrical installation and extraneous metalwork which is within reach and which is likely to transmit a potential, including earth potential. It will be a matter for judgement as to whether individual sections of metal pipework isolated from the general pipe run by plastic fittings are, in fact, liable to transmit a potential, in which case bonding will be necessary if they are simultaneously accessible with exposed electrical metalwork. There will also be requirements for additional bonding in bathrooms, possibly resulting in individual metal taps having to be bonded even though the associated pipework is of plastic material.

Plumbers' Responsibilities

It is obviously unwise to remove any seemingly superfluous bonding connections when repairing, altering or replacing pipework, unless one has expert knowledge of the electrical installation or has consulted an electrical contractor.

Do not interfere with any conductors which form bonds between water pipes, gas pipes, metal sinks, baths, boilers, radiators, tanks and earthing systems, and if these have to be removed for any reason, ensure they are replaced on completion of the work and a continuity test is carried out by a qualified electrician.

Co-operation

Both electrical installers and plumbers could benefit from an exchange of views and information on bonding as applied to domestic plumbing systems, if only in the interests of public safety.

Figure M3 Requirements for bonding in domestic situations
(not intended to be used as a design for a plumbing installation)

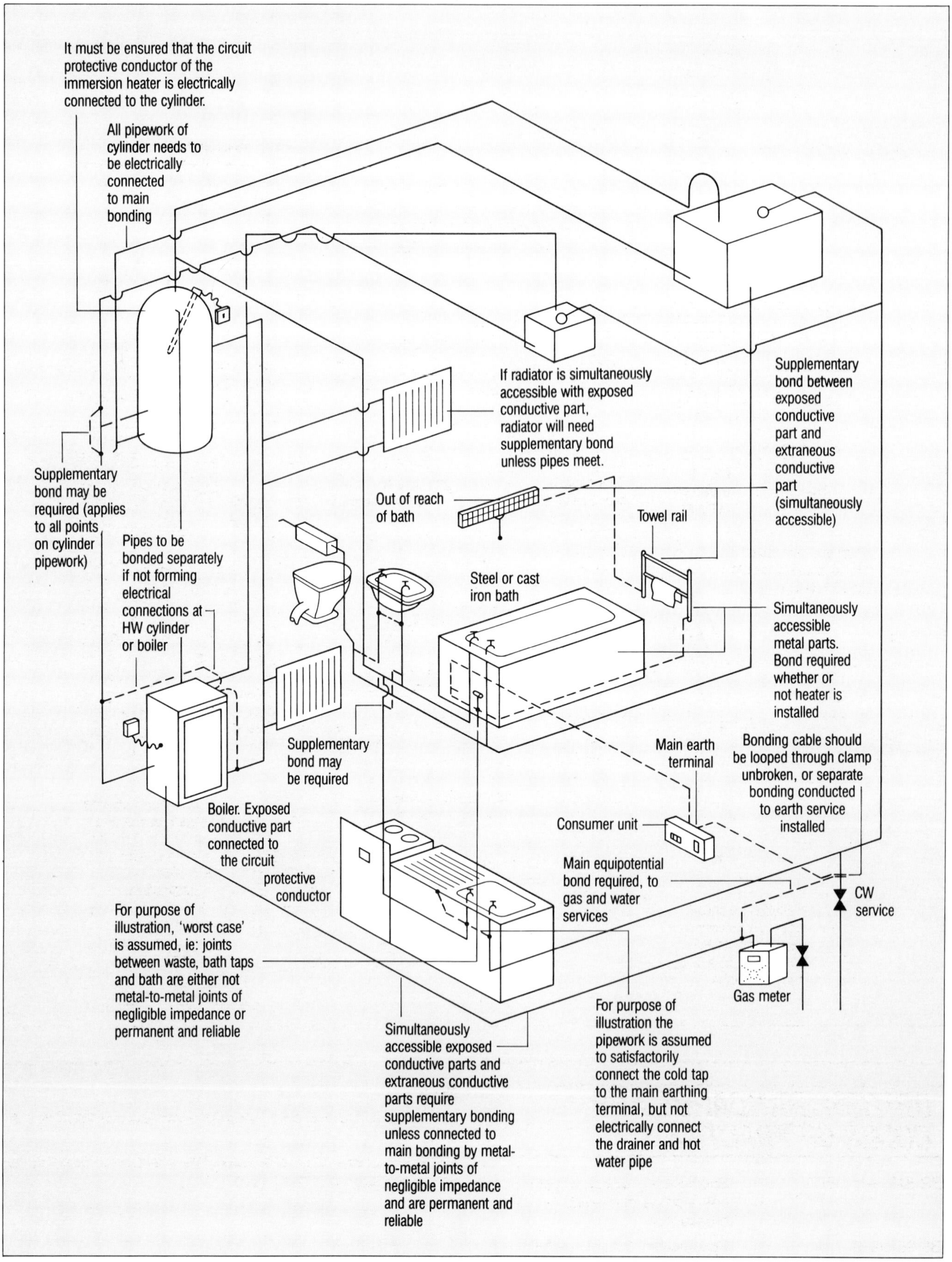

Figure M4 Sockets and fixed equipment outside main equipotential zone (exposed conductive parts of equipment) (diagrammatic only; not intended to be used as a design)

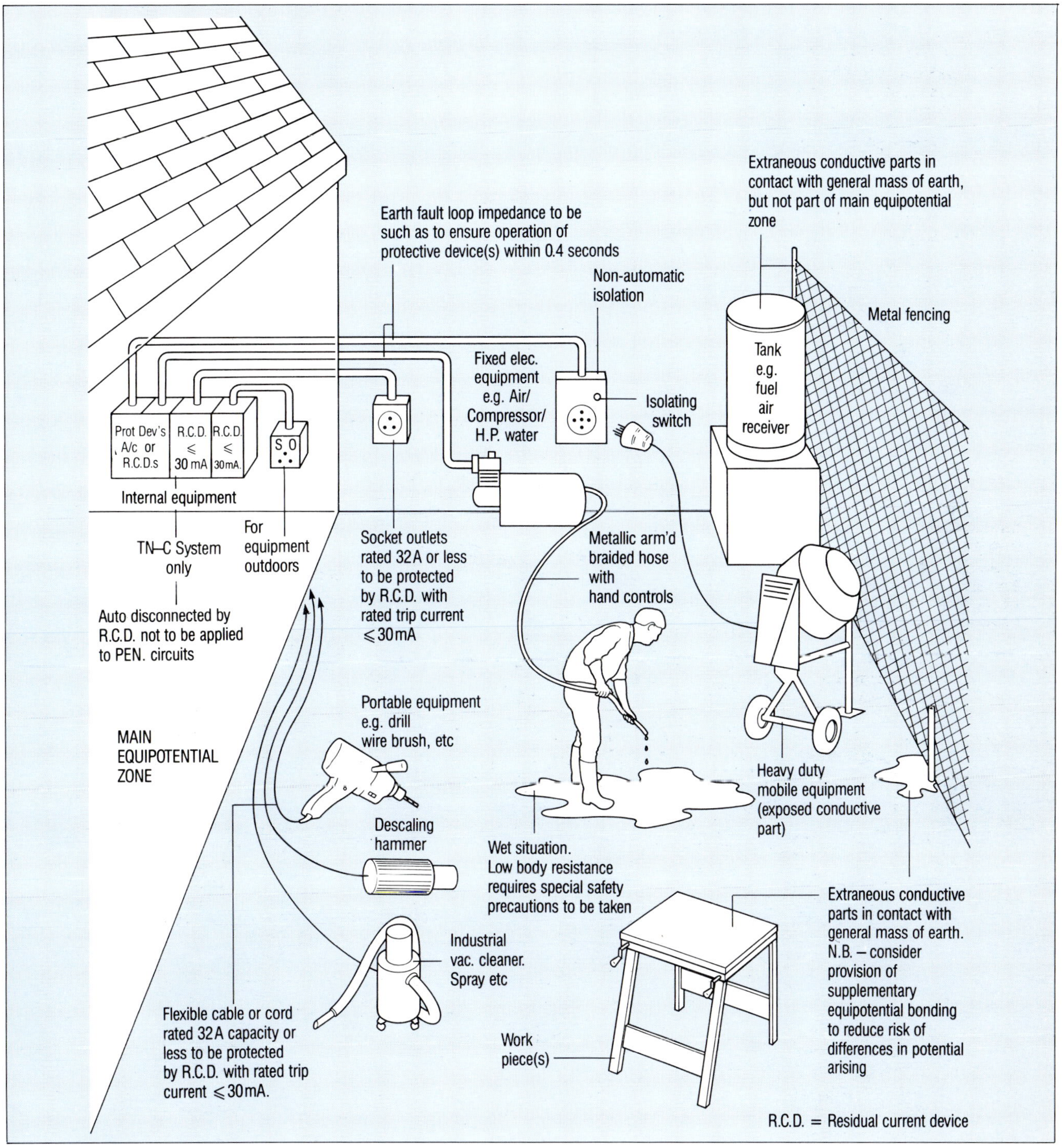

BRITISH STANDARDS AND CODES OF PRACTICE

BS 842:1965 (1967)	Specification for AC voltage-operated earth leakage circuit breakers.
BS 951:1986	Specification for clamps for earthing and bonding purposes.
BS 4293:1983 (1985)	Specification for residual current-operated circuit-breakers.

Note

(1) Year after the Standard's number indicates the latest date of issue.

(2) Year in brackets gives the date of the latest amendment at the time of publication of this Guide.

SECTION N
Conversion Factors and Miscellaneous Data

CONVERSION FACTORS

Table N1 Imperial to SI metric/SI metric to imperial

Unit	Imperial to metric		Metric to imperial	
Length	1 inch	= 25.4 mm	1 mm	= 0.03937 inches
	1 foot	= 0.3048 m	1 m	= 3.281 feet
	1 yard	= 0.9144 m	1 m	= 1.094 yards
	1 mile	= 1.609 km	1 km	= 0.6214 miles
	1 thou	= 25.4 μm	1 μm	= 0.03937 thou
Area	1 in^2	= 645.2 mm^2	1 mm^2	= 0.00155 in^2
	1 ft^2	= 0.0929 m^2	1 m^2	= 10.76 ft^2
	1 yd^2	= 0.8361 m^2	1 m^2	= 1.196 yd^2
	1 acre	= 0.4047 hectare (ha)	1 ha	= 2.471 acres
Volume	1 in^3	= 16390.0 mm^3		
	1 in^3	= 16.39 cm^3	1 cm^3	= 0.006102 in^3
	1 ft^3	= 0.02832 m^3	1 m^3	= 35.31 ft^3
	1 yd^3	= 0.7646 m^3	1 m^3	= 1.308 yd^3
	1 gallon	= 4.546 litres (l)	1 l	= 0.22 gallons
	1 pint	= 0.5683 litres	1 l	= 1.76 pints
	1 ft^3	= 28.32 litres	1 l	= 0.03531 ft^3
Velocity	1 ft/min	= 0.00508 m/s	1 m/s	= 196.9 ft/min
	1 ft/sec	= 0.3048 m/s	1 m/s	= 3.281 ft/s
	1 ft/sec	= 1.097 km/hr	1 km/hr	= 0.9113 ft/s
	1 mile/hr	= 0.4470 m/s	1 m/s	= 2.237 mile/hr
Mass and Density etc	1 lb	= 0.4536 kg	1 kg	= 2.205 lb
	1 oz	= 0.02835 kg	1 kg	= 35.27 oz
	1 oz	= 28.35 g	1 g	= 0.03527 oz
	1 ton	= 1016.0 kg	1 tonne	= 0.09842 ton
		= 1.016 tonne		
	1 cwt	= 50.80 kg	1 kg	= 0.01968 cwt
	1 lb/ft	= 1.488 kg/m	1 kg/m	= 0.672 lb/ft
	1 lb/ft^2	= 4.883 kg/m^2	1 kg/m^2	= 0.2048 lb/ft^2
	1 lb/in^2 (psi)	= 703.1 kg/m^2	1 kg/m^2	= 0.001422 lb/in^2 (psi)
	1 lb/ft^3	= 16.02 kg/m^3	1 kg/m^3	= 0.06243 lb/ft^3
	1 lb/m^3	= 27680 kg/m^3		
Flow Rate	1 gal/min	= 0.07577 l/s	1 l/s	=13.20 gal/min
	1 gal/hr	= 0.001263 l/s	1 l/s	= 791.9 gal/hr
	1 ft^3/min	= 0.4719 l/s	1 l/s	= 2.119 ft^3/min
	1 ft^3/min	= 1.699 m^3/hr	1 m^3/hr	= 0.5886 ft^3/hr
	1 ft^3/hr	= 0.007866 l/s	1 l/s	= 127.1 ft^3/hr
			1 m^3/s	= 2119.0 ft^3/min

Table N1 (Cont'd)

Unit	Imperial to metric		Metric to imperial	
Force (1 kgf = 9.807 N)				
	1 lb f	= 4.448 N	1 N	= 0.2248 lb f
	1 ton f	= 9.964 kN	1 kN	= 0.1004 ton f
Pressure, Stress (1 m head H_2O = 9.807 kN/m^2 [kPa]) (1 mb = 100 N/m^2 [Pa]) (1 Bar = 100 kN/m^2 [kPa])				
	1 in WG	= 249.1 N/m^2 (Pa)	1 N/m^2 [Pa]	= 0.004014 in WG
	1 in WG	= 2.491 mBar	1 mBar	= 0.4014 in WG
	1 lb/in^2 (psi)	= 6.895 kN/m^2	1 kN/m^2	= 0.145 lb/m^2 (psi)
	1 lb/in^2 (psi)	= 68.95 mBar	1 mBar	= 0.01450 lb/in^2 (psi)
	1 ft head H_2O	= 2.989 kN/m^2 [kPa]	1 kN/m^2	= 0.3345 ft head H_2O
	1 ft head H_2O	= 0.02989 Bar	1 Bar	= 33.45 ft head H_2O
	1 lb/in^2 (psi)	= 0.06895 Bar	1 Bar	= 14.50 lb/in^2 (psi)
	1 atmosphere (standard)	= 1.013 Bar	1 Bar	= 0.9872 atmosphere
Power, Heat Flow (1 W = 1 J/s = 1 Nm/s) (1 kW = 1 kJ/s = 3.6 MJ/h)				
	1 hp	= 745.7 W	1 kW	= 1.341 hp
	1 btu/hr	= 0.2931 W	1 W	= 3.412 btu/h
	1 btu/hr ft^2	= 3.155 W/m^2	1 W/m^2	= 0.3170 btu/hr ft^2
	1 btu/ft^3	= 0.03726 mJ/m^3	1 mJ/m^3	= 26.84 btu/ft^3
	1 btu in/ft^2 h °F (Thermal Conductivity	= 0.1442 W/m °C = k)	1 W/m °C	= 6.934 btu in/ft^2 hr °F
	1 btu/hr ft^2 °F (Thermal Transmittance	= 5.678 W/m^2 °C = U)	1 W/m^2 °C	= 0.1761 btu/hr ft^2 °F
Heat (kWhr = 3.6 mJ) (1 Therm = 100000 btu) (1 Cal = 4.1868 J)				
	1 btu	= 1055 J	1 J	= 0.0009478 btu
	1 therm	= 29.31 kWhr	1 kWhr	= 0.03412 therms
	1 therm	= 105.5 MJ	1 MJ	= 0.009479 therms
	1 btu	= 0.2520 kCal	1 kCal	= 3.968 btu
Fuel Consumption	1 mpg	= 0.3540 km/l	1 km/l	= 2.825 mpg
	1 mpg	= 2.825 l/km	1 l/km	= 0.3540 mpg

SECTION N

Conversion Factors and Miscellaneous Data

CONVERSION FACTORS

Table N2 Pressure

dyn/cm²	N/m²	lbf/ft²	mbar	mm HG	in H_2O	kN/m²	in Hg	lbf/in²	kgf/cm²	bar	atm
1	0.1000	2.089×10^{-3}	1×10^{-3}	7.501×10^{-4}	4.015×10^{-4}	1×10^{-4}	2.953×10^{-5}	1.450×10^{-5}	1.020×10^{-6}	1×10^{-6}	9.869×10^{-7}
10	1	0.02089	0.0100	7.501×10^{-3}	4.015×10^{-3}	1×10^{-3}	2.953×10^{-4}	1.450×10^{-4}	1.020×10^{-5}	1×10^{-5}	9.869×10^{-6}
478.8	47.88	1	0.4788	0.3591	0.1922	0.04788	0.01414	6.944×10^{-3}	4.882×10^{-4}	4.788×10^{-4}	4.726×10^{-4}
1000	100	2.089	1	0.7501	0.4015	0.1000	0.02953	0.01450	1.020×10^{-3}	1×10^{-3}	9.869×10^{-4}
1333	133.3	2.785	1.333	1	0.5352	0.1333	0.03937	0.01934	1.360×10^{-3}	1.333×10^{-3}	1.316×10^{-3}
2491	249.1	5.202	2.491	1.868	1	0.2491	0.7356	0.03613	2.540×10^{-3}	2.491×10^{-3}	2.458×10^{-3}
1×10^{4}	1000	20.89	10	7.501	4.015	1	0.2953	0.1450	0.01020	0.0100	9.986×10^{-3}
3.386×10^{4}	3386	70.73	33.86	25.40	13.60	3.386	1	0.4912	0.03453	0.03386	0.03342
6.895×10^{4}	6895	144	68.95	51.71	27.68	6.895	2.036	1	0.07031	0.06895	0.06805
9.807×10^{5}	9.807×10^{4}	2048	980.7	735.6	393.7	98.07	28.96	14.22	1	0.9807	0.9678
1×10^{6}	1×10^{5}	2089	1000	750.1	401.5	100	29.53	14.50	1.020	1	0.9869
1.013×10^{6}	1.013×10^{5}	2116	1013	760.0	406.8	101.3	29.92	14.70	1.033	1.013	1

1 torr = 1 mm Hg (to within 1 part in 7 million) 1 N/m × 10² is sometimes called a pascal.

MISCELLANEOUS DATA

Table N3 Multiples and sub-multiples of units

x factor	Prefix		Symbol
1,000,000,000,000	10^{12}	Tera	T
1,000,000,000	10^{9}	Giga	G
1,000,000	10^{6}	Mega	M
1,000	10^{3}	Kilo	k
100	10^{2}	Hecto	h
10	10^{1}	Deca	da
1	10^{0}	Basic unit	
0.1	10^{-1}	Deci	d
0.01	10^{-1}	Centi	c
0.001	10^{-3}	Milli	m
0.000 0001	10^{-6}	Micro	μ

Table N4 Standard wire gauge thickness

Gauge	Thickness (in)	Thickness (mm)
7/0	0.500	12.70
6/0	0.464	11.79
5/0	0.432	10.97
4/0	0.400	10.16
3/0	0.372	9.449
00	0.348	8.839
0	0.324	8.230
1	0.300	7.620
2	0.276	7.010
3	0.254	6.401
4	0.232	5.893
5	0.212	5.385
6	0.192	4.877
7	0.176	4.470
8	0.160	4.064
9	0.144	3.658
10	0.128	3.251
11	0.116	2.946
12	0.104	2.642
13	0.092	2.337
14	0.080	2.032
15	0.072	1.829
16	0.064	1.626
17	0.056	1.422
18	0.048	1.219
19	0.040	1.016
20	0.036	0.914
21	0.032	0.813
22	0.028	0.711
23	0.024	0.610
24	0.022	0.559
25	0.020	0.518
26	0.018	0.457
27	0.0164	0.417
28	0.0148	0.376
29	0.0136	0.345
30	0.0124	0.315
31	0.0116	0.295
32	0.0108	0.274
33	0.0100	0.254
34	0.0092	0.234
35	0.0084	0.213
36	0.0076	0.193
37	0.0068	0.173
38	0.0060	0.152
39	0.0052	0.132
40	0.0048	0.122
41	0.0044	0.112
42	0.0040	0.102
43	0.0036	0.091
44	0.0032	0.081
45	0.0028	0.071
46	0.0024	0.061
47	0.0020	0.051
48	0.0016	0.041
49	0.0012	0.030
50	0.0010	0.025

Table N5 English zinc gauge thicknesses

Gauge	Thickness (in)	Thickness (mm)
12	0.025	0.635
13	0.028	0.711
14	0.031	0.787
15	0.036	0.914
16	0.041	1.041

IDENTIFICATION OF PIPELINES

For more detailed information see BS 1710 – Identification of Pipelines.

According to the complexity of an installation and to the variety of fluids conveyed, pipes should be identified by either:

(1) basic identification colours only for installations where the determination of merely the basic nature of the contents is sufficient or

(2) basic identification colours and code indications for installations where the precise determination of the contents is of importance.

Basic Identification Colours

The basic identification colour may be applied:

(a) over the whole length or

(b) as a band over a length of about 150 mm, depending on the diameter of the pipe.

Where banding is adopted, any decorative or protective colour of the pipe shall not be any of the other basic identification colours.

The basic identification colour shall be placed at all junctions, at both sides of valves, bulkheads, wall penetrations and at any other place where identification is necessary. Valves may be painted with the identification colour except where the pipeline has been coded with the safety colour for fire fighting, when the valves should be painted red.

Table N6 Basic identification of colours

Pipe contents	Basic identification colour	BS colour references BS 4800
Water	Green	12 D 45
Steam	Silver-grey	10 A 03
Mineral, animal and vegetable oils; combustible liquids	Brown	06 C 39
Gases in gaseous or liquefied condition (except air)	Yellow ochre	08 C 35
Acids and alkalis	Violet	22 C 37
Air	Light blue	20 E 51
Other fluids	Black	Black
Electrical services	Orange	06 E 51

Code Indications

Code indications are:

(1) the safety colours, viz:

a. red for fire fighting;

b. yellow, with black diagonal stripes, for warning of danger;

c. yellow, with black trefoil symbol for ionizing radiation;

d. auxiliary blue in conjunction with green basic colour, to denote pipes carrying fresh water, either potable or non-potable.

(2) information regarding the nature of the contents.

Where code indications are used, they should be placed at all junctions, at both sides of valves, service appliances, bulkheads, wall penetrations and at any other place where identification is necessary.

Application of safety colours and code indication colours.

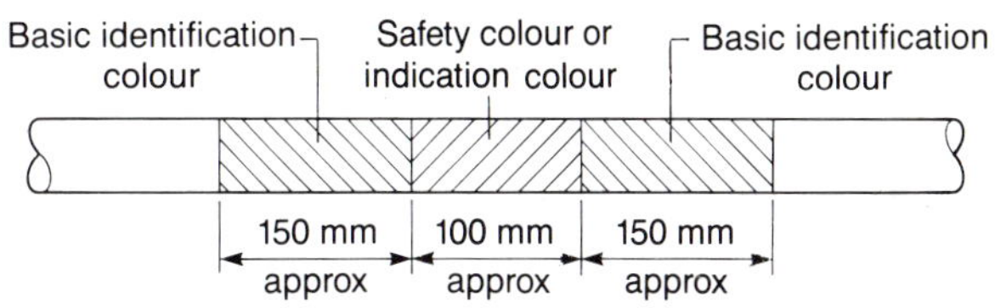

Table N7 Colour references

(a) Safety colour references

Safety colour	BS colour reference BS 4800
Red	04 E 53
Yellow	08 E 51
Auxiliary blue	18 E 53

(b) Code indication colours (if other than safety colours)

Colour	BS 4800
Crimson	04 D 45
Emerald green	14 E 53
Salmon pink	04 C 33
Yellow	10 E 53

(c) Colour code indications for medical gas services

OXYGEN

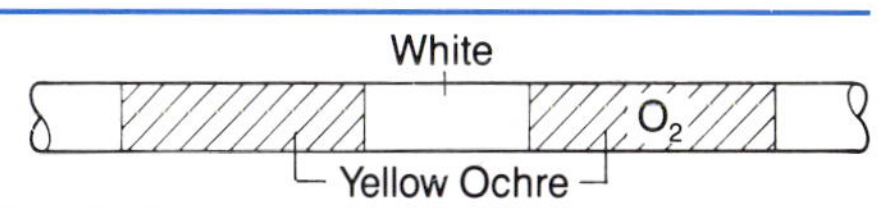

NITROUS OXIDE

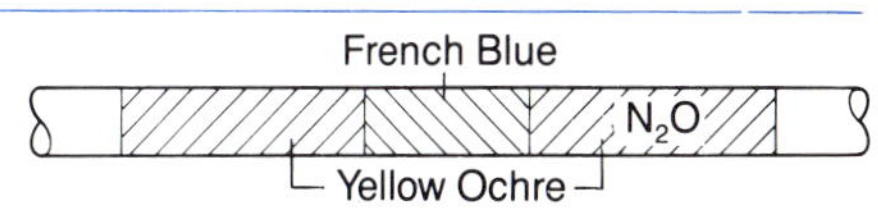

N_2/O_2 MIXTURE

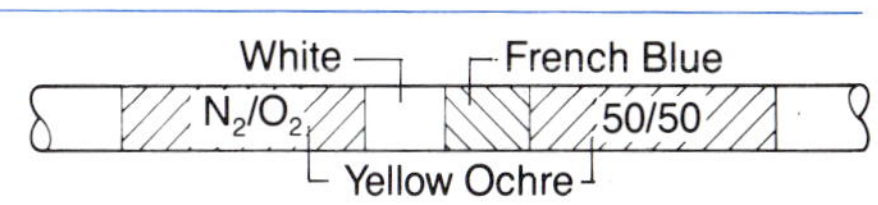

CARBON DIOXIDE

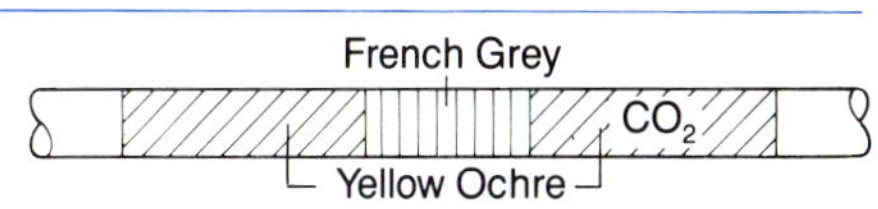

MEDICAL AIR

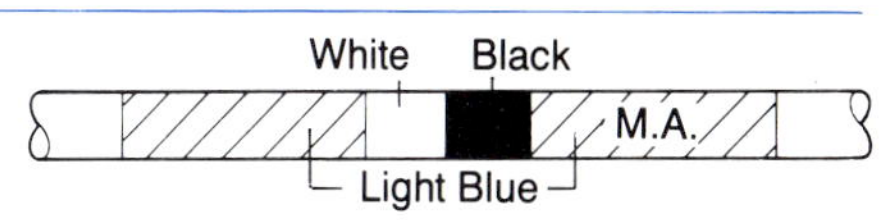

MEDICAL VACUUM

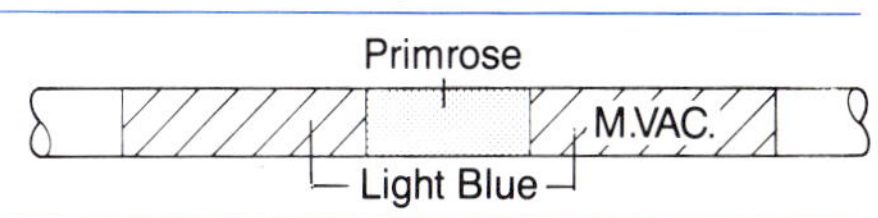

SPARE MEDICAL GAS

All pathological services identified by additional O2 C 37 deep rose pink band.

SECTION N
Conversion Factors and Miscellaneous Data

IDENTIFICATION OF PIPELINES

Table N8 Optional colour code indications for general building services

Pipe contents	Basic colour (approx. 150 mm)		Colour code indication (approx. 100 mm)		Basic colour (approx. 150 mm)
Water					
Drinking	Green		Blue		Green
Cooling (Primary)	Green		White		Green
Boiler feed	Green	Crimson	White	Crimson	Green
Condensate	Green	Crimson	Em. Green	Crimson	Green
Chilled	Green	White	Em. Green	White	Green
Central htg < 100°C	Green	Blue	Crimson	Blue	Green
Central htg > 100°C	Green	Crimson	Blue	Crimson	Green
Cold down service	Green	White	Blue	White	Green
Hot water supply	Green	White	Crimson	White	Green
Hydraulic power	Green		Salmon Pink		Green
Sea, river, untreated			Green		
Fire extinguishing	Green		Safety Red		Green
Compressed air			Light Blue		
Vacuum	Light Blue		White		Light Blue
Steam			Silver Grey		
Drainage			Black		
Electrical conduits and ducts			Orange		
Town Gas					
Manufactured gas	Yellow Ochre		Emerald Green		Yellow Ochre
Natural gas	Yellow Ochre		Yellow		Yellow Ochre
Oils					
Diesel fuel	Brown		White		Brown
Furnace fuel			Brown		
Lubricating	Brown		Emerald Green		Brown
Hydraulic power	Brown		Salmon Pink		Brown
Transformer	Brown		Crimson		Brown
Acids and alkalis			Violet		

BUILDING DRAWINGS

For further information BS 1192 – Building Drawing Practice should be consulted.

Scales

Table N9 Scales for use with metric and imperial systems

Scales for use with SI system	Scales for use with foot/inch system	
1:200	1:192	1/16 in. to 1 ft.
1:100	1:96	⅛ in. to 1 ft.
1:50	1:48	¼ in. to 1 ft.
1:20	1:24	½ in. to 1 ft.
1:10	1:12	1 in. to 1 ft.
1:5	1:4	3 in. to 1 ft.

Where two or more scales are used on the same sheet, these should be particularly clearly indicated.

For sections of drains, sewers, earthworks, etc., it is often convenient to use an exaggerated vertical scale in order to show differences in levels.

Lines

Three thicknesses of lines are used, viz. thick, medium and thin. In general, thick lines are used for site or new building outlines, primary functional elements in section and sectional profiles in component details.

Medium lines are used for general details and secondary elements. Thin lines should be used for dimensional lines, reference grids, leader lines and hatching.

Planes of sections should be indicated by thick broken lines with pointers to show the direction of view, thus:

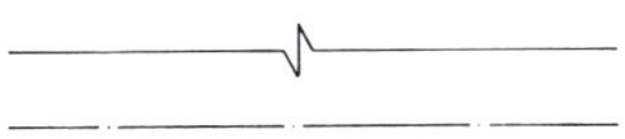
Break line
Centre line

Dimension Lines

Basic dimensions (open arrows)

Tolerance or gap (solid arrows)

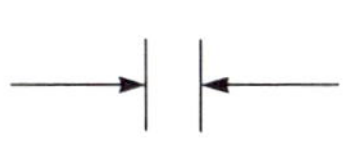

Work size of component (solid arrows)

Dimension Figures

All dimension figures should be written immediately above and along the dimension lines to which they relate, preferably towards the centre.

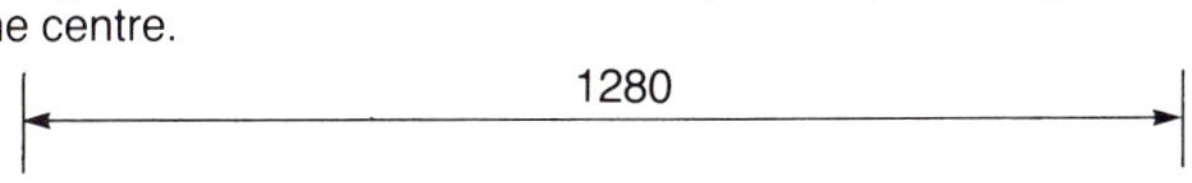

Dimension figures, when not written to be viewed from the bottom of the drawing should be written for viewing only from the right-hand edge.

Sequence of Dimensioning

When dimensions are tabulated, on a drawing or in a schedule, their sequence should be consistent: length, usually the longest, should be given first, width second and height or depth third. Where only two dimensions are given, the same order should be followed.

Symbols

Table N10 Piping Symbols

Position etc.	Symbol
In roof or above ceiling	
At high level	
Vertical drop	
At low level	
Below floor (in screed)	
Direction of flow	
Gradient (rise) or fall	1 in 90

Table N11 Graphical Symbols and Representation
(a) Draughtsmanship

Description	Symbol	Description	Symbol
Centre to centre	C/C	Existing level on plan	X 0.000
Centre line	℄	Existing level on section	⊻ 0.000
Direction of view	→	Finished floor level	FFL
External	ext	Ground level	GL
Internal	int	Required level on plan	X 0.000 (boxed)
Bench mark	BM	Required level on section	⊻ 0.000 (boxed)
Units and Associated Terms			
Approximate	approx	Invert level	IL
Diameter	dia ϕ	Kilometre	km
Diameter, inside	I/D	Metre	m
Diameter	O/D	Millimetre	mm

(b) Components
(i) Drainage

Description	Symbol	Description	Symbol
Discharge pipe	DP	Manhole, surface water	MH
Drainpipe	(See Piping Symbols)	Rainwater head	RWH
Fresh air inlet	FAI	Rainwater outlet	RWO
Grease trap	GT	Rainwater pipe	RWP
Gully	G	Rainwater shoe RWS	
Back inlet gully	BIG	Rodding or cleaning eye	RE or CE
Intercepting trap	IT	Vent pipe	VP
Manhole, soil/foul	MH		

(ii) Sanitary appliances

Description	Symbol	Description	Symbol
Bath (Bth)		Sink (Sk)	S
Bidet (Bt)	BT	Urinal (UR)	

BUILDING DRAWINGS

Table N11 (Cont'd)

Description	Symbol	Description	Symbol
Shower unit (SHR)	S	WC	
		Wash handbasin (WHB)	WB
(iii) Water supply			
Calorifier, horizontal		Hot water tank	HWT
Calorifier, indirect cylinder		Immersion heater	
Cold water storage cistern	CWSC	Mixing valve control (manual)	
Check valve or reflux			
Double check valve assembly			
Combined check valve and vacuum breaker			Direction of flow L–R
Draining tap	DT	Safety valve	
Gas water heater	GWH	Stopvalve	
Header feed and expansion cistern	F&EXPC	Strainer	
Hose bib	HB	Three-way valve	
Hot or cold water draw-off		Water meter	WM
Hot water cylinder	HWC	Water pump	Direction of flow L–R
Anti-vac valve		Towel rail	
(iv) Fire protection			
Fire extinguisher		Hose reel	
Fire hydrant	H	Sprinkler	S
Hose cradle		Fire alarm	
(v) Gas installations			
Gas meter	G	Gas cock	
Gas plug in point	G	Gas point	G
Supply pipe, change of size	25 12	Gas valve	
(vi) Heating and ventilation			
Boiler	B	Radiator	
Surface panel, ceiling type		Thermostat	
Convector heater		Extract fan	
Cooker	C	Humidifier	H

Table N11 (Cont'd)

Description	Symbol	Description	Symbol
		Supply fan	
Electric unit heater			
Unit heater		Unit heater motorised	
Electrically operated valve	3 port	Pneumatically operated valve	3 port
	2 port		2 port

MAXIMUM SPACING OF PIPE SUPPORTS

Table N12 Hot and cold water pipes and gas pipes

Type of piping	Size of pipe (mm)	Spacing for horizontal (m)	Spacing for vertical (m)
Lead	all sizes	0.6*	0.9
Copper (light gauge) and Stainless steel	12	1.2	1.8
	15–28	1.8	2.4
	35–42	2.4	3.0
	54	2.7	3.0
	76–133	3.0	3.6
	159	3.6	4.2
Steel to BS 1387 and Copper (heavy gauge)	10–15	1.8	2.4
	20–25	2.4	3.0
	32	2.7	3.0
	40–50	3.0	3.6
	65–80	3.6	4.5
	100–125	4.0	4.5
	150	4.5	5.4
Grey or ductile iron	50	1.8	1.8
	75–100	2.7	2.7
	150	3.6	3.6

** For pipes conveying hot water and where it is essential to avoid air locks (e.g. primary circulation pipes), continuous support is recommended.*

Table N13 Soil and waste pipes

Type of piping	Size of pipe (mm)	Spacing for horizontal (m)	Spacing for vertical (m)
§Cast iron	all sizes	2.0	3.0
Copper	as for water pipes		
Steel to BS 1387	as for water pipes		
Lead (single tacks)	up to 50	0.6	–
	50 and over	0.9	–
Lead (double tacks)	all sizes	–	1.2
uPVC	32–40	0.5	1.2
	50	0.6	1.2
	75–100	0.9	1.8
	150	1.2	1.8
ABS	32–40	0.5	1.2
	50	0.7	1.2
Polypropylene	32–40	0.5	1.2
	50	0.7	1.2

§At least one fixing must be provided for each unit length.

SPACING OF PIPES

The accompanying table gives the allowable minimum distances between (a) two pipes, (b) pipe and wall and (c) pipe and furring for two pipes with standard malleable or cast iron screwed fittings.

The table has been developed on the basis of the following analysis: between two pipes of differing diameter there will be two values for the minimum distance which will allow free turning of an elbow or tee. The value used will depend upon which pipe is installed first.

In order that there will be no confusion as to which pipe should be installed first, the condition giving the greater distance between pipes will be considered. It is fairly obvious that turning the fitting on the larger pipe will require the large space between pipes for clearance, therefore less space will be required if the larger pipe is installed first.

It is evident that the minimum distance from the wall will be that distance which will allow turning of a fitting on the pipe. This will be the distance K shown in the first column of the table.

The minimum distance from furring (dimension F in the table) will not be less than 6 mm larger than the radius of insulated pipe, because furring is constructed after pipes have been installed.

It should be noted that, for pipes smaller than 25 mm, the distance for insulation is larger than that required for turning fittings. Therefore, on such small pipes, only the allowance for insulation has been considered in the table.

Example

Find the minimum space required for 32 mm and 100 mm pipes if the smaller pipe is located near the wall.

Solution

The table shows, under 32 mm and opposite 100 mm pipes, that 152 mm is the minimum distance between 100 mm and 32 mm pipes.

To the left of 32 mm in the second column of the table, find 54 mm as the minimum distance from the wall of the smaller pipe.

The distance from the furring of the larger pipe is found (above 100 mm) to be 95 mm.

The total space is thus 152 + 54 + 95 = 301 mm.

In the same manner it can be shown that the total space if the pipes were reversed would be 152 + 124 + 51 = 327 mm, so it would be advisable to locate the smaller pipe near the wall or column.

Table N14 Minimum distance allowable between centre lines of screwed steel and copper pipes

Minimum distance from wall (K)	Nom. pipe size	Minimum distance between furring (F), mm											
		38	41	48	51	54	64	70	79	95	114	127	152
		Nominal pipe size, mm											
		15	20	25	32	40	50	65	80	100	125	150	200
mm	mm	Distance between pipes, mm											
216	200	323	238	241	248	251	257	267	276	289	305	321	356
168	150	187	191	194	200	203	210	219	229	241	260	273	
146	125	165	168	171	178	181	187	197	206	219	235		
124	100	140	143	146	152	156	165	171	181	197			
98	90	117	120	124	130	133	140	149	159				
86	65	101	108	111	117	120	127	133					
73	50	92	95	98	101	108	114						
60	40	79	83	86	92	95							
54	32	76	79	83	89								
48	25	73	76	79									
41	20	70	73										
38	15	67											

HEAD AND PRESSURE OF WATER

Note: Pascal Unit of Pressure

The use of various units to describe pressure is creating some confusion, and the increasing use of the Pascal (symbols Pa) in lieu of the N/m² or the bar may lead eventually to it becoming the most commonly used unit of pressure.

This guide does not use the Pascal as a unit of pressure but generally refers to the SI unit for pressure which is the N/m². However, as one Pa is equal to one N/m² it presents no difficulty in relating Pa to N/m² and the symbol has been added to certain tables in this section to enable the user to more fully understand its function.

Table N15 Head and Pressure of Water

Head in metres	Pressure kN/m² or kPa	Pressure mbar & bar	Pressure kN/m² or kPa	Pressure mbar & bar	Head in metres
1	9.81	98 mbar	10	100 mbar	1.02
2	19.61	196 mbar	20	200 mbar	2.04
3	29.42	294 mbar	30	300 mbar	3.06
4	39.23	392 mbar	40	400 mbar	4.08
5	49.03	490 mbar	50	500 mbar	5.10
6	58.84	588 mbar	60	600 mbar	6.12
7	68.65	686 mbar	70	700 mbar	7.14
8	78.45	785 mbar	80	800 mbar	8.16
9	88.26	883 mbar	90	900 mbar	9.18
10	98.07	981 mbar	100	1.00 bar	10.20
11	107.87	1.08 bar	110	1.10 bar	11.22
12	117.68	1.18 bar	120	1.20 bar	12.24
13	127.49	1.27 bar	130	1.30 bar	13.26
14	137.29	1.37 bar	140	1.40 bar	14.28
15	147.10	1.47 bar	150	1.50 bar	15.30
16	156.91	1.57 bar	160	1.60 bar	16.32
17	166.71	1.67 bar	170	1.70 bar	17.34
18	176.52	1.77 bar	180	1.80 bar	18.36
19	186.33	1.86 bar	190	1.90 bar	19.38
20	196.13	1.96 bar	200	2.00 bar	20.40
25	245.17	2.45 bar	250	2.50 bar	25.49
30	294.20	2.94 bar	300	3.00 bar	30.59
35	343.23	3.43 bar	350	3.50 bar	35.69
40	392.27	3.92 bar	400	4.00 bar	40.79
45	441.30	4.41 bar	450	4.50 bar	45.89
50	490.33	4.90 bar	500	5.00 bar	50.99
60	588.40	5.88 bar	600	6.00 bar	61.18
70	686.47	6.86 bar	700	7.00 bar	71.38
80	784.53	7.85 bar	800	8.00 bar	81.58
90	882.60	8.83 bar	900	9.00 bar	91.77
100	980.66	9.81 bar	1 MN/m²	10.00 bar	101.97
200	1.96 MN/m²	19.61 bar	2 MN/m²	20.00 bar	203.94
300	2.94 MN/m²	29.42 bar	3 MN/m²	30.00 bar	305.92
400	3.92 MN/m²	39.23 bar	4 MN/m²	40.00 bar	407.89
500	4.90 MN/m²	49.03 bar	5 MN/m²	50.00 bar	509.86

HYDROSTATIC TABLES

SI Metric Units

1 litre of water weighs 1 kg.
1 cubic metre equals 1000 litres.
Head of water in metres × 9810 = pressure in N/m².
Head of water in metres × 9.81 = pressure in kN/m².
Pressure in kN/m² × 0.102 = head of water in metres.
1 bar equals 1000 mbar = 10^5 N/m² = 10 Pa.
1 millibar (mbar) = 100 N/m² = 100 Pa.

Atmospheric Pressure at Sea Level

Atmospheric pressure = 101 kN/m² = 101 kPa
= 10.33 m head of water
= 762 mm mercury.

Imperial Units

1 gallon of water weighs 10 lb.
1 gallon equals 0.16 cu. ft.
6.25 gallons equal 1 cu. ft.
Head of water in ft. × 62.5 = pressure in lbf/ft².
Head of water in ft. × 0.434 = pressure in lbf/in².
Pressure in lbf/in² × 2.3 = head of water in ft.

Atmospheric Pressure at Sea Level

Atmospheric pressure = 14.7 lbf/in²
= 34 ft head of water
= 30 in in mercury.

RELATION OF SERVICES TO EACH OTHER

Table N16 Relation to Other Services

Pipe or duct conveying	Relation to other services: Cold water	Natural gas	Electricity	Telecom-munications	Hot water and steam	Ventilation and air conditioning	Drainage	Oil	Flammable gases and liquids	Compressed air
Hot water and steam	A	O	A	A	—	F	C	C	L	O
Ventilation and air conditioning	F	L	A	A	F	—	C	C	L	O
Cold water	—	O	A	A	F	F	C	C	C	O
Electricity	C	F	—	M	B	B	C	C	L	O
Telecommunications	C	F	M	—	B	B	C	C	L	O
Natural gas	C	—	F	F	C	L	A C	C	L	G
Drainage	D	D	A	A	D	D	—	D	L	D
Oil	C	J	A	A	B	B	C	—	O	O
Flammable gases and liquid	C	L	L	L	L	L	L	O	—	L
Compressed air	O	G	O	A	O	O	C	O	L	—

Note: It is on occasions necessary to run different services to the same position to operate equipment connected directly to a number of services, eg. motor-driven pumps, thermostatic valves, electrically operated gas valves, etc. The above precautions should be taken in each instance as far as practicable.

Key to Symbols in Above Table

O. No precautions necessary.
A. Not to be run adjacent to.
B. Not to be exposed to temperature or leakage from.
C. Not to be exposed to leakage from.
D. Not to be exposed to effluent from.
F. Shall be prevented by spacing, insulation or other means from coming into contact with.
G. Not to be run in such a position as would cause any escape of air to circulate a gas leakage, or to make the leakage become more dangerous.
J. Not to be run where an escape of gas could cause risk of ignition in presence of oil.
L. Not to be run in same duct.
M. Not to be run adjacent to without further precautions.

APPROXIMATE WEIGHT OF LEAD REQUIRED FOR CAULKED LEAD JOINTS

The following tables give the approximate weight of molten lead required for caulked lead joints. Exact amounts will depend upon the operator and depth of gaskin introduced. Joints made with spun lead or lead rope would require much greater weights.

Table N17 Cast iron soil pipes to BS 416

Dia. of pipe mm		50	65	75	90	100	150
Weight of Lead	lb.	1.5	2.0	2.5	3.0	3.5	5.0
	kg	0.7	0.9	1.1	1.4	1.6	2.3

Table N18 Cast iron pipes for water, gas and sewage to BS 4622 and spun pipes to BS 1211

Dia. of pipe	in.	3	4	*5*	6	*7*	8	*9*	10
	*mm	75	100	125	150	175	200	225	250
Weight of lead	lb	4.0	5.0	6.0	7.0	8.0	10.5	11.5	13.0
	kg	1.8	2.3	2.7	3.2	3.6	4.8	5.2	5.9
Dia. of pipe	in.	12	*14*	15	*16*	18	*20*	21	24
	mm	300	350	375	400	450	500	525	600
Class A–B	lb	15.0	19.0	20.5	22.0	30.5	34.0	36.0	45.0
Weight of lead	kg	6.8	8.6	9.3	10.0	13.8	15.4	16.3	20.4
Class C–D	lb	15.5	19.5	21.0	22.5	31.5	35.0	37.0	46.0
Weight of lead	kg	7.0	8.8	9.5	10.2	14.3	15.9	16.8	20.9

**Sizes of pipes given in mm are nominal. Figures in Italic* type are not preferred diameters.

CORROSION OF METALS

The farther apart two metals appear in the table the more actively will the two metals react when placed in contact in a slightly acid aqueous solution.

Such conditions can occur when dissimilar metals are used in roof work, especially in towns or other manufacturing centres with high atmospheric pollution, which might lead to failure of one of the metals.

Similarly, failure might occur by electrolytic action when dissimilar metals are employed in plumbing services, but this will depend to a large extent on the nature of the water conveyed.

Where any doubt exists as to the suitability of any particular metal or combination of metals, guidance should always be sought from the local water authority.

Table N19 The Electro Chemical Series

Metal	Chemical Symbol	Normal Electrode Potential (volts)
'Noble' or Cathodic, ie. protected end		
Gold	Au	+ 1.42
Platinum	Pt	+ 1.20
Silver	Ag	+ 0.80
Mercury	Hg	+ 0.80
Copper	Cu	+ 0.345
Lead	Pb	− 0.125
Tin	Sn	− 0.135
Nickel	Ni	− 0.250
Cadmium	Cd	− 0.40
Iron	Fe	− 0.44
Chromium	Cr	− 0.71
Zinc	Zn	− 0.76
Aluminium	Al	− 1.67
Magnesium	Mg	− 2.34
Sodium	Na	− 2.71
Potassium	K	− 2.92
Lithium	Li	− 3.02
'Base' or Anodic, ie. corroded end		

COEFFICIENTS OF LINEAR EXPANSION AND MELTING JOINTS

Table N20 Metals and glass

Material		Coefficient of linear expansion per °C	Melting point °C
Aluminium	cast	25.56×10^{-6}	655
	sheet	23.00	655
Brass	cast	18.77	926
	sheet	19.35	1010
Copper	sheet	17.40	1050
	tube	16.90	1050
Iron	cast	10.22	1200
	wrought	11.88	2000
Lead	sheet and pipe	29.30	325
Steel	mild	11.34	1300
	nickel chrome	15.3	—
Tin		21.42	228
Zinc	sheet	26.1	438
Glass	tube	33×10^{-7}per cm	600

Table N21 Plastics

Material		Coefficient of linear expansion per °C	Melting point °C
Acrylonitrile butadiene styrene (ABS)		$83–95 \times 10^{-6}$	—
Cellulose acetate butyrate (CAB)		80–170	
Nylon		70–100	
Polyethylene	low density	280	—
	high density	110–130	—
Polypropylene		110	—
Polyvinyl chloride (uPVC)	normal impact	50	—
	high impact	80	—

DEFINITION OF PLUMBING TERMS

Cold Water Supply

Air vessel	A closed chamber which utilizes the compressibility of contained air, either (i) to promote a more uniform flow of water when connected to the delivery pipe or suction pipe of a reciprocating pump, or to the delivery pipe of a hydraulic ram, or (ii) to minimize shock due to water hammer when connected to a high pressure system.
Air-water storage vessel	A tank or cylinder in which water is stored under the pressure of air compressed in its upper part and forming part of a pumped water supply system. Variations in the amount of water stored are accompanied by variations in the pressure of the air and water and can be used to control the starting and stopping of the pumps.
Backflow	Flow in a direction contrary to the natural or intended direction of flow.
Backnut	1. A locking nut provided on the screwed shank of a tap, valve or pipe fitting for securing it to some other object. 2. A thin threaded nut, dished on one face to retain a grommet, used to form a watertight joint on a long-threaded pipe.
Capacity **Actual capacity**	1. The volume contained in a vessel. 2. Of a storage cistern. The capacity of the cistern measured up to the water line.
Cavitation	A phenomenon in the flow of water consisting in the formation and collapse of cavities in the water bringing about pitting on float operated valve outlets and seatings of valves.
Cistern	A fixed container for water in which the water is at atmospheric pressure. The water is usually supplied through a float operated valve.
Storage cistern **Water Byelaw definition:**	A cistern for storing water. Any cistern, other than a flushing cistern, having a free water surface under atmospheric pressure but does not include a drinking trough or drinking bowl for animals including poultry.
Stopvalve	A valve fitted in a pipeline for isolating the flow of water.
Cross connection	1. A connection between two normally independent pipelines which permits flow from either pipeline into the other. Note: Such a connection may be a possible source of contamination of a water supply. 2. An irregular connection between a foul drainage system and a surface water drainage system.
Cupro-solvency	The ability of some waters to dissolve copper.
Disk feed	A method of regulating the rate of flow into an automatic flushing cistern by the use of a thin disk, with a calibrated orifice, mounted in a union in the feed pipe.
Diversity factor **Probability of simultaneous demand.** **Probability of simultaneous discharge**	A factor used in the design of pipework to determine the maximum rate of flow to allow for in a pipe. When there is a large number of draw-off points it is improbable that all will be in use at the same time. Similarly, where a large number of outlets discharge into one outlet pipe or channel it is improbable that all will be used at the same time. The diversity factor is the probable rate of flow divided by the possible total rate of flow.
Hardness	(of water) That property of water, caused by certain mineral matter dissolved in it, which resists the lathering of soap. Hardness is assessed by expressing the particular mineral content concerned as parts per million (by weight) or by milliequivalents per litre. Formerly it was expressed as 'degrees Clark' which was the mineral content in grains per gallon.
Temporary hardness	That part of the hardness of water which can be removed by boiling, when part of the mineral content is precipitated as solid matter.
Permanent hardness	That part of the hardness of water which remains after the water has been boiled.
HEAD **Pressure**	The vertical height of a column of water which would produce at its base, by virtue of its own weight, a pressure equal to that to which the water is actually subjected.
Static	The pressure head when the water is at rest.
Total	The sum of the position, pressure and velocity heads.
Working head	The head of water at which a pipe, vessel or pump is designed to work.
JOINT **Autogenous welded**	A welded joint in which two parts made of the same metal are welded together with or without the use of a filler rod of the same metal.
Brazed	A joint in which the parts are united with molten brass. (Silver solder) (Many proprietary brands of rod on market.)
Bronze welded	A welded joint in which the parts are united with molten bronze. The process is distinguished from brazing by the local building up of the welding material.
Capillary	A fine clearance spigot and socket joint into which molten solder is caused to flow by capillary action.
Manipulative type compression	A compression joint which requires the end of the pipe to be shaped outwards before assembly to enable the pipe fitting to grip the pipe.
Non-manipulative type compression	A compression joint which does not require the end of the pipe to be shaped before assembly because the pipe fitting is made to grip the outside wall of the plain pipe by means of a compression ring.

DEFINITION OF PLUMBING TERMS

Fusion	A joint made on certain types of plastics piping, eg. polyethylene in which the surfaces which are to be in contact in a closely fitting spigot and socket are separately heated. While both are still hot, the spigot is inserted in the socket, whereupon the two become fused together. Alternatively, pipes in larger sizes can be butt jointed.
Shrunk rubber joint ring	An 'O' ring joint used on certain types of plastics piping, eg. PVC, in which, by the application of heat, the socket is shrunk on to the spigot thereby entrapping the rubber ring.
Solvent welded **Solvent cement**	A joint made on certain types of plastics piping, eg. PVC, in which the surfaces which are to be in contact in a closely fitting spigot and socket are separately smeared with a solvent, after which the spigot is inserted in the socket, whereupon the two become united.
Loading unit	A unit so chosen that the relative water requirements of sanitary appliances can be expressed as multiples of that unit. The loading unit rating of an appliance depends on the duration and rate of flow required at the appliance and on the interval between usage.
PIPES **Standing overflow**	An overflow pipe, consisting of a vertical tube, standing in a cistern and passing through its base.
Distributing pipe– **Water Byelaw definition**	Any pipe (other than an overflow pipe or a flushing pipe) conveying water from a storage cistern, or from a hot water apparatus supplied from a feed cistern, and under pressure from that cistern.
Overflow pipe	A pipe connected to a vessel, sanitary appliance, sewer or chamber to discharge excess water.
Service pipe– **Water Byelaw definition**	So much of any pipe, for supplying water from a main to any premises, as is subject to water pressure from that main, or would be so subject but for the closing of some stopvalve.
Warning pipe– **Water Byelaw definition**	An overflow pipe so fixed that its outlet, whether inside or outside a building, is in a conspicuous position where the discharge of any water therefrom can be readily seen.
Pipe ring **Pipe ring, single** **Split pipe ring**	A ring-shaped clamp, made in halves for screwing or bolting together, which forms part of an assembly for supporting a pipe. One half of the ring incorporates an integral socket screwed internally to receive the threaded end of the supporting unit.
Pipe union	A union for connecting together two pipes. It is made in three parts, two of which are for fixing to the pipe ends, the third being a coupling nut.
Plumbo-solvency	The ability of some waters to dissolve lead.
Back-siphonage	Backflow caused by siphonage of a liquid from an appliance or storage vessel into the pipe feeding it.
Elbow action tap	A quarter turn tap with a lever handle which can be operated by the forearm.
Wrist action tap	A quarter turn tap with a short lever handle which can be operated by the wrist.
Delayed action float operated valve	A float operated valve with its float riding on the surface of the water in an open-topped chamber fitted within a cistern. The opening of the valve is delayed until the water level in the cistern has fallen through a fixed distance, when the chamber empties itself into the cistern through a non-return valve. When the water level rises again, the chamber is refilled by flow over its top edge, thus quickly closing the float operated valve.
Equilibrium float operated valve	A float operated valve so designed that the hydraulic forces acting on the closing disk or piston are hydraulically balanced.
Reverse action float operated valve	A float operated valve, for use in large automatic flushing cisterns, which is open when the float is at top water level and closed when the float is at bottom water level. A pet cock on the supply side of the valve initiates the operation.
Water hammer	A hammering sound caused by violent surges of pressure in water pipes.
Water line	A line marked inside a cistern to indicate the highest water level at which the float operated valve should be adjusted to shut off.
Working pressure	The pressure at which a pipe, vessel or pump is designed to work.

Hot Water Supply

Bower-barffing	A process for rust-proofing cast iron or mild steel in which the metal is raised to red heat and treated with live steam.
Tongue tee	A tee having an internal projection so placed as to divert part of the flow from the main pipe into the branch pipe.
Breech fitting **Breeches fitting** **Breeching fitting**	A symmetrical pipe fitting in which two parallel pipes unite into one pipe. The flow may be in either direction according to the purpose for which the fitting is used.
Calorifier	A storage vessel, not open to the atmosphere, in which a supply of water is heated. The vessel contains an element, such as a coil of pipe, through which is passed a supply of hot water or steam, in such a way that the two supplies do not mix, heat being transferred through the walls of the element.

Conversion Factors and Miscellaneous Data

Cathodic protection	A method of protecting metal pipes or vessels exposed to corrosion. The corrosive action is modified so that the metal to be protected becomes the cathode instead of the anode, and this is done either (i) by attaching a block of other metal, higher in the electrochemical series, such as magnesium, which then becomes the anode and suffers corrosion, or (ii) by applying a small electric current from an external source.
Primary circuit	A circuit in which water circulates between a boiler or other water heater and a hot water storage vessel.
Secondary circuit	A circuit in which water circulates in distributing pipes from and back to a hot water storage vessel.
Closed circuit	A system of pipes and fittings in which the same water is circulated and from which no water is drawn off for use, as in a hot water space heating system.
Water Byelaw definition	Any system of pipes and other water fittings through which water circulates and from which water is not drawn for use, and includes any vent pipe fitted thereto but not the feed cistern or the cold feed pipe.
Feed cistern	A cistern for supplying cold water to a hot water system.
Water Byelaw definition	Any storage cistern used for supplying cold water to a hot water apparatus, cylinder or tank.
Expansion cistern	A cistern forming part of a hot water heating system, which accommodates the increase in volume of the water when hot; it can also act as the feed cistern for the system.
Cradle	A support shaped to fit the underside of a pipe, cylinder or appliance.
Creep	A slow plastic deformation or movement of a material under stress.
Cylinder, hot water Water Byelaw definition	A cylindrical closed vessel for storing hot water under pressure greater than atmospheric pressure.
Indirect cylinder	A cylindrical calorifier with a heating element of annular shape, inside.
Dead leg	A length of hot water pipe leading to a draw-off point and not forming part of a circuit.
Expansion loop	A loop in a pipeline which, by flexing, can accommodate expansion and contraction movement.
Expansion joint	A joint which permits relative movement of the jointed parts caused by expansion and contraction due to temperature change.
One-pipe circulation	The circulatory flow of a stream of hot water in one direction and a stream of cooler water in the opposite direction at the same time and in the same pipe.
Parasitic circulation	In a domestic hot water system the unwanted circulation of hot water in a pipe circuit, intended to be temporarily inactive, or in a vent pipe.
Cold feed pipe Feed pipe	A distributing pipe conveying cold water from a cistern to a hot water apparatus.
Vent pipe Exhaust pipe Expansion pipe	A pipe, in a hot water appliance, for the escape of air and for the same discharge of any steam generated.
Plumbing unit	A pre-fabricated plumbing assembly of pipes and fittings with a supporting framework.
Direct system	A hot water supply system in which the water supplied to the draw-off points is heated by a primary source of heat such as solid fuel, gas, electricity or oil.
Indirect system	A hot water system in which the water supplied to the draw-off points is heated by means of a calorifier.
Spray mixing tap	A tap, supplied with hot and cold water and incorporating a mixing device operated by the user. The mixed water is delivered at a restricted rate of flow in the form of a spray.
Spray tap	A tap supplied with water at a predetermined temperature which it delivers at a restricted rate of flow in the form of a spray.

Sanitation

Air admittance valve	A check valve which admits air into a soil/waste stack system.
Inverted branch	A branch fitting, for a soil or waste ventilating pipe, on which the branch, which may terminate in either a socket or spigot, is either inclined below the horizontal or is parallel to the main pipe and facing downwards. (The branch is intended for the connection of a branch ventilation pipe.)
Chair	A metal frame for building into a thin partition wall and the floor so as to provide means of supporting a wash basin, W.C. pan or other sanitary appliance clear of the floor.
Flushing cistern	A cistern provided with a device for rapidly discharging the contained water and used in connection with a sanitary appliance for the purpose of cleansing the appliance and carrying away its contents into a drain. Note: The nominal size of a cistern is the quantity of water discharged per flush.
Dual flush cistern	A W.C. flushing cistern which gives flushes of two different volumes, either of which can be selected by the user.
Cross vent Yoke vent	A short relief vent between the main soil or waste pipe and a main ventilating pipe.
Crown	1. The highest point of the internal surface of a pipe or culvert at any cross section. 2. Of a trap. The topmost point of the outlet leg of a trap.

DEFINITION OF PLUMBING TERMS

Discharge unit — A unit so chosen that the relative load-producing effect of sanitary appliances can be expressed as multiples of that unit. The discharge unit rating of an appliance depends on its rate and duration of discharge and on the interval between discharges.

Flushing trough
Trough cistern — A flushing apparatus which combines several discharging units in one long cistern body to allow the more frequent flushing of the W.C. pans in a range than is possible with individual flushing cisterns.

Garchey system — A system for the conveyance by water carriage, and for the disposal at a central plant, of domestic refuse from multi-storey buildings.

Group connector — A cast iron or uPVC double or triple branch fitting, on a drain, with socketed branches to receive waste pipes and a spigoted outlet for connecting to the socketed inlet of a gully or grease trap.

Loop vent — A branch ventilating pipe which, after being carried above the spillover level of the sanitary appliance it serves, is connected at a lower level to a soil or waste pipe, which is not freely ventilated to atmosphere, or to the drain serving the sanitary appliance.
Note: In the United States the term is used with the following meaning: 'A branch ventilating pipe which serves the topmost range of not more than 8 sanitary appliances. The upper end is connected to a stack vent and the lower end to the horizontal branch soil or waste pipe between the two appliances most remote from the discharge stack.'

Soil pipe
Soil conductor — A pipe which conveys to a drain the discharge from a water closet, urinal or slop hopper. In the 'one pipe' and 'single stack' systems the soil pipe also conveys to a drain the discharges from baths, wash basins, sinks and similar appliances.

Soil-waste discharge pipe — A pipe for conveying both soil and waste water.

Ventilating pipe
Soil ventilating pipe — A pipe in a soil and/or waste pipe system which facilitates the circulation of air within the system and protects trap seals from excessive pressure fluctuation.

Waste pipe
Waste conductor
Waste, deprecated — A pipe which conveys to a drain or soil pipe the discharge from a sanitary appliance used for ablutionary, drinking or culinary purposes.

Waste ventilating pipe — A ventilating pipe in a waste pipe system.

Receiver, diluting — A vessel, into which a laboratory sink waste pipe discharges. It is arranged so that its contents start to drain only when it is full, thus ensuring that, if a corrosive liquid is emptied into a laboratory sink connected to a diluting receiver there is a chance of that corrosive liquid being diluted before discharge into the waste pipe or drainage system. Sometimes the outlet is trapped.

Relief vent — An additional ventilating pipe connected to a soil or waste pipe at any point where excessive pressure fluctuation is likely to occur.

Induced-siphonage — The extraction of water from a trap by siphonage set up by reduction of pressure at the outlet of the trap.

Self-siphonage — The extraction of water from a trap by siphonage set up by the momentum of the discharge from the sanitary appliance to which the trap is attached.

Stack vent — The extension of a soil or waste stack, above the highest branch soil or waste pipe connection, terminating in an end open to atmosphere.

Two-pipe system — A soil and waste system comprising two independent pipes, namely, a soil pipe conveying soil directly to the drain and a waste pipe conveying waste water to the drain through a trapped gully. The system may also require ventilating pipes.

One-pipe system — A soil and waste system in which a single soil-waste pipe conveys soil and waste water directly to the drain. The system may also require ventilating pipes.

Single-stack system — A one-pipe system from which, subject to the observance of certain stringent rules, all or most of the trap ventilating pipes are omitted.

Spray tap basin — A wash basin designed to receive the discharge from a spray tap without splashing. It normally has one taphole and no overflow and is fitted with a grated waste.

Performance test — A test for the stability of trap seals in soil and/or waste pipe systems.

Catchpot trap — A bottle trap with a removable lower part large enough to retain waste for examination or recovery.

Deep seal trap — A trap having a water seal 3in. or more deep.

Resealing trap — A trap designed to retain an effective water seal after relieving excessive pressure fluctuations either at the inlet or outlet of the trap.

Running trap — A tubular trap having the inlet and outlet in horizontal alignment.

Vent stack — A main ventilating pipe.

Wash basin — A sanitary appliance intended for washing the upper parts of the body. It has a waste connection and a piped water supply and is generally of ceramic ware but can also be of metal or of plastics.

Grated waste — A waste fitting with an integral grating flush with its inlet end and hence incapable of accepting a waste plug.

Depth of water seal — The depth of water which would have to be removed from a fully charged trap before air could pass through the trap.

Wet venting — The use of the lower portion of the ventilating pipe, in a waste pipe system or a soil/waste pipe system, as a waste pipe.

Drainage

Aerobic — Pertaining to the action of micro-organisms in the presence of oxygen.

Anerobic — Pertaining to the action of micro-organisms in the absence of oxygen.

Anti-splash shoe — A rainwater fitting fixed at the lower end of a rainwater pipe and so shaped as to reduce splashing when rainwater is discharging into the open air.

Back-drop
Tumbling bay — The vertical fall in a back-drop connection.

Back-drop connection — A vertical or steeply sloping connection to or near the invert level of a manhole from a sewer or drain at a higher level.

Balcony outlet — A fitting intended to be interposed in a vertical rainwater pipe passing through a balcony, and providing an inlet for the drainage of rainwater from the balcony.

Benching — Sloping surfaces so constructed on either side of channels at the base of a manhole as to discourage the accumulation of deposits.

Barron's bend
Slipper bend — A short three-quarter section branch bend.

Knuckle bend — A short radius bend.

Trapping bend — An unsocketed 90° cast iron bend, with one long arm for use as a dip pipe in, for example, a petrol intercepting trap.

Catchpit — A pit excavated or a chamber constructed in and below the normal level of a ditch, drain or sewer to trap silt and solid matter and facilitate its removal.

Cesspool — 1. An underground chamber for the reception and storage of foul water, the contents of which are periodically removed for disposal.
2. A box-shaped receiver constructed in a roof or gutter for collecting rainwater which then passes into a rainwater pipe connected thereto.

Foul water drain — A drain intended to convey foul water.

Process drain — A drain intended to convey process or trade wastes only.

Subsoil drain
Subsoil water drain — 1. A drain intended to collect and carry away subsoil water.
2. A drain intended to disperse into the subsoil the effluent from a septic tank or an overflowing cesspool.

Surface water drain — A drain intended to convey surface water only.

Drain chute — A tapered drain fitting fitted to the inlet or outlet of a manhole to facilitate rodding.

Drop connection — A length of drain laid vertically immediately before its connection to a sewer or to another drain.

Dry weather flow — 1. The foul water, together with infiltration, if any, flowing in a sewer or drain in dry weather.
2. The rate of flow of foul water, together with infiltration, if any, in a sewer or drain in dry weather.

Puddle flange
Puddling flange — A flange on a pipe, at the point at which it will pass through a water retaining structure, to increase the resistance to leakage along the exterior of the pipe.

Hydraulic gradient — The loss of head in liquid flowing in a pipe or channel expressed per unit length of the pipe or channel. In a channel the hyraulic gradient is equal to the slope of the free surface of the flowing liquid.

Hydraulic mean depth — A factor used in calculating the rate of flow of a liquid in a pipe or channel. It is obtained by dividing the cross-sectional area of the liquid by the length of the wetted perimeter of the pipe or channel.

Invert — 1. The lowest point of the internal surface of a pipe or channel at any cross section.
2. The lower portion of the internal surface of a drain, sewer or channel.

Joinder junction — A junction pipe in which the branch is manufactured with a closed end designed to be cut off when the branch is required for use.

Flushing manhole — A manhole at which water for flushing is introduced into a drainage system.

Rainwater shoe — 1. A rainwater fitting at the foot of a rainwater pipe to discharge rainwater into the open air clear of the building surface to which the pipe is fixed.
2. A drain fitting, fixed horizontally at the foot of a rainwater pipe, having a vertical or horizontal inlet and an inspection opening with either a grating or a sealed cover.

Raising piece
Gully riser
Rising piece — A fitting for extending the height of a gully or of a rainwater shoe. A gully raising piece may have branch inlets.

Ramp — A short length of a pipeline or channel laid at a considerably steeper gradient than the adjoining portions.

DEFINITION OF PLUMBING TERMS

Rest bend **Duck foot bend**	A cast iron or stoneware bend, having a foot formed integrally in its base, used to receive a vertical pipe or line of pipes.
Self-cleansing velocity	The velocity of a flowing liquid in a pipe or channel necessary to prevent the deposition of solids in suspension.
Soffit **Crown**	The highest point of the internal surface of a pipe or culvert at any cross section.
Combined system	A drainage system in which foul water and surface water are conveyed by the same pipes.
Separate system	A drainage system in which foul water and surface water are conveyed in separate pipes.
Concentric taper **Concentric taper pipe**	A taper for connecting together pipes having their axes in the same straight line.
Level invert taper **Level invert taper pipe**	A taper for connecting together pipes having their inverts in the same straight line.
Buchan trap	(Scottish) An intercepting trap with a horizontal inlet having also a vertical socket on the inlet side of the trap for the connection of an access pipe. It is also obtainable with means of access on the outlet side of the trap.
High invert trap **Low back trap (Scottish)**	A trap with a vertical inlet in which the level of the outlet invert is high in relation to the inlet.
Intercepting trap with reverse clearing arm	An intercepting trap designed to be fixed on the inlet side of an interceptor chamber.
Reverse action intercepting trap	An intercepting trap which is used when the water from a subsoil drainage system needs to be brought into the main drainage system of a building. A ventilating pipe is joined to the trap at the end of the subsoil drain.
Low invert trap	A trap with a vertical inlet in which the level of the outlet invert is low in relation to the inlet.
Anti-flooding valve	A non-return valve, for use on a drain, usually consisting of a floating ball and seating, preventing the backing up of flood water.
Wetted perimeter	The line of contact between a pipe or channel, and the liquid flowing therein, at a cross section.

Note: Where more than one title is given to a definition, the first is the preferred term.

USEFUL ADDRESSES

Addresses are classified as follows:

1 Employers' Trade Associations
2 Manufacturers' and other Trade Bodies
3 Professional Bodies
4 Research and Advisory Bodies
5 Regional Water Authorities and Water Companies in England and Wales
6 Fuel and Energy Interests
7 Government Departments
8 Training Organisations
9 Trade Unions
10 Safety Organisations
11 Overseas

1 Employers' Trade Associations

Association of Boiler Setters, Chimney and Furnace Constructors
64 Queen Street, London EC4 Tel: 01-248 1755

British Constructional Steelwork Association
35 Old Queen Street, Westminster, London SW1H 9HZ
Tel: 01-222 2254

Building Employers' Federation
82 New Cavendish Street, London W2 4JG Tel: 01-229 2488

Committee of Associations of Specialist Engineering Contractors (CASEC)
ESCA House, 32-34 Palace Court, London W2 4HY
Tel: 01-229 1266

Electrical Contractors' Association (ECA)
ESCA House, 32-34 Palace Court, London W2 4HY
Tel: 01-229 1266

Electrical Contractors' Association of Scotland
23 Heriot Row, Edinburgh EH3 6EW Tel: 031-225 7221/2

Federation of Associations of Specialists and Sub-Contractors (FASS)
19-20 Charter House, Lord Montgomery Way, Portsmouth PO1 2SB Tel: 0705-861 361

Federation of Civil Engineering Contractors
Cowdray House, 6 Portugal Street, London WC2A 2HH
Tel: 01-404 4020

Federation of Master Builders
33 John Street, London WC1N 2BB Tel: 01-242 7583

Heating and Ventilating Contractors' Association (HVCA)
ESCA House, 32-34 Palace Court, London W2 4JG
Tel: 01-229 2488

House Builders' Federation
82 New Cavendish Street, London W1M 8AD Tel: 01-580 4041

Metal Roofing Contractors' Association
c/o Drake and Scull Ltd, Hamlyn House, Highgate Hill, London N19 Tel: 01-272 0233

National Association of Plumbing, Heating and Mechanical Services Contractors (NAPHMSC)
6 Gate Street, London WC2A 3HX Tel: 01-450 2678

National Federation of Roofing Contractors
15 Soho Square, London W1V 5FB Tel: 01-439 1743

Scottish & Northern Ireland Plumbing Employers' Federation (SNIPEF)
2 Walker Street, Edinburgh, EH3 7LB Tel: 031-225 2255

2 Manufacturers' and Other Trade Associations

Aluminium Federation
Broadway House, Calthorpe Road, Five Ways, Birmingham B15 1TN Tel: 021-455 0311

Asbestos Cement Manufacturers' Association
602 Castle Lane West, Bournemouth, Dorset BH3 9YF
Tel: 0202-301 260

Association of British Roofing Felt Manufacturers
69 Cannon Street, London EC4N 5AB Tel: 01-248 4444

Association of Bronze and Brass Founders
c/o Heathcote & Coleman, 136 Hagley Road, Birmingham, B16 9PN Tel: 021-454 4141

Association of Industrial Building Component Manufacturers
26 Store Street, London WC1E 7BT Tel: 01-636 5400

Association of Installers of Unvented Hot Water Supply Systems (Scotland and Northern Ireland)
2 Walker Street, Edinburgh EH3 7LB Tel: 031-225 2255

Boiler and Radiator Manufacturers' Association
Fleming House, Renfrew Street, Glasgow C3 6TG
Tel: 041-332 0826

British Bath Manufacturers' Association
Fleming House, Renfrew Street, Glasgow C3 6TG
Tel: 041-332 0826

British Combustion Equipment Manufacturers' Association
The Fernery, Market Place, Midhurst, West Sussex
Tel: 073-081 2782

British Compressed Gases Association
Alembic House, 93 Albert Embankment, London SE1 7TU
Tel: 01-735 3001-8

British Electrical & Allied Manufacturers' Association Ltd
Leicester House, 8 Leicester Street, London WC2H 7BN
Tel: 01-437 0678

British Industrial Measuring and Control Apparatus Manufacturers' Association
20 Peel Street, London W8 7PD Tel: 01-727 2614

British Insurance Association
PO Box 538, Aldermary House, Queen Street, London EC4
Tel: 01-248 4477

British Malleable Tube Fittings Association
The Chestnuts, Otley, Ipswich, Suffolk IP6 9NS
Tel: 047339 340

British Mechanical Engineering Confederation
112 Jermyn Street, London SW1Y 4UR Tel: 01-930 0865

British Non-Ferrous Metals Association
6 Bathurst Street, London W2 2SD Tel: 01-723 7465

British Plastics Federation
5 Belgrave Square, London SW1X 8PH Tel: 01-235 9483

British Plumbing Fittings Manufacturers' Association
10 Greenfield Crescent, Birmingham, B15 3AU
Tel: 021-456 3322

British Pump Manufacturers' Association
3 Pannel Court, Chertsey Street, Guildford, Surrey GU1 4EU
Tel: 0483 37997

British Steel Corporation
PO Box 403, 33 Grosvenor Place, London SW1
Tel: 01-235 1212

British Valve Manufacturers' Association
Artillery House, Artillery Row, London SW1P 1RT
Tel: 01-799 2241

British Effluent & Water Association
5 Castle Street, High Wycombe, Bucks HP13 6RN
Tel: 0494 444544

British Bathroom Council
Federation House, Stoke-on-Trent, Staffordshire ST4 2RT
Tel: 0782 747074

Building Societies Association
14 Park Street, London W1Y 4AL Tel: 01-629 0515

Confederation of British Industry
Centre Point, 103 New Oxford Street, London WC1A 1DU
Tel: 01-379 4700

Contractors' Plant Association
28 Eccleston Street, London SW1 Tel: 01-730 7117

Energy Systems Trade Association (ESTA)
11 Colinsmith Drive, Grove, Nr Wantage, Oxon
Tel: 02357 67940

Energy Users Research Association (EURA)
PO Box 97, Altrincham, Cheshire WA14 5HT Tel: 02357 67940

Engineering Employers' Federation
Broadway House, Tothill Street, London SW1H 9NQ
Tel: 01-222 7777

Engineering Equipment Users' Association
20 Grosvenor Gardens, London SW1 Tel: 01-235 5316

Eurisol-UK, Association of British Manufacturers of Mineral Insulating Fibres
39 High Street, Redbourn, Herts AL3 7LW Tel: 058285 4624

Fire Extinguishing Trade Association
Beech House, 35-37 London Road, Reigate, Surrey RH2 9QU Tel: 07372 41841

Foundry Equipment & Supplies Association Ltd
21 John Adam Street, London WC2N 6JH Tel: 01-839 6171

Galvanizers' Association
34 Berkeley Square, London W1X 6AJ Tel: 01-499 6636

Hevac Association
Sterling House, 6 Furlong Road, Bourne End, Bucks, SL8 5DG Tel: 06285 31186

High Performance Pipe Association
Service House, 61-63 Rochester Road, Aylesford, Maidstone, Kent ME20 7BS Tel: 0622 882093

Hydraulic Association
Glen House, Stag Place, London SW1 Tel: 01-834 8740

Industrial Warm Air Heater Manufacturers' Association
Plum Tree Court, London EC4 Tel: 01-583 3000

Lead Contractors' Association
34 Berkeley Square, London W1X 6AJ Tel: 01-499 8422

Manufacturers of Domestic Unvented Supply Systems Equipment (MODUSSE)
c/o Building Centre, 26 Store Street, London WC1E 1BT
Tel: 01-637 1022
or The MODUSSE Director Tel: 0732 850744

Metal Sink Manufacturers' Association
Fleming House, Renfrew Street, Glasgow G3 6TG
Tel: 041-332 0826

National Association of Lift Makers
Leicester House, 8 Leicester Street, Leicester Square, London WC2 Tel: 01-437 0678

National Brassfoundry Association (NBA)
5 Greenfield Crescent, Edgbaston, Birmingham B15 3BE
Tel: 021-454 2177

National Council for Building Material Producers
33 Alfred Place, London WC1E 7EN Tel: 01-580 3344

National Home Improvement Council
26 Store Street, London WC1E 7BT Tel: 01-637 9709

National House Building Council
58 Portland Place, London W1N 4BU Tel: 01-387 7201

USEFUL ADDRESSES

National Economic Development Office (NEDO)
Millbank Tower, 21-41 Millbank, London SW1P 4QX
Tel: 01-211 3000

Oil Appliance Manufacturers' Association
Oil & Chemical Plant Constructors' Association
Suite 41-48 Kent House, 87 Regent Street, London
W1R 7HF Tel: 01-734 5246

Plastic Pipe Manufacturers' Society
c/o Wenham Major & Co, 89 Cornwall Street, Birmingham 3
Tel: 021-236 1866

Pump Distributors Association of Great Britain
Artillery House, Artillery Row, London SW1P 1RT
Tel: 01-799 2257

Scientific Instrument Manufacturers' Association
Leicester House, 8 Leicester Street, London WC2H 7BN
Tel: 01-437 0678

Swimming Pools & Allied Trade Associations
Farraday House, 17 Essendene Road, Caterham, Surrey,
CR3 5PB Tel: 0883 40110

Timber Trade Federation
Clareville House, Whitcombe Street, London WC1
Tel: 01-839 1891

Water Authorities Association
1 Queen Anne's Gate, London SW1H 9BT Tel: 01-222 2111

Water Companies Association
14 Great College Street, Westminster, London SW1P 3RX
Tel: 01-222 0644

Water Heater Manufacturers' Association
56 Oxford Street, Manchester M1 6EV Tel: 061-236 0384

Water-Tube Boilermakers' Association
7th Floor, 197 Knightsbridge, London SW7 1RS
Tel: 01-581 2621

3 Professional Bodies

Architectural Association
36 Bedford Square, London WC1 Tel: 01-636 0974

Association of Consulting Engineers
12 Caxton Street, London SW1H 0QL Tel: 01-222 6557

Association of London Boroughs Engineers and Surveyors
128-142 High Road, Ilford, Essex IG1 1DD Tel: 01-478 302
Ext. 145

Association of Supervisory and Executive Engineers
Wix Hill House, West Horsley, Surrey KT24 6DZ
Tel: 0483-222 383

Association of Water Officers
15 Market Place, South Shields, Tyne & Wear NE33 1JQ
Tel: 0914 563882

Chartered Institute of Building
Englemere, Kings Ride, Ascot, Berkshire SL5 8BJ
Tel: 0990 23355

Chartered Institution of Building Services Engineers
Delta House, 222 Balham High Road, London SW12 9BS
Tel: 01-675 5211

Construction Industry Research & Information Association
6 Storeys Gate, London SW1P 3AU Tel: 01-839 6881

Engineering Council
Canberra House, 10-16 Maltravers Street, London
WC2R 3ER Tel: 01-240 7891

Incorporated Association of Architects and Surveyors
Billing Brook Road, Weston Favell, Northampton
Tel: 0644 404121

Institute of Clerks of Works of Great Britain Incorporated
41 The Mall, Ealing, London W5 3TJ Tel: 01-579 2917

Institute of Marine Engineers
76 Mark Lane, London EC3R 7JN Tel: 01-481 8493

Institute of Materials Handling
St Ives House, St Ives Road, Maidenhead, Berks
Tel: 0628 28011

Institute of Measurement and Control
20 Peel Street, London W8 Tel: 01-727 0083

Institute of Petroleum
61 New Cavendish Street, London W1M 8AR Tel: 01-636 1004

Institute of Plumbing
64 Station Lane, Hornchurch, Essex RM12 6NB
Tel: 04024 72791

Institute of Refrigeration
7 Mill Lane, Carshalton, Greater London SM5 2JR
Tel: 01-647 7033

Institute of Water Pollution Control
Ledson House, 53 London Road, Maidstone Kent
ME16 8JH Tel: 0622-62034

Institution of British Engineers
Regency House, 3 Marlborough Place, Brighton BN1 1UB
Tel: 0273 61399

Institution of Chemical Engineers
165-171 Railway Terrace, Rugby, Warwick CV21 3HQ
Tel: 0926 78214

Institution of Civil Engineers
1 Great George Street, London SW1 Tel: 01-222 7722

Institution of Corrosion, Science and Technology
14 Belgrave Square, London SW1X 8PZ Tel: 01- 245 91896

Institution of Electrical Engineers
Savoy Place, London WC2 Tel: 01-240 1871

Institution of Gas Engineers
17 Grosvenor Crescent, London SW1X 7ES Tel: 01-245 9811

Institution of Mechanical Engineers
1 Birdcage Walk, London SW1 Tel: 01-222 7899

Institution of Municipal Engineers
22 Eccleston Square, London SW1 Tel: 01-834 5082

Institution of Plant Engineers
138 Buckingham Palace Road, London SW1W 9SG
Tel: 01-730 0469

Institution of Production Engineers
Rochester House, 66 Little Ealing Lane, London W5 4XX
Tel 01-579 9411

Institution of Structural Engineers
11 Upper Belgrave Street, London SW1 Tel: 01-235 4535

Institution of Water and Environmental Management
15 John Street, London WC1N 2EB Tel: 01-430 0899

Plastics and Rubber Institute
11 Hobart Place, London SW1 Tel: 01-245 9555

Royal Incorporation of Architects in Scotland
15 Rutland Square, Edinburgh 1, Scotland Tel: 031-229 7205/6

Royal Institute of British Architects
66 Portland Place, London W1N 4AD Tel: 01-580 5533

Royal Institute of Chartered Surveyors
12 Great George Street, London SW1 Tel: 01-222 7000

Royal Society of Health
13 Grosvenor Place, London SW1 Tel: 01-235 9961

Royal Town Planning Institute
26 Portland Place, London W1N 4BE Tel: 01-636 9107

4 Research & Advisory Bodies

Acoustical Investigation & Research Organisation Ltd
26/28 Market Place, St Albans, Herts AL3 5DG Tel: 0727 31176

British Board of Agrément
PO Box 195 Bucknalls Lane, Garston, Watford, Herts
WD2 7NG Tel: 0923 70844

British Coal Utilization Research Association
Randalls Road, Leatherhead, Surrey Tel: 0372 379222

British Compressed Air Society
Leicester House, 8 Leicester Street, London WC2 7BN
Tel 01-437 0678

British Non-Ferrous Metals Federation
6 Bathurst Street, London W2 2SD Tel: 01-723 7465

British Standards Institution
2 Park Street, London W1 Tel: 01-629 9000

British Standards Institution Testing
Maylands Avenue, Hemel Hempstead, Herts HP2 4SQ
Tel: 0442 3111

British Standards Institution Certification & Assessment Services
PO Box 375, Milton Keynes MK14 6LL Tel: 0908 220908

British Steel Corporation
PO Box 403, 33 Grosvenor Place, London SW1X 7JG
Tel: 01-235 1212

Building Centre
26 Store Street, London WC1E 7BT.
Administration – Tel: 01-637 1022;
Information – Tel: 0344 884999;
Bookshop – Tel: 01-637 3151

Regional Offices
Bristol
Building Centre, Colston Avenue, Bristol BS1 4TW
Administration – Tel: 0272 22953; Information – 0272 27002.

Glasgow
Building Centre, McData Unit, Weighhouse Close, 47 High Street, Paisley PA1 2AL Tel: 041-840 1199

Manchester
Building Centre, 113-115 Portland Street, Manchester M1 6FB
Administration – Tel: 061-236 9802; Information – Tel: 061-236 6933

Building Conservation Trust
Apartment 39, Hampton Court Palace, East Molesley, Surrey
KT8 9BS Tel: 01-943 2277

Building Cost Information Service
85-87 Clarence Street, Kingston-upon-Thames, Surrey
KT1 1RB Tel 01-549 2542

Building Research Establishment
Garston, Watford, Herts Tel: 09273 74040

Building Research Advisory Service
Building Research Station, Garston, Watford WD2 7JR
Tel: 0923 76612

Fire Research Station, Borehamwood, Herts WD6 2BL
Tel: 01-953 6177

Princes Risborough Laboratory, Princes Risborough, Aylesbury, Bucks HP17 9PX Tel: 08444 3101

BRE Scottish Laboratory, Kelvin Road, East Kilbride, Glasgow G75 0RZ Tel: 03552 33941

Building Services Research and Information Association (BSRIA)
Old Bracknell Lane, Bracknell, Berks RG12 4AH
Tel: 0344 26511

Central Office of Information
Hercules Road, Westminster Bridge, London SE1 7DU

Construction Industry Research & Information Association
6 Storeys Gate, London SW1P 3AU Tel: 01-839 6881

Consumers' Association
14 Buckingham Street, London WC2N 6DS Tel: 01-839 1222

Copper Development Association (CDA)
Orchard House, Mutton Lane, Potters Bar, Herts EN6 3AP
Tel: 0707 50711

District Heating Association
(Combined Heat and Power Association)
Bedford House, Stafford Road, Caterham, Surrey
CR3 6JA Tel: 0883 42323

Electrical Research Association
Cleeve Road, Leatherhead, Surrey Tel: 0372 74151

Energy Conservation and Solar Centre
Midland Road, London NW1 2AH Tel: 01-380 1002

Engineering Industry Training Board
PO Box 148, Clarendon Road, Watford, Herts WD1 1HS
Tel: 0923 44322

Fire Protection Association
Aldermary House, Queen Street, London EC4N 1TJ
Tel: 01-248 5222

Fire Research Station
Melrose Avenue, Borehamwood, Herts Tel: 01-953 6177

Glass Advisory Council
6 Mount Row, London W1Y 6DY Tel: 01-629 8334

Hydraulics Research Station
Wallingford, Oxon Tel: 049 132381

Industrial Society
PO Box 1BQ, Robert Hyde House, 48 Bryanston Square, London W1 Tel: 01-262 2401

International Solar Energy Society (UK Section)
King College, London, Atkins Building South, Camden Hill Road, London W2 7AH Tel: 01-938 2919

Lead Development Association
34 Berkeley Square, London W1X 6AJ Tel: 01-499 8422

National Economic Development Council
Millbank Tower, Millbank, London SW1 Tel: 01-211 3000

National Home Improvement Centre
26 Store Street, London WC1E 7BT Tel: 01-437 9709

National House Builders' Council
58 Portland Place, London W1N 4BU Tel: 01-387 7201

National Industrial Fuel Efficiency Service Ltd
2 Darnley Road, Edgbaston, Birmingham B16 8TE
Tel: 021-454 4471

National Inspection Council for Electrical Installation Contracting (NICEIC)
237 Kennington Lane, London SE11 5QJ Tel: 01-582 7746

National Society for Clean Air
134/137 North Street, Brighton BN1 1RG Tel: 0273 26313

Office of Fair Trading
Chancery House, 53 Chancery Lane, London WC2A 1SP
Tel: 01-242 2858

Plastics and Rubber Institute
11 Hobart Place, London SW1 Tel: 01-245 9555

Royal Society of Health
13 Grosvenor Place, London SW1X 7EN Tel: 01-235 9961

Rubber & Plastics Research Association of Great Britain
Shawbury, Shrewsbury SY4 4NR Tel: 0743 94383

Science Research Council
PO Box 18, North Star Avenue, Swindon, Wilts SN2 1ET

Society of British Gas Industries (SBGI)
36 Holy Walk, Leamington Spa, Warwickshire CV32 4LY
Tel: 0926 34357

USEFUL ADDRESSES

Society of Chemical Industries
14 Belgrave Square, London SW1X 8PS Tel: 01-235 3681

Solid Fuel Advisory Service
Hobart House, Grosvenor Place, London SW1X 7AE
Tel: 01-235 2020

Stainless Steel Advisory Service
PO Box 161, Shepcote Lane, Sheffield S9 1TR
Tel: 0742 440060

Timber Research and Development Association
Hugenden Valley, High Wycombe, Bucks Tel: 024 024 3091

Tin Research Institute
Fraser Road, Perivale, Greenford, Middlesex Tel: 01-997 4254

Town & Country Planning Association
17 Carlton House Terrace, London SW1Y 5AS
Tel: 01-930 8903/4/5

Vitreous Enamel Development Council
New House, Tilehurst, Wadhurst, East Sussex TN5 7AL
Tel: 0580 200152

Water Research Centre, Headquarters
John L van der Post Building, Henley Road, Medmenham, PO Box 16, Marlow, Bucks SL7 2HD Tel: 0491 571531

Water Research Centre, Water Byelaws Advisory Service
660 Ajax Avenue, Slough, Berkshire SL1 4BG Tel: 0753 37277

Welding Institute
Abington Hall, Abington, Cambridge Tel: 0223 891162

Zinc Development Association (ZDA)
34 Berkeley Square, London W1X 6AJ Tel: 01-499 6636

5 Regional Water Authorities and Water Companies in England & Wales

Anglian Water Authority
Ambury Road, Huntingdon PE18 6NZ Tel: 0480 56181

Water Companies

Cambridge Water Company
Rustat Road, Cambridge, CB1 3QS Tel: 0223 47351

East Anglian Water Company
163 High Street, Lowestoft Tel: 0502 2406

Essex Water Company
342 South Street, Romford, Essex RM1 2AL
Tel: 0708 46076

Tendring Hundred Waterworks Company
Mill Hill, Manningtree, Essex CO11 2AZ Tel: 0206 392155

Northumbrian Water Authority
Regent Centre, Gosforth, Newcastle upon Tyne NE3 3PX
Tel: 091 3843151

Water Companies

Hartlepools Water Company
3 Lancaster Road, Hartlepool, Co Durham TS24 8LW
Tel: 0429 4405/6

Newcastle and Gateshead Water Company
PO Box 10, Allendale Road, Newcastle upon Tyne, NE6 2SW
Tel: 0632 654144

Sunderland and South Shields Water Company
29 John Street, Sunderland SR1 1JT Tel: 0783 57123

North West Water Authority
Dawson House, Great Sankey, Warrington WA5 3LW
Tel: 092572 4321

Severn Trent Water Authority
2297 Coventry Road, Sheldon, Birmingham B26 3PU
Tel: 021 743 4222

Water Companies

Cheadle Water Works Company Ltd
43 Chapel Street, Cheadle, Staffordshire Tel: 05384 2388

East Worcestershire Waterworks Company
47 New Road, Bromsgrove B60 2JT Tel: 0527 75151

South Staffordshire Waterworks Company
50 Sheepcote Street, Birmingham B16 8AR
Tel: 021 643 8131

Southern Water Authority
Guildbourne House, Worthing Sussex BN11 1LD
Tel: 0903 205252

Water Companies

Eastbourne Waterworks Company
14 Upperton Road, Eastbourne, Sussex BN21 1EP
Tel: 0323 21371

Folkestone and District Water Company
The Cherry Garden, Cherry Garden Lane, Folkestone, Kent CT19 9QB Tel: 0303 76951

Mid Kent Water Company
PO Box 45 High Street, Snodland, Kent ME6 4AH
Tel: 0634 240313

Mid Sussex Water Company
6 Boltro Road, Hayward Heath, Sussex RH16 1BA
Tel: 0444 2662

Portsmouth Water Company
Brockhampton Springs, PO Box 8, West Street, Havant, Hants PO9 1LG Tel: 07012 6333

West Kent Water Company
Cramptons Road, Sevenoaks, Kent TN14 5DG
Tel: 0732 52307

South West Water Authority
Penninsula House, Rydon Lane, Exeter, Devon EX2 7HR
Tel: 0392 219666

Thames Water Authority
Nugent House, Vastern Road, Reading, Berks RG1 8DB
Tel: 0734 593333

Water Companies

Colne Valley Water Company
Blackwell House, Aldenham Road, Watford WD2 2EY
Tel: 0923 23333

East Surrey Water Company
London Road, Redhill, Surrey RH1 1LJ Tel: 0737 66333

Lee Valley Water Company
PO Box 48 Bishop's Rise, Hatfield, Herts A10 9HL
Tel: 07072 64311

Mid Southern Water Company
Frimley Green, Camberley, Surrey, GU16 6HZ
Tel: 0252 5031/7

North Surrey Water Company
The Causeway, Staines, Middlesex TW18 3BX
Tel: 0784 55464

Rickmansworth and Uxbridge Valley Water Company
Batchworth, Rickmansworth, Herts WD3 1LB
Tel: 0923 76633

Sutton District Water Company
41 Carshalton Road, Sutton, Surrey SM1 4LQ
Tel: 01-643 8050

Welsh Water Authority
Cambrian Way, Brecon, Powys LD3 7HP Tel: 0874 3181

Water Companies

Chester Waterworks Company
15 Newgate Street, Chester CH1 1DR Tel: 0244 20501

Wrexham and East Denbighshire Water Company
21 Egerton Street, Wrexham LL11 1ND Tel: 0978 2259

Wessex Water Authority
Wessex House, Passage Street, Bristol, Avon BS2 0JG
Tel: 0272 290611

Water Companies

Bournemouth & District Water Company
Alderney Waterworks, West Howe, Bournemouth BH11 8N Tel: 02016 2261

Bristol Waterworks Company
PO Box 218 Bridgwater Road, Bristol BS99 7AU
Tel: 0272 665881

West Hampshire Water Company
Knapp Hill, Mill Road, Christchurch, Dorset BH23 2LU
Tel: 0202 3361

The Weld Estate
Lulworth Castle, Dorset Tel: 092 941 352

The Cholderton and District Water Company
Estate Office, Cholderton, Salisbury, Wiltshire
Tel: 098064 203

Yorkshire Water Authority
West Riding House, 67 Albion Street, Leed LS1 5AA
Tel: 0532 448201

Water Company

York Waterworks Company
Lendal Tower, Yorks YO1 2DL Tel: 0904 22171

Water Services in Northern Ireland
Water Services Headquarters
Parliament Buildings, Stormont, Belfast BT4 3SS
Tel: 0232 63210

6 Fuel and Energy Interests

Atomic Energy Authority United Kingdom
11 Charles Street, London SW1Y 4QP Tel: 01-930 5454

British Gas PLC
59 Bryanston Street, London W1A 2AZ Tel: 01-723 7030

Central Electricity Generating Board
Sudbury House, Newgate Street, London EC1
Tel: 01-248 1202

Electricity Council
30 Millbank, London SW1 Tel: 01-834 2333

Energy Technology Support Unit (ETSU)
AERE, Harwell, Abingdon, Oxon Tel: 02357 834621

National Coal Board
Hobart House, Grosvenor Place, London SW1
Tel: 01-235 2020

National Gas Consumers' Council
4th Floor, 162-168 Regent Street, London W1R 5TB
Tel: 01-439 0012

Petrofina (UK) Ltd
Petrofina House, York Road, London SE1 7NT
Tel: 01-928 8000

Shell UK Oil Ltd
Shell-Mex House, Strand, London WC2 Tel: 01-430 3000

Solid Fuel Advisory Service
Hobart House, Grosvenor Place, London SW1
Tel: 01-235 2020

7 Government Departments

Department of Education and Science
Elizabeth House, York Road, London SE1 7PH
Tel: 01-928 9222

Department of Employment
Caxton House, Tothill Street, London SW1H 9NA
Tel: 01-213 3000

Department of Energy
Thames House South, Millbank, London SW1P 4QJ
Tel: 01-211 3000

Department of the Environment
2 Marsham Street, London SW1 3EB Tel: 01-276 8607

Departments of Health and Social Security
Alexander Fleming House, Elephant Castle, London SE1
Tel: 01-407 5522

Department of Prices and Consumer Protection
Millbank Tower, Millbank, London SW1P 4QS Tel: 01-211 3000

Department of Trade and Industry
1 Victoria Street, London SW1 Tel: 01-215 7877

Energy Business Centre
3 Hans Crescent, London SW1X 0LN Tel: 01-225 0233

Energy Conservation Regional Offices

London South & South East Region
Room 583 Charles House, 375 Kensington High Street, London W14 8QH Tel: 01-603 2060/2070 Ext. 245

Eastern Region
Room 582 Charles House, 375 Kensington High Street, London W14 8QH Tel: 01-603 2060/2070 Ext. 248

West Midlands Region
Ladywood House, Stephenson Street, Birmingham B2 4DT
Tel: 021-632 4111 Ext. 539

East Midlands Region
Severns House, 2 Middle Pavement, Nottingham NG1 7DW
Tel: 0602 56181 Ext. 282

Yorkshire and Humberside Region
Room 30, Priestly House, 3-5 Park Road, Leeds LS1 5LF
Tel: 0532 443171 Ext. 345

Northern Region
Stanegate House, Groat Market, Newcastle upon Tyne NE1 1YN Tel: 0632 324722 Ext. 218

South West Region
Room 535, The Pithay, Bristol BS1 2PB Tel: 0272 291071 Ext 204

Wales Region
Welsh Office, Industry Department, Cathays Park, Cardiff CF1 1NQ Tel: 0222 823126

Scotland Region
Scottish Economic Planning Department, Energy Division, Room 6/39, New St Andrew's House, Edinburgh
Tel: 031-556 8400

Northern Ireland Region
Department of Economic Development, Energy Conservation Division, Netherleigh House, Massey Avenue, Belfast
Tel: 0232 63244

Her Majesty's Stationery Office (Government Bookshops)
49 High Holborn, London WC1
Tel: 01-211 5656

Home Office
Whitehall, London SW1 Tel: 01-213 3000

Meteorological Office
London Road, Bracknell, Berks RG12 2SZ Tel: 0344 20242

Ministry of Agriculture, Fisheries and Food
Whitehill Place, London SW1 Tel: 01-839 7711

USEFUL ADDRESSES

Scottish Development Department
New St Andrew's House, Edinburgh 1 Tel: 031-556 8400

8 Training Bodies

Business Technician Education Council
Central House, Upper Woburn Place, London WC1 0HH
Tel: 01-388 3288

City and Guilds of London Institute
76 Portland Place, London W1 Tel: 01-580 3050

Construction Industry Training Board (CITB)
Bircham Newton, Nr King's Lynn, Norfolk PE31 6RH
Tel: 0553 776677

CITB Regional Offices

North London
56 Park Street, Luton, Beds LU1 3JD Tel: 0582 27462

South London and Publicity Dept
Cowdray House, Clements Lane, London EC2A 2HH

CITB Training Centres

Bircham Newton, Nr King's Lynn, Norfolk PE31 6RH
Tel: 0553 776677

Edison Street, Hillington, Glasgow C52 4XN
Tel: 041-882 6455

Education Council for Heating and Ventilating
ESCA House, 34 Palace Court, London W2 4JG
Tel: 01-229 2488

Engineering Industry Training Board
PO Box 148, Clarendon Road, Watford WD1 1HS
Tel: 0923 44322

Heating, Ventilating Domestic Engineers' National Joint Industrial Council
ESCA House, 34 Palace Court, London W2 4JG
Tel: 01-229 2488

Joint Industry Board for Plumbing Mechanical Engineering Services in England and Wales
Brook House, Brook Street, St Neots, Huntingdon, Cambs
PE19 2BP Tel: 0480 76925-8

Manpower Services Commission
Moorfoot, Sheffield S1 4PQ Tel: 0742 753275
(Will supply lists of local Skills Training Agencies)

National Joint Council for the Building Industry
11 Weymouth Street, London W1 Tel: 01-580 1740

Polytechnic of the South Bank
Faculty of Environmental Science and Technology, Borough Road, London SE1 Tel: 01-928 8989

Scottish and Northern Ireland Joint Industry Board for Plumbing Mechanical Engineering Services
2 Walker Street, Edinburgh, EH3 7LB Tel: 031-225 2255

Scottish Technician and Vocational Education Council (SCOTVEC)
38 Queen Street, Glasgow G1 3TY Tel: 041-204 2271

Training Services Division
162-168 Regent Street, London W1R 6OE Tel: 01-214 6000

Water Industry Training Association
Tadley Court, Tadley Common Road, Tadley Nr Basingstoke
RG26 6TB Tel: 07356 3011

and at:

Melvin House, Dairy Road, Kilwinning, Ayrshire
KA13 7HW Tel: 0294 54621

Burn Hall, Huby, Sutton-on-Forest, Yorkshire, YO6 1JB
Tel: 03473 484

Millis House, Hilltop, Oakwood, Derby DE2 4FZ
Tel: 0332 831 681

9 Trade Unions

Amalgamated Society of Boilermakers, Shipwrights, Blacksmiths and Structural Workers
Lifton House, Eslington Road, Newcastle upon Tyne 2
Tel: 0632 813205

Amalgamated Union of Engineering Workers (Constructional Section)
190 Cedars Road, Clapham, London SW4 0PP
Tel: 01-622 4451

Amalgamated Union of Engineering Workers (Technical, Administrative & Supervisory Section)
Onslow Hall, Little Green, Richmond, Surrey Tel: 01-948 2271

Electrical, Electronic, Telecommunication and Plumbing Union (EETPU)
Hayes Court, Bromley, Kent Tel: 01-462 7755

10 Safety Organisations

British Red Cross Society
9 Grosvenor Crescent, London SW1X 7EJ Tel: 01-741 1231

British Safety Council
62/64 Chancellors Road, London W6 9RS Tel: 01-741 1231

Confederation for the Registration of Gas Installers (CORGI)
St Martin's House, 140 Tottenham Court Road, London
W1P 0AS Tel: 01-387 9185

Health & Safety Commission (Secretariat)
Regina House, Old Marylebone Road, London NW1
Tel: 01-723 1262

Health & Safety Executive
Baynards House, 1 Chepstow Place, Westbourne Place, London W2 4TF Tel: 01-243 6000

Heating & Ventilating Joint Safety Committee
ESCA House, 34 Palace Court, London W2 4JG
Tel: 01-229 2488

Institution of Industrial Safety Officers
222 Uppingham Road, Leicester LE5 0QG

Merseyside and North West Safety Centre
c/o Government Training Centre, Stopgate Lane, Aintree, Liverpool L9 6AW Tel: 051-525 0702

ROSPA Industrial Safety Training Centre
22 Summer Road, Acocks Green, Birmingham 27
Tel: 021-706 8121

Royal Life Saving Society
Desborough House, 14 Devonshire Street, Portland Place, London WC1N 2AT Tel: 01-589 5678

Royal Society for the Prevention of Accidents (ROSPA)
Cannon House, The Priory, Queensway, Birmingham B4 6BS
Tel: 021-233 2461

St John's Ambulance Association and Brigade
1 Grosvenor Crescent, London SW1 7EF Tel: 01 235 5231

11 Overseas

European Economic Community

EEC Commission, Directorate General for Internal Market, Rue de la loi 200, B-1049 Brussels, 7350040/735 80 40

Australia

Building Services Corporation
PO Box 667 Crows Nest, New South Wales 2065.

Master Plumbers Association of Victoria
525, King Street, Melbourne 3000, Victoria

Master Plumbers Mechanical Contractors Association of New South Wales
121, Bland Street, Ashfield 2131, New South Wales

Master Plumbers & Mechanical Services Association of SA Inc
220, Henley Beach Road, Torrensville, South Australia

Austria

Bundesinnung der Sanitar - und Heizungsinstallateure
Wiedner Hauptstrasse 63, Postfach 352, A - 1045 Wien

Bundesinnung der Spengler und Kupferschmiede
Bauernmarkt, 13, A - 1011 Wien

Belgium

Federation Nationale des Associations de Patrons Installateurs Sanitaires et de Chauffage au Gaz, Plombiers, Zingueurs et Ardoisiers Couvreurs de Belgique FBIC

National Federation of Belgian Associations of Master Sanitary Engineers, Gas Heating Engineers, Plumbers, Zinc Workers and Slaters
Rue Willems 14/16, 1040 Bruxelles

Denmark

Dansk Varme-, Ventilations-och Sanitets Installator Forening

Danish Heating, Ventilation and Sanitation Installers Association
Hojnaesvej, 75, DK - 2610 Rodovre

France

Union Internationale de la Couverture, Plomberie, Installations Sanitairies, Gez et Hydraulique Generale

International Union for Roofing, Plumbing, Sanitary Installations, Gas and Water.
9, rue La Perouse, 75784 Paris Cedex 16 France Tel: (1) 47 20 10 20
Telex: Fedebat 611 975 F

Union National des Chambres Syndicales de Couverture et de Plomber de France

National Union of French Roofing and Plumbing Trade Associations
9, Rue La Perouse, 75784 Paris Cedex 19 France Tel: (1) 47 20 10 20
Telex: Fedebat 611 975 F

Germany

Zentralverband Sanitar Heizungs-Kilma

Central Association for Plumbing and Heating Technology
Rathausallee, 6. 5205 Saint Augustin 1 (RFA)

Holland

Algemene Vezeniging Van Ordernemers in het Loodgieters Sanitar en Gasverwarmings – Installatatiebedrijf
Nieuwe Sarklaan 55, Postbus 90725, 2509LS's, Gravenhage/La Haye, (Pays-Bas)

Italy

Association Nazionale Installatori di Impianti Termici e de Ventilazione, Idrici, Sanitari, Electtrici, telefonici ed Affini.

National Association for Installers of Heating, Ventilation, Plumbing, Sanitary, electrical and telephone Services
Via giorgio Jan 5, 20120 Milano, Italy.

Luxembourg

Federations Reunies des Patrons Installateurs Sanitaires et Installateurs de Chaffage du grande Duche de Luxembourg.

Association of Sanitary and Central Heating Engineers of the Grand Duchy of Luxembourg.
41, rue Glesener, 1631 Luxembourg

Federation des Gatrons-Ferblantiers et Calorifugeurs du Grand Duche de Luxembourg.
41, rue Glesener, 1631 Luxembourg

New Zealand

New Zealand Plumbers, Gasfitters and Drainlayers Board
PO Box 11-422, Wellington

Spain

Confederacion Espanola de Instaladores y Mantenedores
Grincipe de Vergera 74, 28006 – Madrid (Espagne)

Sweden

Rorfirmornas Riksforbunnd
Norttullsgatan 6, 11329 Stockholm, (Suede)

Switzerland

Association Suisse des Maitzes Ferblantiers et Appareilleurs

Swiss Association of Plumbing and Installation Contractors
Auf der Mauer 11, 8023, Zurich, Switzerland.

United States

American Backflow Prevention Association
PO Box 1563, Akron, Ohio 44309 -1563

American Society of Plumbing Engineers
3617 Thousand Oaks Boulevard, 210 Westlake, CA 91362

American Society of Sanitary Engineering
PO Box 40362, Bay Village, Ohio 44140

National Association of Plumbing, Heating and Cooling Contractors
3416 Robards Court, Louisville, KY40218

SECTION A HOT AND COLD WATER SUPPLIES 1-49

Page Number

Pipe Sizing 1-6

Assessment of probable demand 1
Assessment of required flow rate 1
Equivalent pipe lengths 1
Loading units 1
Loss of head at point of delivery 2
Loss of head through ballvalves and stopvalves 2
Relative discharging power of pipes 2

Tables

A1 Recommended minimum rates of flow of various appliances or fittings 1
A2 Loading units 1
A3 Equivalent pipe length 1
A4 Loss of head through draw-off taps expressed as equivalent pipe lengths 2
A5 Loss of head through draw-off taps (metres) 2
A6 Loss of head through taps and equivalent pipe lengths 2
A7 Flow rates at H/L = 1 2
A8 Number of outlets at 100% demand at full flow conditions 2
A9 Average water consumption of domestic and commercial appliances 2

Graphs

A1 Pipe sizing chart, copper tube to BS 2871 Table X 3
A2 Pipe sizing graphs, heavy galvanized steel tubes to BS 1387 4
A3 Head loss through ballvalve orifices 5
A4 Head loss through stopvalves 6

Hot Water Storage 7-8

Method of assessing recovery/storage relationships for hot water storage vessels 7

Figures

A1 Demand pattern histogram 7
A2 Storage make-up ratio curve 8

Tables

A10 Cold water storage in buildings 7
A11 Assessment of hot water demand and storage requirements 7
A12 Peak hourly loads 8

Sizing of Primary and Secondary Circulations 8-9

Circulating pressure 8
Hot water distributing pipes 9
Maximum permissible lengths of uninsulated hot water draw-off pipes 9
Primary circulation pipes 8
Secondary circulation pipes 9

Table

A13 Density of water 8

Pumped Hot Water Systems 9

Figure

A3 Recommended pump positioning for hot water circulations 9

Water Quality 9-12

Acoustic and vibration in pipework 12
Deionised water 10
Distributed loads on steel joists 11
Operation of a base exchange water softener 10
Softened water 10

Figure

A4 Re-circulation deionisation system 10

Tables

A14 Distributed loads on steel joists (i) 11
A15 Distributed loads on steel joists (ii) 11
A16 Distributed loads for spans up to 3 metres 11

Outward Thrust in Pipelines Due to Internal Pressure 12-13

Values of K for various bends 12

Table

A17 Bearing capacity of soils 13

Graph

A5 External thrust developed in pipelines 13

Page Number

Acoustic and Vibration Control 13-15

Acoustic 13
Acoustic bridge 15
Acoustic transmission 15
Aeration of water pipes 15
Float operated valve oscillation 14
Float operated valve water noise 15
Impact noise in sanitary fittings 15
Pump noise 14
Tap water oscillation (valve bounce) 14
Thermal movement noise 15
Water hammer (impulsive noise) 14
Water flow noise 14
Water terminal fittings 14

Water Hammer 15-24

Basic design principles 16
Shock pressure rise 16
Sizing of hydro-pneumatic accumulators 17
System protection 16

Figures

A5 Hydro-pneumatic accumulator 16
A6 Accumulator sizing 18

Tables

A18 Accumulator sizing (i) 19
A19 Accumulator sizing (ii) 19
A20 Flow rates from copper tubes 19

Graphs

A6 Velocity of shock wave/pipe wall thickness 21
A7 Hydro-pneumatic accumulator selection chart (single units) 22
A8 Hydro-pneumatic accumulator selection chart (multiple units) 23
A9 Hydro-pneumatic accumulator selection chart (5-50mm pipe run) 24

Inspection and Testing of Water Mains and Water Supply Systems 25

Cold water systems 25
Hot water systems 25
Sterilizaton of cold water systems 25

Prevention of Contamination of Water by Backsiphonage or Cross Connection 25-26

Backflow Prevention Devices 26-28

Check valve 27
Pipe interrupter 27
Type A air gap 27
Type B air gap 27
Vacuum breaker 28
Upstands on supply and distributing pipes 28

Figures

A7-A12 Examples of backflow prevention 26
A13 Type A air gap 27
A14 Type B air gap 27
A15 Spring loaded check valve 26
A16 Pipe interrupter 26
A17 Vacuum breaker 27
A18-A19 Backflow protection by the use of upstands 28

Tables

A21 Requirements for backflow protection 26
A22 Required dimensions for Type A air gaps 27

Protection of Draw-off Taps, Hose Connections, Bidets, Clothes and Dishwashing Machines and Cisterns 28-32

Acceptable methods of backflow protection at points of use or delivery of water 30
Bidets 29
Draw-off taps 28
Primary heating circuits and feed cisterns 30
Secondary backflow protection 30
Shower hose connections 29
Storage cisterns 29
Washing machines and dishwashers 29

Figures

A20 Arrangement of a domestic cold water cistern complying with Water Byelaw requirements for preventing contamination of stored water 30

Page Number

A21 Secondary backflow protection on a common supply pipe serving two or more dwellings 31
A22 Secondary backflow protection on a common distributing pipe serving two or more dwellings 32

Tables

A23 Air gap dimensions 29
A24 Class 1 risk of contamination from a substance continuously or frequently present 31
A25 Class 2 risk of contamination from a substance which may be present 31
A26 Class 3 risk of contamination 31

Unvented Hot Water Systems 32-35

Application of Building Regulations 34
Application of Water Byelaws 33
Check valve 33
Discharge pipework from relief valves 34
Expansion relief valve 33
Expansion vessel 33
Inlet pressure reducing/limiting valve 32
Line strainer 32
Stopvalve 32
Temperature relief valve 33

Figures

A23 Schematic layout of an unvented domestic hot water installation 32
A24 Pressure reducing and limiting valves 33, 34
A25 Expansion relief valve 33
A26 Temperature and pressure relief valve 33
A27 Accommodation of expansion water in an expansion vessel; unvented installation 34
A28 Accommodation of expansion water in a cistern 34
A29 Accommodation of expansion water in an expansion vessel 34
A30 Example method of providing safe disposal of hot water discharge from safety devices 34

Corrosion Causes, Effects and Preventative Measures 35-38

Bacteria 37
Basic theory 35
Contact with dissimilar materials 37
Differential aeration 36
Dissolved gases 36
Dissolved salts 36
Flow, erosion, impingement and cavitation 37
Ion concentration 36
Organic matter 38
pH 36
Solution potential 35
Stray current corrosion 38
Surface effects 36
Temperature 35

Figure

A31 Pourbaix diagram of Fe-H_2O systems 36

Table

A27 Galvanic series of metals and alloys in natural waters 37

Effects of Corrosive Environments 39-41

Aluminium and aluminium alloys 38
Brasses 39
Bronzes 39
Cadmium 38
Cast iron 38
Copper 38
Copper and copper alloys 38
Cupro nickel alloys 40
Lead 40
Magnesium 40
Plastics and rubber components 41
Soft solders 40
Stainless steels 40
Steels 40
Zinc 41

Table

A28 Maximum recommended water velocities for copper tube at different temperatures 39

Prevention of Corrosion 41-44

Assessing the corrosivity of the local environment 41
Chilled water systems 42, 44
Cleaning and descaling boilers and associated plant 45
Cooling waters 42, 44

Page Number

Corrosion control and prevention 45
External and internal systems 41, 44
Potable (supply), hot and cold domestic waters 41, 44
Primary and secondary heating waters 42, 44
Steam raising plant 42, 44

Tables

A29 Assessment of condition of recirculating systems from analytical data 42
A30 Recommended water characteristics for evaporative cooling towers 42
A31 Recommended water characteristics for shell boilers 43
A32 Recommended water characteristics for water-tube boilers 43
A33 Recommended water characteristics for once-through boilers 44
A34 Cleaning and descaling boilers 45

Welding/Brazing and Soft Soldering/Adhesive Bonding 46-47

Adhesive bonding 47
Brazing 46
Soft soldering 47
Welding 46

Table

A35 Composition of typical soft solders 47

References 48

British Standards and Codes of Practice 48-49

SECTION B DOMESTIC CENTRAL HEATING 50-77

Design Considerations 50-58

Flow of water in pipes 53
Heat emission from insulated pipes 53
Heat emission from pipes 50
Heat losses through building structures 58
Vertical pipes 50

Tables

B1 Temperature and ventilation rates 50
B2 Percentage of heat emission from vertical copper pipes as compared with horizontal pipes 51
B3 Heat emission from insulated copper pipes 53
B4 Resistance to hot water flow through copper pipes 53
B5 Resistance to hot water flow at 0.61m/s through fittings 58
B6 Heat transmittance coefficients (U values) 58

Graphs

B1 Heat emission from painted single horizontal uninsulated copper tubes to BS 2871 51
B2 Heat emission from unpainted single horizontal uninsulated steel tubes to BS 1387 52

Heating Controls 59

Sensing elements 59
Timing controls 59
Valves 59

Systems of Control 60-65

Boiler air supply 65
External temperature-compensating controls 62
Frost thermostats 63
Fuel storage 65
Full control 60
Fully pumped systems 60
Fully pumped system with 2-port control valve 63
Fully pumped system with 3-port control valve 61
Optimum start control systems 63
Partial control 60
Sealed systems 63
Ventilation of boiler compartments 65
Ventilation of living rooms 65
Zone control 62

Figures

B1 Programmer or time switch control 60
B2 Two-port motorised valve added to system 61
B3 Fully pumped system with 3-port valve control 61
B4 Fully pumped system with 2-port motorised valves 62
B5 System with external temperature control 62
B6 Three stages of operation of an expansion vessel 63

INDEX

Page Number

B7 Arrangement of sealed system components 63
B8 Connection direct from water supply pipe in accordance with Model Byelaw 14 64

Tables

B7 Capacities of expansion vessels 64
B8 Air opening sizes 65
B9 Minimum free area of ventilation to rooms where heating appliances are fitted 65
B10 Solid fuel storage capacities 65
B11 Recommended minimum capacity of oil storage tanks 65

Electric Central Heating Storage Boilers 66-69

Assessing the existing system 66
Commissioning 67
Control sequence 67
Converting the existing system 67
Design considerations 66
Dry core boiler 66
Dry core boiler capacity 67
Economy 7 tariff 66
Electrical supply 66
Existing pipework 67
Heat loss calculations 66
Installation-siting the boiler 67
Plumbing considerations 67
System controls-user 67
Water storage boiler 69

Figures

B9 Dry core boiler 66
B10 Typical existing system 68
B11 System conversion with dry core boiler and Economy 7 water heating 69
B12 Total system schematic 69

Domestic Electric Water Heating 69-76

Central storage system 69
Cylinder insulation 70
Cylinder size 70
Design principles 70
Economy 7 controller 72
Economy 7 cylinders and combination units 72
Economy 7 systems 72
Economy 7 tariff 70
Electric installation 76
Immersion heaters 72
Immersion heater arrangements 70

Figures

B13 Domestic electricity tariff 70
B14 Immersion heater arrangements 71
B15 Recommended Economy 7 systems 72
B16 Economy 7 water heating controller 73
B17 Electricity Council specification for a direct cylinder 73
B18 Electricity Council specification for an indirect cylinder 74
B19 Economy 7 cylindrical combination units 74
B20 Economy 7 rectangular combination units 74
B21 Instantaneous water heater 74
B22 Undersink and oversink water heaters 74
B23 Wall kettle 75
B24 Wiring installation for twin side entry arrangements 75
B25 Wiring installation for dual element arrangement 75

Table

B12 Specification for immersion heaters 75

Graph

B3 Profile of hot water for a size 7 (120 litre) cylinder 71

References 76

British Standards and Codes of Practice 77

SECTION C SANITARY PLUMBING AND DRAINAGE 78-136

Design of Sanitary Pipework Systems 78-89

Admission of rainwater into discharge stacks 84
Air admittance valves 82
Bends and branches at base of stacks 83
Branch to stack connections 88
Capacities of stacks 83

Page Number

'Collar Boss' system 81
Connections from washing machines 80
Depth of trap seals 80
Design of single branches and fittings 82
Discharge pipe systems 78
Discharge pipes 80
Discharge units 84
Large branch connections 81
Modified single stack system 79
Prevention of cross flow 81
Single stack system 79
Sizing procedure for discharge unit method 86
Small branch connections 81
Stack offsets 81
Stub stacks 79
Termination of discharge and ventilating stacks 83
Ventilating pipes and stacks 83
Ventilated stack system 78
Ventilated system 78
Waste disposal units 80
Waste traps and discharge pipes 80

Figures

C1 Ventilated system 78
C2 Ventilated stack system 78
C3 Single stack system 79
C4 Modified single stack system 79
C5 Connection of washing appliances without venting 80
C6 Connection of washing appliances with venting 81
C7 Restricted connection area on stack (i) 81
C8 Examples of permitted connections 81
C9 Restricted connection area on stack (ii) 81
C10 Examples of permitted connections (50mm parallel branch) 82
C11 'Collar Boss' fitting 82
C12 Length and fall of basin waste 82
C13 Air admittance valve 83

Tables

C1 Minimum internal diameter of traps 81
C2 Maximum capacities of stacks flowing quarter full 84
C3 Discharge units 86
C4 Maximum number of discharge units allowed on branch discharge pipes 86
C5 General guide for the sizing of ventilating pipes and stacks 86
C6 Discharge stack sizing (a) 88
C7 Discharge stack sizing (b) 88
C8 Design of branch discharge pipes 89

Graphs

C1 Probability graph 85
C2 Discharge units and flow rates 87

Kitchen and Laboratory Drainage 90-91

Kitchen Drainage 90
Access 90
Backflow problems 90
Branch discharge pipes 90
Drainage (above ground) 90
Floor drainage 90
Grease traps 90
Materials 90
Pumped installations 90
Specific requirements 90
Traps 90
Laboratory Drainage 91
Dilution 91
Distribution 91
Incompatible wastes 91
Radioactive waste 91
Replacement 91

Materials for Above Ground Drainage Systems 91

Selection 91
Types of materials 91

Inspection and Testing of Discharge Pipes 92-93

Air test 92
Commercial or public buildings 93
Dwellings 93
Performance of testing systems 92

Tables

C9 Number of appliances to be discharged simultaneously for testing stability of trap seals-dwellings 92
C10 Number of appliances to be discharged simultaneously for testing stability of trap seals-commercial or public buildings 93

	Page Number
Vacuum System	**93**
Methods of Waste Collection	**93**
Sanitary Accommodation	**93-100**
Sanitary accommodation for disabled people	100
Scale of provision of sanitary appliances	94
Tables	
Scale of provision of appliances:	
C11 Office buildings and shops	94
C12 Dwellings	95
C13 Accommodation for elderly people	95
C14 Residential homes for elderly people	96
C15 Factories	96
C16 Schools and higher educational establishments	97
C17 Cinemas, concert halls, theatres etc	98
C18 Hotels	98
C19 Restaurants and canteens	99
C20 Swimming pools	99
C21 Art galleries, libraries, museums, places of public assembly	99
C22 New public houses	100
C23 Marinas and yacht harbours	100
The Design of Drainage Systems	**100-109**
Computer programs	101
Discharge units	103
Gravity flow pipe design charts	104
Hand calculations	102
Model Water Byelaws	103
Percentage gradients	102
Rational Lloyd-Davies Method	101
Theory	104
Typical computer program	101
Figure	
C14 Typical computer program	102
Table	
C24 Average water consumption for varying households	103
Graphs	
C3 Design chart, 100mm pipe (K_s = 0.6)	105
C4 Design chart, 100mm pipe (K_s = 1.5)	106
C5 Design chart, 150mm pipe (K_s = 0.6)	107
C6 Design chart, 150mm pipe (K_s = 1.5)	108
C7 Design chart, 250mm pipe (K_s = 1.5)	109
Sewage Pumping Systems	**110-115**
Designing pipework systems to resist seismic disturbances	112
Drainage charts for obtaining the HMD	112
Drainage definitions	111
Drainage formulae	111
Pipework suspension	113
Precautionary considerations	113
System integrity	113
Ventilation of pumping chambers	110
Figures	
C15 Pipework suspension, co-axial	112
C16 Concrete duct (radioactive protection)	113
C17 Spring hammer	113
C18 Sway brace	114
Table	
C25 Proportional areas and depths at HMD's	112
Below Ground Drain Renovation	**114-116**
Cast iron drainage	115
Closed circuit television	116
High pressure jetting	115
Renovation	114
Renovation or Renewal?	114
Methods of Pipe Renovation	**116-118**
Cement grouting with pistons	116
Drain liners	117
Insitu liners	117
Liquid resin grouting	116
Pipe bursting	118
Preformed pipes	118
Sleeved uPVC pipes	118
Small man entry 'egg and barrel' shaped sewers	118
Special grouting to individual joints and cracks	116
Spraying	118
Two-part chemical grouting	116

	Page Number
uPVC spiral relining	117
Figures	
C19 Cement and resin grouting	116
C20 Insertion of insitu liner	117
C21 Curing insitu liner	117
C22 'Permaline' liner	117
C23 'Renoline' sleeve	117
Rainwater Systems	**118-133**
Calculation examples	119
Methods of design	118
Gutter outlets	126
Prevention of ice on roofs	133
Rainfall intensity tables	123
Sizing charts	128
Sources of data	118
Use of gutter sizing charts	124
Figures	
C24 Flow chart showing calculation procedures	119
C25 UK map, key to rainfall tables	122
C26 'Ice-stop' self regulating heating system	133
Tables	
C26(a-g) Rainfall tables	123
C27 Estimated capacities of outlets for flat roofs	124
C28 Categories of risk	124
C29 Flow rates, square/rectangular ungrated gutter outlets	126
C30 Flow rates, circular ungrated gutter outlets	126
Graphs	
C8 Gutter sizing nomogram	125
C9 Vertical rainwater pipe sizing chart	128
C10 Rectangular box gutter sizing chart	129
C11 Trapezoidal gutter sizing chart, 1 to 1 side slope	130
C12 Trapezoidal gutter sizing chart, 1 to 1.5 side slope	131
C13 Trapezoidal gutter sizing chart, 1 to 2 side slope	132
Solid Waste Refuse Disposal	**133-134**
Design considerations	133
On-site treatment	134
Pipeline systems	134
Refuse collection	133
Scope	133
References	**135**
British Standards and Codes of Practice	**135-136**

SECTION D PUMPS AND PUMPING 137-142

Definitions and Descriptions	**137-138**
Absolute pressure	137
Atmospheric pressure	137
Capacity	137
Cavitation	137
Efficiency	137
Friction head	137
Gauge pressure	137
Net positive suction head (NPSH)	137
Power	137
Power and efficiency	137
Self-priming	137
Static delivery head	137
Static head	138
Suction head (negative)	138
Suction head (positive)	138
Total delivery head	138
Total head on pump	138
Total suction head (positive)	138
Total suction lift (negative)	138
Vacuum	138
Vapour pressure	138
Velocity head	138
Figure	
D1 Pumping definitions	138
Pump Types	**139**
Applications	139
Centrifugal pump	139

Page Number

Positive displacement pump 139

Graph

D1 Typical characteristic curves 139

Systems 139-141

Effects of pump on system 140
Packaged systems 139
Unpackaged component systems 139

Figures

D2(a) Open circuit 140
D2(b) Open circuit pressure diagram 140
D3(a) Closed circuit 140
D3(b-e) Closed circuit pressure diagrams for various pump positions 140

Graph

D2 Pump graph 141

Pump Selection 141

Noise 141

Pump Installation 142

Discharge non-return valve 142
Electric motors 142
Foot valve 142
Location 142
Piping 142
Relief valve 142
Suction strainer 142
Switch gear 142

British Standards 142

SECTION E PIPED GAS SERVICES 143-169

Natural Gas 143-146

Factors affecting pressure loss 143
Flow of gas in pipes 143
Gas flow tables 143
Meter house dimensions 145
Pipe sizing tables 144
Pipe support tables 145

Tables

E1 Pipe sizing, horizontal copper tube 144
E2 Pipe sizing, horizontal steel pipe 144
E3 Pipe sizing, effect of elbows, tees or bends 144
E4 Pipe supports, iron/steel pipes 145
E5 Pipe supports, copper pipes 145
E6 Industrial gas meter sizes 145
E7 Typical equipment gas consumption figures 146

Liquefied Petroleum Gas Installations 146-152

Building control and planning permission 146
General properties 146
LPG bulk tank location and safety distances 147
LPG storage in cylinders 150
Pipe sizing 151
Pipe sizing tables 152
Storage tanks and fittings 149

Figures

E1 Small bulk vessel adjacent to a building 149
E2 Small bulk vessel at domestic premises 149
E3 Storage tank dimensions 150
E4 Cylinder storage 150

Tables

E8 Minimum recommended safety distances for LPG storage vessels 147
E9 Cylinder sizes 150
E10 Guide to pipe sizes for appliances 151
E11(a-d) Pipe sizing tables 151

Compressed Air 153-161

Closed cycle cooling 154
Compressing cooling 153
Compressing the air 153
Cooling the air 154
Dryers 156

Page Number

Layout 156
Main pressure reduction 156
Pump assisted circulation 153
Receivers 155
Removing moisture 154
Rotary compressors 154
Separators 156
Sizing compressed air mains 156
Thermo-syphon circulation 153

Figure

E5 Typical layout of a compressed air plant 161

Tables

E12 Final temperature of adiabatic compression from free air 153
E13 Effect of altitude on compressor volumetric efficiency 153
E14 Cooling tank capacities 154
E15 Ratio of compression 157
E16 Resistance of pipe fittings 157
E17 Formula for converting volume of compressed air to volume of free air 157
E18 Relative discharging capacities of steel tubes to BS 1387 157
E19 Discharge of air through orifices 158
E20 Receivers for compressed air systems 159
E21 Vapour chart 159
E22 Equivalent volume of compressed air at common pressure 160
E23 Volume of compressed air carried by medium grade steel pipes 160
E24 Typical equipment consumption of compressed air 160

Graphs

E1 Moisture content of air 155
E2 Compressed air sizing nomogram 158
E3 Compressed air – laboratories 160

Vacuum 161-167

Bench outlets 164
High vacuum systems 164
Hospital systems 162
Laboratory systems 163
Pipelines 163
Pipe sizing 165
Pipe sizing for individual chamber work 165
Pumping units 164
Siting of pumps 163

Table

E25 Comparative vacuum scale 162

Graphs

E4 Pump sizing 165
E5 Vacuum pipe sizing 167

Medical Gases 166

References 168

British Standards and Codes of Practice 168-169

SECTION F FIRE PROTECTION SERVICES 170-188

Principal Causes of Fire in Buildings 170

Figure

F1 Fire triangle 170

Classification of Fire Risks 170

Table

F1 Classification of fire risks 170

Insurance Companies, Role and Influence 170

Sprinkler Installations 171-181

Alarms 180
Area covered by sprinklers 176
Automatic sprinklers; design principles 172
Classes of system 172
Combined high velocity systems 180
Coverage of a sprinkler system 176
Duplicate water supplies 174
Fire brigade inlets 176
General description 171
Grading of systems 173

Page Number

High velocity water spray system 179
Medium velocity systems 180
Other installations using water drenchers 177
Pressure and flow requirements 175
Pressure tank 174
Principal components 171
Projectors 179
Proving of water supplies 175
Risk categories 172
Siting of sprinkler heads 176
Superior water supplies 173
Systems using steam 180
Types of water spray systems 179
Water spray projector systems 178
Water supplies 173

Figures

F2 Typical sprinkler installation 171
F3 Pressure tank and town main with back pressure valves 175
F4 Diagrammatic layout of the pipework of a sprinkler system 176
F5 Multiple control system 176
F6(a) Automatic control 177
F6(b) Open sprayer 177
F7 Typical drencher system 178
F8 Types of drencher 178
F9 High velocity water spray projectors 179
F10 Water spray projector systems 179
F11 Arrangement of projectors to protect a transformer 180

Tables

F2 Fire Officers' Committee risk categories 172
F3 Working air pressure for pressure tanks 175
F4 Pressure and flow requirements 175
F5 Maximum area covered by a sprinkler 176

Hose Reel Installations 181-183

Automatic 'on and manual off' 183
Automatic 'on and off' 183
Fire protection discharge from jets 181
Pipework 182
Positioning of hose reels 181
Special provision-water supplies 182
Statutory regulations 181
Testing and approval 183
Types of hose reels 181
Water supplies 182

Figure

F12 Typical arrangement of hose reel system 182

Tables

F6 Height of jets 181
F7 Discharge from jets 182

Dry Risers 183-184

Inlets 183
Installation 183
Outlets 183
Size and positioning of risers 183
Statutory regulations 183
Testing and approval 184

Figure

F13 Typical arrangement of a dry riser 183

Wet Risers 184

Installation 184
Outlets 184
Pumping equipment 184
Size and positioning 184
Statutory regulations 184
Testing and approval 184
Water supplies 184

Figure

F14 Typical arrangement of a wet riser 185

Foam Systems 184-185

Application 185
Types of system; Design data 185
Fixed foam units 185
Foam branch pipes 185
Foam inlets 185
High expansion foam 185
Mobile foam units 185
Portable extinguishers 185

Page Number

Carbon Dioxide Systems 185-186

Application 185
Location application 185
Non-conductivity 185
Safety precautions 186
Storage 186
Testing and maintenance 186
Total flooding 186
Types of system 186
Uses and limitations 185

Halogenated Extinguishing Agent Systems 187

Private Hydrants 187

By-pass valves 187
Connections 187
Markings 187
Outlets 187

References 187

British Standards and Codes of Practice 187

SECTION G STEAM SUPPLIES 189-200

Steam 189-200

Condensate 193
Design considerations 189
Expansion of a steam pipeline 197
Pipeline velocity in steam mains 193
Separators 196
Strainers 196
Tables 189
Traps 194
Trap checking 196

Figures

G1 Diagrammatic steam main 189
G2 Condensate removal from distribution main 193
G3 Air venting on distribution line 193
G4 Condensate removal from branch lines 193
G5 Float trap with thermostatic air valve 194
G6 Inverted bucket trap 196
G7 Operation of bimetal steam trap 196
G8 Strainer 196
G9 Steam trap checking equipment 196
G10 Expansion loop 197
G11 Alternative loop 197
G12 Expansion bellows 197
G13 Sliding joint 197

Tables

G1 Steam tables 190
G2 Pressure factors for pipe sizing 191
G3 Pipeline capacity and pressure drop factors 192
G4 Flow of water in steel pipes 194
G5 Flow of water in heavy steel pipes 195
G6 Expansion of pipes 197
G7 Heat emission from pipes 198
G8 Pipeline capacities at specific velocities 199

Graph

G1 Sensible, latent and total heat in steam 189

British Standards and Codes of Practice 200

SECTION H PIPEWORK EXPANSION 201-207

Considerations for Design and Installation 201-207

Anchor design 202
Cold draw 202
Expansion bellows 207
Flanged units 202
Guides along the remaining pipe run 207
Guides nearest the bellows 206
Pipe guides and supports 205
Plastic soil and waste pipes 207

INDEX

Page Number

Screwed end units 202
Welded end units 202

Figures

H1 Typical anchor positions 201
H2 Expansion loop 202
H3 Expansion bellows 202
H4 Flanged unit cold draw 202
H5 Welded end unit cold draw 202
H6 Anchor for copper pipework (light load) 202
H7 Anchor for steel pipework (light load) 203
H8 Light to medium load anchor 203
H9 Typical duct application 203
H10 Typical riser application 203
H11 Alternative light to medium load anchor 203
H12 Application using angle welded to pipe 203
H13 Typical heavy load anchor 203
H14 Typical service walkway application 203
H15(a) Friction coefficient – point contact 204
H15(b) Friction coefficient – edge contact 204
H15(c) Friction coefficient – line contact 204
H15(d) Friction coefficient – face to face contact 204
H16 Pipeline failure may occur without pipe guides 205
H17(a) Bellows positioned centrally in an axial installation 205
H17(b) Bellows positioned at end of line in an axial installation 205
H18 Strap-type guide 205
H19 Strap-type guide with tube roller 205
H20 Strap-type guide with roller and chair 205
H21 Guide using angle rollers on box section 206
H22 Strap-type guide using angle rollers on box section 206
H23 Tee support with angle guide 206
H24 Tee support with roller and angle guide 206
H25 Guides for large bore and/or high pressure pipes 206
H26 Tube-type guide 206
H27 Tube-type guide with rods 206
H28 Slip-on flange type guide 206
H29 Guides near bellows 206
H30 Installation of additional guides 207

Table

H1 Coefficients of linear expansion 201

British Standards and Codes of Practice 207

SECTION I DESIGNING FOR THE DISABLED 208-216

Introduction 208

Approaches to Buildings 208

Automatic entrance doors 208
Dropped kerbs 208
Entrance doors 208
Handrails 208
Level/ramped access 208
Parking spaces 208
Stepped access 208

Figures

I1 Disabled logo 208
I2 Handrail dimensions 208

Inside Buildings 208-209

Door handles 209
Double swing doors 208
Floors 209
Glass doors and other glass areas 208
Hazards 209
Internal circulation areas 209
Internal doors 208
Kick plates 208
Level changes 209
Passage ways 209
Sliding doors 208
Spring closers 208
Switches 209
Thresholds 208

Figure

I3 Dimensions – internal doors and adjacent areas 209

WC Compartments in Public Buildings 209

Access through lobbies 209

Page Number

Access to WC compartments 209
Commercial buildings 209
Entertainment facilities 209
Points to consider for WC's 209
Travel facilities 209

Facilities for Wheelchair Disabled 210-211

Additional facilities 211
Dimensions 210
Doors 210
Hand dryers 210
Hand rinse basin 210
Mirrors 211
Restricted space 211
Support rails 210
Toilets 210
Toilet paper holder 210
WC macerating boxes and WC pans 211
WC seats 211

Figure

I4 Dimensions of WC compartments and facilities 210

WC Facilities for Ambulant Disabled 211

Washbasin Fittings 211

WC Levels in Private Facilities 211

Considerations 211
Selection of raised toilet seats 211

Facilities for Private Use and Nursing Homes 212-214

Bathing 212
Baths and accessories 213
Bath inserts 213
Bath lifts 213
Baths with built-in seats 213
Bathroom layout 212
Bidets 214
Disposal of incontinence and sanitary towels 214
Doors 212
Drainage 214
Fittings 212
Flooring 212
Grabrails 212, 214
Heating 212
Showering 213
Space 212
Urinals 214

Advanced WC Cubicle Design for Wheelchair Disabled 214

Central Heating for the Disabled 214-215

Automatic controls 215
Boilers 215
Design criteria 214
Gas/oil fired 215
Heat emitters 215
Off-peak electric storage heaters 215

References 216

British Standards and Codes of Practice 216

SECTION J LEGIONNAIRES' DISEASE 217-220

Engineering Implications 217

Approved materials 219
Cold water storage systems and tanks 217
Cooling towers 219
Continuous chlorination plant 218
Domestic hot water calorifiers 219
Duties of the maintenance department 217
Humidifiers 219
Outbreak control 217
Stagnant water positions – occupied 218
Stagnant water positions – unoccupied 219
Thorough testing of all outlets 218
Water softening and hot water storage systems 218

Page Number

Figure

J1 Hot water warning notice 217

References 220

SECTION K FOUNTAIN AND WATER DISPLAY TECHNIQUES 221-224

Design Considerations 221-223

Construction principles 221
Fountain installations 223
Ten important rules 221
Types of fountains and water displays 221
Watercourse installations 223

Figures

K1 A simple fountain unit using a submersible pump 222
K2 Fountain installation incorporating a dry mounted centrifugal pump 222
K3 Fountain installation with an open underground reservoir 223

The Art of Water Display 224

Cascades 224
Construction and maintenance 224
Important aspects of waterfall design 224
Small fountains 224
Water sculptures 224
Waterfall arrangements 224

SECTION L AIRFLOW IN DUCTS 225-231

Duct Sizing 225-229

Duct fittings resistance 225
Extract ventilation 225
Ideal system 229

Table

L1 Ventilation head resistance factors (K) for bathroom and WC ventilation ducts 225

Graphs

L1 Relationship of capacity of rectangular and circular ducts 226
L2 Relationship of K, V and head loss for duct fittings resistances 227
L3 Sizing chart – ventilation ducts for bathrooms and WC's 228

Fans 229-231

Axial flow fan 230
Centrifugal fan 229
Propeller fan 230
Regulations and requirements 231

Figures

L1 Centrifugal fan 229
L2 Paddle blade 229
L3 Forward curved blade 229
L4 Backward curved blade 229
L5 Propeller fan 230
L6(a) Axial flow fan 230
L6(b) Axial flow fan mounting, non guide vane type 230
L6(c) Axial flow fan with guide vanes 230
L7 Roof mounted supply duct with recirculating damper control 231
L8 Roof mounted extract fan – natural inlet 231
L9 Recommended proportions for a plain canopy hood 231
L10 Combined supply and extract system with propeller fans 231
L11 Cross ventilation for public rooms 231

British Standards and Codes of Practice 231

SECTION M ELECTRICAL EARTHING AND BONDING IN WATER INSTALLATIONS 232-235

Page Number

Introduction 232-233

Bonding 232
Drainage systems 233
New regulations 233
Plumbers' responsibilities 233
Problems with plastics 232
Protective multiple earthing 232
Supplementary bonding 233

Figures

M1 Protective multiple earthing, method of bonding 232
M2 Protective multiple earthing, area electricity board's requirements for main bonding 232
M3 Requirements for bonding in domestic situations 234
M4 Sockets and fixed equipment outside main equipotential zone 235

SECTION N CONVERSION FACTORS AND MISCELLANEOUS DATA 236-259

Conversion Factors 236-238

Tables

N1 Imperial to SI metric/SI metric to imperial 236
N2 Pressure 238

Miscellaneous Data 238-239

Tables

N3 Multiples and sub-multiples of units 238
N4 Standard wire gauge thickness 238
N5 English zinc gauge thickness 239

Identification of Pipelines 239-240

Tables

N6 Basic identification of colours 239
N7 Colour references 239
N8 Optional colour code indications for general building services 240

Building Drawings 241-242

Lines 241
Sequence of dimensioning 241

Tables

N9 Scales for use with metric and imperial systems 241
N10 Piping symbols 241
N11 Graphical symbols and representation 241

Maximum Spacing of Pipe Supports 242-243

Tables

N12 Hot and cold water pipes and gas pipes 242
N13 Soil and waste pipes 243

Spacing of Pipes 243

Table

N14 Maximum distance allowable between centre lines of screwed steel and copper pipes 243

Head and Pressure of Water 244

Table

N15 Head and pressure of water 244

Hydrostatic Tables 244

Relation of Services to Each Other 245

Table

N16 Relation to other services 245

INDEX

Page Number

Approximate Weight of Lead Required for Caulked Lead Joints 245

Tables

N17 Cast iron soil pipes to BS 416 245
N18 Cast iron pipes for water, gas and sewage to BS 4622 and spun pipes to BS 1211 245

Corrosion of Metals 246

Tables

N19 The electro-chemical series 246

Coefficients of Linear Expansion and Melting Points 246

Tables

N20 Metals and glass 246
N21 Plastics 246

Page Number

Definitions of Plumbing Terms 247-252

Cold water supply 247
Hot water supply 248
Sanitation 249
Drainage 251

Useful Addresses 252-259

Employers' trade associations 252
Manufacturers' and other trade associations 253
Professional bodies 254
Research and advisory bodies 255
Regional Water Authorities and Water Companies in England and Wales 256
Fuel and energy interests 257
Government departments 257
Training bodies 258
Trade unions 258
Safety organisations 258
Overseas 258